MATHEMATICAL LOGIC AND ITS APPLICATIONS

MATHEMATICAL LOGIC AND ITS APPLICATIONS

Edited by
Dimiter G. Skordev
Sofia University
Sofia, Bulgaria

PLENUM PRESS • NEW YORK AND LONDON

Library of Congress Cataloging in Publication Data

Advanced International Summer School and Conference on Mathematical Logic and Its Applications (1986: Druzhba, Varna, Bulgaria)
Mathematical logic and its applications.

Bibliography: p.
"Proceedings of an Advanced International Summer School and Conference on Mathematical Logic and Its Applications, in honor of the 80th anniversary of Kurt Gödel's birth, held September 24–October 4, 1986, in Druzhba, Bulgaria"–T.p. verso.
1. Logic, Symbolic and mathematical–Congresses. 2. Gödel, Kurt–Congresses. I. Skordev, Dimiter Guenchev. II. Gödel, Kurt. III. Title.
QA9.A1A38 1986 511.3 87-20243
ISBN 0-306-42599-8

Conference organized by:
Bulgarian Academy of Sciences
Sofia University "Kliment Okhridski"
International Foundation "Lyudmila Zhivkova"

Conference sponsored by:
Union of Bulgarian Mathematicians: Polytechnics, Gabrovo, Assenovgrad, Shumen, Vratsa, and Mathematical Center divisions
Dimitrov Youth Communist League
Youth Movement for Technology and Science

Proceedings of an Advanced International Summer School and Conference on Mathematical Logic and Its Applications, in honor of the 80th anniversary of Kurt Gödel's birth, held September 24–October 4, 1986, in Druzhba, Bulgaria

A Division of Plenum Publishing Corporation
233 Spring Street, New York, N.Y. 10013

Printed in the United States of America

Kurt Gödel (1906–1978)

Photograph taken by Dr. Veli Valpola on May 9, 1958, in Princeton

ORGANIZING COMMITTEE OF THE MEETING

A. Buda, G. Gargov/Secretary, N. Georgieva,
L. Ivanov/Treasurer, D. Skordev/Chair, D. Vakarelov

PROGRAM COMMITTEE

Henk Barendregt (Nijmegen)
Johan van Benthem (Amsterdam)
Jan Bergstra (Amsterdam)
Douglas Bridges (Buckingham)
Osvald Demuth (Prague)
Albert Dragalin (Debrecen)
Yurij Ershov (Novosibirsk)
Jens Erik Fenstad (Oslo)
Nadezhda Georgieva* (Sofia)
Robert Goldblatt (Stanford)
Sergej Goncharov (Novosibirsk)
Herman Ruge Jervell (Oslo)
Nikolaj Nagornyj (Moscow)
Valerij Nepomniaschy (Novosibirsk)
Rohit Parikh (New York)
Helena Rasiowa (Warsaw)
Krister Segerberg (Auckland)
Dimiter Skordev/Chair* (Sofia)
Andrzej Skowron (Warsaw)
Boris Trakhtenbrot (Tel Aviv)
Dimiter Vakarelov* (Sofia)

* Selection Committee Member

OTHER REFEREES

George Gargov (Sofia)
Ljubomir Ivanov (Sofia)
Jan Klop (Amsterdam)
Evangelos Kranakis (Amsterdam)
Dag Normann (Oslo)
Solomon Passy (Sofia)
Petio Petkov (Sofia)
Slavian Radev (Sofia)
Adrian Rezus (Nijmegen)
Ivan Soskov (Sofia)
Jordan Zashev (Sofia)

PREFACE

The Summer School and Conference on Mathematical Logic and its Applications, September 24 - October 4, 1986, Druzhba, Bulgaria, was honourably dedicated to the 80-th anniversary of Kurt Gödel (1906 - 1978), one of the greatest scientists of this (and not only of this) century. The main topics of the Meeting were: Logic and the Foundation of Mathematics; Logic and Computer Science; Logic, Philosophy, and the Study of Language; Kurt Gödel's life and deed. The scientific program comprised 5 kinds of activities, namely:

a) a Gödel Session with 3 invited lecturers
b) a Summer School with 17 invited lecturers
c) a Conference with 13 contributed talks
d) Seminar talks (one invited and 12 with no preliminary selection)
e) three discussions

The present volume reflects an essential part of this program, namely 14 of the invited lectures and all of the contributed talks. Not presented in the volume remained six of the invited lecturers who did not submit texts:
Yu. Ershov - The Language of Σ-expressions and its Semantics;
S. Goncharov - Mathematical Foundations of Semantic Programming;
Y. Moschovakis - Foundations of the Theory of Algorithms;
N. Nagornyj - Is Realizability of Propositional Formulae a Gödelean Property;
N. Shanin - Some Approaches to Finitization of Mathematical Analysis;
V. Uspensky - Algorithms and Randomness - joint with A. N. Kolmogoroff, and - To the Notions of a Problem and of an Algorithmic Size; one of the lectures of D. Bridges - Constructive Complex Analysis since 1970, the invited seminar talk of D. Siefkes - How Richard Büchi Worked with Terms, the three discussions (one on Gödel, one on recursion, and one on mathematical logic and its applications), and the non-selected seminar talks.

For the Conference, 26 submitions arrived on time (2 came too late) of which the Selection Committee, based on the referees' reports (between 2 and 4 for each paper) accepted the half, all being included in the volume.

The invited paper of C. Christian and contributions of E. Kranakis and M. Zimand were not orally presented at the Meeting by their absent authors.

The Meeting enjoyed participants and contributors with wide variety of nationalities and from all generations of logicians. Out of 83 ones (8 actually did not attend), we had 37 from Bulgaria and 46 from 18 other countries in Eurasia, America and the Pacific, the youngest participants being students, while among the older were authors of classical results in logic and mathematics. This colourful variety surely reflects on the contents of the present volume. Indeed, in the 27 contributions included one can find (not mentioning the papers on Gödel) both, papers elaborating in classical areas of logic, and such directed towards revision of these last.

As a whole the papers are not only with different intensions, but in many different fields of logic: in constructive mathematics and intuitionistic logic, set theory and proof theory, abstract computability, modal and dynamic logic, lambda calculus, philosophy and history, and in the large and rapidly enlarging area known nowadays as logic in computer science. Thus one can trace here the fashionable wave of computerizing the mathematics and the logic, in particular. But, on the other hand, the majority of the contributions are devoted to the sound mathematical logic of notions and phenomena of intrinsic interest. So this volume gives some idea of what the logic was yesterday, what it is like today, and, hopefully, some hints on what it will be tomorrow.

Obviously, organizing such a meeting, selecting its scientific program, and editing its Proceedings, is not a job supposed to be done by a single person. That is why I would like to take this opportunity and to share the responsibility with all colleagues and friends whose contribution to this deed was more than simply "a help", and to whom I would like to express here more than simply "acknowledgements". To start with the institutions, besides to the Organizers, Co-Sponsors and the Institute of Youth Studies, special thanks are also due to the Central Council of Bulgarian Trade Unions, and to the administration of the Trade Union House "Zarya", where the Meeting took place. Personally, first of all to all members of the Program and Selection Committees, whom I have the pleasure to consider as co-editors of these Proceedings, and to all other referees mentioned above.

Thanks to all members of the Organizing Committee, for the endless consultations on how the job has to be done, and especially to those who did most of it:
to George Gargov, who, jointly with V. Goranko, did his best and it was much for best arrangement of accomodation of all the participants in the Meeting;
to Ljubomir Ivanov, who pursued the democratic financial policy of our enterprise, for his expedition and active cooperation;
to Dimiter Vakarelov, for arranging extremely fruiful contacts with helpful institutions.

Thanks to all members of the Sector of Mathematical Logic (including the students) who voluntarily joined the Organizing Committee sharing its duties:
to Moni Passy, who helped very much in carrying on the huge correspondence with authors, referees and Publisher and who did a lot of technical and other work for the very preparation of the manuscript of the volume;
to Petio Petkov, who – among many other things – organized the Gödel Session as a pillar of the whole Meeting and, jointly with V. Sotirov, the printing of our beloved Gödel poster, for his renaissance enthusiasm;
to Valentin Goranko, who efficiently served as Local Arrangements Chair;
to Miroslav Genov, who arranged the excursion in the vicinities of Varna.

That is not all. I cannot help mentioning also (or again) the names of Angel Ditchev, Slavian Radev, Ivan Soskov, Alexandra Soskova, Vladimir Sotirov and Tinko Tinchev for their readiness and responsiveness in doing the things. And of Elka Dimitrova, Rumjana Draganova, Petko Evtimov, Valja Ficherova, Mitko Janchev, Vera Kaneva, Garo Keshishian, Pavlina Mileva, Nevjana Nikolova, Stela Nikolova, Lili Petrova, Valja Petrova, Rossina Petrova, and of Roza Vassileva – for their real help.

Three more persons should be also thanked to for well done jobs: the artist Stoil Mirchev – for the Gödel poster, the photographer Vassil Plougchiev – for the masterpieces, and the editor Elena Vidinska of Jusautor Agency – for successful negotiations with Plenum Press. And one more institution: the Plenum Publishing Corporation – obviously what for.

My last and foremost thanks are to all the participants in the Meeting and contributors to the volume, without whom neither the Meeting would take place, nor the Proceedings might appear.

December 1986

Dimiter Skordev

CONTENTS

GÖDEL SESSION (INVITED PAPERS)

* indicates the lecturer invited

CONFERENCE (CONTRIBUTED PAPERS)

G Ö D E L   S E S S I O N

(invited papers)

REMARKS CONCERNING KURT GÖDEL'S LIFE AND WORK

Curt C. Christian

Institute for Logistic
University of Vienna
Vienna, Austria

Gödel was born on April 28th, 1906, in Brno. His mother Marianne, also born in Brno, stemmed on her father's side from the family Handschuh in Rheinland, Germany. Gödel's father Rudolf came from Vienna and was managing director and partner of one of the leading textile companies in Brno. He died in the age of 54 and left his family well situated. So Gödel's mother, an extremely subtle and distinguished lady, became the center of the family. The elder son Rudolf decided to study medicine and was soon a famous radiologist.

Kurt Gödel attended the secondary school in Brno and went to Vienna in 1924 where he stayed till 1939 - apart from some short stays abroad. On other occasions I have already talked about Gödel's years of study, his teachers (O.Prof. Furtwängler, Hahn, Wirtinger, Ao.Prof. Menger, Doz. Lense, Helly, W. Mayer, Vietoris, Hornich, Thirring, Schlick) and his relations to the Vienna Circle. In 1929 he submitted his thesis "Über die Vollständigkeit des Funktionenkalküls". The presentation of his famous habilitation paper "Über formal unentscheidbare Sätze der Principia Mathematica und verwandter Systeme" followed already 1932. He obtained the venia legendi on March 11th, 1933. His "Habilitationsvater" Hahn recognized the great import of this paper as can be seen from the expert's report, which is reprinted in my paper "Leben und Wirken Kurt Gödels" (Monatshefte für Mathematik). Not everybody was willing to accept immediately Gödel's contribution. This means Zermelo, but also Hilbert.

1938 Gödel married Adele Porkert. After the annexation of Austria by the German Reich the Austrian standards of habilitation were annulled, so Gödel had to apply for appointment to be "beamteter Dozent neuer Ordnung"; this application was granted in 1940. But Gödel didn't make use of that because already in the end of 1939 - <u>after</u> beginning of war - he emigrated to Princeton, USA, where on the occasion of two foregoing flying visits he had lectured about his work on the relative consistency of the axiom of choice and the continuum hypothesis, which had been outlined at Vienna. Gödel - citizen of the United States since 1948 - got his professorship at the famous Institute for Advanced Study not before 1953 in the age of 47.

In my Paper cited above I already mentioned, that apart from his most famous papers Gödel elaborated some other treatises in logic, physics and philosophy. Of course he had many scholars who gained from his inspirations. He received many awards of honour in the USA, the United Kingdom and Austria. The Austrian Academy of Sciences elected him to its honorary member on occasion of his 60th birthday. In connection with this matter a letter from Gödel, addressed to the Academy is reprinted in my paper cited above. Unfortunately Gödel did not make use of a professsorship by title at the University of Vienna, bestowed on February 23rd, 1966. The faculty of natural sciences, emerged from the separation of the old philosophical faculty into three parts, proposed the bestowal of the honorary doctorship of natural sciences to Kurt Gödel, after he had not made any objection against it referring to my inquiry in this connection. Shortly afterwards Gödel died so with the agreement of his widow the graduation was performed posthumously.

Gödel was a reserved, lovable and co-operative person, who in former times liked to be in company with other people, but in an uncommunicative way. Later he had a small circle of friends, e.g. Einstein and Morgenstern. He had an unstable nervous system, suffered formerly from a "Duodenalulcus", later from presumably nervous stomach trouble and depressions, several times he was under medical treatment in hospital. On the occasion of a nervous breakdown subsequent to his most famous paper the number theorist Furtwängler asked, whether Gödel's nervous constitution would be a consequence of his creative power or conversely. Cesare Lombroso, author of "Genie und Irrsinn" and Ernst Kretschmer, author of "Geniale Menschen", would reply to that question in the sense of an interaction and not in the sense of a one-sided causal relationship.

Gödel died on January 14th, 1978, ostensibly of "Herzversagen", presumably partially of "chronische Inanition" caused by his special diet.

Doubtless the meritorious book "Gödel, Escher, Bach" by Hofstadter increased the acquaintance of Gödel's results enormously. Therefore I don't want to go into much detail, but I would like to raise the question: Is there a change in the appraisal of Gödel's results 8 years after his death?

Doubtless there are people speaking about a so-called "Gödel-myth" and uttering some reservations. I want to give some remarks concerning three objections in connection with the two incompleteness theorems. My remarks concerning the first incompleteness theorem refer to the question: Gödel's diagonalization - Cantor's diagonalization, a translation?, Gödel's self reference - Eubulides' self reference - a translation?; concerning the second incompleteness theorem I ask: Bernays' conditions: yes - no, an overlooking of conditions? As everybody knows, Gödel proved his incompleteness theorem (Every consistent, recursively axiomatized extension of the Peano system has at least one undecidable - i.e. neither provable nor refutable - sentence) by specifying a formal sentence, which is true under standard interpretation if and only if it is unprovable in the number-theoretic formalism. That sentence can be called "diagonalized Gödel formula". It emerges from the Gödel formula (a sentence-formula with exactly one free variable) by substituting the formal Gödel number of the Gödel formula for the one free variable.

It is well known that there is an injective function from formulas into the natural numbers. The range of this function are the Gödel numbers. Why do we talk about a diagonalized Gödel formula as a diagonalized formula? The set of all number-theoretic formulas is recursively enumerable - especially the set of all such formulas with exactly one free variable. We can talk about the first, second, third, .. formula. From each of these formulas with one free variable we can get infinitely many sentences substituting the variable successively by the formal Gödel number of the first second, third, .. formula. Arranging these sentences in a quadratic scheme we have a diagonal consisting of the sentences

$F_i \frac{x_1}{\ulcorner F_i \urcorner}$, which originate from the sentence-formulas with one free variable by substituting their formal Gödel numbers for the variable. Gödel did <u>not</u> construct his sentence in the manner of Cantor's method by mixing the formulas

of the diagonal together, but he chose the one formal sentence which is true under standard interpretation if and only if it is unprovable in Peano formalism.
Assuming the consistency of number theory he showed, that the assumption of provability of that sentence would cause its unprovability, which implies the unprovability according to the principium in contrarium. In the same way he showed, that assuming the ω-consistency the refutability of his sentence would imply its unrefutability, which implies the unrefutability like before. The most important basis in this proof is the Gödelization-monomorphism from syntactical entities and the relations between them to the natural numbers and the relations between them corresponding to the former in a homomorphical way. Another important basis is the representation theorem concerning recursive relations and functions.

A totally different thing is Cantor's diagonalization method developped to show the uncountability of the real numbers, or the interval $[0,1)$ (including 0, excluding 1). Assuming the countability of $[0,1)$, we can talk about the first, second, third, ... real number of this interval. Each of these numbers is characterized by an infinite sequence of decimals. Arranging these decimals into a quadratic scheme the position in column k and line i contains the k-th decimal of the i-th real number in the enumeration. Cantor built a new real number - let us call it "Cantor number" - using <u>all</u> decimals of the diagonal. The integer part of the Cantor number is 0 and the decimals are given as follows: the i-th decimal is 0 if the i-th decimal of the i-th real number is different from 0, otherwise it is 1. Clearly the Cantor number must occur in $[0,1)$, so it has a number - say k - in the enumeration. The k-th decimal of the Cantor number would be 1 if it should be 0 according to the definition and it should be 0 if it should be 1 according to the definition. We have a contradiction. The assumption of the countability of $[0,1)$ resp. the real numbers is false.
Cantor's diagonalization method, which has many applications - e.g. in the proof of the uncountability of the set of functions with domain $\mathbb{N}$ and range $\{0,1\}$ - doesn't have anything to do with Gödel's method. The only common thing is the indirect proof method. If one wants to raise the question, whether Cantor's noncountabilityresult or Gödel's first incompleteness result would be more fruitful for mathematics, so maybe the one or other mathematician not interested in formal mathematics would prefer Cantor's theorem concerning the real numbers. Doubtless Gödel's result is of much more intricacy. Students can understand Cantor's result just in the beginning of their university education, but to comprehend Gödel's result in full detail one has to make many preparations. Taking for granted that one has mastered the involved presuppositions the proof of Gödel's first incompleteness theorem can clearly be done in the same shortness as Cantor's proof. Cantor's diagonalization is in no way the predecessor of Gödel's diagonalization. It is often difficult to reconstruct the way how mathematicians get to their results. Many results are reached after searching in vain something totally different for a long time. We cannot be sure that the liar's paradox of Eubulides gave rise to the first incompleteness theorem. But we can show that the incorrect self reference contained in the liar's pseudo-sentence differs totally from the correct self reference in the diagonalized Gödel formula. Therefore there is no reason to think little of Gödel's contribution. Certainly today we have still other proof methods for Gödel's first incompleteness theorem. In order to prove the undecidability and incompleteness of number theory and set theory, one shows

1. Every recursively representative and consistent theory is undecidable.
   A theory is recursively representative means that all recursive relations (resp. all recursive functions) are negation-preserving (resp. functionally) representable in it.

The proof of this sentence requires one of the following non-representation-lemmata:

α) In a consistent theory there is no negation-preserving representation of the set of numbers having diagonals being theorem numbers.

β) In a consistent theory with functionally representable diagonal function there is no negation-preserving representation of the set of theorem numbers.

2. Every recursively axiomatized, recursively representative and consistent theory is not only undecidable but also incomplete. The proof of this sentence requires apart from the just mentioned first sentence a further lemma:
   Every recursively axiomatized and maximally consistent theory is decidable
   This sentence is equivalent with:
   Every recursively axiomatized, undecidable and consistent theory is incomplete.

3. If reduced number theory means a formal system, arising from Peano formalism (with $x<y :\leftrightarrow \exists z_{\neq 0}(y=x+z)$) replacing the scheme of mathematical induction by singular sentences just sufficient to get recursively representativity - e.g. replacing the scheme of mathematical induction by the dichotomy principle $(x=0 \vee \exists y(x=y'))$, the commutative form of the first addition axiom $(0+x=x)$ and the principle of uniqueness of remainder $(\exists q(y=x\cdot q+z_1) \wedge z_1 < x \wedge \exists q(y=x\cdot q+z_2) \wedge z_2 < x \dot{\rightarrow} z_1=z_2)$ - the following sentence holds by the fact that reduced number theory as well as set theor are recursively representative theories:
   Every theory which is a recursively axiomatized, consistent extension of reduced number theory or is a recursively axiomatized, consistent extensio of set theory, is undecidable and incomplete.
   This sentence contains Gödel's result of incompleteness of number theory and Church's result of undecidability of number theory.

Church's theorem means that the set of theorem numbers of number theory is not recursive, i.e. there is no algorithmic procedure to determine derivability or non-derivability. This meaning of "decidable" differs from the one mentioned above which deals with sentences of formal theories. I only mention that completeness in the sense of the completeness theorem is also not the same as completeness in the sense of the incompleteness theorem. According to the completeness theorem the Peano formalism is complete, i.e. every formula valid in all models of Peano formalism is derivable; but it is <u>not</u> complete in the sense of the incompleteness theorem, because not every irrefutable sentence is derivable.
Furthermore the opinion is maintained sometimes in the literature that Gödel had overlooked some conditions - today known as Bernays' conditions - necessary for the proof of his second incompleteness theorem. This is incorrect, too.
The second incompleteness theorem states the undecidability (i.e. non-derivability and irrefutability) of a sentence which is true under standard interpretation if and only if number theory is consistent. As a result, the consistency of Peano formalism cannot be proved using only means from itself whereas for some subsystems this <u>can</u> be done.
Last year I have given in Leipzig a two-hour-lecture on this topic. +)

In 1963 Cohen strengthened Gödel's results on relative consistency to another undecidability result, namely that of the axiom of choice and the

+) This lecture will be published in the "Sitzungsberichte der Österreichischen Akademie der Wissenschaften" under the title "Bemerkungen zu drei Einwänden gegen Kurt Gödel".

continuum hypothesis. Assuming the consistency of basic set theory, then the axiom of choice is undecidable in basic set theory and the continuum hypothesis is undecidable in basic set theory with axiom of choice.

Gödel himself tended to the opinion that in the future one will find a sentence in conformity with intuition which gives rise to the refutation of the continuum hypothesis. Unfortunately we don't have any such sentence up to now. Many derivations using the special continuum hypothesis can also be done by using Martin's axiom, which follows from the continuum hypothesis but not vice versa. But Martin's axiom cannot be the desired intuitive sentence: If Martin's axiom would imply the negation of the continuum hypothesis in ZF+CH, then this theory would be inconsistent, so ZF would be inconsistent by Gödel's relative consistency result. Hence the consistency of ZF implies that the negation of the continuum hypothesis cannot follow from Martin's axiom. Furthermore Martin's axiom also cannot imply the continuum hypothesis because the system ZF+MA+ $\neg$CH is relative consistent, as you know. The continuum hypothesis therefore is undecidable within the extension of basic set theory by Martin's axiom. On the present state one is inclined to accept the continuum hypothesis. But Gödel's challenge still remains in force. The acceptance of the generalized continuum hypothesis, by which there is no immediate-sized set between an infinite set and its powerset of course does not represent a risk on consistency, yet it is a strong assertion if you consider e.g. that the union of a set may result into a such one, that between these two sets there are infinitely many sets of different sized.
In connection with Gödel's relative consistency result I would like to stress, that Gödel was familiar with the importance of the infinitesimal calculus very early. He himself essentially contributed to the justification of Leibniz's idea of infinitesimals by his result, as I demonstrated in a paper which was published in America.
A former assistent of Hans Hahn, Mrs. Olga Taussky-Todd, who was on friendly terms with Kurt Gödel, often invited him to a cup of tea and attended Schlick's seminar with him, kindly sent to me a letter of Kurt Gödel to Mrs. Julia Robinson where Gödel emphasized the great importance of non-standard-analysis.

As a summary we can say that even eight years after Gödel's death nobody can undermine Gödel's extraordinary position justified by the diversity, the degree of difficulty and the great import with regard to the contents of his results. There is no reason to complain of a "Gödel myth".

FACETS OF INCOMPLETENESS

John W. Dawson, Jr.

Pennsylvania State University
York, PA U.S.A.

Great intellectual discoveries, like fine gems held up before the eyes, continue to fascinate and to reveal new facets as they are beheld in the eye of the mind. As an example of this phenomenon, we may consider the Gödel incompleteness theorems, which, during the 55 years since their discovery, have repeatedly yielded new insights in the course of ongoing reinterpretation.

I. TRADITIONAL PERSPECTIVES

Elsewhere (Dawson 1984) I have discussed in detail the circumstances surrounding Gödel's first public announcement of his discovery of arithmetically undecidable statements. There I stressed that the event occurred but one day after Gödel had lectured on the results of his dissertation, in which he had established the completeness of first-order logic.

In the context of the time – that is, from the perspective of Hilbert's program – Gödel's completeness theorem came as an expected, but nonetheless notable, result, for it seemed to confirm that first-order logic proves all that it "ought" to: in modern terms, that the theorems of any first-order theory coincide with those of its statements true in all its models. Consequently, the adequacy of any such theory must depend only on its _non_-logical axioms, which may be defective in various ways. In particular:

(1) they may be _inconsistent_, in which case _every_ statement of the theory (including, perhaps, the mendacious assertion of its own consistency) will be a theorem:

(2) they may be unsound, that is, not true of the intended objects of study; or

(3) they may not be recursive (or even recursively enumerable), and so be effectively unrecognizable.

It is natural to ask whether a theory free from all these defects (such as, presumably, Peano Arithmetic) must provide an adequate formalization of its subject matter – in particular, whether it must prove all true statements formalizable within it. And as Gödel so perceptively noted in the introduction to his dissertation (in remarks deleted from the published version), the non-existence of undecidable statements was a necessary presumption for those who, like Hilbert, equated the existence of mathematical objects with the consistency of formal axioms describing them and sought to establish the consistency of theories by "finitary" means.

It is important to understand that, for Hilbert, the consistency of arithmetic was not in doubt. His hope was rather that if a consistent theory was sufficiently strong, it should be able to demonstrate that consistency. That idea, however, was refuted by Gödel's second incompleteness theorem, which may be viewed as a profound converse to the trivial fact stated in (1): any recursively axiomatizable extension of (even a small part of) PA that does prove (a suitably encoded assertion of) its own consistency must be inconsistent. At the time of the Königsberg conference Gödel had only his first incompleteness theorem, exhibiting a particular, somewhat artificial statement of PA that was neither provable nor refutable within it; but he obtained the second theorem soon thereafter, and because of its impact on Hilbert's program it is understandable that it should have overshadowed the first theorem.

According to the commonly accepted view (as disseminated, for example, in most logic textbooks), Gödel's theorems demolished Hilbert's program – contrary to Hilbert's and Gödel's opinions at the time. The received view also tends to emphasize the syntactic diagonalization employed in Gödel's proof and the seemingly paradoxical nature of his self-referential statement. As such, it reflects attitudes of the time (1930), when syntactic formalism held sway and when Gödel's theorems were often conflated with other quasi-paradoxical results (such as the Löwenheim-Skolem theorem) that were thought to be antinomial, pathological or artifactual (due to unnecessary cardinality restrictions or the restriction to first-order logic).

In this regard, it is worth noting that the incompleteness theorems can be stated (as above) without reference to semantic notions and that, in his proofs, Gödel argued purely syntactically. But he preceded his formal demonstrations with an informal discussion in which he invoked the truth of his undecidable statement. It is now clear that Gödel was led to his discovery of incompleteness by semantic considerations, but that he avoided semantic arguments in his published proofs because he believed there was a prevailing prejudice against them; and indeed, it was to be more than twenty years before logicians began to make significant use of model-theoretic notions. Quite recently, Kripke has found a way to use such notions to prove the incompleteness theorems without appeal to self-reference (i.e., without any use of fixed-point theorems). An account of his results is given in Quinsey 1980.

In the meantime, as is well known, the incompleteness theorems came to be subsumed within the theory of recursive functions that was developed by Church, Kleene, Turing, and Post. That development has now been chronicled a number of times, so I will say no more about it here, except to note Gödel's conspicuous absence from it (see Davis 1982 for more on this point) and to remark that whereas the computer revolution has caused the practical importance of Turing's work to be widely recognized, the air of paradox surrounding Gödel's theorems and the tendency to view them in the context of Hilbert's program has given rise to the notion, widespread even today, that they are essentially irrelevant to mathematical practice.

II. THE QUEST FOR "NATURAL" UNDECIDABLE STATEMENTS

Considered together, the completeness and incompleteness theorems could have been seen to imply the existence of non-standard models of arithmetic, models not only not isomorphic to the natural numbers (such as Skolem was to construct just two years later), but not elementarily equivalent to them. Yet, remarkably, no one was to observe this until the 1950's (except for a passing and somewhat misleading remark by Gödel himself in his review of Skolem's work; see the commentary by Vaught, pp.376-379 in Gödel 1986).

Had the study of non-standard models been pursued more vigorously, it is possible that more "natural" (i.e., mathematically relevant) undecidable statements might have been discovered earlier. As it was, not until the 1970's did Kirby and Paris develop a theory of cuts in nonstandard models of arithmetic, a breakthrough that led shortly to the celebrated Paris-Harrington theorem (1977), which states that the finite Ramsey theorem of combinatorics (a statement provable in PA) remains true, but becomes unprovable in PA, if the homogeneous set H is required to have cardinality greater than its least element. (The finite Ramsey theorem asserts that given natural numbers n, m, and s, there is a natural number k such that for any partition of the unordered n-tuples of natural numbers $< k$ into m classes, there is a set H of size $\geq s$, all of whose unordered n-tuples belong to the same partition classes.)

Although Ramsey's theorem plays an important role in modern combinatorial studies, it nevertheless arose in the context of a logical investigation, as did the Paris-Harrington modification of it; and it must be admitted that the latter seems ad hoc from a combinatorial point of view. One might therefore seek to find an arithmetically undecidable statement that arose outside of logical investigations. An example of such is Goodstein's theorem, which concerns so-called complete base b expansions of integers. (That is, if the usual base b expansion of an integer n, say as

$$n = c_1 b^{e_1} + c_2 b^{e_2} + \ldots + c_k b^{e_k},$$

is modified by repeated expansion of the exponents e_i to the base b, n will eventually be (uniquely) expressed as a super-exponential polynomial in which all coefficients and exponents involve only numbers $\leq b$. For example, the complete base 3 expansion of 172 is $172 = 2\cdot 3^{3+1} + 3^2 + 1$, and the complete base 2 expansion of 100 is $100 = 2^{2^2+2} + 2^{2^2+1} + 2^2$.) For any integers $n > 0$ and $b \geq 2$, let $c_b(n)$ be the complete base b expansion of n, and let $R_{b+1}(n)$ be the result of replacing each b in the expansion $c_b(n)$ by $b+1$. Then define the Goodstein sequence starting at n by $n_o = c_2(n)$, $n_{k+1} = R_{k+3}(n_k) - 1$ if $n_k > 0$, and $n_{k+1} = 0$ otherwise. In 1944, R.L. Goodstein proved by infinitary methods that every such sequence eventually converges to 0. In 1982, Kirby and Paris showed that Goodstein's theorem is not provable in PA.

In both these examples (and a welter of others proposed as "natural" undecidable statements) there is an important connection with rapid functional growth. A detailed survey of results in this area has been given by Simpson (1986); here we mention only a result of Solovay concerning the Paris-Harrington theorem. Toward that end, suppose that $\forall x \exists y\, P(x,y)$ is a statement of PA true of the natural numbers, and define f(x) to be the <u>least</u> y such that P(x,y). According to a theorem of Kreisel (1952), if P(x,y) can be proved in PA to be a <u>computable</u> relation, then if $\forall x \exists y\, P(x,y)$ is provable in PA it must eventually be dominated by a certain very rapidly growing computable function g; on the other hand, if f eventually dominates g, then $\forall x \exists y\, P(x,y)$ must be unprovable in PA. Now, given n and m as in the finite Ramsey theorem, let $s = n+1$ and define f(n) to be the least k satisfying the Paris-Harrington theorem. Solovay's theorem then states that f eventually dominates g and is not <u>provably</u> computable in PA. (See <u>Ketonen and Solovay 1981</u>, <u>Smoryński 1980</u>, or <u>Spencer 1983</u>.)

It is worth noting that, in contrast to set theory, almost all the statements so far shown to be undecidable in PA have been <u>informally</u> decided. Smoryński (<u>1977</u>, pp.862-864) mentions an artificial example, due to Kreisel, whose truth value remains unknown; are there "natural" examples here too?

III. PROOF-LENGTH CONSIDERATIONS AND ALGORITHMIC INFORMATION THEORY

Since first-order logic does not suffice to identify those arithmetic statements that are true of the natural numbers (but only the smaller set of those that are true in <u>all</u> models of PA), it is natural to try to augment the means of first-order logic so that the theorems obtained <u>will</u> coincide with the statements true of the natural numbers. Hilbert himself suggested how to do so, at almost the same time that the incompleteness theorems were announced: he proposed adding a restricted version of the <u>ω-rule</u>, the infinitary rule of inference that permits $\forall x\, P(x)$ to be deduced from the premises $P(\underline{0})$, $P(\underline{1})$, $P(\underline{2})$, ... It is an interesting question whether Hilbert made his proposal in an effort to rescue his program from the effects of Gödel's results; see the discussion by Feferman, pp.208-213 in <u>Gödel 1986</u>.

Gödel himself felt that Hilbert's program would be compromised if the ω-rule were to be added. But the fact remains that the theorems of ω-logic <u>do</u> coincide with the statements true of the natural numbers, and this suggests another interpretation of Gödel's results, namely, that incompleteness arises because PA is not <u>ω-complete</u>, that is, because there are unprovable universal statements $\forall x\, P(x)$, each of whose numerical instances $P(\underline{n})$ is nonetheless provable (and thus true). (It should be noted that if PA is formalized in the language containing only + and × as function symbols, then Gödel's undecidable statement is not of such purely universal form. But it follows from Matijasevič's negative resolution of Hilbert's 10th Problem (see Davis, Matijasevič, and Robinson 1976) that there are unsolvable Diophantine equations that are not provably unsolvable in PA, and such unsolvability can be expressed in the form

$$\forall x_o \ldots \forall x_n (P(x_o, \ldots, x_n) \neq Q(x_o, \ldots, x_n)),$$

where P and Q denote polynomials with natural numbers as coefficients.) Note that a purely universal statement is finitely refutable, so if it can be proved consistent with PA, it must in fact be true.

To appreciate the motivation behind Hilbert's introduction of the ω-rule, consider the question: How could $\forall x\, P(x)$ fail to be provable despite the provability of every numerical instance $P(\underline{n})$? The answer is, presumably, because the various instances hold for an infinite variety of reasons. In particular, it could be that the length of the shortest proof in PA of each $P(\underline{n})$ (as measured by the number of lines (not symbols) in each such proof) increases without bound as n increases. This in turn suggests the converse question, first raised by Kreisel: If there is a uniform bound to the lengths of the shortest proofs in PA of the instances $P(\underline{n})$, must $\forall x\, P(x)$ be provable in PA?

This question became a well-known unsolved problem. In 1973, Parikh answered affirmatively the corresponding question for the system PA^*, a modification of PA in which + and × are expressed as ternary relations rather than binary functions. For PA itself, the question remained open until 1985, when it was settled by William Farmer in his doctoral dissertation at the University of Wisconsin. Somewhat disappointingly, the answer turns out to depend critically on the particular formalization chosen. As an example of one of Farmer's results, the answer is "yes" if PA is formalized by using the schema "$\exists x(x = \tau)$, provided x does not occure in τ" in place of the usual substitution schema.

The proof-length measure described above was introduced by Gödel himself, who, in his 1936, obtained the first of what are now called "speed-up" theorems. (See Arbib 1966.) Gödel cannot, however, be considered a founder of computational complexity theory, since the modern notion of program (or proof) complexity depends upon the later concept of universal Turing machine. But Gödel's incompleteness theorems, interpreted in computation-theoretic terms, have recently given rise to the development of algorithmic information theory.

In this reinterpretation, an important ingredient is a striking new definition of the probabilistic notion of randomness, due independently to Kolmogorov and Chaitin. According to this definition, we consider a fixed universal Turing machine $\mathcal{U}$ that accepts finite binary strings as inputs (regarded as programs) and generates binary strings as outputs. A program is said to compute the finite string s if it causes $\mathcal{U}$ to halt after generating s. The string s is then defined to be random if the shortest binary programs that compute s have about as many bits as s itself. (The phrase "about as many" can be made precise in any of various ways, among which the choice is somewhat arbitrary.) The length of any such minimal program is defined to be the complexity of s.

A number that satisfies the Kolmogorov-Chaitin criterion for randomness can be shown to satisfy the usual statistical criteria for randomness as well. It can also be shown that, given an axiom system S, there is a constant C such that no string can be proved in S to have complexity greater than C. In fact, this statement is just a variant of the incompleteness theorem; its proof exploits the fact that any of the well-known logical paradoxes can be employed, in the manner of Gödel, to construct an undecidable statement. In particular, whereas Gödel noted that his undecidable statement could be regarded as a formal version either of the Paradox of the Liar or of Richard's paradox, the proof of the statement above relies on a variant of Berry's paradox. ("There is a smallest positive integer whose definition requires more characters than are in this sentence.") For suppose the formal system S is employed to determine the complexity of strings. We may then ask the universal machine $\mathcal{U}$ to "use the axioms of S to find the first string (in some fixed enumeration) of complexity greater than C", where the quoted command is expressed by a binary program of length C. By the definition of "complexity $> C$", $\mathcal{U}$ cannot halt when given this input.

We may also define $a(n)$ to be the greatest natural number whose binary representation has complexity $\leq n$ (if there are any such numbers; otherwise $a(n)$ is undefined). Then, given any formal system S (whose axioms are encoded as a binary string), there is a constant c, depending only on the rules of inference of S, such that if S proves every true statement of the form "s has complexity k", for each $k \leq n + c$, then either the string encoding the axioms of S has length greater than $a(n)$ bits or some of those proofs must be of length $> a(n)$. This is another instance of rapid functional growth, since it can be shown that $a(n)$ eventually dominates any partial recursive function.

Finally, we may suppose that the machine $\mathcal{U}$ is designed so as to request input bits whenever it requires them, and that the input bits so requested are generated by a random process such as coin-tossing. We may then ask for the probability, before any input bits are fed in, that $\mathcal{U}$ will eventually halt. This probability is a well-defined real number Ω that can be shown to be random according to the algorithmic definition. As such it is irrational, whence it follows that the first n bits of its binary expansion encode enough information to solve the halting problem for all programs of length $\leq$ n (because, since Ω is the sum of the probabilities of those strings that cause $\mathcal{U}$ to halt, any string of length k that does cause $\mathcal{U}$ to halt must contribute $1/2^k$ to the value of Ω ; and, since Ω is irrational, $\Omega_n < \Omega < \Omega_n + 1/2^n$, where Ω_n denotes the number defined by the first n digits of Ω). Hence the digits of Ω, could they be determined, would suffice to settle all finitely refutable conjectures. (See Gardner 1979, from which this discussion is adapted. For more detailed surveys of algorithmic information theory, see Chaitin 1974 or 1977).

IV. MODAL CONSIDERATIONS

Modal logic — the study of the logic underlying the notions "necessary" and "possible" — has been fraught with conceptual difficulties since its inception. Modern axiomatizations, the most popular of which is S4 , were introduced by C.I. Lewis in his book A Survey of Symbolic Logic (1918). When construed as quantifiers (denoted respectively by $\Box$ and $\Diamond$), some syntactic properties (such as the equivalence of $\Diamond A$ with $\neg \Box \neg A$) of the intuitive notions of necessity and possibility seem evident, but others are problematical. Lewis himself gave inconsistent explanations of the meaning of his modal operators, and modal semantics was not significantly clarified until the work of Kripke in the 1960's.

In the interim, two papers appeared that were to lead ultimately to an almost serendipitous reinterpretation of the modal calculi. The first, by Gödel (1933), pointed out that Heyting's intuitionistic propositional calculus could be axiomatized in terms of a provability operator B and that, incidentally as it were, the same formal statements also provided a new axiomatization of Lewis' S4 if B were everywhere replaced by $\Box$. Thus, with a quite different aim in mind, Gödel introduced the modal axiom scheme consisting of

(1) all propositional tautologies of ordinary logic;

(2) all formulas of the form $\Box(A \to B) \to (\Box A \to \Box B)$;

(3) all formulas of the form $\Box A \to \Box\Box A$; and

(4) all formulas of the form $\Box A \to A$,

together with the inference rules modus ponens and necessitation ("from A deduce $\Box A$").

The provability interpretation of modal logic that Gödel thus broached was largely ignored at the time, both because the paper 1933 was overshadowed by Gödel's greater contributions and because, as Gödel himself noted, the second incompleteness theorem showed that the operator B could not be construed as representing provability in a fixed formal system; in particular, the axioms (4), so interpreted, would express a (formally unprovable) soundness schema. Indeed, as Löb later showed in his 1955 (the second of the two papers alluded to above), the soundness principle $\Box A \to A$ is provable if and only if A itself is. Moreover, just as Gödel demonstrated the provability (and thus the truth) of a sentence asserting its own unprovability (in modal terms, a sentence S such that $S \leftrightarrow \neg \Box S$), Löb demonstrated the provability (hence also the truth) of a sentence asserting its own provability (a sentence H such that $H \leftrightarrow \Box H$).

Despite Löb's investigations, the general question of how provability within a formal system might be given a modal axiomatization was not pursued until the 1970's. Then Solovay and others established two fundamental results for the system obtained from Gödel's system by omitting axiom schema (4) and adding Löb's rule: from $\Box A \to A$ deduce A (or alternatively, by taking modus ponens as the only inference rule and adjoining to axioms (1) - (3) all formulas $\Box A \to \Box\Box A$, together with all "boxes" of them, and all formulas $\Box(\Box A \to A) \to \Box A$, together with all their "boxes"). Specifically, if we assume that to each atom p of this system G is correlated some statement p^* of PA, then to every formula A of G we may correlate a statement A^* of PA by replacing each atom in A by its star and each occurrence of $\Box$ in A by Gödel's provability predicate Bew. Then:

I. A is a theorem of G if and only if for every correlation $*$, A^* is a theorem of PA.

II. If G, in its alternative formulation, is extended by adding as axioms all formulas of the form $\Box A \to A$ (but not their boxes), then A is a theorem of the extended system if and only if for every correlation $*$, A^* is true in the natural numbers.

(For further details, see Boolos 1979 or 1984 or Solovay 1976.)

In sharp contrast to these completeness theorems for propositional provability logic are the following recent results concerning predicate provability logic PG (in which the modal operators are adjoined to first-order logic without identity, constants, or function symbols):

I'. The set of sentences A of PG for which each A^* is a theorem of PA is Π_2^0-complete (Vardanyan 1985) — in particular, the set is not recursively axiomatizable.

II'. The set of sentences A of PG for which each A^* is true in the natural numbers is not arithmetical (Artemov 1985); indeed, it is complete Π_1^0 in the set of true sentences of arithmetic (Boolos 198?).

V. THE MIND/MECHANISM DEBATE

Do the incompleteness theorems say anything about the nature of our minds, and if so, what? An extensive philosophical literature has grown up around this question during the past twenty years, largely in response to the claim by J.R. Lucas (in his 1964) that the incompleteness theorems, together with Church's Thesis, refute mechanism.

In outline, Lucas' argument is that given any Turing machine, we can apply Gödel's construction to generate a problem that the machine is incapable of solving, yet whose solution we are able to apprehend; hence the power of our minds exceeds that of any Turing machine, and so, by Church's Thesis, the operation of the mind cannot be reduced to any effective (mechanizable) procedure.

It is easy to attack Lucas' reasoning, and philosophers have delighted in doing so. A full discussion of all the issues that have been raised is beyond the scope of this survey (see Webb 1980 for a penetrating analysis), but two salient weaknesses in Lucas' argument are worth noting. First, Gödel's theorems assume the consistency of the underlying system, a premise that (on empirical grounds!) our minds may but rashly be presumed to satisfy. Second, to carry out Gödel's construction, we must know which particular Turing machine (in some given enumeration, say) we are given; yet if our own minds are Turing machines, there is no reason to believe we have the means to tell which one(s).

Some critics (notably Judson Webb) have gone further, arguing not only that Lucas' reasoning is flawed, but that his conclusion is untenable as well; indeed, Webb contends that Church's Thesis is itself "the principal bastion protecting mechanism" (ibid., p. vii). A more cautious view was put forward by Gödel in his unpublished Gibbs Lecture of 1951. There he argued that "the following disjunctive conclusion is inevitable: Either mathematics is incompletable in the sense that its evident axioms can never be comprised in a finite rule, that is to say, the human mind (even within the realm of pure mathematics) infinitely surpasses the powers of any finite machine, or else there exist absolutely unsolvable Diophantine problems". Corresponding to this "disjunctive incompletability of mathematics", Gödel saw a disjunction of philosophical implications, in which "either alternative [stands] decidedly opposed to materialistic philosophy. Namely, if the first alternative holds, [then it] seems ... that the working of the human mind cannot be reduced to the working of the brain, which to all appearances is a finite machine", while "On the other hand, the second alternative ... seems to disprove the view that mathematics is only our own creation." In fact, Gödel did not believe that the mind could be identified with the physical brain (a view that, accoring to Wang, he considered "a prejudice of our time") and, though he accepted Turing's analysis of the notion of mechanizable procedure, Gödel rejected Turing's argument "that mental procedures cannot carry any farther than mechanical procedures"; according to Gödel, Turing erred in supposing "that a finite mind is capable of only a finite number of distinguishable states ... disregard[ing] completely the fact that mind, in its use, is not static, but constantly developing." (Wang 1974, p.325).

The debate over mind versus mechanism will surely continue. Will Gödel's view prevail? Or will Gödel one day be regarded, with Copernicus and Darwin, as having dealt one of the three most devastating blows to man's self-esteem?

REFERENCES

Arbib, Michael A.
1966 "Speed-up theorems and incompleteness theorems", in E.R. Caianiello (ed.), Automata Theory, Academic Press, pp.6-24.

Artemov, S.N.
1985 "Nearifmeticnostj istiinostnych predikatnych logik dokazuemosti", Soviet Math. Doklady 284:2, pp. 270-271.

Boolos, George
1979 The Unprovability of Consistency, Cambridge U. Press.
1984 "The logic of provability", American Mathematical Monthly 91, pp. 470-480.
198? "The sentences of predicate provability logic true under every interpretation" (preprint), to appear.

Chaitin, Gregory J.
1974 "Information-theoretic limitations of formal systems", Journal of the Association for Computing Machinery 21, pp.403-424.
1977 "Algorithmic information theory", IBM Journal of Research and Development 21, pp.350-359, 496.

Davis, Martin
1982 "Why Gödel didn't have Church's thesis", Information and Control 54, pp.3-24.

Davis, Martin, Yu. Matijasevič and J. Robinson
1976 "Hilbert's tenth problem. Diophantine equations: positive aspects of a negative solution", in F.E. Browder (ed.), Mathematical Developments Arising from Hilbert Problems, pp.323-378.

Dawson, Jr., John W.
1984 "Discussion on the foundation of mathematics", History and Philosophy of Logic 5, pp.111-129.
1986 "The reception of Gödel's incompleteness theorems", to appear in PSA 1984, vol.2 (Proceedings of the 9th Biennial Meeting of the Philosophy of Science Association).

Farmer, William
1984 "Length of proofs and unification theory", Ph.D. dissertation, University of Wisconsin, Madison.

Gardner, Martin
1979 "Mathematical Games", Scientific American, November 1979, pp.20-34.

Gödel, Kurt
1931 "Über formal unentscheidbare Sätze der Principia Mathematica und verwandter Systeme, I", Monatshefte für Mathematik und Physik 38, pp. 173-198 . Reprinted and translated in Gödel 1986, pp. 144-195.
1933 "Eine Interpretation des intuitionistischen Aussagenkalkuls", Ergebnisse eines mathematischen Kolloquiums 4, pp.39-40. Reprinted and translated in Gödel 1986, pp. 300-303.
1936 "Über die Länge von Beweisen", Ergebnisse eines Mathematischen Kolloquiums 7, pp. 23-24. Reprinted and translated in Gödel 1986, pp.396-399.
1986 Collected Works, vol. I, Oxford U. Press.

Ketonen, Jussi and Robert M. Solovay
1981 "Rapidly growing Ramsey functions", Annals of Mathematics 113, pp.267-314.

Kirby, L. and J. Paris
1982 "Accessible independence results for Peano arithmetic", Bulletin of the London Mathematical Society 14, pp.285-293.

Kolmogorov, A.N.
1968 "Logical basis for information theory and probability theory", IEEE Transactions in Information Theory IT-14, pp.662-664.

Kreisel, Georg
1952 "On the interpretation of non-finitistic proofs, II", Journal of Symbolic Logic 17, pp.43-58.

Löb, M.H.
1955 "Solution of a problem of Leon Henkin", Journal of Symbolic Logic 20, pp. 115-118.

Lucas, John R.
1964 "Minds, machines, and Gödel", in Alan Ross Anderson (ed.), Minds and Machines, Prentice-Hall, pp.43-59.

Parikh, Rohit
1973 "Some results on the length of proofs", Transactions of the American Mathematical Society 177, pp.29-36.

Paris, Jeff, and Leo Harrington
1977 "A mathematical incompleteness in Peano arithmetic", in J. Barwise (ed.), Handbook of Mathematical Logic, North-Holland Pub. Co., pp. 1133-1142.

Quinsey, Joseph
1980 "Some problems in logic", Ph.D. dissertation, St. Catherine's College, Oxford University.

Simpson, Stephen G.
1986 "Unprovable theorems and fast-growing functions", to appear in S.G. Simpson (ed.), Logic and Combinatorics, American Mathematical Society.

Smoryński, Craig
1977 "The incompleteness theorems", in J. Barwise (ed.), Handbook of Mathematical Logic, North-Holland Pub. Co., pp.821-865.

1980 "Some rapidly-growing functions", The Mathematical Intelligencer 2, pp.149-154.

Solovay, Robert M.
1976 "Provability interpretations of modal logic", Israel Journal of Mathematics 25, pp.287-304.

Spencer, Joel
1983 "Large numbers and unprovable theorems", American Mathematical Monthly 90, pp.669-675.

Vardanyan, V.A.
1985 "O predikatnoj logike dokazuemosti", (preprint) Naucnyj sovet po kompleksnoj probleme "Kibernetika" AN SSSR, Moscow.

Wang, Hao
1974 From Mathematics to Philosophy, Routledge and Kegan Paul, pp. 298-328.

Webb, Judson
1980 Mechanism, Mentalism, and Metamathematics, D. Reidel Publishing Co.

GODEL'S LIFE AND WORK

P. P. Petkov*

Sector of Mathematical Logic
Faculty of Math. and Mech.
Boul. Anton Ivanov 5
Sofia 1126, Bulgaria

Since the end of last century an investigation of the methods of mathematical reasoning has been gaining momentum; due to that development the field of mathematical logic has originated and advanced. Kurt Gödel was one of those who contributed most to its forming and progress. Information about Gödel's life is still comparatively scanty. Most likely that is due to the secluded and simple way of life he led.

The list of Gödel's works presented here is not an exhaustive one. A more comprehensive list of his scientific writings is to be found in the union of (Dawson, 1983) and (Dawson, 1984). A list of 25 of Gödel's works was published in the collection (Foundations of Mathematics, 1969), but it was not an exhaustive one. From (Dawson, 1984) one is left with the impression that the list presented there was, most probably, made by Gödel himself. If so then it is very likely to include those of Gödel's works which he considered, at the time, to be worthy of more attention. As is well-known, an editorial board consisting of: S. Feferman, J. W. Dawson, Jr., S. C. Kleene, G. H. Moore, R. M. Solovay and J. van Heijenoort is taking steps to publish the complete works of Gödel. The first volume of his works, containing all his publications in the period 1929-1936, has already come out (Gödel, 1986). It also includes notes providing the historical context for these works and pointing out their relation to and influence on other studies. Gödel's "Nachlass" has been catalogued by J. W. Dawson, Jr. and is now accessible to research workers (as has already been reported by several sources) at Firestone Library at Princeton University. E. Köhler, W. Schimanovich and P. Weibel are preparing a volume on Gödel's life and work (in German). As Feferman's preface to (Gödel, 1986) reads "their volume is in many respects intended to be complementary" to Feferman-Dawson-Kleene-Moore-Solovay-van Heijenoort's edition.

The present article is a completely revised version of the report (Petkov, 1986). My Bulgarian colleagues L. Avramov, D. Vakarelov, G. Gargov, L. Ivanov, S. Passy, S. Radev, D. Skordev, V. Sotirov, J. Stojanov and T. Tinchev were of great help to me while I was preparing the above mentioned report. They called my attention to some facts and pointed out, let me use or supplied bibliography essential to my work. E. Dimitrov and

---

* English translation by Snezhana Anastassova

V. Mihova placed their Bulgarian translations of (Christian, 1980) and of the German text in (Grattan-Guinness, 1979) at my disposal. John W. Dawson, Jr. looked through the original version of this article and made many useful remarks. Here I should like to express my thanks to them and to all other colleagues who helped me with their attention and comments.

There are no less than 125 sources in Bulgarian (original or translation texts) which mention Gödel. Since the sources in Bulgarian dealing with the problems of mathematical logic are few in number and are either informative, philosophical or elementary in nature, the already mentioned number of 125 sources again shows the importance attached to Gödel's research work. Few mathematical achievements stir up such great interest, meet with such a wide response and come to be quoted almost everywhere due to their crucial nature.

1. CHILDHOOD, BOYHOOD AND SECONDARY EDUCATION

The great logician was born on 28 April 1906 in Brünn, Moravia, which was then part of the Austro-Hungarian empire (it is now the town of Brno, Czechoslovakia). His father Rudolf's Catholic family came to Brünn from Vienna. The father was managing director and part owner of one of the leading textile firms in Brünn. His mother Marianne, whose maiden name was Handschuh, came from a Protestant family of weavers. At the beginning of his career her father was a weaver in the Rhineland region. Later he moved to Brünn, continuing in the textile trade. The child Kurt was baptized in the German Lutheran Church and named Kurt Friedrich. Kurt Gödel was the second child in the family. He had an elder brother called Rudolf. According to his biographers the family lived in comparatively easy circumstances. The child is said to have been very bright and to have possessed a great deal of inquisitiveness. Naturally, most children like to ask questions, but the fact that little Kurt's nickname "Mr. Why" (Christian, 1980) was kept long in the family memory goes to show that his questions, probably, made a great impression on them in contrast to those of the other children.

Gödel received his elementary and secondary education in his native town. From 1912 to 1916 he attended the Evangelische Privat-Volks-und Bürgerschule and from 1917 to 1924 the Staats Realgymnasium mit deuscher Unterrichtssprache. Latin was obligatory at his secondary school. French was also taught as a modern foreign language. Gödel took two extra courses besides the obligatory subjects - his extras were shorthand and English. The records preserved show that he had the highest possible marks in all his subjects. At the same time it is evident from the records that he was often excused from school, which speaks of his poor health. During 1915-1916 and 1916-1917 school years he was excused from his sports classes. He suffered from rheumatic fever and he had a severe attack probably during the 1915-1916 school year. In 1975 Gödel filled in a questionnaire made up by the sociologist Burke D. Grandjean. Some of his answers are discussed in (Dawson, 1984), (Feferman, 1986) and (Dawson, 1986). According to one of his answers Gödel remembered that he had become aware of his interest in mathematics about the age of fourteen.

In 1924, after his graduation from the Realgymnasium, he took up physics at the University of Vienna. Another of his answers in Grandjean's questionnaire read that until then he had had a hazy notion of the intellectual and cultural life in Vienna at that time, except for the information he received through the Neue freie Presse.

2. UNIVERSITY YEARS

At the beginning of his studies Gödel was greatly impressed by Philipp Furtwängler's lectures on number theory (Wang, 1981). In 1926 his interests turned from physics to mathematics. In (Wang, 1981) it was pointed out that the reason for that was "his interest in precision". As will be seen further in some of his later articles, Gödel again dealt with physics problems. At the same time, of course, his striving to achieve the greatest possible precision and clarity never gave way.

During his studies Gödel drew closer to Professor Hans Hahn (1879-1934). Hahn was one of the mathematicians who exercised considerable influence on the development of contemporary functional analysis. He was the author of a very good book on the theory of real functions. In its second edition (Hahn, 1932) he expressed his acknowledgements to Gödel, Hausdorff, Hurewicz, Knaster, Lindenbaum, Menger, Nöbeling, Radakovic, Sierpiński and Szpilrajn "for their benevolent support".

About 1926 Gödel, under Hans Hahn's influence, started attending the meetings of Moritz Schlick's (1882-1936) circle. As a student Schlick studied physics. He took his doctoral degree under Max Planck's supervision, his dissertation dealing with the physics problems of light. From 1922 to 1936 he was holder of the chair for Philosophy of the Inductive Sciences, becoming in that way Ernst Mach and Ludwig Boltzmann's successor. M. Schlick's circle (also known under the name of the "Vienna Circle") was a significant phenomenon in West European philosophy. United under one name - logical (neo)-positivists - but still having divergent views, the members of that circle sought to transfer the methods of the exact sciences to philosophy.

The article (Wang, 1981) - the bulk of its text having been prepared under Gödel's supervision and approval - shows that, in spite of his accepting part of the neo-positivistic criticism of the state in which philosophy was at that time, he was against their negation of objective reality and their stamping of metaphysical problems as meaningless. When Gödel took part in discussions his stand was always non-positivistic. The only thing he approved of was their method of analyzing philosophical concepts through the application of the methods of mathematical logic.

Another mathematician who used to attend the Vienna Circle meetings was Karl Menger (1902-1985). He wrote in the fields of topology, mathematical logic and mathematical economy, as well as mathematical didactics. Later Gödel published quite a number of his works in the proceedings of the mathematical colloquim organized and directed by Menger.

In (Feferman, 1986), W. Wirtinger, E. Helly, W. Mayer and L. Vietoris were also mentioned as mathematics lecturers at the University of Vienna at that time.

Thus we get to the year 1928, the time when Gödel had to enroll at the University for his senior year. That year Hilbert and Ackermann's famous book on mathematical logic came out and Gödel acquainted himself with its contents. The book posed the problem of the completeness of the classical predicate calculus as open. It was time for Gödel to choose a theme for his doctoral thesis and he chose exactly that problem. His thesis, in which he arrived at a positive solution of the problem, was accepted for defense in the autumn of 1929. In that dissertation Gödel made his first discovery in the field of mathematical logic which, as is well-known, is taught now in almost all university courses in that discipline.

3. ACTUAL INFINITY ABSTRACTION; SET THEORY; PARADOXES; HILBERT'S PROGRAM

The way things stood in mathematics and mathematical logic during the 30's has been described, time and again, in many works. It may be said that it was a period of crisis as well as of development.

Let us assume that there is a set consisting of a small number of elements - for example 4-5 - and we are considering the simple properties of these elements and the simple relations arising between them. The term "simple property" means that one can easily tell whether a certain element has it or not. The other term "simple relation" bears the same meaning - one can tell for every group of elements whether they enter into the relation under consideration or not. Reasoning about such properties and relations in the case of the above mentioned sets is usually very easy to check, and arguments questioning their real meaning are difficult to find. Things stand differently if the set of objects under consideration has a large number of elements - for example

$$\left.10^{10^{\cdot^{\cdot^{\cdot^{10}}}}}\right\} \text{10 times}$$

How, for instance, is an assertion of the following type then to be understood: (A) for every element x of this set there is an element y of the same set so that the relation P(x,y) is satisfied? The number of elements considered is too great. Does the assertion (A) say that all possible cases have been checked separately? Obviously not. What does it say then?

The case does not look simpler when the set of objects considered is infinite. (Yet the very assumption of the existence of infinite sets is an idealization, meant to simplify, to a certain degree, theoretical reasoning.) The following question arises here: in what way (i.e., by means of what logical laws) is it advisable to construct one's reasoning in cases when infinite sets are being considered? That question can be given different, non-equivalent answers. One of them is related to the idealization called actual infinity abstraction. The acceptance of that idealization lies in the following: operations with infinite sets are done, if possible, by pretending that they are finite sets having a small number of elements. With complex properties and relations arising between the objects of infinite sets, operations are done as if we were operating with simple properties and relations like the ones mentioned above. Of course that can be done only within certain limits - every mathematician knows that many of the properties of the finite sets are not valid for infinite sets and vice versa. At any rate the acceptance of the actual infinity abstraction means disregarding, as much as possible, the fact that infinite sets cannot be "surveyed at a glance", that we cannot have command of all the elements of an infinite set at the same time. To accept the actual infinity abstraction means, in figurative language, to accept a type of reasoning characteristic of people possessing great mind-power, people capable of envisaging the infinite sets in the same way as others can envisage sets containing 4-5 elements. A typical example of an assertion for whose proof the actual infinity abstraction is used considerably is the so-called D. König's lemma (König, 1926): a tree in which there is a finite number of branches from every vertex and which contains finite paths of arbitrary length, contains also an infinite path.

The theory of sets created by George Cantor (1845-1918) at the end of the preceding century is grounded on concepts related to the actual infinity abstraction. It is common knowledge that in 1895 Cantor noticed a certain contradiction in his theory and later other mathematicians also arrived at contradictions in the same theory. Since there had already been a well established general feeling that the set theory could underlie all of mathematics (or almost all of it), these contradictions attracted a great deal of interest. Different remedies for treating the disease were suggested. Following Bertrand Russell's (1872-1970) suggestion and using Gottlob Frege's (1848-1925) achievements, a hierarchical set theory was created in which no contradictions were detected at the time, but it could not be stated with absolute certainty that such would not be spotted later on. The case of axiomatic set theory, founded by Ernst Zermelo (1871-1953) in 1908 and improved by Abraham Fraenkel (1891-1965) about 1925, was similar. In addition to that, another question arose, the question about the methods of reasoning used for determining that a certain mathematical theory was free of contradictions. It seems obvious that if we want to prove the consistency of a theory, then the methods we use to demonstrate this should be more reliable than the ones used by the theory itself - otherwise the demonstration of the consistency would appear to be, on the whole, too unconvincing.

Moreover, the question about inconsistency was not the only one that mattered, as Luitzen E. J. Brouwer (1881-1966) started vigorously arguing at the time. No less important was the question about the intuitive convincingness (we could probably add - about the use in practice) of mathematical achievements, in which there always is, nolens volens, some kind of abstraction. Brower leveled his splashing and peremptory criticism at the methods of reasoning employed then, expressed mainly in Cantor's set theory and its applications. To put it mildly, he considered their value to be quite problematic; according to him the results obtained by means of those methods could be, if not contradictory, then at least misleading.

It is common knowledge that David Hilbert (1862-1943) took it upon himself to repulse Brower's attack through carrying out a program of research on mathematical logic. The principle ideas constituting Hilbert's program can be summed up in the following four items:

3.1 Disregarding the Meaning of the Predicates under Consideration

Instead of operating with predicates, operations are performed on the formal expressions (i.e., strings of symbols) which express them. When considering a certain mathematical theory, it has only to be specified which strings of symbols are regarded as its axioms and which rules for operating with strings of symbols correspond to the logical laws used in driving other assertions from the ones belonging to that theory. The description of the axioms and the rules affects only the written form of the predicates but not their meaning.

3.2 Finitism

It is suggested that axiomatic mathematical theories should be formulated in such a way that:

(a) there should be a finite number of axioms and rules of inference;

(b) every rule of inference should be relevant to a finite number of objects, i.e., in deriving a new string of symbols it should use only a finite number of string symbols;

(c) there should be a clearly stated method by means of which, without any major difficulties (and by performing only a finite number of elementary operations), every string of symbols could be identified as being an axiom or not; also there should be a clearly defined method by means of which it could be stated, for every string of symbols, whether it is derivable from other given strings of symbols by means of a given rule of inference.

It was believed (by Hilbert and his followers) that every mathematical theory (or at least a substantial part of every mathematical theory) could be reduced to an axiomatic theory satisfying requirements (a), (b) and (c).

Naturally, in the further reasoning about a certain theory there should be absolute certainty. For that purpose it was suggested that all similar reasoning should follow the same lines so that, if need arose, it could also be made into a theory (called the metatheory of the first one), satisfying requirements (a), (b) and (c). In that way the actual infinity abstraction would prove to be isolated. However, after the strings of symbols in a certain axiomatic system were invested with meaning again, the possibility remained for them to turn out to be assertions, properties and relations having direct bearing on infinite sets; also, after such investing with meaning, the axioms and the rules of inference might prove to be in agreement with the actual infinity abstraction. It was expected that those circumstances would afford an opportunity for theories, that used so-called "dubious" abstractions, to regain their previously well-respected status.

3.3 Principles of the Ideal Elements; Conservative Extensions of Mathematical Theories

Critics regarded those assertions, properties and relations, as well as methods of demonstration, in which any involvement of the actual infinity abstraction could be detected, as not being sufficiently "real". What could be the use of such assertions? For example: König's lemma claimed that in every tree of a certain kind an infinite path could be traced. What was the "real" meaning of that assertion? Did the proof of the lemma provide a method by means of which, in every similar tree, an infinite path could "really" be traced, i.e., did it provide a rule which gave for each number n, after a finite number of steps, the nth point of the path? It was easy to see that the proof of the lemma did not do that. Then what could the use of assertions like König's lemma be?

Hilbert's answer was that such assertions had an auxiliary part to play; their use made the study of "real" assertions easier. In the same way having a knowledge of complex numbers made it easier to solve problems arising in the theory of real numbers. In relation to mathematical theories, Hilbert's idea can be summed up as follows: think of a theory R whose objects, predicates and rules of inference are recognized as sufficiently "real". That theory could be extended to another theory I through the addition of new - "ideal" - objects, predicates and rules of inference, whose reality might seem problematic, at least on the face of it. If every assertion of the theory R which could be proved for I could also be proved to hold true for R, then we would rightfully say that the theory I is a conservative extension of R. Conservative extensions could be useful, because it is a more or less easier task to find proofs in them than in the initial theories. Hilbert obviously believed that Cantor's set theory (with certain corrections made to it to eliminate the existing contradictions) was an example of a very successful theory which could be accepted as I in relation to the "real" part of mathematics.

3.4 Reducing the Conservation Problem to the Consistency Problem

Hilbert thought that if the theory I was consistent then it should be a conservative extension of R. His reasoning concerning that matter can be found in his report of July 1927, published in (Hilbert, 1928). An account of his reasoning can be found in (Smoryński, 1977) and in (Smoryński, 1985).

Here we end our outline of Hilbert's program for the logical study of mathematical theories (and for rehabilitating some of them).

4. THE COMPLETENESS THEOREM FOR THE PREDICATE CALCULUS

A set of "logical laws" might turn out to be suitable for a certain group of mathematical theories and unsuitable for another. In Hilbert and Ackermann's book, mentioned above, a description was given of a formal axiom system for the first-order predicate calculus in accordance with the actual infinity abstraction. The axioms of that system were valid and the rules of inference led from valid formulas to valid formulas, provided that the universe was a finite set and the atomic formulae were interpreted by means of simple predicates (with "simple" bearing the meaning elucidated at the beginning of paragraph 3). Hilbert and Ackermann wrote that they did not know of a proof from which it could be seen that the axiom system under consideration provided an exhaustive account of all logical laws (i.e., of the logical laws expressible in the language of the first-order predicate calculus and in accord with the actual infinity abstraction). Of course, it was not yet clear whether a formal system of some kind (i.e., an axiom system constructed according to the terms of paragraph 3.1 and satisfying the requirements of paragraph 3.2) which could provide such an exhaustive account could be found at all.

In his doctoral dissertation Gödel proved that the axiom system of the predicate calculus, formulated in essence by Russell, was complete. (Here he had in mind the predicate calculus with identity as well as the predicate calculus without identity.) The last pages of his dissertation were devoted to proving that every (at most denumerable) system of formulae of the predicate calculus was either satisfiable or contained a finite subsystem whose conjunction was refutable. The results of the dissertation were published in (Gödel, 1930), where the compactness theorem was also stated in its final form for the denumerable case: "for a denumerably infinite system of formulas to be satisfiable it is necessary and sufficient that every finite subsystem be satisfiable (Theorem X)". In (Gödel, 1930) the introduction to the (Gödel, 1929) version was omitted. An argument, bearing a strong resemblance to König's lemma, constituted an important step in the proof of the completeness theorem. Use of the actual infinity abstraction in the metatheory here was, to a certain extent, obligatory. With other naturally arising definitions of logical validity, a theorem similar to Gödel's completeness theorem proved to be impossible - see for example (Plisko, 1977).

Many versions of the proof of the completeness theorem have since been given. At present, similar theorems have been proved for a number of logical systems differing from the predicate calculus in that in them the basic operations also express other notions not present in the predicate calculus. Notes about the historical situation and the connections between Gödel's completeness theorem and the works of Emil Post (1897-1954). Thoralf Skolem (1887-1963), Paul Bernays (1888-1978) and some others can be found in (Zygmunt, 1973), (Kleene, 1976) and (Dreben and van Heijenoort, 1986).

In (Gödel, 1932), in answer to a question of Menger, a theorem was deduced, according to which every consistent set of propositional formulae has a model, including the case of indenumerably many formulae.

In (Maltsev, 1936), Anatolii Maltsev (1909-1967) extended the compactness theorem to include the case when the set of predicate formulae under consideration was non-denumerable. The generalization thus obtained has many applications. One of the simplest, and perhaps the most impressive, of its applications is the proof of the existence of infinitesimals. Since 1960, Abraham Robinson (1918-1974) and his followers have been developing non-standard analysis on the basis of that existence.

5. FROM 1930 to 1933; WITHOUT ACADEMIC RANK OR POST

Gödel graduated from the university at the time of the Great Depression. That probably was the main reason he could get neither a suitable post nor an academic rank. His father died on 23 February 1929 - about a year before Gödel defended his dissertation. After his father's death his mother went to live with him and his brother, an already established radiologist, in Vienna (Christian, 1980). According to several biographical sources, after the father's death the family lived in comparatively easy circumstances and could meet Kurt's needs.

In (Menger, 1979) the following moving story is told. In the spring of 1932, on Menger's initiative, the university lecturers - the chemist H. Mark, the physicist H. Thirring, H. Hahn, K. Menger and also G. Nöbeling - delivered a series of five popular lectures entitled "Crisis and Reconstruction in the Exact Sciences". The lectures aroused great interest and the money collected from the entrance fees (the tickets were as expensive as those for the Vienna Opera House) was intended to help young mathematicians who had no academic post.

It was at the beginning of that period of unsettled social standing, in 1930, that Gödel proved his famous incompleteness theorems.

6. INCOMPLETENESS

In the period 1925-1930 intensive logico-mathematical research was carried out in connection with Hilbert's program. The main task was that of proving the consistency of some "big" formal system such as one of the systems for set theory or analysis. Proofs had already been published achieving essential though humbler objectives - namely, demonstrating the consistency of some fragments of the arithmetic of natural numbers. What remained to be done was to proceed to the more complex theories; for that purpose, advanced ideas were at hand, and their realization was considered to be only a matter of overcoming certain technical difficulties.

Gödel also tackled the problem of the consistency of analysis - as (Wang, 1981) shows - in the summer of 1930. At that time Hilbert had suggested a "direct" finitistic method for proving consistency, but Gödel thought his method inadequate. He believed that difficulties should be overcome one by one and that in that case, first the consistency of number theory should be established, and only after that should the relative consistency of analysis to number theory be proved. Later on, while trying to solve the second problem, he arrived at the conclusion that in formal axiom systems similar to the axiom systems then known for set theory, assertions neither provable nor refutable were inevitably present. In that way he arrived at the first incompleteness theorem (not for arithmetic, but for a somewhat different formal system). In (Dawson, 1984b) we read that on 26 and 29 August 1930, Gödel told Carnap, Feigl and Waismann about his discovery while talking to them at the Café Reichsrat in Vienna. Gödel's first recorded announcement of the discovery of that theorem can be found in the proceedings of a discussion on the foundations of mathematics which took place on 7 September 1930 in Königsberg. The following persons took part in the discussion (listed in the order they first took the floor): Hahn, Carnap, von Neumann, A. Scholz, Heyting, Gödel and Reidemeister. At that time, according to (Wang, 1981), von Neumann was greatly interested in Gödel's results and had a talk with him, as a consequence of which Gödel was stimulated to arrive at an undecidable assertion in arithmetic. At the same time, but completely independently of that result, Gödel also deduced the second incompleteness theorem. Hahn presented an abstract of these results to the Vienna Academy of Sciences on 23 October 1930. A detailed account of the results was published in (Gödel, 1931). Unlike the completeness theorem, the proofs of the incompleteness theorems satisfy rather strict requirements for constructivity. The excitement that the incompleteness theorems were met with is easy to explain, considering their negative effect on the expectations that Hilbert's program would be successful.

On 15 September 1931, Gödel gave a lecture on the results he had obtained about incompleteness, at the autumn meeting of Deutsche Mathematiker-Vereinigung at Bad Elster. In this connection, a short-term correspondence started between him and Zermelo, of which an account has been given in (Grattan-Guinness, 1979) and (Dawson, 1985). Zermelo was wrong to think that there was a mistake in Gödel's work, but at the same time, Zermelo's critical attitude towards finitism, expressed in these letters, cannot be regarded as completely unjustified. In 1936 J. B. Rosser improved the incompleteness theorems by disposing of the ω-consistency condition. In 1950 R. M. Robinson suggested a simple arithmetical system for which the incompleteness theorems were valid. Considerable simplification of the structure of undecidable propositions derived from the incompleteness theorems was achieved by the famous result in (Matiyasevich, 1970). Taking into account all these achievements, the incompleteness theorems would read as follows:

First Incompleteness Theorem: let T be a formal system which is a recursively axiomatizable extension of Robinson's system. Then a Diophantine equation can be specified such that:

(1) if T is consistent, then there are no natural numbers which are solutions of the equation; and

(2) if T is consistent, then in T we can neither prove nor refute the fact that the equation under consideration has no solution in the set of natural numbers.

Second Incompleteness Theorem: let T be a formal system which is a recursively axiomatizable extension of Robinson's system. Then a Diophantine equation could be worked such that:

(1) the assertion that T is consistent is equivalent to the assertion that the equation has no solution in the domain of natural numbers; and

(2) if T is consistent, then it cannot be proved in T that the equation has no solutions in the set of natural numbers.

The incompleteness theorems have been the object of many research studies. Here we can name, for example, the following: (Kleene, 1950), (Smoryński, 1985), (Kanovic, 1977) and (Kanovic, 1984)

7. FROM 1933 TO 1940

In June 1932, Gödel presented his research work on incompleteness as his Habilitationsschrift at the University of Vienna, and on 11 March 1933 he became Privatdozent. In (Christian, 1980) some details about that event can be found; for example, excerpts are quoted both from Gödel's autobiography and Hahn's report (written on the occasion). In his report, Hahn strongly recommended that Gödel should be conferred an academic title. The title Privatdozent did not bring Gödel a regular salary, but at least it gave him the right to deliver lectures. Before proceeding we must say that this paper will not discuss some of Gödel's articles published before 1933. They dealt with the intuitionistic propositional calculus, intuitionistic arithmetic and modal logic.

During the academic year 1933-1934 Gödel was invited to lecture at the newly founded Institute for Advanced Study in Princeton, USA. Up to the academic year 1938-1939 he occasionally delivered lectures, sometimes at the University of Vienna, sometimes at the Institute for Advanced Study in Princeton. From February to May 1934 he lectured on the incompleteness theorems at Princeton (Gödel, 1934); these lectures have gone down in history as one of the series of episodes which led to the specification of the mathematical notion of "algorithm". More information on the history of that problem can be found in (Kleene, 1952), (Kleene, 1967) and (Kleene, 1986).

During that period Gödel also studied the axiom of choice and the continuum hypothesis. It is common knowledge that there have been many attempts to prove the continuum hypothesis, but all of them have been unsuccessful. It was exactly the problem of proving the continuum hypothesis that occupied first place in Hilbert's famous 1900 list of problems. As regards the axiom of choice, it seemed, on the face of it, as if it should not have aroused any doubt (at least from the point of view of the actual infinity abstraction). However, its formulation and application in 1904 by E. Zermelo stirred up a violent controversy in which J. Hadamard (1865-1963), R. Baire (1874-1932), F. Bernstein (1878-1956), E. Borel (1871-1956), H. Poincaré (1854-1912), B. Russell, H. Lebesgue (1875-1941) and some other mathematicians took part. Despite the fact that in the set-theoretic axiom systems created at that time no inconsistencies were detected, some research workers arrived at consequences in those systems that were interpreted as being either too strong or paradoxical. In the proofs of the controversial theorems the presence of the axiom of choice was constantly noticed (other axioms, of course, also were there, but they did not attract attention). The question of whether the axiom of choice could be added to the other axioms of the set-theoretical axiom system of the time without resulting in inconsistency was a crucial one.

According to (Wang, 1981), Gödel started thinking about the continuum problem in 1930, after he had heard of Hilbert's suggestion for a proof of the continuum hypothesis. Probably while studying set theory Gödel was at the same time considering those two problems - that of the continuum hypothesis and that of the axiom of choice. According to (Dawson, 1984a) in October 1935, he told von Neumann of his proof of the relative consistency of the axiom of choice. Unfortunately, a month later Gödel started having problems with his health, and up to 1937 he was not able to do any serious work. A shorthand note found in Gödel's Nachlass (extended by Cheryl Dawson) states that on the night of 14/15 June, Gödel finally succeeded in proving, in essence, the relative consistency of the generalized continuum hypothesis. His first publication of those findings was (Gödel, 1938).

Much later, during 1963 and 1964, P. J. Cohen proved the independence of the axiom of choice and the continuum hypothesis as well - see (Cohen, 1963), (Cohen, 1964), (Cohen, 1966).

On 20 September 1938 Gödel married Adele Porkert. The ceremony took place in Vienna. Two weeks after the wedding he went back to the USA for the 1938-1939 academic year while Adele remained in Austria. Before that, on 13 March 1938, Austria's annexation to Nazi Germany was declared.

During the fall term of the 1938-1939 academic year, Gödel delivered lectures on the consistency of the axiom of choice and the generalized continuum hypothesis at the IAS. During the spring term he was at the University of Notre Dame where, together with Karl Menger, he conducted a seminar on elementary logic.

In the summer of 1939 Gödel rejoined his wife in Vienna. After Austria's annexation to Germany the academic rank of Privatdozent was abolished. Because of that, Gödel applied for the new academic rank of Dozent neuer Ordnung. At the same time, husband and wife took steps to arrange their emigration to Princeton. They were permitted to depart, and from 18 January to 4 March 1940 they traveled through the USSR, Manchuria and Japan. From then on till the end of their lives they lived in Princeton, where Gödel worked at the Institute for Advanced Study ((Christian, 1980), (Dawson, 1984), (Dawson, 1986)).

8. PRINCETON, NEW JERSEY. THE INSTITUTE FOR ADVANCED STUDY

In 1940, as we read in the (Encyclopedia Americana, 1949), the town of Princeton, New Jersey, USA, had 7719 inhabitants. Princeton is about 30 miles from New York and about 10 miles from the state capital, Trenton. The town also is the site of a famous university.

The Institute for Advanced Study was created in 1930 and opened in 1933 to be a research center on the supra-university level. According to (Kaysen, 1976) its first director, Abraham Flexner, made it the Institute's object to study fundamental problems which challenge the intellect, without fear of flying in the face of major difficulties. Immediate application of the results obtained to the needs of the moment was not to be expected. The Institute's staff consisted of two distinct groups: the first one was the considerably small but fairly unchangeable body of its permanent members, and the second was made up of its temporary members, who got a one-year appointment. The latter were, in general, young, chosen people who had recently defended their dissertations, and the Institute afforded them the opportunity to work in close contact with the members of its whole research staff. Due to the competent choice of members the Institute soon gained international prestige. The School of Mathematics was its oldest and largest division. Its group of physicists was also a representative one - for example, Albert Einstein worked there from 1933 until the end of his life. The Institute was generally well liked by the young mathematicians as a place for creative work - see, for example, (Bott, 1985).

9. FROM 1940 TO 1978

In 1941, according to (Wang, 1981), Gödel arrived at a new proof of the relative consistency of the axiom of choice, still unpublished. On 15 April 1941, he delivered a lecture at Yale University; a brief publication of the ideas there was (Gödel, 1958). About 1943, according to (Wang, 1981), but in the summer of 1942, according to (Dawson, 1986), Gödel found a proof of the independence of the axiom of choice in finite type theory.

However, he did not continue his work along these lines and did not prepare a publication. At that time his interests were directed more and more to philosophy. In 1946 Gödel became a permanent member and in 1953 a professor at the Institute for Advanced Study. His biographers draw attention to his strong sense of responsibility, which made him devote much time to the works of the junior members of the Institute. "He would pore interminably over the writings of candidates for the Institute for Advanced Study, even though they were only candidates for a year's membership and the writings were remote from his field" wrote (Quine, 1979). See also (Feferman, 1986).

In 1949 and 1952 Gödel published articles on Einstein's theory of relativity. Gödel's and Einstein's biographers mention that the two scientists were often seen together, especially on their way home or riding together on the bus. Some of Einstein's visitors report having talked to Gödel at Einstein's. After Einstein's death Gödel (together with Bruria Kaufman) took part in arranging his scientific manuscripts. (See, for example, (Pais, 1982).) According to (Wang, 1981) Gödel's interest in the theory of relativity, however, derived from his interest in Kant's philosophy of space and time rather than his conversation with Einstein.

For a comparatively long period of time Gödel took an active interest in Leibniz's scientific heritage. Later he also started studying Husserl's works. More information about that period can be found, for example, in (Feferman, 1986).

All of Gödel's biographers write about his health problems. Gödel followed a strict diet and was very frail-looking. He was susceptible to cold and used to wear warm clothes all the time, despite Princeton's mild climate. "He could be seen on a warm day trudging along a Princeton street in an overcoat" (Quine, 1979). In 1977 Adele underwent a serious operation. Her poor health prevented her from looking after her husband and on 29 December 1977, Gödel was hospitalized. He died on 14 January 1978.

REFERENCES

Bott, R., 1985, On topology and other things, Notices of the Amer. Mathem. Soc., 32, No. 2 (March):152-158.

Christian, C., 1980, Leben und Wirken Kurt Gödels, Monatshefte für Mathem., 89:261-273.

Christian, C., 1983, Der Beitrag Gödels für die Rechtfertigung der Leibnizschen Idee von den Infinitesimalen, Sitzungsberichte der Osterreichischen Akademie der Wissenschaften, Mathem. Naturw. Klasse, 192, Heft 1-3:25-44.

Christian, C., 1984, Gödel's contribution to the justification of Leibniz's notion of the infinitesimals, Historia Mathematica, 11:215-219.

Cohen, P. J., 1963, The independence of the continuum hypothesis I, Proc. Nat. Acad. Sci. USA., 50:1143-1148.

Cohen, P. J., 1964, The independence of the continuum hypothesis II, Proc. Nat. Acd. Sci. USA., 51:105-110.

Cohen, P. J., 1966, "Set Theory and the Continuum Hypothesis", Benjamin Inc., New York.

Davis, M., 1977, "Applied Nonstandard Analysis", Wiley, New York.

Dawson, J. W. Jr., 1983, The published work of Kurt Gödel: an annotated bibliography, Notre Dame JFL., 24:255-284.

Dawson, J. W. Jr., 1984, Addenda and corrigenda to "The published work of Kurt Gödel", Notre Dame JFL., 25:283-287.

Dawson, J. W. Jr., 1984a, Kurt Gödel in sharper focus, Mathem. Intelligencer, 6, No. 4:9-17.

Dawson, J. W. Jr., 1984b, Discussion on the foundations of mathematics, History and Philosophy of Logic, 5:111-129.
Dawson, J. W. Jr., 1985, Completing the Gödel-Zermelo correspondence, Historia Mathematica, 12:66-70.
Dawson, J. W. Jr., 1986, A Gödel chronology, in: (Gödel, 1986).
Dreben, B., and van Heijenoort, J., 1986, Introductory note to 1929, 1930 and 1930a, in: (Gödel, 1986).
"Encyclopedia Americana. Complete in 30 volumes", 1949, Americana Corporation.
Feferman, S., 1986, Gödel's life and work, in: (Gödel, 1986).
"Foundations of Mathematics. Symposium Papers Commemorating the Sixtieth Birthday of Kurt Gödel", 1969, J. Bulloff, T. C. Holyoke, S. W. Hahn, eds., Springer, Berlin.
Gödel, K., 1929, On the completeness of the calculus of logic, in: (Gödel, 1986).
Gödel, K., 1930, Die Vollständigkeit der Axiome des logischen Funktionenkalküls, Monatshefte für Mathem. und Physik, 37:349-360.
Gödel, K., 1930a, Eininge metamathematische Resultate über Entscheidungsdefinitheit und Widerspruchsfreiheit, Anzeiger der Akademie der Wiss. in Wien, 67:214-215.
Gödel, K., 1931, Uber formal unentscheidbare Sätze der Principia mathematica und verwandter Systeme I, Monatshefte für Mathem. und Physik, 38:173-198.
Gödel, K., 1932, Eine Eigenschaft der Realisierungen des Aussagenkalküls, Ergebnisse eines mathematischen Kolloquiums, 2:27-28.
Gödel, K., 1934, "On undecidable propositions of formal mathematical systems" (mimeographed lecture notes, taken by S. C. Kleene and J. B. Rosser), Princeton. Reprinted with revisions in: "The Undecidable", M. Davis, ed., Raven Press, Hewlett, N.Y.
Gödel, K., 1938, The consistency of the axiom of choice and of the generalized continuum hypothesis, Proc. Nat. Acad. Sci. USA., 24:241-242.
Gödel, K., 1949, An example of a new type of cosmological solutions of Einstein's field equations of gravitation, Rev. of Modern Physics, 21:447-450.
Gödel, K., 1952, Rotating universes in general relativity theory, in: "Proceedings of the Intern. Congress of Mathematicians, Cambridge, Massachusetts, USA, 30 August - 6 September 1950, I", Amer. Math. Soc., Providence, R.I.
Gödel, K., 1958, Uber eine bisher noch nicht benützte Erweiterung des finiten Standpunktes, Dialectica, 12:280-287.
Gödel, K., 1986 "Collected Works, Vol. I, Publications 1929-1936", Eds. by S. Feferman (ed. in chief), J. W. Dawson, Jr., S. C. Kleene, G. H. Moore, R. M. Solovay, J. van Heijenoort, Oxford University Press, New York; Clarendon Press, Oxford.
Grattan-Guinness, I., 1979, In memoriam Kurt Gödel: his 1931 correspondence with Zermelo on his incompletability theorem, Historia Mathematica, 6:294-304.
Hahn, H., 1932, "Reele Funktionen, Erster teil, Punktfunktionen", Akad. Verlagsgesellschaft MBH., Leipzig.
Hilbert, D., 1928, Die Grundlagen der Mathematik, Abhandlungen aus dem math. Seminar der Hamburgischen Universität, 6:65-85.
Kanovic, M. I., 1977, Otsenka slozhnosti nepolnoti arifmetiki, Doklady Akademii Nauk SSSR., 238:1283-1286.
Kanovic, M. I., 1984, O nezavisimosti invariantnich predlozhenii, Doklady Akademii Nauk SSSR., 276:27-31.
Kaysen, C., 1976, "The Institute for Advanced Study. Report of the Director, 1966-1976", Princeton.
Kleene, S. C., 1950, A symmetric form of Gödel's theorem, Indagationes Mathematicae, 12:244-246.

Kleene, S. C., 1952, "Introduction to Metamathematics", van Nostrand, New York.
Kleene, S. C., 1967, "Mathematical Logic", Wiley, London.
Kleene, S. C., 1976, The work of Kurt Gödel, Journ. Symb. Logic., 41:761-768.
Kleene, S. C., 1978, An addendum to "The work of Kurt Gödel", Journ. Symb. Logic., 43:613.
Kleene, S. C., 1986, Introductory note to 1930b, 1931 and 1932b, in: (Gödel, 1986).
König, D., 1926, Sur les correspondances multivoques des ensembles, Fundamenta mathematicae, 8:114-134.
Kreisel, G., 1980, Kurt Gödel, 28 April 1906 - 14 January 1978, Biographical Memoirs of Fellows of the Royal Society, 26:148-224 (corrections ibid. 27:697, and 28:718).
Maltsev, A. I., 1936, Untersuchungen aus dem Gebiete der mathematischen Logik, Matematicheskii sbornik, 1:323-336.
Matiyasevich, Y., 1970, Diophantovost perechislimih mnozhestv, Doklady Academii Nauk SSSR., 191:279-282.
Menger, K., 1979, The new logic, in: "Selected Papers in Logic and Foundations, Didactics, Economics", D. Reidel, London.
Pais, A., 1982, "'Subtle is the Lord ...', The Science and the Life of Albert Einstein", Clarendon Press, Oxford; Oxford University Press, New York.
Petkov, P. P., 1986, Zhivotat i deloto na Kurt Gödel, in: "Proceedings of the Fifteen Spring Conference of the Union of Bulgarian Mathematicians, 2 - 6 April", BAN., Sofia.
Plisko, V. A., 1977, Nearifmetichnost klassa realisuemih predikatnih formul, Izvestya Academii Nauk SSSR., ser. Math., 41:483-502.
Quine, W. W., 1979, Kurt Gödel (1906-1978), Year Book of the Amer. Philos. Soc., 1978, 81:84.
Read, C., 1970, "Hilbert", Springer, New York.
Robinson, A., 1966, "Non-Standard Analysis", North-Holland, Amsterdam.
Robinson, R. M., 1950, An essentially undecidable axiom system, in: "Proceedings of the International Congress of Mathematicians, Cambridge, Massachusetss, USA., 30 August - 6 September 1950, I", Amer. Math. Soc., Providence, R.I.
Rosser, J. B., 1936, Extensions of some theorems of Gödel and Church, Journ. Symb. Logic, 1:87-91.
Smoryński, C., 1977, The incompleteness theorems, in: "Handbook of Mathematical logic", J. Barwise, ed., North-Holland, Amsterdam.
Smoryński, C., 1985, "Self-Reference and Modal Logic", Springer, New York.
van Heijenoort, J., 1986, see Dreben, B., and van Heijenoort, J.
Wang, H., 1981, Some facts about Kurt Gödel, Journ. Symb. Logic., 46:653-659.
Zygmunt, J., 1973, A survey of the methods of proof of the Gödel-Malcev's completeness theorem, in: "Studies in the History of Mathematical Logic", S. J. Surma, ed., Ossolineum, Wroclaw.

S U M M E R   S C H O O L

(invited papers)

CATEGORIAL GRAMMAR AND LAMBDA CALCULUS

Johan van Benthem

Mathematical Institute
University of Amsterdam
The Netherlands

1. INTRODUCTION

Categorial Grammar and logical Type Theory stem from the same historical source, viz. the Fregean and Russellian idea of a pervasive function/argument structure in Language. Nevertheless, the two fields have developed in quite different ways, one becoming a more linguistic enterprise, the other a more mathematical one. (For the former, see Bach et al., 1986, Buszkowski et al., 1986 - for the latter, Gallin 1975, Barendregt, 1981.) Even so, there is also a more theoretical logical component to Categorial Grammar, which has been studied recently by various authors (cf. Buszkowski, 1982, Došen, 1986, van Benthem, 1986a,e). And in that direction, various connections have emerged with research in Type Theory and Lambda Calculus - exploiting analogies between categorial grammars, Gentzen calculi for implication and fragments of typed lambda-languages. In this paper, we shall survey this development, adding various new results on definability and preservation.

2. CATEGORIAL GRAMMAR

The Standard System

Categorial Grammar is based on a correspondence between grammatical *categories* and semantic *types*. The latter are generated, starting from certain basic types, by successive pairing. For instance, basic types could be e ('entity') and t ('truth value'), with compounds (a,b) denoting *functions* from a-type objects to b-type ones. Now, simplex expressions in natural language will be assigned one or more types, while complex expressions may, or may not, acquire a type via a rule of Function Application:

if X has type (a,b) and Y has type a,

then the concatenation XY has type b.

If no preferred position is encoded for the argument expression, then XY will be admitted too, with type b.

Example (simple natural language sentences).

(1) Mary toils

$$\frac{e \quad (e,t)}{t}$$

(2) Mary hears Lily

$$\frac{e \quad \dfrac{(e,(e,t)) \quad e}{(e,t)}}{t}$$

(3) A child smiled

$$\frac{\dfrac{((e,t),\ ((e,t),t)) \quad (e,t)}{((e,t),t)} \quad (e,t)}{t}$$

These sentences illustrate a few common types: e (proper names), t (sentences), (e,t) (intransitive verbs, common nouns), $(e,(e,t))$ (transitive verbs), $((e,t),t)$ (noun phrases), $((e,t),\ ((e,t),t))$ (determiners). Other examples are (t,t) (one-place connectives ("not")), $(t,(t,t))$ (two-place connectives ("and", "or")), $((e,t),\ (e,t))$ (adjectives ("quiet"), adverbs ("quietly"), or - about the most complex case - $(((e,t),t),\ ((e,t),(e,t)))$ (prepositions).

If a complex linguistic expression receives a type through successive function application steps, starting from some initial type assignment to its simplex components, then it has been recognized as belonging to the corresponding category. In particular, initial assignments generate *languages* as sets of expressions receiving type t. The main attraction here is that recognition implies semantic interpretability, by the order of function application in the process of calculating the final type. Moreover, varying orders of recognition (for the same type) may model ambiguities of meaning:

Example (propositional formulas without brackets).

(1) $\neg$ p $\wedge$ q

$$\frac{\dfrac{(t,t) \quad t}{t} \quad \dfrac{(t,(t,t)) \quad t}{(t,t)}}{t}$$

("$(\neg p \wedge q)$")

(2) $\neg$ p $\wedge$ q

$$\frac{(t,t) \quad \dfrac{t \quad \dfrac{(t,(t,t)) \quad t}{(t,t)}}{t}}{t}$$

("$\neg(p \wedge q)$")

Flexible Rules

The standard framework has gradually been enriched with so-called rules of *type change*. One central example is the 'Geach Rule':

'expressions occurring in type (a,b) can also occur in type $((c,a),(c,b))$, for arbitrary types c'.

This rule accounts for the multiple uses of negation (as sentence negation: (t,t), or predicate negation: $((e,t),(e,t))$, or noun phrase negation, etc.), without having to postulate an infinity of initial types. For various other applications of this principle, see van Benthem (1986e). Another kind of type change is given in the 'Montague Rule':

'expressions occurring in type a can also occur in type $((a,b),b)$, for arbitrary types b'.

This rule allows us to use proper names (type e) as noun phrases (type $((e,t),t)$), whenever convenient. For other applications, see Zwarts (1986) or Keenan & Faltz (1985) (where all intransitive verbs have been raised from (e,t) to $(((e,t),t),t)$). Nowadays, most categorial grammars incorporate some such rules, to allow for smoother linguistic description.

Behind these rules of type change, a logical mechanism may be observed. The above categorial trees resemble logical derivations, with Function Application corresponding to Modus Ponens:

$$\frac{a \quad (a,b)}{b} \qquad\qquad \frac{\mathrm{a} \quad \mathrm{a}\to\mathrm{b}}{\mathrm{b}}$$

This analogy was discovered already in Lambek (1958). Type change rules correspond to valid principles of *constructive implication*, and the process of linguistic recognition for a sequence A of initial types to some outcome type b may be viewed as a deduction of the sequence $A \Rightarrow b$. (Thus, "Parsing as Deduction" was implemented by Lambek long before the popularity of this slogan in the seventies.) To capture the relevant transitions, a Gentzen-type natural deduction calculus may be employed. E.g., both the Geach Rule and the Montague Rule can be derived by iterating Modus Ponens and Conditionalization.

Example (deriving the Geach Rule).

$$\dfrac{\dfrac{\dfrac{\dfrac{\overset{1}{c} \quad \overset{2}{(c,a)}}{a} \quad (a,b)}{b}}{(c,b)}\ \text{(withdrawing 1)}}{((c,a),(c,b))}\ \text{(withdrawing 2)}$$

Still, not every valid principle of constructive implication reflects a correct law of natural language type change. For instance, there is no evidence for a transition $t \Rightarrow (e,t)$ (sentences do not transmute into intransitive verbs), or for a general contraction $(a,(a,b)) \Rightarrow (a,b)$. To derive these, vacuous or multiple occurrences of premises are to be withdrawn. And so,

Lambek's calculus differs from the usual logical one in its *structural* rules: in each use of Conditionalization, one and only one occurrence of the antecedent type may be withdrawn. Thus, we are doing a kind of relevant logic, making sure that each occurrence of a piece of evidence is used.

There can be different Lambek calculi, arising from varying stipulations on the structural rules. For instance, a decision is still needed as to the order of premises: arbitrary *permutations* might be permitted (as is usual in logic), but this is negotiable. (See van Benthem, 1986e, for some discussion.) Also, linguistic arguments have been put forward for liberalizing Conditionalization after all. E.g., "and" can jump from type $(t,(t,t))$ (sentence conjunction) to $((e,t),((e,t),(e,t)))$ (predicate conjunction): a transition whose derivation requires using some e-premise twice. (Technically, the structural rule of *Thinning* is to be invoked here.) Thus, the proper perspective is that of a *spectrum* of categorial calculi, ranging from the standard system (with Modus Ponens only) to full constructive conditional logic.

Remark There are several variants and extensions of the above categorial framework. For instance, Montague Grammar employs an additional *basic type* (s), to model intensional constructions. Moreover, the internal structure of linguistic categories may necessitate the use of *subtypes*, and other devices to separate expressions of different sorts. And finally, syntactic peculiarities may enforce the use of *directed* types $a\backslash b$ (left-searching), b/a (right-searching), instead of the undirected (a,b). Such further possibilities will be disregarded here.

3. RECOGNIZING POWER

The various categorial calculi each represent a mechanism for recognizing natural (and artificial) languages. For the standard format, with directed types, the resulting family is precisely that of the *context-free* languages. With richer combination modes, the question is still open. (See Buszkowski, 1982, and Buszkowski et al., 1986, for what is known of the recognizing power of the Lambek format.) For the case with undirected types, results are equally scarce (cf. van Benthem, 1986e).

One problem here is that, in the more abstract derivational setting, determining which language is recognized by a given initial assignment to basic expressions (i.e., symbols in the alphabet), amounts to a possibly hard question of proof analysis. Moreover, the effect of adding rules of type change, while always increasing the number of possible readings ('strong capacity'), can be either an increase or a decrease in the family of languages recognized ('weak capacity'). For instance, adding the structural rule of Permutation to the standard format results in a loss of recognition: as not all context-free languages are permutation-closed. It also leads to a gain in recognition: as, e.g., the permutation closure of the regular (and hence context-free) language $(a\ b\ c)^*$ becomes recognizable. And the latter is the non-context-free set of all sequences having equal numbers of symbols a,b,c.

There is a two-fold connection with logical *Proof Theory* in all this. First, logical systems or rules of inference can be used as mechanisms for linguistic recognition ('Deduction as Parsing'). E.g., one could use full constructive conditional logic as a categorial grammar format too (although the class of languages recognized turns out to be rather poor: see van Benthem, 1986e). Or more sensitively, one can study the recognizing effects of various structural rules. E.g., when adding Thinning to the standard format, we can no longer recognize such languages as the propositional formulas (as is easy to show). Do we also gain non-context-free languages in this format? More generally, one gets various questions concerning a logical motivation for traditional divisions in mathematical linguistics. E.g., is there a natural proof system which recognizes precisely the *regular* languages? Finally, the second connection is this. Given the above perspective of recognition as deduction, general techniques from Proof Theory (such as normalization of derivations) can be brought to bear upon investigations of linguistic strength: surely, a surprising application.

4. CATEGORIAL SEMANTICS AND TYPE THEORY

Categorial grammars are often used as a mainly syntactic device of description, producing convenient distinct constituent structures in equal quantity to semantically felt distinct readings. But, by itself, syntactic diversity does not guarantee the correct semantic distinctions: for that, one needs to know the *meaning* of the various modes of categorial combination, i.e., a categorial semantics. Another motive for the latter study is the desire to provide a semantics for the various calculi of type change. This has been attempted by adapting general techniques from ordinary logic, whether algebraic (see Buszkowski, 1982) or more model-theoretic (see Došen, 1986, van Benthem, 1986e, on Kripke models for Lambek calculi). The most enlightening perspective for present purposes, however, is the following.

Meaning Instructions

The standard world picture of Categorial Grammar is that of basic domains D_e (individual entities) and $D_t = \{0,1\}$, with a function hierarchy on top: $D_{(a,b)} = (D_b)^{D_a}$. Derivations in the standard system describe objects in these domains, employing only the *application* fragment of a type-theoretical language, via the following correspondence:

$$a \quad + \quad (a,b) \quad \Rightarrow b$$

$$\mathrm{a} \qquad \mathrm{a} \rightarrow \mathrm{b} \quad \Rightarrow \mathrm{b}$$

$$x_a \quad + \quad y_{(a,b)} \quad \Rightarrow y(x)$$

Next, the richer Lambek modes of combination induce meaning instructions involving *lambda abstraction* as well.

Example The canonical meaning for the Geach Rule can be read off systematically from its natural deduction tree in Section 2:

$$\cfrac{\cfrac{\cfrac{\cfrac{x_c \quad y_{(c,a)}}{y(x)} \quad z_{(a,b)}}{z(y(x))}}{\lambda x \bullet z(y(x))}}{\lambda y \bullet \lambda x \bullet z(y(x))}$$

Here, abstractions encode conditionalizations. (This is the idea known to logicians as 'formulae-as-types': cf. Barendregt, 1981, Appendix I.)

The full application/lambda language corresponds to derivations in full constructive conditional logic. But, the Lambek calculus employs only a fragment of this, in which each symbol λ binds *exactly one* free variable occurrence - while, moreover, no subterm is closed. The proof of this result is in van Benthem (1983), which provides the relevant, effective conversions between Lambek derivations and lambda terms. Incidentally, the conversion from derivations to terms is a paradigmatic case of what has been advocated as 'type-driven translation' in the linguistic literature.

Of course, different Lambek calculi will be 'calibrated' by different fragments of the lambda language. E.g., raising sentence conjunction to predicate conjunction would require double binding:

$$\lambda x_{(e,t)} \bullet \lambda y_{(e,t)} \bullet \lambda z_e \bullet \text{"and"}_{(t,(t,t))}(y(z))(x(z)).$$

These variations can have significant semantic effects.
Notably, the above standard Lambek fragment is *logically finite*, in the sense that only finitely many logically distinct readings can be produced for a given sequence of items reducing to some single type (cf. van Benthem, 1983, 1986a). This reflects the linguistic intuition that complex expressions cannot be infinitely ambiguous. By contrast, in the full lambda lan-

uage, logical finiteness fails. E.g., a sequence

$$a_{(e,t)} \qquad b_{((e,t),(e,t))}$$

as infinitely many distinct readings, of the forms

$$\lambda x_{((e,t),(e,t))} \bullet x(x(\ldots(x(a))\ldots))(b).$$

<u>Remark</u> One could also envisage rules of type change without any lambda-explanation at all. With such arbitrary rules, all type 0 languages become recognizable (Buszkowski, 1986): but, the semantic attraction has gone.

<u>Extensions</u>

Although the full application/abstraction language may not be involved when expressing natural language meanings, there could still be type changes requiring the next logical level: being the language of Type Theory with an additional *identity*. For instance, some linguists claim that proper names (type e) can also function as one-place predicates (type (e,t); as in the German phrase "der Heinrich"). Since the inference from e to e → t is not constructively valid, no pure lambda explication suffices here. Instead, the meaning instruction is the formula $\lambda y_e \bullet y_e \equiv x_e$. (By the way, the explicit linguistic occurrence of the verb "to be" itself is not a good argument for taking such a step. This will simply be a transitive verb, not an 'emergent' notion - whose general behaviour can in fact be derived via the Lambek calculus: cf. van Benthem, 1986a.) Unfortunately, the language of type theory has excessive power. For instance, all the usual logical constants are definable in it (see Gallin, 1975), whereas these certainly do not represent mere modes of categorial combination. Admissible type changes would now include all sequents of the form $A \Rightarrow t$, using conjunctions of trivial identities. (But e.g., $t \Rightarrow e$ would remain forbidden.) So, in any case, reasonable *fragments* will have to be studied.

A more natural extension, perhaps, is motivated by a difference between Boolean operators and their logical companions, the *quantifiers*. Higher-order uses of the former can be accounted for as consequences of their initial types, via the Lambek rules. But, this is not true for quantifiers. For instance, the first-order universal quantifier has type $((e,t),t)$, with meaning

$$\lambda x_{(e,t)} \bullet \forall y_e \bullet x(y).$$

A second-order universal quantifier has type $(((e,t),t),t)$, with an associated meaning

$$\lambda x_{((e,t),t)} \bullet \forall y_{(e,t)} \bullet x(y).$$

The second meaning has no Lambek derivation from the first, not even in the widest sense. For $(e \to t) \to t \Rightarrow ((e \to t) \to t) \to t$ is not a valid principle of constructive implication. And yet, the two meanings are closely related - be it by a different mechanism, viz. *substitution* of types (in the above case: (e,t) for e). Any substitution σ for basic types induces a syntactic transformation on type indices in terms, such that terms of type a become terms of type $\sigma(a)$. (This observation fails in the full language with identity.) The latter kind of transition points at genuine *polymorphism* in a second-order lambda calculus with variables over types. Then, the quantifiers could be of type $((x,t),t)$, where x ranges over all possible types.

Remark For the Boolean operators, such polymorphism is not appropriate. For, define the *width* of a type as follows:

width $(a) = 1$ for basic types a (including variables),

width $((b,c))$ = width $(c) + 1$.

Any single polymorphic schema with a right-most type constant t (the appropriate format for negation) has some fixed finite width, which is not changed by substitutions for type variables. But, occurrences of negation may have types of arbitrary finite width (consider n-ary predicate negation). On the other hand, the Lambek rules do produce all of these occurrences. Of course, in general, type change and polymorphism could be *combined*.

Digression What is the significance, in categorial terms, of the transition from constructive to *classical* conditional logic? As is well-known, the latter logic arises from the former by adding all instances of Peirce's law: $((a \to b) \to a) \to a$. In fact, it suffices to add all non-constructive *atomic* instances of this principle (cf. van Benthem, 1986b). So, in our case adding two constant terms will be enough:

$$c_{(((e,t),e),e)} \quad \text{and} \quad c_{(((t,e),t),t)}.$$

Only the first seems to admit of a natural meaning, however: it represents some canonical transition from choice functions (type $((e,t),e)$) to individual objects (type e) - such as $\lambda x_{((e,t),e} \bullet x(\lambda y_e \bullet \bot)$ (apply x to the empty set).

5. INTERACTION OF LOGIC AND LINGUISTICS

Through the above connections, ideas from logic and linguistics can interact. For instance, as appeared in Section 3, Proof Theory can be applied to mathematical linguistics. One further illustration, in the light of Section 4, is this. Categorial meanings for a compound expression must always be *predicative*, in the sense that only domains are used representing subtypes of the initial types of the simplex components, or proper subtypes of the outcome type. This follows from the *Subformula Property* of normalized derivations in Lambek calculi. And many other topics from the Lambda Calculus generally could be relevant. (E.g., W. Pohlers has remarked that the embedding of typed calculi into untyped ones might reflect the 'opportunistic' character of natural language expressions: acquiring types only as required by context.) This also holds for the equivalent enterprise of *Combinatory Logic*, which provides another way of classifying categorial meanings (cf. Steedman, 1985).

Remark Logically 'equivalent' formalisms can still provide quite different ways of cutting the cake of natural language. E.g., the Combinatory Logic framework suggests a hierarchy of type change calculi based on various sets of combinators, which need not be the natural one in our Lambek perspective. Thus, matters of notation, usually disregarded in logic, can be vital in applications. In particular, what counts as a reasonable 'fragment' of a language is a highly intensional question, depending on the intended application. This also holds for formalisms inside logic itself. E.g., the predicate-functor logic of Quine (1971) arranges the language of first-order logic quite differently from the usual patterning in layers of quantifier complexity - leading to new perspectives on what count as large decidable fragments (cf. Bacon, 1985). Likewise, uses of logical formalisms in natural language semantics suggest fragments defined by certain restrictions on multiplicity and depth of variable bindings, which have not been studied traditionally. (For instance, the linguistic phenomenon of 'bounded dependencies' can be modelled in a predicate logic in which quantifiers can only bind variables separated from them by at most some fixed number of intervening quantifiers. Are such fragments decidable?)

Also to be observed nowadays is the emergence of *new* questions on the border between categorial semantics and ordinary logic. These often arise from a *specialization* of the general perspective in the Lambda Calculus. Instead of having arbitrary basic types, one starts with the specific e and t, endowing the latter domain with *Boolean structure*. For instance, general Boolean structure in natural language has been investigated at length in Keenan & Faltz (1985). In addition, *intensional* basic types can be introduced with various special effects (van Benthem, 1986d). But also, *specific* types can be studied for their semantic contents, witness the logical study of determiners/generalized quantifiers in van Benthem (1986a).

Starting from such special cases, semanticists try to arrive at more general notions, applicable across all categories - and hence, at the level of generality of the Lambda Calculus.

One example is the notion of *logicality* in all types as invariance for permutations of the individual domain (suitably lifted to higher types). A sample result is this. Do Lambek type changes, when performed on a logical item in some type, again yield a logical item in the new type? This seems a reasonable constraint on type change. An in fact, it is satisfied: all lambda-terms having only logical parameters in the above sense define denotations which are still permutation-invariant (van Benthem, 1986a, Section 7.5). Further possible constraints on type change will be considered below, as well as other general questions for type theory suggested by the linguistic connection.

6. BOOLEAN STRUCTURE AND MONOTONICITY

The Boolean particles "not", "and", "or" occur, in one form or another, across all human languages. So, there is a semantic case for studying Boolean structure in a general type theory. To begin with, every type ending in a final occurrence of t has an associated domain with a natural Boolean structure. (E.g., think of $(a_1,(\ldots(a_n,t)\ldots))$ as the power set of the cartesian product $D_{a_1}\times\ldots\times D_{a_n}$.) This again leads to a general *inclusion* structure $\sqsubseteq$ on all such domains, definable as follows:

on D_t, $\sqsubseteq$ is $\leq$; on D_e, it is the identity,

on $D_{(a,b)}$, $f \sqsubseteq g$ iff, for all $x \in D_a$, $f(x) \sqsubseteq g(x)$ in D_b.

This represents a generalized implication in all linguistic categories, not just sentences.

Now, various questions arise. For instance, should Boolean structure be preserved absolutely, in the sense that all functions used as linguistic denotations should be Boolean *homomorphisms*? This seems too strong in general. E.g., even Lambek changes on homomorphic denotations need not remain homomorphic (cf. van Benthem, 1986a). But, a more plausible constraint would be that the meaning instructions τ of Section 4 preserve *logical implication*, in the following sense (with some harmless abuse of notation):

'if $x \sqsubseteq y$ in a type domain D_a,

then $\tau(x) \sqsubseteq \tau(y)$ in the domain of the changed type a'.

This form of preservation is not guaranteed, however. E.g., the Geach instruction has it, but the Montague instruction - reading $\lambda y_{(a,b)} \cdot y(x_a)$, lacks it. (Consider the case of $a = (e,t)$). Thus, there is a general question of *preservation*. Which syntactic forms for lambda-terms are necessary and sufficient to guarantee preservation of logical implication for one of their parameters in the above sense?

A sufficient condition for the above semantic *monotonicity* is *positive occurrence* of the relevant variable in the term τ; where

- x occurs positively in x,
- if an occurrence of x is positive in τ,

then it is also positive in compounds $\tau(\tau')$, $\lambda y . \tau$.

(Note that the relevant variable occurred positively in the Geach instruction of Section 4, but not in the Montague instruction.) But, the converse does not hold. E.g., given the trivial nature of $\sqsubseteq$ on types ending in a final e, parameters with those types may occur in arbitrary syntactic positions. And even with 't-types', exceptions occur.

Example. The term $\tau = x_{(t,t)}(x_{(t,t)}(y_t))$ is monotone in the parameter $x_{(t,t)}$ (1), without having a definition in which this parameter occurs only positively (2).

Proof of (1). The domain $D_{(t,t)}$ is a 4-element Boolean algebra with atoms $\lambda x_t . x_t$ and $\lambda x_t . \neg x_t$. It suffices to check the cases where, for some fixed y, τ has value 1, replacing x by some $\sqsubseteq$-larger function: τ will always retain the value 1.

Proof of (2). Let τ' be any term with free variables $x_{(t,t)}$, y_t, in which x occurs only positively. By an earlier observation (Section 5), τ' can be made predicative, having only bound variables of types t, (t,t). (The lambda-conversions producing such a form will not turn positive occurrences into non-positive ones.) Such a predicative τ' cannot have an initial λ (its type being t), so it must be an application. The latter's head cannot have a prefix λ, τ' being in normal form, nor can it be an application itself (which would yield the non-functional type t) - and so it must be $x_{(t,t)}$. But then, the argument term cannot have a prefix λ either, etc.: τ' must consist of some sequence of $x_{(t,t)}$, followed by a final y_t. If x occurs only positively here, then the only possible form is $x_{(t,t)}(y_t)$: which is evidently not equivalent to the original τ. □

This example illustrates the possible complexity of preservation results in *higher-order* settings - which need not be the obvious counterparts of those in first-order logic (where , e.g., semantic monotonicity and positive occurrence are simply related). Indeed, no elegant syntactic characterization need exist here at all! But, for the special *fragment* corresponding to the standard Lambek calculus ('single-bind lambda terms'; cf. Section 4), there is such a result.

Theorem A single-bind lambda term τ is semantically monotone with respect to some parameter x_a with a ending in t, if and only if τ is logically equivalent to a term all of whose occurrences of x_a are positive.

Proof The crux is to prove the assertion from left to right. Consider any lambda *normal form* for τ. If the (!) occurrence of x_a is not positive here, then the following procedure will find a counter-example to semantic monotonicity for τ.

We define two assignments A_1, A_2, starting with

$A_1(x_a) = 0_a$, $A_2(x_a) = 1_a$;

where 0_a , 1_a are the constant zero- and one-functions in the type $a = (a_1, \ldots (a_n, t) \ldots)$. (I.e., $0_a = \lambda x_1 \bullet \ldots \bullet \lambda x_n \bullet 0$, etc.) So, $A_1(x_a) \sqsubseteq A_2(x_a)$. Now, we extend A_1 , A_2 by making identical choices on further free variables, working upward by stages in the construction tree of τ. As x_a does not occur positively, some first governing functor must be encountered, where we will disturb the inclusion between the A_1 - and A_2 - values. This non-inclusion is then preserved until τ is reached: which is the desired counter-example.

The following facts about normal forms will be used.

(i) Each application $\tau_1(\tau_2)$ is of the form 'leading variable, followed by a string of suitable argument terms of which τ_2 is the last'. (For, function heads followed by arguments can never be λ-forms.)

(ii) Each subterm τ_1 occurs in a maximal *context* of the form $\lambda y_1 \bullet \ldots \bullet \lambda y_k\ [\tau_1(\tau_2) \ldots (\tau_m)]$, which is either the whole term τ or the argument of some application.

Now, by the assignment made to x_a, its context γ will also denote the constant zero- or one-function in its type (say, b), where A_1 , A_2 can be extended arbitrarily on the relevant free variables. Next, a governing function is found, with some leading variable u. Case 1: the type of u ends in t. Then assign any function to u which yields value 1 for all tuples of arguments having 0_b at the position of γ, and value 0 for all others. Case 2: the type of u ends in e. Then assign any function which marks 0_b at the position of γ by some value $e_1 \in D_e$, and 1_b by some e_2 distinct from e_1. Other free variables in the scope of u can be given arbitrary values. As a result, the A_1 - value of the u-term is no longer included in its A_2 - value. And the same will hold for the full context of u, which may have some initial lambdas in front. (By the definition of inclusion for functions, $A_1(\lambda y \bullet \sigma) \sqsubseteq A_2(\lambda y \bullet \sigma)$ would imply $A_1(\sigma) \sqsubseteq A_2(\sigma)$. Note that the domain of the assignment may be taken to decrease in this step.)

Finally, if there are further higher contexts, the non-inclusion can be preserved by making appropriate stipulations for leading variables, using $0, 1, e_1, e_2$, but now without the earlier reversal between the Boolean values.

This procedure relies crucially on the single-bind property of Lambek terms. For, the latter ensures that no clashes can occur between assignments made at some stage and those necessary in a wider context, as no free variable in any subterm occurs twice. □

Remark Another approach to preservation of logical implication would be to insist that *all* terms have this property, and then construct some class of models for the language where this is true. (This 'Scott strategy' was suggested by U. Mönnich.)

In preservation results in first-order logic, however, the formulation is somewhat different. There, the language has constants for the Boolean operators, and positive occurrences in argument positions of conjunction and disjunction are counted as positive too. (Moreover, one maintains an alternating 'positive/negative' count with negations.) So, we have a

Question: To prove the appropriate generalization of the above preservation theorem to a language with added constants for the Boolean operators.

Actually, van Benthem(1986c) presents a systematic calculus for obtaining monotonicity indications in this setting. It consists of a Lambek natural deduction calculus with additional rules for 'monotonicity marking', deriving both from general type-theoretic effects and from monotonicity effects of special lexical items (such as the Boolean constants). So, the question is if this calculus is *complete*, up to logical equivalence.

The explicit addition of Boolean constants to the type-theoretic language also affects other questions mentioned in the above. For instance, will the logical finiteness of the Lambek fragment change with this addition? (It probably does not.)

7. FIRST-ORDER TYPES

It has often been remarked by linguists that natural language expressions tend to occur with relatively simple types. E.g., Chierchia (1985) claims a universal restriction to an *order* of at most three. Here, the measure of complexity is this:

order $(x) = 0$ for basic types x,

order $((a,b))$ = maximum(order $(a) + 1$, order (b)).

Thus, n-ary truth functions or operations on individuals are first-order, while e.g., $((t,t), t)$ or $((e,t), ((e,t), t))$ are second-order. Actually, *first-order* types seem to be predominant: especially when predicates (e,t) are re-categorized in a new basic type p (as has been advocated occasionally). How is this low complexity to be explained? Zwarts (1986) points at the proof of Gaifman's Theorem (mentioned in Section 3), which actually shows that each context-free language has a recognizing categorial grammar with first-order types only. Thus, for combinatorial syntactic purposes, there may be just no need to go higher up. A more semantic explanation might be to show how denotations at lower orders already suffice for *defining* (all) higher-order objects. And indeed, there is a global (be it rather weak) sense in which the expressions of first-order types already populate all higher types:

Theorem For every type a, there exists a sequence A of first-order types such that the sequent $A \Rightarrow a$ is Lambek derivable.

Proof By induction on the complexity of a. The basic case, with $a = e$ or t, is obvious. Next, let $a = (a_1, (\ldots(a_n, t)\ldots))$. (The case with final e is analogous.) We want first-order types A such that $A \Rightarrow a$ is derivable, or equivalently, $A, a_1, \ldots, a_n \Rightarrow t$. Now, if a_i is not basic itself, it has the form, say, $(a_i^1, (\ldots(a_i^k, e)\ldots))$. By the inductive hypothesis, there exist first-order sequences X_i $(1 \leq i \leq k)$ such that $X_j \Rightarrow a_i^j$ is Lambek-derivable. So, gathering all these sequences together, for all compound types a_i, into one large X, we obtain:

$X, a_1, \ldots, a_n \Rightarrow Y$ is Lambek derivable;

where Y is some sequence of basic types e,t. Adding one more first-order 'bridge-type' to X results in the required sequence A deriving a E.g., if $Y = \langle e,t,e \rangle$, then add $(e, (t, (e,t)))$. □

In the next Section, further definability results will be proved - showing that, in some cases, first-order objects contain all information about the higher orders already. Nevertheless, another strategy is possible too, viz. to reduce our standard function hierarchies to models which do not grow so fast (or, not at all) in higher orders. For instance, if attention were to be restricted to structures where all functions should respect available Boolean structure, then the hierarchy might start like this. $D_{(e,t)}$ consists of *all* functions from individuals to truth values (assuming that the former have no prior Boolean structure to be preserved). But then, functions at the next level $D_{((e,t),t))}$ will have to be homomorphisms with respect to the Boolean structure of $D_{(e,t)}$: which leads to severe reductions in size. E.g., when D_e is finite, there is a natural *bijection* between D_e and these homomorphisms (i.e., ultrafilters on $D_{(e,t)}$). So, the hierarchy would 'flip-flop', in a manner reminiscent of Scott's recent categorial models.

The linguistic issue of 'low type complexity' of natural language is a not uncontroversial one. The aim of this Section has been merely to show how logical points can be relevant to such discussions.

8. DEFINABILITY OF DENOTATIONS

One question which has received much attention in recent semantic literature is that of the *expressive power* of natural language. Which mathematically possible denotations can be defined by ordinary linguistic expressions? (See Keenan & Faltz, 1985, or van Benthem, 1986a.) Notably, this has been studied for type-theoretic structures starting from a *finite* individual domain D_e. (With infinite base domains, a combinatorial explosion occurs, outgrowing our countable supply of expressions.) In terms of meaning instructions for sentences, the question becomes one of expressive power for various fragments of the lambda/application language. (Once *identity* is added, higher-order objects can be defined by trivial enumeration of their graphs. But, this full language is too rich to be of general interest: cf. Section 4.)

Truth-functional Types

First, consider types involving t only, with a fixed hierarchy starting from $D_t = \{0,1\}$. If natural language "and", "or" and "not" are truly fundamental, we would like to see a general *functional completeness* result; extending the usual one concerning n-ary (first-order) connectives to all higher types, such as $((t,t),t)$, etc. Note again that we allow only applications and lambdas, as in the definitions of

material implication: $\lambda x_t \bullet \lambda y_t \bullet \neg (x \wedge \neg y)$

material equivalence: $\lambda x_t \bullet \lambda y_t \bullet \neg (x \wedge \neg y) \wedge \neg (y \wedge \neg x)$.

__Theorem__ All pure t-type objects are definable by means of application/abstraction terms using the Boolean operators.

Proof By induction on t-types, we construct terms τ_A for every object A, having A as their constant denotation (under all assignments). In the basic domain D_t, 0 is defined by $x_t \wedge \neg x_t$, 1 by $\neg(x_t \wedge \neg x_t)$. Then, a compound type a may be viewed as a relation in $D_{a_1} \times \dots \times D_{a_n}$:

$$(a_1,(a_2,\dots,(a_n, t)\dots)).$$

Now, any relation of this kind is a finite union of singleton cases. So, using *disjunction*, it suffices to consider the latter. Here, we need a term of the form

$$\lambda x_{a_1} \bullet \dots \bullet \lambda x_{a_n} \bullet \text{'}x_{a_1} = A_1\text{'} \wedge \dots \wedge \text{'}x_{a_n} = A_n\text{'},$$

for any n-tuple $< A_1,\dots,A_n >$ in $D_{a_1} \times \dots D_{a_n}$. Thus, it remains to find formulas '$x_{a_i} = A_i$' of type t which are true, under any assignment α, if and only if $\alpha(x_{a_i}) = A_i$. Of course, the point is to do this non-trivially: without using identity. Now, if $a_i = t$, use

- x_t, if $A_i = 1$
- $\neg x_t$, if $A_i = 0$.

If a_i is itself complex, however: say, $a_i = (b_1, (b_2, t))$, then use the conjunction of all terms

- $x_{a_i}(\tau_{B_1})(\tau_{B_2})$, for all B_1, B_2 in the 'argument domains' of a_i such that $A_i(B_1)(B_2)$ holds
- $\neg x_{a_i}(\tau_{B_1})(\tau_{B_2})$, for the other couples in these domains.

(These τ-terms exist by the main inductive hypothesis.)

It is easy to check that this procedure produces the correct outcomes, also for higher types. □

As a special case, consider *monotone* Boolean functions (cf. Section 6), which, in the above terms, define relations closed under inclusion of tuples, defined in the obvious pointwise sense. Such relations can be defined by disjunctions of terms of the form:

$$\lambda x_{a_1} \bullet \dots \bullet \lambda x_{a_n} \bullet \text{'}x_{a_1} \sqsupseteq A_1\text{'} \wedge \dots \wedge \text{'}x_{a_n} \sqsupseteq A_n\text{'}.$$

Here, the inclusion formulas can be defined as follows. If $a_i = t$, then use x_t if $A_i = 1$. (Otherwise, the requirement is vacuously true.) If a_i is complex, then a conjunction is needed of positive formulas like the earlier $x_{a_i}(\tau_{B_1})(\tau_{B_2})$ - where the τ_{B_1}, τ_{B_2} themselves might involve negations too (as *their* denotations need not be monotone). The resulting definitions are *positive* in the extended sense of Section 6. Thus, in a sense, conjunction and disjunction are indeed characteristic for *all* monotone truth-functional objects.

For the more restricted *single-bind* fragment associated with Lambek meaning instructions, however, functional completeness fails. Not even all binary truth functions will be definable now. For, the available (normal form) schemata

$$\lambda x_t \bullet \lambda y_t \bullet \quad . \ldots x_t \ldots y_t \ldots$$

using $\neg$, $\wedge$, $\vee$ can be enumerated quite simply - yielding only those truth functions having exactly one or three entries 1 in their truth table. It would be of interest to study minimal fragments having full functional completeness, measuring their distance from the Lambek formalism.

Adding Individual Types

Next, the same questions can be raised for e,t - type hierarchies - starting from a base domain D_e whose individuals all have proper names in the language. This time, compound types can be either relational as above (ending in a final t), or more functional (ending in a final e). Still, the above expressive completeness argument will go through, provided that we add two features to the language: *identity between individuals* (type $(e,(e,t))$) as well as a *description operator*.

Theorem All e,t - objects are definable in the application/ abstraction language with individual identity and definite description.

Proof The main new points are these.
In the basic case, one uses the names for D_e - objects. To define a functional object f, with type $(a_1, (a_2, \ldots, (a_n, e)\ldots))$, one first defines its *graph* G_f with type $(a_1, (a_2, \ldots, (a_n, (e,t))\ldots))$, as in the preceding proof. Then f itself may be extracted using one description operator:

$$\lambda x_{a_1} \bullet \ldots . \lambda x_{a_n} \bullet \iota x \bullet G_f (x_{a_1}, \ldots, x_{a_n}, x).$$

In the auxiliary formulas 'x = A', this requires the following additions. The basic case with A of type e, can be handled using individual identity. A compound case like $(b_1,(b_2, e))$ is described by clauses again using identity ($\equiv$) for individual names E:

$$x_{b_1,(b_2,e)}(\tau_{B_1})(\tau_{B_2}) \equiv E,$$

enumerating the graph of the relevant A. □

Remark In these definitions, no unrestricted description operator is ever used. In fact, it suffices to have one choice function "the" of type $((e,t),e)$ (in addition to the single "is" of type $(e,(e,t))$). Without these enrichments of the language, however, not even all operations on individuals (type (e,e)) would be definable.

Discussion Although natural language does possess the above expressions "is" and "the", it still falls short of the full expressive power of the Theorem. There are several reasons for this. First, with Lambek restrictions on variable occurrences, many of the above schemata will be inadmissible as natural language meanings. Moreover, there remain additional *syntactic* restrictions on natural language expressions which make some of the latter forms inexpressible. E.g., there is no coherent sequence "is a and b is ..., or ... is c and d is", unless one counts the more artificial "x,y are, respectively, a and b or c and d". Finally, natural language categories are also subject to certain *denotational constraints* (see van Benthem, 1986a, chapters 1,2,3), which make certain a priori possible denotations inaccessible.

One drawback of the earlier definitions is their being *local* in specific models. E.g., the universal quantifier "everyone" will be defined by a conjunction "John and Mary and..." enumerating the specific domain of individuals. No *global* uniform definition is forthcoming, correct in all models. (See also Keenan & Stavi, 1981, on this theme.)

Example (1) The universal quantifier "everyone" (∀) of type $((e,t),t)$ is not globally definable using Boolean operators only. This is because the transition from (t,t) and $(t,(t,t))$ to $((e,t),t)$ is not even correct in classical conditional logic (cf. Section 3).
(2) But even adding names for individuals, as well as individual "be" will not produce a correct definition:

> the universal quantifier is not uniformly definable in this enriched application/abstraction language.

For, consider a defining schema in normal form (cf. Section 5), being $\lambda x_{e,t} \bullet B_t$, where all subterms exemplify subtypes of (e,t); t; $(t,(t,t))$; (t,t); e; $(e,(e,t))$. By some syntactic analysis of possible forms, reflecting the simple character of first-order derivations, B_t must be equivalent to some Boolean assertion about (non-) membership of certain individuals in $x_{e,t}$: which is clearly insufficient. □
(3) Finally, adding the descriptor "the" will make the universal quantifier definable as follows ('empty complement'):

$\lambda x_{e,t} \bullet x_{e,t}$ ("the" $(\lambda y_e \bullet \neg x_{e,t}(y_e))$).

But, by familiar arguments, even this richer language cannot globally define such *higher-order* quantifiers as "most (individuals in the universe)".

Question Is there at least a natural global defining set for all *logical* denotations in the sense of Section 5?

9. AXIOMATIZATION AND DECIDABILITY

One central reason for doing formal semantics has been to account for *inference* in natural language. In this light, the inferential behaviour of the various lambda languages of the preceding Sections deserves closer attention. Here, we shall merely point out some results and questions.

Consider identities between terms in the application/abstraction language, interpreted on our standard function hierarchies. For the pure *application* language, valid consequence between such identities is axiomatizable (and decidable) in a simple 'equational calculus'. For the full language with *abstraction* (but without identity), such valid consequence is non-axiomatizable and indeed quite complex. (This follows from theorem 5 of Friedman, 1975.) On the other hand, validity of single equations, without premises, is axiomatized by the usual typed extensional lambda calculus. (This is theorem 3 from the same paper.)
In the language with *identity*, however, even the latter notion becomes non-axiomatizable (cf. Gallin, 1975).

These general results do not transfer directly to the *fragments* that are of interest to us. For instance, Gallin found an interesting axiomatizable 'extensional' fragment of the full type-theoretic language. And in our case, one would like to know if valid consequence *is* axiomatizable for the single-bind Lambek fragment.
That this case may be much simpler is suggested by the result in van Benthem (1986b) that single-bind *predicate logic* is *decidable* (as opposed to general predicate logic). In these cases, the *fine-structure* of proofs requires more care - as not all usual basic rules of inference respect single-bondage. (E.g., Boolean Associativity does, but Distributivity does not: $x \wedge (y \vee z)$ becomes $(x \wedge y) \vee (x \wedge z)$.)

Remark In addition to ordinary extensionality ('equality of graphs implies equality of functions'), the following inference is also valid on standard models:

'from $\lambda x_{a,b} \bullet x(y_a) \equiv \lambda x_{a,b} \bullet x(z_a)$ to $y_a \equiv z_a$ '.

This says, in other words, that the Montague Rule of Section 2 is one-one on denotations of type a. But, not all Lambek transitions have this feature. E.g., consider the sequent $(((e,t),t),t) \Rightarrow (e,t)$, with associated term $\lambda x_e \bullet y_{(((e,t),t),t)} (\lambda z_{e,t} \bullet z(x))$. Two properties of noun phrases (type $((e,t),t)$) can agree on all individual noun phrases, and yet be different.

Finally, in line with earlier remarks, lambda calculi could be studied enriched with certain central logical constants from natural language. For instance, when the *Boolean operators* are added, does Friedman's completeness theorem still hold with respect to some suitable axiomatic system? (One needs at least the typed extensional lambda calculus plus the Boolean axioms. But, there are also additional principles of extensionality; such as the implication from $x_{(t,t)}(1) \equiv y_{(t,t)}(1)$, $x_{(t,t)}(0) \equiv y_{(t,t)}(0)$ to $x_{(t,t)} \equiv y_{(t,t)}$.) This would be an interesting generalization of the usual completeness theorem for Boolean Algebra. Moreover, the case of valid *identities* is quite central now, because implications between type t expressions can be reduced to categorial validities (equality with Boolean 1) in the usual way.

Similar questions arise when *quantifiers* are added, say $\exists, \forall$ as constants of type $((e,t),t)$. Will, e.g., the logic of the existential quantifier be exhausted by the obvious 'modal' distribution principle $\exists(P \vee Q) \equiv \exists P \vee \exists Q$? A final addition might be individual *identity* ("be") and descriptors ("the"), as introduced in Section 8. This would give us quite a working system of 'natural logic' for ordinary language.

10. CONCLUSION

The semantics of natural language generates many questions on the boundary with logic. In particular, the theory of Categorial Grammar interacts with Lambda Calculus and Type Theory. That the latter are important in the study of *programming* languages has been known for some time already. The general picture ought to be that logic provides a general theory of semantics for *all* kinds of language: formal languages, natural languages and programming languages. That it pays to look for general analogies here was realized already by Montague (and has been illustrated vividly, e.g., in Janssen, 1983). Moreover, these applications of logic may also enrich that discipline itself.

REFERENCES

1. E. Bach, R. Oehrle and D. Wheeler, eds, 1986, Categorial Grammars and Natural Language Structures, Reidel, Dordrecht.

2. J. Bacon, 1985, 'The Completeness of a Predicate-Functor Logic', Journal of Symbolic Logic 50, 903-926.

3. H. Barendregt, 1981, The Lambda Calculus: its Syntax and Semantics, North-Holland, Amsterdam.

4. J. van Benthem, 1983, The Semantics of Variety in Categorial Grammar, report 83-26, Department of Mathematics, Simon Fraser University, Burnaby (B.C.). (To appear in Buszkowski, Marciszewski & van Benthem (eds), 1986.)

5. J. van Benthem, 1986a, Essays in Logical Semantics, Reidel, Dordrecht.

6. J. van Benthem, 1986b, 'Logical Syntax', manuscript, Mathematisch Instituut, Universiteit van Amsterdam.

7. J. van Benthem, 1986c, 'Meaning: Interpretation and Inference', to appear in Synthese, (M-L dalla Chiara and G. Toraldo di Francia, eds, 'Proceedings Conference on Theories of Meaning, Florence 1985').

8. J. van Benthem, 1986d, 'Strategies of Intensionalisation', to appear in Filozófiai Figyelö, (L. Pólos, ed, 'Festschrift for Imre Ruzsa').

9. J. van Benthem, 1986e, 'The Lambek Calculus', report 86-06, Mathematical Institute, University of Amsterdam. (To appear in Bach, Oehrle & Wheeler (eds), 1986.)

10. W. Buszkowski, 1982, Lambek's Categorial Grammars, Instytut Matematicki, Adam Mickiewicz University, Poznan, Poland.

11. W. Buszkowski, W. Marciszewski and J. van Benthem, eds, 1986, Categorial Grammar, John Benjamin, Amsterdam.

12. G. Chierchia, 1985, 'Formal Semantics and the Grammar of Predication', Linguistic Inquiry 16:3, 417-443.

13. K. Došen, 1986, 'Sequent Systems and Groupoid Models', Mathematical Institute, University of Beograd.

14. J-E Fenstad, ed, 1971, Proceedings of the Second Scandinavian Logic Symposium, North-Holland, Amsterdam.

15. H. Friedman, 1975, 'Equality between Functionals', in Dold and Eckmann, eds, Logic Colloquium, Boston 72-73, Springer, Heidelberg, (Lecture Notes in Mathematics, vol. 453), 22-37.

16. D. Gallin, 1975, Intensional and Higher-Order Modal Logic, North-Holland, Amsterdam.

17. T. Janssen, 1983, Foundations and Applications of Montague Grammar, dissertation, Mathematical Center, Amsterdam.

18. E. Keenan and Y. Stavi, 1981, 'A Characterization of Natural Language Determiners', to appear in Linguistics and Philosophy.

19. E. Keenan and L. Faltz, 1985, Boolean Semantics for Natural Language, Reidel, Dordrecht.

20. J. Lambek, 1958, 'The Mathematics of Sentence Structure', American Mathematical Monthly 65, 154-170.

21. J. Lambek, 1980, 'From λ-Calculus to Cartesian-Closed Categories', in J. Seldin and J. Hindley, eds, 1980, To H.B. Curry: Essays in Combinatory Logic, Lambda Calculus and Formalism, Academic Press, New York, 375-402.

22. W.V.O. Quine, 1971, 'Predicate-Functor Logic', in J. Fenstad (ed), 309-315.

23. M. Steedman, 1985, 'Combinators, Categorial Grammars and Parasitic Gaps', School of Epistemics, Edinburgh.

24. F. Zwarts, 1986, Categoriale Grammatica en Algebraïsche Semantiek, proefschrift, Nederlands Instituut, Rijksuniversiteit, Groningen.

A CONSTRUCTIVE MORSE THEORY OF SETS

Douglas S. Bridges

University of Buckingham
Buckingham MK18 1EG, England

1. Introduction

In this paper I shall outline a foundational system for constructive mathematics analogous to that given in [11] for classical mathematics. By 'constructive mathematics' I shall mean mathematics as understood by Errett Bishop and his followers [2, 3], the mathematics of which

> the primary concern ... is number, and this means the positive integers ... Everything attaches itself to number, and every mathematical statement ultimately expresses the fact that if we perform certain computations within the set of positive integers, we shall get certain results. [2, pp. 2-3]

In other words, mathematics as we understand it is characterised by *numerical content* and *computational meaning*.

Among the interesting features of our formal system are the following. First, there are very precise rules for the classification and manipulation of symbols and expressions. These rules describe a high-level language for the expression and development of constructive mathematics, and a formal proof within the system is essentially a program written in that language. There is therefore a real possibility, which I hope to pursue in later papers, that the system outlined below will provide an alternative to existing approaches to computer-based theorem proving, such as those described in [7] and [10].

Perhaps the most unusual feature of our formal system is that it draws no distinction between terms and formulae: *every mathematical object can be regarded as either a set or a proposition.* Thus we shall interpret $x \in P$ in the Fregean manner, as 'the concept P applies to x', and we shall draw no distinction between, for example,

for all x, <u>u</u>x

and

the intersection, as x runs, of the sets <u>u</u>x.

(Note that Morse uses <u>u</u>x, <u>v</u>y, ... for unary predicates, <u>u</u>'xy, <u>v</u>'st, ... for binary predicates, <u>u</u>"xyz, <u>v</u>"rst, ... for ternary predicates, and so on.) However, we do not entirely equate mathematics with logic: as well as the principles of logic itself, our development requires certain construction rules, by the application of which we can be sure that at least part of our mathematics has constructive content and meaning; we also require the axiom of dependent choice. Among the most important of the construction rules is that which allows us to construct the set of all natural numbers, and thence the class of recursive functions.

The use of the phrase 'part of our mathematics' requires further explanation. Within the formal system we can talk about a very large class of mathematical objects, but not all of those will be constructively defined. This exactly parallels the situation in informal mathematics, where we can, for example, talk about the least upper bound of a bounded set of real numbers without necessarily recognising this as an object which has been, or ever can be, constructed. In this context, it is worth noting that a mathematical statement may have constructive significance without being, as a term, a well-constructed object: for example,

every natural number is constructively well-defined

is certainly a statement with constructive significance, but as a term it is not a well-constructed object (it actually equals the universe of our formal system).

For background material on the classification and construction of expressions, and on rules of inference, we refer the reader to [11, Ch. 0], with one warning: for reasons discussed in [5], where Morse places the symbol Λ among his expressions of Class 1, we place it among the expressions of Class 3.

The classical logic and set theory corresponding to the work of this paper can be found in sections 1.0 - 1.6, 2.0 - 2.102, 2.130 - 2.139, and 2.191 - 2.195 of [11]. It should be understood that if a term appears below without explicit definition, then its classical definition carries over unchanged into our constructive setting. Also, if a statement, other than a definition or an axiom, appears below without proof, either its constructive proof is essentially the same as its classical one, or the constructive proof is comparatively routine.

2. Logic

Our formal logic is based on the three primitive connectives $\rightarrow$, $\wedge$, $\vee$ and the two quantifiers $\bigwedge$, $\bigvee$, all of which have their customary constructive interpretations. We also require the definitions

.0 $(U \equiv \bigvee xx)$

.1 $(0 \equiv \bigwedge xx)$

.2 $(\neg p \equiv (p \rightarrow 0))$

Note that we use $\neg$ for negation, where Morse uses $\sim$.

The axioms of logic divide into the *propositional axioms*

.3 $(p \rightarrow (p \wedge p))$

.4 $((p \wedge q) \rightarrow (q \wedge p))$

.5 $((p \rightarrow q) \rightarrow ((p \wedge r) \rightarrow (q \wedge r)))$

.6 $((p \rightarrow q) \rightarrow ((q \rightarrow r) \rightarrow (p \rightarrow r)))$

.7 $(q \rightarrow (p \rightarrow q))$

.8 $((p \wedge (p \rightarrow q)) \rightarrow q)$

.9 $(p \rightarrow (p \vee q))$

.10 $((p \vee q) \rightarrow (q \vee p))$

.11 $(((p \rightarrow r) \wedge (q \rightarrow r)) \rightarrow ((p \vee q) \rightarrow r))$

and the *predicate axioms*

.12 $((y \rightarrow \underline{u}x) \rightarrow (y \rightarrow \bigwedge x\, \underline{u}x))$

.13 $((\underline{u}x \rightarrow y) \rightarrow (\bigvee x\, \underline{u}x \rightarrow y))$

.14 $(\bigwedge x\, \underline{u}x \rightarrow \underline{u}x)$

.15 $(\underline{u}x \rightarrow \bigvee x\, \underline{u}x)$

.16 $(y \rightarrow \bigwedge xy)$

.17 $(\bigvee xy \rightarrow y)$

The propositional axioms are the first eight of those due to Heyting [9, pp. 105-106]. The main difference between our approach and Heyting's is that whereas he takes negation as a primitive concept, for us it is defined by 2.1, 2.2, and the explanatory definition

.18 $(\text{not } p \equiv \neg p)$

In fact, it is easy to derive Heyting's two axioms of negation as theorems:

.19 $(((p \rightarrow q) \wedge (p \rightarrow \neg q)) \rightarrow \neg p)$

.20 $(\neg p \rightarrow (p \rightarrow q))$

Our predicate axioms are essentially those described in [15, page 11]. Bearing in mind the definition

.21 $(\text{omniscience} \equiv \bigwedge x(x \vee \neg x))$

.22 $(\text{the law of excluded middle} \equiv \text{omniscience})$

we have been particularly careful to avoid the appearance of such theorems as (omniscience), $(\neg\neg p \rightarrow p)$, and $(\neg\bigwedge x\ \underline{u}x \rightarrow \bigvee x\neg\underline{u}x)$ as consequences of our axioms.

Amongst the theorems that we can deduce within our system are

.23 $\neg 0$

.24 U

Thus the universe U corresponds to the truth value 'true', and the empty set 0 to the truth value 'false'.

3. Set theory

We begin with the *axiom of definition*

.0 $((x \equiv y) = (x = y))$

The first group of set-theoretic axioms is concerned with truth and certain rules for the manipulation of sets:

.1 $(x \leftrightarrow (0 \in x))$

.2 $((t \in U) \rightarrow ((t \in (x \in y)) \leftrightarrow (x \in y)))$

.3 $((t \in a) \rightarrow ((t \in (x \rightarrow y)) \leftrightarrow ((t \in x) \rightarrow (t \in y))))$

.4 $((t \in \bigwedge x\ \underline{u}x) \leftrightarrow \bigwedge x(t \in \underline{u}x))$

.5 $((t \in \bigvee x\ \underline{u}x) \leftrightarrow \bigvee x(t \in \underline{u}x))$

.6 $((t \in (x \wedge y)) \leftrightarrow ((t \in x) \wedge (t \in y)))$

.7 $((t \in (x \vee y)) \leftrightarrow ((t \in x) \vee (t \in y)))$

Among the elementary deductions we can make from these axioms and the definitions and axioms for logic, we have

.8 $(0 \in U)$

.9 $(x \subset y \rightarrow (x \rightarrow y))$

.10 $(x \in a \rightarrow x \in U)$

.11 $(x \in 0 \rightarrow x \in a)$

.12 $(x \in 0 \rightarrow 0)$

.13 $(U = \neg\neg U)$

.14 $(0 = \neg\neg 0)$

Our choice of definition for U was dictated by a desire to obtain 3.10. Had we made the definition $(U \equiv \bigvee x\neg x)$ - perhaps closer to that of [11, 1.0.6] - the nearest we could have come to 3.10 would have been

$$((x \in \neg a \rightarrow x \in U) \wedge (x \in \neg\neg a \rightarrow x \in U))$$

Note that in spite of 3.13 and 3.14, we trust that $(x = \neg\neg x)$ is not a theorem of our system.

We now introduce two *axioms of equality*:

.15 $((x \in U) \rightarrow ((x = y) \leftrightarrow \bigwedge t((x \in t) \rightarrow (y \in t))))$

.16 $((x = y) \rightarrow (\underline{u}x = \underline{u}y))$

from which we readily obtain such results as

.17 $(x = y \rightarrow (x \in t \rightarrow y \in t)$

.18 $(x = y \rightarrow \underline{v}x = \underline{v}y)$

.19 $(x = y \wedge s = t \rightarrow \underline{u}'xs = \underline{u}'yt)$

4. <u>Singletons</u>

Singletons can be approached in two ways. From the first of these

.0 $(\text{sng } x \equiv \bigwedge y(y \rightarrow (x \in y)))$

.1 $(\text{singleton } x \equiv \text{sng } x)$

we derive the fundamental theorem

.2 $(y \in \text{sng } x \leftrightarrow y = x \wedge y \in U)$.

The second approach

.3 $(\text{sngl } x \equiv \bigwedge y((x \in y) \rightarrow y)$

.4 $(\text{single } x \equiv \text{sngl } x)$

produces the corresponding criterion

.5 $(x \in U \rightarrow y \in \text{sngl } x \leftrightarrow y = x)$

and is linked to the first by

.6 $(x \in U \rightarrow \text{sng } x = \text{sngl } x)$

For a more general characterisation of singletons we turn to

.7 $(\text{singleton is } a \equiv (\Pi a = \nabla a \in a)$

where

.8 $(\Pi a \equiv \bigwedge y(y \in a \rightarrow y))$

.9 $(\nabla a \equiv \bigvee y(y \in a \wedge y))$

We then have

.10 $(\text{singleton is } a \rightarrow y \in a \leftrightarrow y = \Pi a = \nabla a)$

and, not unexpectedly,

.11 $(x \in U \rightarrow \text{singleton is sng } x \wedge \text{singleton is sngl } x)$

.12 (singleton is a $\rightarrow$ a = sng ∇a = sngl Πa)

Morse makes the definition

$$(\text{singleton is } a \equiv (\Pi a = \nabla a))$$

from which, with the aid of the law of excluded middle, he is able to deduce that

$$(\text{singleton is } a \leftrightarrow \Pi a = \nabla a \in a)$$

We now show that Morse's use of excluded middle in this context is essential. To this end, suppose that

$$(\Pi a = \nabla a \rightarrow \Pi a \in a)$$

and consider the case where $a \equiv (0 \in (p \vee \neg p)) \cap \{0\}$. We have

$$\begin{aligned} t \in \Pi a &\leftrightarrow \bigwedge y((p \vee \neg p) \wedge y = 0 \rightarrow t \in y) \\ &\leftrightarrow (p \vee \neg p) \rightarrow t \in 0 \\ &\leftrightarrow \neg(p \vee \neg p) \\ &\leftrightarrow 0 \\ &\leftrightarrow t \in 0 \end{aligned}$$

Thus $\Pi a = 0$; similarly, $\nabla a = 0$. So, by our assumption, $0 = \Pi a \in a$; whence $0 \in (p \vee \neg p)$, and therefore $p \vee \neg p$, by axiom 3.1. Thus our assumption entails the law of excluded middle.

Singletons appear in many contexts, such as classification:

.13 $(\mathrm{E}x\ \underline{u}x \equiv \bigvee x(0 \in \underline{u}x \wedge \text{sng } x))$

.14 (The set of points x such that $\underline{u}x \equiv \mathrm{E}x\ \underline{u}x$)

.15 $(\{x : \underline{u}x\} \equiv \mathrm{E}x\ \underline{u}x)$

With these definitions it is routine to establish

.16 $(x \in U \rightarrow x \in \mathrm{E}x\ \underline{u}x \leftrightarrow \underline{u}x)$

5. Axioms of construction

We believe that a mathematical object is constructively well-defined if and only if it belongs to the universe. Accordingly, we are led to the following *axioms of construction*.

.0 $(\omega \in U)$

.1 $(((A \in U) \wedge \bigwedge x((x \in A) \rightarrow (\underline{u}x \in U))) \rightarrow (\bigvee x((x \in A) \wedge \underline{u}x) \in U)$

.2 $(((x \in U) \wedge (y \in U)) \rightarrow ((x \vee y) \in U))$

.3 $((x \in U) \leftrightarrow (\text{sngl } x \in U))$

.4 $(A \in U \rightarrow \text{sb } A \in U)$

Here, ω, the *set of natural numbers*, and sb x, the *power set of x*, are introduced by the definitions

.5 $(\omega \equiv \wedge A; (0 \in A \wedge \wedge x \in A((x \vee \text{sng } x) \in A))A)$

.6 $(\text{sb } A \equiv \text{E}x(x \subset A))$

Axiom 5.0 places ω firmly at the base of our mathematics, and is very much in keeping with the quotation from [2] at the beginning of this paper. The power set axiom (5.4) has been criticised by some authors (for example, [13], [1, pp. 190-191]); nevertheless, in the absence of any convincing argument that it is nonconstructive, and being aware of the simplifications its use makes in, for example, measure theory, we prefer to adopt it at this stage.

Morse derives 5.1, 5.2, and several important theorems of construction from one powerful *axiom of replacement*

$$((\vee x((A \in U) \wedge (x \subset A) \wedge ((\underline{u}x \in U) \wedge \underline{u}x)) \in U),$$

via the consequent *theorem of replacement*

$$(\vee x((A \in U) \wedge (x \in A) \wedge (\underline{u}x \in U) \wedge \underline{u}x) \in U).$$

His derivation of this theorem rests heavily on the use of omniscience.

Among the elementary theorems of construction that follow from axioms 5.0 - 5.4 are

.7 $(B \subset A \in U \rightarrow B \in U)$

.8 $(A \in U \rightarrow \triangledown A \in U)$

.9 $(x \in U \rightarrow \text{sng } x \in U)$

.10 $(\neg(x \in U) \rightarrow \text{sng } x = 0 \wedge \text{sngl } x = U)$

Note that we can deduce classically from 5.10 that

$$(\text{sng } x = \text{sngl } x \rightarrow x \in U)$$

and from 5.9 and 5.10 that

$$(\text{sng } x \in U).$$

6. Ordered pairs

The classical theory of ordered pairs carries over with only minor changes from [11, 2.55-2.63]. In turn, this leads to various theorems of substitution, of which the most important are

.0 $(\wedge x,y;\ \underline{u}'xy\ \underline{v}'xy = \wedge x \wedge y;\ \underline{u}'xy\ \underline{v}'xy)$

and

.1 $(\vee x,y;\ \underline{u}'xy\ \underline{v}'xy = \vee x \vee y;\ \underline{u}'xy\ \underline{v}'xy)$

For constructive proofs of 6.0 and 6.1, we need the following lemmas.

.2 $(\bigwedge x \bigwedge y \bigwedge z(\underline{w}'xy = (0 \in \underline{u}'xy \rightarrow \underline{v}'xy) \wedge \underline{u}z = (0 \in \mathrm{st}\ z(x,y)\ \underline{u}'xy \rightarrow$
$\mathrm{st}\ z(x,y)\ \underline{v}'xy) \rightarrow \bigwedge x \bigwedge y\ \underline{w}'xy = \bigwedge z\ \underline{u}z))$

<u>Proof</u> For the meaning of such expressions as $\mathrm{st}\ zy\ \underline{u}'xx'$ see [11, 0.58].
We have

$$
\begin{aligned}
(c = a,b \rightarrow \bigwedge z\ \underline{u}z &\subset \underline{u}c = \underline{w}'ab \\
&\rightarrow \bigwedge z\ \underline{u}z \subset \underline{w}'ab)
\end{aligned}
$$

whence

$$(\bigwedge a \bigwedge b(\bigwedge z\ \underline{u}z \subset \underline{w}'ab)$$

and therefore

$$(\bigwedge z\ \underline{u}z \subset \bigwedge x \bigwedge y\ \underline{w}'xy).$$

On the other hand,

$$
\begin{aligned}
(t \in \bigwedge x \bigwedge y\ \underline{w}'xy &\rightarrow \bigwedge x \bigwedge y(0 \in \underline{u}'xy \rightarrow t \in \underline{v}'xy) \\
&\rightarrow (0 \in \mathrm{st}\ z(x,y)\ \underline{u}'xy \rightarrow \mathrm{st}\ z(x,y)\ \underline{u}'xy \\
&\qquad \rightarrow \bigvee x \bigvee y(z = x,y \wedge \underline{u}'xy) \\
&\qquad \rightarrow \bigvee x \bigvee y(z = x,y \wedge 0 \in \underline{u}'xy \wedge \\
&\qquad\qquad (0 \in \underline{u}'xy \rightarrow t \in \underline{v}'xy)) \\
&\qquad \rightarrow \bigvee x \bigvee y(z = x,y \wedge t \in \underline{v}'xy)) \\
&\rightarrow (0 \in \mathrm{st}\ z(x,y)\ \underline{u}'xy \rightarrow t \in \mathrm{st}\ z(x,y)\ \underline{v}'xy) \\
&\rightarrow t \in (0 \in \mathrm{st}\ z(x,y)\ \underline{u}'xy \rightarrow \mathrm{st}\ z(x,y)\ \underline{v}'xy) \\
&\rightarrow t \in \underline{u}z)
\end{aligned}
$$

whence

$$(\bigwedge x \bigwedge y\ \underline{w}'xy \subset \underline{u}z)$$

Thus

$$(\bigwedge x \bigwedge y\ \underline{w}'xy \subset \bigwedge z\ \underline{u}z \subset \bigwedge x \bigwedge y\ \underline{w}'xy)$$

from which the result follows. □

.3 $(\bigwedge x \bigwedge y \bigwedge z(\underline{w}'xy = (0 \in \underline{u}'xy \rightarrow \underline{v}'xy) \wedge$
$\underline{u}z = (0 \in \mathrm{st}\ z(x,y)\ \underline{u}'xy \rightarrow \mathrm{st}\ z(x,y)\ \underline{v}'xy) \rightarrow \bigvee x \bigvee y\ \underline{w}'xy = \bigvee z\ \underline{u}z)$

We now give the

<u>Proof of 6.0.</u> From the theory of notation [11, Ch. 0] and 6.2 above, we have

$$
\begin{aligned}
(\bigwedge x,y;\ \underline{u}'xy\ \underline{v}'xy &= \bigwedge z;\ \mathrm{st}\ z(x,y)\ \underline{u}'xy\ \mathrm{st}\ z(x,y)\ \underline{v}'xy \\
&= \bigwedge z(0 \in \mathrm{st}\ z(x,y)\ \underline{u}'xy \rightarrow \mathrm{st}\ z(x,y)\ \underline{v}'xy) \\
&= \bigwedge x \bigwedge y(0 \in \underline{u}'xy \rightarrow \underline{v}'xy) \\
&= \bigwedge x \bigwedge y;\ \underline{u}'xy\ \underline{v}'xy) \qquad \square
\end{aligned}
$$

We omit the proof of 6.1.

Morse's classical proofs of theorems 6.0 and 6.1 depend on the statement

$$(\text{orderedpair is } p \vee \neg\text{orderedpair is } p)$$

Although, at first sight, it seems reasonable to assume that we can recognise whether or not an object is an ordered pair, the following argument shows that this is not, in fact, the case.

Let

$$(z = (\text{sng } 0 \,\text{"}\, \text{ss } x) \cup (\text{sng sng } 0 \,\text{"}\, ((x = y) \wedge \text{ss } y)))$$

Then

$$\begin{aligned}(z = a,b &\rightarrow \text{sng sng } 0 \,\text{"}\, \text{ss } b \;=\; \text{sng sng } 0 \,\text{"}\, ((x = y) \wedge \text{ss } y) \\ &\rightarrow \text{ss } b = ((x = y) \wedge \text{ss } y) \\ &\rightarrow 0 \in ((x = y) \wedge \text{ss } y) \subset (x = y) \\ &\rightarrow 0 \in (x = y) \\ &\rightarrow x = y)\end{aligned}$$

the third last line holding because $(0 \in \text{sng } 0 \in \text{ss } b)$. On the other hand,

$$\begin{aligned}(x = y \rightarrow (x = y) = U \\ &\rightarrow z = (\text{sng } 0 \,\text{"}\, \text{ss } x) \cup (\text{sng sng } 0 \,\text{"}\, \text{ss } y) \\ &\rightarrow \text{orderedpair is } z)\end{aligned}$$

so that

$$(\neg\text{orderedpair is } z \rightarrow \neg(x = y))$$

Hence

$$\begin{aligned}(\bigwedge p(&\text{orderedpair is } p \vee \neg\text{orderedpair is } p) \\ &\rightarrow \bigwedge x \bigwedge y((x = y) \vee \neg(x = y)) \\ &\rightarrow ((t \in x) = U) \vee \neg((t \in x) = U) \\ &\rightarrow (t \in x \vee \neg(t \in x)) \\ &\rightarrow \bigwedge t(t \in (x \vee \neg x) \leftrightarrow t \in U) \\ &\rightarrow x \vee \neg x \in U \\ &\rightarrow x \vee \neg x \\ &\rightarrow \text{omniscience})\end{aligned}$$

While on the topic of substitution, we note that of the miscellaneous notational theorems in Section 2.70 of [11], all but (2.70.10) and (2.70.18) can be proved constructively. It is easy to see that the two which fail in our constructive setting entail

$$(\neg\bigwedge x \; ux \leftrightarrow \bigvee x \neg ux)$$

and

$$(\bigwedge x(y \vee ux) \leftrightarrow y \vee \bigwedge x \; ux)$$

respectively, and so are essentially nonconstructive.

7. Relations and functions

The definitions and main results in 2.76 - 2.91 of [11] carry over largely unchanged into our constructive setting. Thus we have such theorems of construction as

.0 $(R \in U \rightarrow \text{dmn } R \in U \wedge \text{rng } R \in U)$

.1 $(\text{function is } f \wedge \text{dmn } f \in U \rightarrow f \in U)$

.2 $(A \in U \wedge B \in U \rightarrow A,,B \in U)$

An important difference between Morse's classical approach and our constructive one occurs at the introduction of the expression λ. Where Morse places λ among his expressions of Class 1 (see [11, p. 22]) and introduces λ by the definition

.3 $(\lambda x\ \underline{u}x \equiv \mathcal{E}x,y(y = \underline{u}x))$

we consider λ as an expression of Class 3, and make the definitions

.4 $(\lambda x;\underline{v}x\ \underline{u}x \equiv \mathcal{E}x,y(\underline{v}x \wedge y = \underline{u}x))$

.5 $(\text{lambda } x \text{ with } \underline{v}x,\ \underline{u}x \equiv \lambda x;\underline{v}x\ \underline{u}x)$

.6 $(\lambda x\ \underline{u}x \equiv \lambda x;(x = x)\ \underline{u}x)$

We then obtain 7.3 as a simple consequence of definition 7.6 and the theory of notation. We also obtain such important theorems as

.7 $(f = \lambda x;\underline{v}x\ \underline{u}x \rightarrow \text{function is } f \wedge \text{dmn } f = \mathcal{E}x(\underline{v}x \wedge \underline{u}x \in U) \wedge$
$(x \in \text{dmn } f \rightarrow .fx = \underline{u}x))$

.8 $(f = \lambda x\ \underline{u}x \rightarrow \text{function is } f \wedge \text{dmn } f = \mathcal{E}x(\underline{u}x \in U) \wedge$
$(x \in \text{dmn } f \rightarrow .fx = \underline{u}x))$

and

.9 $(f = \lambda x \in A\ \underline{u}x \rightarrow \text{function is } f \wedge \text{dmn } f = \mathcal{E}x \in A(\underline{u}x \in U) \wedge$
$(x \in \text{dmn } f \rightarrow .fx = \underline{u}x))$

Note Morse's notation for the value of a function f at x:

.10 $(.fx \equiv \Pi\text{vs } fx)$

where

.11 $(\text{vs } fx \equiv \mathcal{E}y\vee x(x,y \in f))$

In Morse's classical theory, it readily follows that

$$(\lambda x;\underline{v}x\ \underline{u}x = \lambda x(0 \in \underline{v}x \rightarrow \underline{u}x))$$

is a theorem, from which he derives 7.7. Unfortunately, this procedure is essentially nonconstructive: as we showed in [5], if 7.3 and

$$(\text{dmn } \lambda x(0 \in \underline{v}x \rightarrow \underline{u}x) = \mathcal{E}x(\underline{v}x \wedge \underline{u}x \in U))$$

are both theorems of our system, then we can derive the law of excluded middle. It is for this reason that we adopt our different approach to the expression λ.

8. Unicity and unique choice

Morse's definitions

.0 $(\text{One } x\ \underline{u}x \equiv \bigvee y \bigwedge x(\underline{u}x \leftrightarrow x = y))$

.1 $(\text{The } x\ \underline{u}x \equiv \bigwedge x;(\text{One } x\ \underline{u}x \wedge \underline{u}x)x)$

easily lead to the theorems

.2 $(\text{One } x\ \text{u}x \rightarrow \underline{u}x \leftrightarrow x = \text{The } x\ \underline{u}x)$

.3 $(\neg\text{One } x\ \underline{u}x \rightarrow \text{The } x\ \underline{u}x = U)$

and the *theorem of unique choice*

.4 $(\bigwedge x \in A \text{ One } y\ \underline{u}'xy \rightarrow \text{One } f(\text{on } A \text{ is } f \wedge \bigwedge x \in A\ \underline{u}'x.\dot{f}x)$

In 8.4, the choice function is, of course, $\lambda x \in A \text{ The } y\ \underline{u}'xy$.

9. Wellfounded sets

Our notion of wellfounded set has its origins in Richman's definition of 'constructive ordinal' [14].

.0 $(\text{wellfounded is } A \equiv \bigwedge S \subset A(\bigwedge x \in A(x \cap A \subset S \rightarrow x \in S) \rightarrow S = A))$

It is not hard to prove

.1 $(\text{wellfounded is } A \rightarrow \neg\bigvee f(\text{on } \omega \text{ to } A \text{ is } f \wedge \bigwedge n \in \omega(.f \text{ scsr } n \in .fn)))$

where

.2 $(\text{scsr } x \equiv (x \vee \text{sng } x))$

We also have

.3 $(\text{wellfounded is } \omega)$

and

.4 $(n \in \omega \rightarrow \text{wellfounded is } n)$

More generally, defining

.5 $(\text{transitive is } A \equiv \bigwedge x \in A \bigwedge y \in A \bigwedge z \in A(x \in y \wedge y \in z \rightarrow x \in z))$

we have

.6 $(\text{wellfounded is } A \wedge \text{transitive is } A \wedge B \subset A \rightarrow \text{wellfounded is } B)$

In view of the foregoing, it would seem reasonable to adopt an axiom that all sets are wellfounded, especially as such an axiom would be classically equivalent to the standard *axiom of foundation*

$$(x \in A \rightarrow \bigvee y(y \in A \wedge y \cap A = 0))$$

and to Morse's axiom (2.5.9). We prefer not to do this, but to introduce the hypothesis of wellfoundedness where it is appropriate or necessary.

It may well be asked why we do not simply adopt the standard classical axiom of foundation, and be done with it. The answer is simple: as was first observed by Myhill [12], the classical axiom is essentially nonconstructive. To see this, suppose that the classical axiom obtains, and apply it to the set $\{1\} \cup (x \vee \neg x)$, to conclude that $(0 \in (x \vee \neg x))$, from which we obtain the law of excluded middle.

10. Recursion

As in Morse's classical approach, so in our constructive one we obtain powerful and general recursion theorems from which the familiar elementary recursion theorems follow as special cases. All this is made possible by the following definitions.

.0 (inducive is A ≡ $\bigwedge$x ∈ scsr A(wellfounded is x ∧ ∇x ⊂ x))

.1 (Induced Rxy u'xy on A ≡ (relation is R ∧ dmn R ⊂ A ∧ $\bigwedge$x ∈ A(vs Rx = st strc Rx y u'xy)))

.2 (R is induced on A by u'xy in x and y ≡ Induced Rxy u'xy on A)

.3 (Ndc Axy u'xy ≡ The R(induced Rxy u'xy on A))

.4 (ndc HA ≡ Ndc Axy sng.Hy)

.5 (ndc'ha ≡ ndc ⅄g(g = 0 ∧ a ∨ g ≠ 0 ∧ .h.g∇dmn g)ω)

.6 (ndc"Sa ≡ ndc ⅄g(g = 0 ∧ a ∨ g ≠ 0 ∧ ..S∇dmn g.g∇dmn g)ω)

We begin with two lemmas.

.7 (wellfounded is A ∧ Induced Rxy u'xy on A ∧ Induced Sxy u'xy on A → R = S)

Proof. Let T = Ɛx ∈ A(vs Rx = vs Sx). Then

(x ∈ A ∧ x ∩ A ⊂ T
→ (y ∈ strc Rx → $\bigvee$p$\bigvee$q(y = p,q ∈ R ∧ p ∈ x ∩ dmn strc Rx ⊂ x ∩ A ⊂ T)
→ $\bigvee$p$\bigvee$q(y = p,q ∈ R ∧ vs Rp = vs Sp ∧ p ∈ x)
→ $\bigvee$p$\bigvee$q(y = p,q ∈ S ∧ p ∈ x)
→ y ∈ strc Sx)
→ strc Rx ⊂ strc Sx)

Similarly,

(x ∈ A ∧ x ∩ A ⊂ T → strc Sx ⊂ strc Rx)

so that

$$\begin{aligned}(x \in A \wedge x \cap A \subset T \ &\rightarrow \ \mathrm{strc}\ Rx = \mathrm{strc}\ Sx\\ &\rightarrow \ \mathrm{vs}\ Rx = \mathrm{st}\ \mathrm{strc}\ Rx\ y\ \underline{u}'xy\\ &\qquad = \mathrm{st}\ \mathrm{strc}\ Sx\ y\ \underline{u}'xy = \mathrm{vs}\ Sx\\ &\rightarrow \ x \in T)\end{aligned}$$

Thus $T = A$ and

$$(x \in A \rightarrow \mathrm{vs}\ Rx = \mathrm{vs}\ Sx)$$

from which the result is almost immediate. □

.8 $(\text{wellfounded is } A \wedge \text{Induced } Rxy\ \underline{u}'xy \text{ on } A \rightarrow R = \mathrm{Ndc}\ Axy\ \underline{u}'xy)$

<u>Proof</u> Use (10.7) and (8.2). □

We can now prove the *general recursion theorem*:

.9 $(\text{wellfounded is } A \wedge \text{inducive is } A \rightarrow \text{Induced } Rxy\ \underline{u}'xy \text{ on } A \leftrightarrow$
$R = \mathrm{Ndc}\ Axy\ \underline{u}'xy)$

<u>Proof</u> Let

$$((R = \mathrm{Ndc}\ Axy\ \underline{u}'xy) \wedge \bigwedge a(\underline{u}a = \mathrm{Ndc}\ axy\ \underline{u}'xy) \wedge$$
$$(S = \mathrm{E}x,y(x \in A \wedge y \in \underline{u}'x\ \underline{u}x)))$$

Then

$$(\text{relation is } S \wedge \mathrm{dmn}\ S \subset A \wedge \bigwedge x \in A(\mathrm{vs}\ Sx = \underline{u}'x\ \underline{u}x))$$

Next, we have

$$\begin{aligned}(a \in A \wedge x \in a \ &\rightarrow \ x \subset a\\ &\rightarrow \ \mathrm{strc}(\mathrm{strc}\ Sa)x = \mathrm{strc}\ S(a \cap x) = \mathrm{strc}\ Sx)\end{aligned}$$

In view of this and the easily proved statement

$$(x \in a \rightarrow \mathrm{vs}(\mathrm{strc}\ Sa)x = \mathrm{vs}\ Sx)$$

it follows that

$$\begin{aligned}&(a \in A \wedge \bigwedge x \in a(\mathrm{vs}\ Sx = \underline{u}'x\ \mathrm{strc}\ Sx)\\ &\rightarrow \text{relation is } \mathrm{strc}\ Sa \wedge \mathrm{dmn}\ \mathrm{strc}\ Sa \subset a \wedge\\ &\quad \bigwedge x \in a(\mathrm{vs}(\mathrm{strc}\ Sa)x = \underline{u}'x\ \mathrm{strc}(\mathrm{strc}\ Sa)x)\\ &\rightarrow \text{Induced } (\mathrm{strc}\ Sa)xy\ \underline{u}'xy \text{ on } a))\end{aligned}$$

Thus, by (10.7) and (8.2),

$$(a \in A \wedge \bigwedge x \in a(\mathrm{vs}\ Sx = \underline{u}'x\ \mathrm{strc}\ Sx) \rightarrow \mathrm{strc}\ Sa = \underline{u}a)$$

We now let

$$(T = \mathrm{E}a \in A \bigwedge x \in A(\mathrm{vs}\ Sx = \underline{u}'x\ \mathrm{strc}\ Sx)$$

Then, as $(a \in A \rightarrow a \subset A)$, we have

$$\begin{aligned}(a \in A \wedge a \cap A \subset T \wedge x \in a \ &\rightarrow \ x \in a \subset T\\ &\rightarrow \ x \in A \wedge \bigwedge y \in x(\mathrm{vs}\ Sy = \underline{u}'y\ \mathrm{strc}\ Sy)\end{aligned}$$

$$\rightarrow\ x \in A \wedge \mathrm{strc}\ Sx = \underline{u}x$$
$$\rightarrow\ \mathrm{vs}\ Sx = \underline{u}'x\ \underline{u}x = \underline{u}'x\ \mathrm{strc}\ Sx)$$

Hence

$$(a \in A \wedge a \cap A \subset T \ \rightarrow\ \bigwedge x \in a(\mathrm{vs}\ Sx = \underline{u}'x\ \mathrm{strc}\ Sx)$$
$$\rightarrow\ a \in T)$$

from which we deduce that T = A. In turn, this and the foregoing yield

$$(a \in A \ \rightarrow\ \mathrm{strc}\ Sa = \underline{u}a$$
$$\rightarrow\ \mathrm{vs}\ Sa = \underline{u}'a\ \underline{u}a = \underline{u}'a\ \mathrm{strc}\ Sa)$$

whence

$$(\mathrm{Induced}\ Sxy\ \underline{u}'xy\ \mathrm{on}\ A)$$

Reference to (10.7) and (8.2) completes the proof. □

With the general recursion theorem at hand, there is little problem in the derivation of the familiar recursion theorems in the forms

.10 $(\mathrm{on}\ A\ \mathrm{to}\ A\ \mathrm{is}\ h \wedge a \in A$

$\rightarrow\ \mathrm{on}\ \omega\ \mathrm{to}\ A\ \mathrm{is}\ f \wedge .f0 = a \wedge \bigwedge n \in \omega(.f\mathrm{scsr}\ n = .h.fn) \leftrightarrow f = \mathrm{ndc}'ha)$

.11 $(\mathrm{on}\ \omega,,A\ \mathrm{to}\ A\ \mathrm{is}\ h \wedge a \in A$

$\rightarrow\ \mathrm{on}\ \omega\ \mathrm{to}\ A\ \mathrm{is}\ f \wedge .f0 = a \wedge \bigwedge n \in \omega(.f\mathrm{scsr}\ n = .h(n,.fn))$

$\leftrightarrow\ f = \mathrm{ndc}''\text{ʎ}n \in \omega\ \text{ʎ}x \in A.h(n,x)a)$

The guarantee that the functions defined by recursion in 10.10 and 10.11 are constructively well-defined is provided by axiom 5.0:

.12 $(\mathrm{on}\ A\ \mathrm{to}\ A\ \mathrm{is}\ h \wedge a \in A \rightarrow \mathrm{ndc}'ha \in U)$

.13 $(\mathrm{on}\ \omega,,A\ \mathrm{to}\ A\ \mathrm{is}\ h \wedge a \in A \rightarrow \mathrm{ndc}''\text{ʎ}n \in \omega\ \text{ʎ}x \in A.h(n,x)a \in U)$

To end this section, we prove Morse's *axiom of infinity*:

.14 $(U = \bigvee c \bigwedge x(c \wedge (c \in U) \wedge ((x \in c) \rightarrow \mathrm{sng}\ x \in c))))$

<u>Proof</u> By 5.9, 7.8, 10.10, and 10.12,

$$(f = \mathrm{ndc}'\text{ʎ}x\ \mathrm{sng}\ x\ 0$$
$$\rightarrow\ \mathrm{on}\ \omega\ \mathrm{is}\ f \wedge .f0 = 0 \wedge \bigwedge n \in \omega(.f\ \mathrm{scsr}\ n = \mathrm{sng}.fn) \wedge f \in U),$$

whence, by 7.0,

$$(c = \mathrm{rng}\ \mathrm{ndc}'\text{ʎ}x\ \mathrm{sng}\ x\ 0$$
$$\rightarrow\ 0 \in c \wedge c \in U \wedge (x \in c \rightarrow \mathrm{sng}\ x \in c)$$
$$\rightarrow\ \bigwedge x(0 \in c \wedge c \in U \wedge (x \in c \rightarrow \mathrm{sng}\ x \in c)))$$

Thus

$$(\bigvee c \bigwedge x(0 \in c \wedge c \in U \wedge (x \in c \rightarrow \mathrm{sng}\ x \in c)),$$

from which the result readily follows. □

The purpose of the axiom of infinity in Morse's system is to prove 5.0 and 5.9. We prefer to introduce 5.0 as one of our axioms of construction, and then derive 5.9 as a theorem.

11. Axioms of choice

For the purposes of constructive analysis, the strongest choice axiom that appears to be needed is the *axiom of dependent choice*:

.0 $(a \in A \wedge \bigwedge x(x \in A \rightarrow \bigvee y(y \in A \wedge \underline{u}'xy))$
$\rightarrow \bigvee f(\text{on } \omega \text{ to } A \text{ is } f \wedge .f0 = a \wedge \bigwedge n(n \in \omega \rightarrow \underline{u}'.fn\ .f \text{ scsr } n)))$

A simple consequence of this is the *principle of countable choice*:

.1 $(\bigwedge n \in \omega \bigvee x \in A\ \underline{u}'nx \rightarrow \bigvee f(\text{on } \omega \text{ to } A \text{ is } f \wedge \bigwedge n \in \omega\ \underline{u}'n.fn))$

We also have the more general choice principle

.2 $(\text{on } \omega \text{ is } T \wedge a \in .T0 \wedge \bigwedge n \in \omega \bigwedge x \in .Tn \bigvee y \in .T\text{scsr } n\ \underline{u}'xy$
$\rightarrow \bigvee f(\text{on } \omega \text{ is } f \wedge .f0 = a \wedge \bigwedge n \in \omega(.fn \in .Tn \wedge \underline{u}'.fn.f \text{ scsr } n))$

A well-known theorem of Goodman and Myhill [8] states that the full axiom of choice

$$(\bigwedge x \in A \bigvee y \in B\ \underline{u}'xy \rightarrow \bigvee f(\text{on } A \text{ to } B \text{ is } f \wedge \bigwedge x \in A\ \underline{u}'x.fx))$$

implies the law of exluded middle.

12. Families of sets

In the following, note that in contrast to Bishop's usage, the empty set is both subfinite and countable.

.0 $(\text{subfinite is } A \equiv \bigvee f(\text{function is } f \wedge \text{dmn } f \in \omega \wedge \text{rng } f = A)$

.1 $(\text{subfnt} \equiv \mathrm{E}A \text{ subfinite is } A)$

.2 $(\text{countable is } A \equiv \bigvee f(\text{upon } \omega \text{ onto } A \text{ is } f))$

.3 $(\text{cbl} \equiv \mathrm{E}A \text{ countable is } A)$

.4 $(\text{inhabited is } A \equiv \bigvee x(x \in A))$

.5 $(\text{htd} \equiv \mathrm{E}A \text{ inhabited is } A)$

.6 $(\triangledown' F \equiv \bigvee G \in \text{subfnt} \cap \text{sb } F \text{ sng } \triangledown G)$

.7 $(\pi' F \equiv \bigvee G \in \text{htd} \cap \text{subfnt} \cap \text{sb } F \text{ snf } \pi G)$

.8 $(\triangledown'' F \equiv \bigvee G \in \text{cbl} \cap \text{sb } F \text{ sng } \triangledown G)$

.9 $(\pi'' F \equiv \bigvee G \in \text{htd} \cap \text{cbl} \cap \text{sb } F \text{ sng } \pi G)$

The inhabitedness condition is necessary in 12.7 and 12.9 to avoid the appearance of $\pi 0$ (which equals U).

We believe that, barring the fact that Bishop's definition applies to complemented sets [2, Ch. 3, Sect. 2], the following definition captures the spirit of his inductive approach to Borel sets.

.10 $(\text{Borel } F \equiv \bigwedge B;(F \subset B = \nabla''B = \pi''B)B)$

Of the following theorems, perhaps the most interesting are 12.13 and 12.17, each of which illustrates the inductive nature of definition 12.10.

.11 $(F \subset \text{Borel } F)$

.12 $(F \subset B = \nabla''B = \pi''B \rightarrow \text{Borel } F \subset B)$

.13 $(F \subset S = \nabla''S = \pi''S \subset \text{Borel } F \rightarrow S = \text{Borel } F)$

.14 $(\text{Borel } F = \nabla''\text{Borel } F = \pi''\text{Borel } F)$

<u>Proof</u> $(F \subset B = \nabla''B = \pi''B \rightarrow \text{Borel } F \subset B$

$\rightarrow \nabla''\text{Borel } F \subset \nabla''B = B \wedge \pi''\text{Borel } F \subset \pi''B = B)$

whence

$(\nabla''\text{Borel } F \subset \bigwedge B(F \subset B = \nabla''B = \pi''B \rightarrow B) \wedge$

$\pi''\text{Borel } F \subset \bigwedge B(F \subset B = \nabla''B = \pi''B \rightarrow B))$

The result now follows because $(F \subset \nabla''F \wedge F \subset \pi''F)$. □

.15 $(S \subset \text{Borel } F \rightarrow \nabla''S \subset \text{Borel } F \wedge \pi''S \subset \text{Borel } F)$

.16 $(\text{Borel } F = \bigwedge B;(F \subset B \wedge \bigwedge S \subset B(\nabla''S \subset B \wedge \pi''S \subset B))B)$

.17 $(F \subset X \subset \text{Borel } F \wedge \bigwedge S \subset X(\nabla''S \subset X \wedge \pi''S \subset X) \rightarrow X = \text{Borel } F)$

Note that in what appears to be the definitive constructive approach to integration theory, Borel sets are not used at all ([3], [4]).

13. <u>An example of formalisation</u>

To convince ourselves that the formal system described above is adequate for the presentation of constructive mathematics, it is worth sketching the formal layout of part of an interesting proof that appears in [2, p. 177]. The result whose proof we shall formalise is

.0 *If A is a complete, located subset of a metric space* (X,ρ), *then for each x in X there exists* ξ *in A such that* $\rho(x,\xi) > 0$ *entails* $\rho(x,A) > 0$.

An informal proof of 13.0 goes as follows. Construct an increasing binary sequence (δ_n) such that

$$\delta_n = 0 \rightarrow \rho(x,A) < 1/n$$

and

$$\delta_n = 1 \rightarrow \rho(x,A) > 1/2n.$$

Fix any point a of A, and define a sequence (x_n) in A as follows. If $\delta_1 =$

1, then $x_n \equiv a$ for all n. If $\delta_n = 0$, then x_n is chosen in A so that $\rho(x,x_n) < 1/n$; if $\delta_1 = 0$ and $\delta_n = 1$, then $x_n \equiv x_m$, where m is the unique integer such that $\delta_{m-1} = 0$ and $\delta_m = 1$. Then (x_n) is a Cauchy sequence in A, and converges to the required point ξ of A.

In order to formalise the first part of this proof, we shall assume that appropriate formal definitions of such notions as real number, metric space, $.\rho(x,y)$, and so on, have been given.

Let

$$(\Phi = \lambda m,n \in \{12\},,\omega(m \wedge ((m = 1 \wedge .\rho(\xi,A) < n^{-1}) \vee (m = 2 \wedge .\rho(\xi,A) > (2n)^{-1}))))$$

Then

$$(\text{on } \{12\},,\omega \text{ is } \Phi)$$

Moreover, as

$$(\bigwedge n \in \omega(.\rho(\xi,A) > (2n)^{-1} \vee .\rho(\xi,A) < n^{-1}))$$

$$\bigwedge x \in \mathbb{R} \bigwedge y \in \mathbb{R}((x = y \rightarrow (x = y) = U) \wedge (\neg(x = y) \rightarrow (x = y) = 0))$$

and

$$(\bigwedge x \in \mathbb{R} \bigwedge y \in \mathbb{R}((x < y \rightarrow (x < y) = U) \wedge (\neg(x < y) \rightarrow (x < y) = 0))$$

it is clear that

$$(\bigwedge n \in \omega \bigvee m \in \{12\}(.\Phi(m,n) = m))$$

Thus, by 11.1, there exists g such that

$$(\text{on } \omega \text{ to } \{12\} \text{ is } g \wedge \bigwedge n \in \omega(.\Phi(.gn,n) = .gn))$$

We now set

$$(h = \lambda n,m \in \omega,,\omega \max(.g \text{ scsr } m - 1,n))$$

$$(\delta = \text{ndc}''\lambda n \in \omega\ \lambda x \in \omega\ .h(n,x)(.g0 - 1))$$

and

$$(\alpha = \text{The } m \in \omega(.\delta m = 0 \wedge .\delta \text{ scsr } m = 1))$$

Then

$$(\text{on } \omega \text{ to } \omega \text{ is } \delta \wedge .\delta 0 = .g0 - 1 \wedge \bigwedge n \in \omega(.\delta \text{ scsr } n = \max(.g \text{ scsr } n - 1,.\delta n)))$$

Noting that

$$\begin{aligned}(n \in \omega \wedge .\Phi(1,n) = 1 \rightarrow 1 &= 1 \cap (U \cap (.\rho(\xi,A) < n^{-1})) \cup (0 \cap (.\rho(\xi,A) > (2n)^{-1}))\\ &= 1 \cap (.\rho(\xi,A) < n^{-1})\\ &\rightarrow 0 \in (.\rho(\xi,A) < n^{-1})\\ &\rightarrow .\rho(\xi,A) < n^{-1})\end{aligned}$$

and that, likewise,

$$(n \in \omega \wedge .\Phi(2,n) = 2 \rightarrow .\rho(\xi,A) > (2n)^{-1})$$

we easily prove by induction that

$$(\mathrm{rng}\ \delta \subset \{01\} \wedge \bigwedge n \in \omega(.\delta\ \mathrm{scsr}\ n \geq .\delta n \wedge (.\delta n = 0 \rightarrow .\rho(\xi,A) < n^{-1}) \wedge (.\delta n = 1 \rightarrow .\rho(\xi,A) > (2n)^{-1})))$$

We also have

$$(.\delta 0 = 0 \rightarrow \bigwedge n \in \omega(n > 1 \rightarrow (.\delta n = 1 \rightarrow \alpha \in \omega \wedge .\delta\alpha = 0 \wedge .\delta\ \mathrm{scsr}\ \alpha = 1))$$

and

$$(\bigwedge n \in \omega \bigwedge x \in A \bigvee y \in A((.\delta\ \mathrm{scsr}\ n = 1 \wedge x = y) \vee (.\delta\ \mathrm{scsr}\ n = 1 \wedge .\rho(\xi,y) < (n+1)^{-1})))$$

Now, a simple corollary of 11.2 is

$$(a \in A \wedge \bigwedge n \in \omega \bigwedge x \in A \bigvee y \in A\ \underline{u}''nxy \rightarrow \bigvee f(\mathrm{on}\ \omega\ \mathrm{to}\ A\ \mathrm{is}\ f \wedge .f0 = a \wedge \bigwedge n \in \omega\ \underline{u}''n.fn.f\ \mathrm{scsr}\ n))$$

It should now be clear that with a any point of A, there exists y such that

$$(\mathrm{on}\ \omega\ \mathrm{to}\ A\ \mathrm{is}\ y \wedge ((.\delta 0 = 1 \wedge .y0 = a) \vee (.\delta 0 = 0 \wedge .\rho(\xi,.y0) < 2^{-1})) \wedge \bigwedge n \in \omega((.\delta\ \mathrm{scsr}\ n = 1 \wedge .y\ \mathrm{scsr}\ n = .yn) \vee (.\delta\ \mathrm{scsr}\ n = 0 \wedge .\rho(\xi,.y\ \mathrm{scsr}\ n) < (n+1)^{-1})))$$

To show that such a term y fulfils our requirements for the proof of 13.0, it only remains to prove the statements

$$(.\delta 0 = 1 \rightarrow y = \lambda n \in \omega\ a)$$

$$(n \in \omega \wedge .\delta n = 0 \rightarrow .\rho(\xi,.yn) < n^{-1})$$

and

$$(n \in \omega \wedge .\delta 0 = 0 \wedge .\delta n = 1 \rightarrow .yn = .y\alpha \in A)$$

We omit the simple details of their proofs.

14. Conclusion

The foregoing is merely an outline of the possibilities afforded by a constructive development of set theory in Morse's high-level language. constructive analogues of many parts of Morse's classical text, including, for example, cardinals and their arithmetic, remain to be investigated. However, I am particularly interested in two, not necessarily unrelated, developments: the first of these is an attempt to reflect within the formal system some notion corresponding to 'the set of proofs of p'; the second is the exploration of constructive Morse set theory as a framework for theorem-proving on a computer. Some progress on the first of these topics is discussed in the paper [6].

REFERENCES

[1] M.J. Beeson, *Foundations of Constructive Mathematics*, Springer Verlag, Berlin 1985

[2] Errett Bishop, *Foundations of Constructive Analysis*, McGraw-Hill, New York 1967

[3] Errett Bishop and Douglas Bridges, *Constructive Analysis*, Grundlehren der math. Wissenschaften, Bd 279, Springer Verlag, Berlin 1985

[4] Errett Bishop and Henry Cheng, Constructive Measure Theory, Mem. Amer. Math. Soc 116 (1972)

[5] Douglas Bridges, A note on Morse's lambda notation in set theory, Z. Math. Logik Grundlagen Math., 24, 493-494 (1978)

[6] Douglas Bridges, A constructive Morse set theory II, to appear

[7] R.L. Constable et al., *Implementing Mathematics with the Nuprl Proof Development System*, Prentice-Hall, New Jersey 1986

[8] N. Goodman and J. Myhill, Choice implies excluded middle, Z. Math. Logik Grundlagen Math., 23, 461 (1978)

[9] A. Heyting, *Intuitionism*, 3rd edn., North-Holland, Amsterdam 1971

[10] P. Martin-Löf, Constructive mathematics and computer programming, in: Logic, Methodology and Philosophy of Science VI, pp. 153-175, North-Holland, Amsterdam 1982.

[11] A.P. Morse, *A Theory of Sets*, Academic Press, New York 1965

[12] J. Myhill, Some properties of intuitionistic Zermelo-Fraenkel set theory, in: Cambridge Summer School in Mathematical Logic, pp. 206-231, Lecture Notes in Mathematics 337, Springer Verlag, Berlin 1973

[13] J. Myhill, Constructive set theory, J. Symbolic Logic 40, 347-382 (1975)

[14] F. Richman, The constructive theory of countable abelian p-groups, Pacific J. Math 45, 621-637 (1973)

[15] A.S. Troelstra, *Principles of Intuitionism*, Lecture Notes in Mathematics 95, Springer Verlag, Berlin 1969

DIFFERENTIATION OF CONSTRUCTIVE FUNCTIONS OF A REAL VARIABLE AND RELATIVE COMPUTABILITY

Osvald Demuth and Pavel Filipec

Department of Computer Science, Charles University
Institute of Economics, Czechoslovak Academy of Sciences
Prague, Czechoslovakia

The paper belongs to constructive mathematics of Markov school. The aim of it is to show an introducing of some relativized notions into constructive mathematical analysis (CMA) and to present a few results concerning differentiability of constructive real-valued functions of a real variable as an example of the utilization of the relativization.

Introduction

Constructive mathematics can be described as that sort of mathematics in which, when one proves that there exists a solution to a parametric problem, one is able to extract from the proof an algorithm for computing the solution from values of the parameters. (Cf. Beeson,[1] p. 3, and Beeson.[2])

Considering that Beeson´s book[1] published recently gives an exposition of constructive mathematics in a broad sense and that Šanin´s paper,[3] Kušněr´s book[4] and briefly also Demuth and Kučera´s paper[5] contain an analysis of basic features of constructive mathematics of Markov school, we limit ourselves in this introduction to a few remarks only.

Constructive mathematics of Markov school (CM), founded by A.A. Markov, N.A. Šanin and their followers after 1945, systematically uses, in contrast to intuitionism and Bishop´s constructivism, a mathematical concept of algorithm.

The basic features of CM are as follows.

(i) Objects studied in CM are so-called constructive objects, i.e. above all words (strings) in an alphabet and also objects finitely codable by words in a certain alphabet.

(ii) It is used in the abstraction of potential realizability (but not the abstraction of actual infinity) and the theory of algorithms with Markov's principle.

(iii) It is used as a special algorithmical interpretation of intuitionistic logic (so-called constructive logic) created by Šanin.

Points (i) and (ii) do not require more detailed explanations. We shall return to point (iii) in Section 1.

An introduction to CMA with a detailed bibliography up to 1971 can be found in Kušněr's book.[4] The bibliography of the Prague seminar on constructive mathematics up to 1978 is contained in Demuth and Kučera's paper.[5]

After these remarks we can turn to our topic. Already in Šanin's paper[3] there were considered several constructive analogues of the concept of real number. In addition to the basic analogue (constructive real numbers) let us mention so-called pseudonumbers which were systematically studied afterwards. The situation has been simplified, when it turned out that pseudonumbers can be regarded as codes of "real numbers constructive relative to the jump of empty set."

The introduction of relativized concepts into CMA strengthened applications of recursion theory (RT) in CMA. Further connections between CMA and RT are due to natural relations between real numbers and sets of natural numbers, e.g. via binary expansions of real numbers. These relations make it possible to employ in CMA such notions and methods of RT as degrees and reducibilities. On the other hand, there are results in RT based on applications of methods and concepts of CMA.

The concepts and notations of CMA used in this paper will be introduced in what follows. The phrase "cannot fail to" indicates a double negation. We use the following symbols for logical connectives: $\&$ conjunction, $\vee$ disjunction, $\neg$ negation, $\Rightarrow$ implication, $\Leftrightarrow$ equivalence. The sign $\leftrightharpoons$ will be adopted in place of the word "denotes."

1. Basic Concepts and Facts

The basic objects of constructive mathematical analysis studied in this paper are words in the alphabet Ξ containing, among others, the letters 0, $|$, $-$, $/$, $\Diamond$, $\Box$, Δ, ∇ etc. By Ξ^* we denote the set of all words in the alphabet Ξ. The signs Λ , $\overline{\overline{\cdot}}$, $\simeq$ denote the empty word, graphical equality, and conditional graphical equality, respectively. The symbols U, V play the role of variables for the elements of Ξ^* .

Markov algorithms, which we employ, are the algorithms over the alphabet Ξ. The applicability of an algorithm $\mathfrak{A}$ to a word P is denoted by $!\mathfrak{A}(P)$. The term "set" is understood in the same sense as in Šanin's paper[6] and { ; } is the notation to indicate set formation. Let $\mathfrak{M}$ and $\mathfrak{N}$ be sets, i.e. subsets of Ξ^*. We write $\mathfrak{M} \cap \mathfrak{N}, \mathfrak{M} \setminus \mathfrak{N}, \mathfrak{M} \cup \mathfrak{N}$ and $\mathfrak{M} \hat{\cup} \mathfrak{N}$ for the intersection, the difference, the union, and the quasi-union of $\mathfrak{M}$ and $\mathfrak{N}$, respectively. By the definition, a word P is a member of the union (resp. the quasi-union) of $\mathfrak{M}$ and $\mathfrak{N}$ iff $P \in \mathfrak{M} \vee P \in \mathfrak{N}$ (resp. $\neg\neg(P \in \mathfrak{M} \vee P \in \mathfrak{N})$). By a finite set we mean a set for which a list of all its elements can be given, and by an infinite set we mean a set different from any finite set.

By natural numbers (NNs) we mean the words 0, $0|$, $0||$, ... The set of all NNs will be denoted by N. We use the indexing of finite sets of NNs defined in Rogers' book,[7] p. 70. By $\mathcal{D}_n$ we denote the finite set whose index is n. Let us recall that $\mathcal{D}_0 = \emptyset$. For any word P and NN n we define $P^0 \leftrightharpoons \Lambda$ and $P^{n|} \leftrightharpoons P^n P$. The integers and the rational numbers (RtNs) are introduced as certain words in Ξ. As abbreviational notation for NNs, integers and RtNs we shall employ the standard notation of the form 2, -3, $\frac{2}{3}$ etc. Let us note that the arithmetical operations over these numbers are, in line with requirements of CM, realized algorithmically.

The symbols k, l, m, n, p, q, s, and t play the role of variables for NNs, i and j the role of variables for integers and a, b, c, and d the role of variables for RtNs.

We use the constructive interpretation of mathematical judgements.[6] Its substance is in an algorithmical interpretation of $\exists$ and $\vee$. For example, the formula $\forall U \exists V \; A(U,V)$ holds iff there exists a Markov algorithm $\mathfrak{A}$ such that $\forall U \, (\,!\mathfrak{A}(U) \,\&\, A(U,\mathfrak{A}(U)))$ holds. A disjunctive formula holds iff it is possible to determine its valid member algorithmically. Let us recall the importance of so-called normal formulae.[6] (Normal formulae do not contain $\vee$ and $\exists$ and each of their subformulae is equivalent to its own double negation.) By a normal set we mean any set for which the membership can be given by a normal formula. It should be noted that in this paper we introduce only variables whose domains (of admissible values) are normal sets. The consequence of this fact is that the interpretation of $\forall\exists$-formulae with such variables is analogical to that given above.

In reducibilities and in relativizations we use only normal sets of NNs. The symbols A, B, C play the role of metavariables for normal sets of NNs. $A \oplus B$ denotes $\{m; \exists k(m = 2k \ \& \ k \in A \ \vee \ m = 2k+1 \ \& \ k \in B)\}$.

Let us fix Markov algorithms wd and en establishing one-to-one correspondence between Ξ^* and N , where wd maps N onto Ξ^* and en is an algorithm inverse to wd . With the help of this numbering we carry over concepts introduced initially for sets of NNs (such as recursivity, recursive enumerability etc.) to the sets of words. Also, the well-known equivalence of Markov algorithms and <u>partial recursive functions</u> (PRFs) can be expressed in this way. The use of either Markov algorithms or PRFs depends on practical needs of the context.

Using PRFs we use notations given in Rogers' book.[7] So W_m $(m \in N)$ is a fixed effective enumeration of recursively enumerable (r.e.) sets of NNS; for any NN $k \geq 1$, τ^k is a mapping from the set of all ordered k-tuples of NNs onto N defined in Rogers' book,[7] p. 64, and π_1^k , π_2^k , ... , π_k^k are inverse mappings to τ^k, i.e. such that $\tau^k(\pi_1^k(n), \dots, \pi_k^k(n)) = n$ for any NN n . All these functions are primitive recursive.

As for the relative computability, we use the constructive reformulation of its characterization, given in Rogers' book.[7] The only difference is caused by the constructive interpretation of the existential quantifier. In fact, if B is a (normal) set of NNs and m , n, k are NNs, then by $\langle m\rangle^B(k) \simeq n$ we denote

$$\neg\neg \exists s t \, (\tau^4(k,n,s,t) \in W_{\rho(m)} \ \& \ \mathcal{D}_s \subseteq B \ \& \ \mathcal{D}_t \subseteq N \setminus B)$$

and by $!\langle m\rangle^B(k)$ we denote $\neg\neg\exists p(\langle m\rangle^B(k) \simeq p)$; $\langle m\rangle^B$ we call a <u>B-PRF with the index m</u> . (For notation see Rogers' book,[7] § 9.2.) For details and for the employment of relative computability in CM see Demuth, Kryl and Kučera's paper[8] and Demuth's paper.[9] We consider the relativized PRFs as predicatively defined correspondences. Let us note that these predicates are equivalent to normal formulae. By W_m^B we denote $\{n; !\langle m\rangle^B(n)\}$. It is useful to suppose that $W_0 = \emptyset$, consequently $W_{\rho(0)} = \emptyset$ and $W_0^B = \emptyset$ for any B . B' is <u>the jump of</u> B , i.e. the set $\{n; !\langle n\rangle^B(n)\}$, $B^{(0)}$ denotes B and for any NN n $B^{(n+1)}$ is $(B^{(n)})'$. Frequently, $[n]$ stands in this paper for $\emptyset^{(n)}$.

Let for any set B of NNs and any NNs p and k $(1 \leq k)$ be

$$\langle p\rangle_k^B (m_1, \dots, m_k) \leftrightharpoons \langle p\rangle^B(\tau^k(m_1, \dots, m_k)) \ .$$

Then we can for any NNs k and q $(1 \leq k, q)$ construct a primitive recursive function s^q_k such that for every set B and NN p

$$\langle p\rangle^B_{q+k}(m_1,\ldots,m_q,n_1,\ldots,n_k) \simeq \langle s^q_k(p,m_1,\ldots,m_q)\rangle^B_k(n_1,\ldots,n_k) .$$

$\emptyset$-PRFs are just the PRFs and therefore $\langle m\rangle^{\emptyset}$ is an indexing of PRFs. We usually write $\langle m\rangle$ instead of $\langle m\rangle^{\emptyset}$.

We can construct a primitive recursive function φ_0 of two variables and an NN $list$ such that for any NNs m and k and any set B of NNs $\varphi_0(m,0) = 0$, $\mathcal{D}_{\varphi_0(m,k)} \subseteq \mathcal{D}_{\varphi_0(m,k+1)}$, $W_m = \bigcup_k \mathcal{D}_{\varphi_0(m,k)}$, $\langle list\rangle^B_2$ is a total function, and

$$p \in \mathcal{D}_{\langle list\rangle^B_2(m,k)} \Longleftrightarrow \neg\neg \exists qst(\langle p,q,s,t\rangle \in W_{\varphi_0(\rho(m),k)} \,\&\, \mathcal{D}_s \subseteq B \,\&\, \mathcal{D}_t \cap B = \emptyset).$$

Thus we have a standard listing of B-r.e. sets (of NNs) for any B.

It should be noted that we are interested, owing to the natural connection between concepts of constructive mathematical analysis and arithmetical predicates, mainly in the computability relative to jumps of empty set. It is known from the results of E.M. Gold and P. Putnam that the $\emptyset^{(n)}$-PRFs $(1 \leq n)$ can be represented on the basis of recursive functions by means of non-effective limits. By the use of relative computability we improve our ability to handle effective processes without leaving constructive program. The advantage of the use consists both in substantial simplification and in clearness of formulations.

Relativized algorithms can be introduced on the basis of relativized PRFs. Let B be a set of NNs and let m be a NN. The correspondence $[\![\langle m\rangle^B]\!]$ defined by $[\![\langle m\rangle^B]\!](U) \simeq wd(\langle m\rangle^B(en(U)))$ is called a <u>B-algorithm with the index m</u> . Let us note that $\emptyset$-algorithms are just the correspondences realizable by Markov algorithms.

If B is a set of NNs and F is a B-algorithm, then for any word P we denote by $\tilde{F}_P$ a B-algorithm such that $\forall V(\tilde{F}_P(V) \simeq F(PV))$.

The relativized existential quantifier and the relativized disjunction can be defined on the basis of relativized algorithms in the same way as the constructive interpretation of $\exists$ and $\vee$ is expressed by means of Markov algorithms. Let B be a set of NNs. We write $\exists^B U\ A(U)$ (" B-exists a word U ... ") for $\exists m(!\,[\![\langle m\rangle^B]\!](0) \,\&\, A([\![\langle m\rangle^B]\!](0)))$ and we write $A_1 \overset{B}{\vee} A_2$ for $\exists^B V((V = \Lambda \Rightarrow A_1) \,\&\, (V \neq \Lambda \Rightarrow A_2))$, where A and A_1, A_2 have no occurences of variables m and V, respecively. In such a manner the interpretation of the quantifier $\exists^B$ and of $\overset{B}{\vee}$ is reduced to the interpretation of the quantifier $\exists$. Of course, $\exists$ and $\exists^{\emptyset}$ are equivalent. Further, $\exists$ "implies" $\exists^B$ for any set B , and $\exists^B$ "implies" $\neg\neg\exists$.

Let B be a set of NNs and let $\mathfrak{M}$ be a normal set of words in Ξ. A B-algorithm F is said to be

1) a B-sequence of elements of $\mathfrak{M}$ if $\forall n (!F(n) \;\&\; F(n) \in \mathfrak{M})$,
2) a B-sequence of B-algorithms of a certain type if for any NN n $\tilde{F}_{n\square}$ is a B-algorithm of the type.

In the sequel we present B-sequences of words (or B-algorithms) by their "members" using notation $\{\ldots\}^{B}_{n}$. Of course, e.g. "$\{P_n\}^{B}_{n}$ is a B-sequence of words" means that there exists a B-algorithm F such that $\forall n (!F(n) \;\&\; F(n) \doteq P_n)$. In the sequel the abbreviation B-SNNs (resp. B-SRtNs) means B-sequence of NNs (resp. of RtNs).

The central concept of mathematical analysis is the concept of real number. There exist more mutually non-equivalent constructive formulae that characterize, from the classical point of view, the fundamentality, i.e. cauchyness. Indeed, a sequence F of RtNs is said to be

(i) canonically fundamental if $\forall n\, k\, (n \leq k \Rightarrow |F(k) - F(n)| \leq 2^{-n})$,
(ii) fundamental if $\forall n\, \exists m\, \forall k\, (m \leq k \Rightarrow |F(k) - F(m)| \leq 2^{-n})$,
(iii) pseudo-fundamental if $\forall n\, \neg\neg \exists m\, \forall k\, (m \leq k \Rightarrow |F(k) - F(m)| \leq 2^{-n})$.

Let us note that, in accordance with constructive interpretation of mathematical judgements, the fundamentality of F means the existence of an algorithmical modulus of fundamentality of F, i.e. the existence of an algorithm transforming every NN n into some NN m with the corresponding property. In the case of pseudo-fundamentality of F the formula is equivalent to a normal formula and its interpretation does not differ from the classical one, specifically the existence of algoritmical modulus is not required. As is known, in the case of Specker $[0]$-sequence of RtNs such an algorithmical modulus does not exist, indeed. For sequences of RtNs we obtain the concept of B-fundamentality so that we replace in (ii) $\exists$ by $\exists^{B}$. It will be useful for us that a B-sequence of RtNs is pseudo-fundamental iff it is B'-fundamental.

Let us mention that the similar situation, we have met in the case of fundamentality, is also found in the case of constructive formulations of other concepts such as convergence, continuity, uniform continuity, differentiability etc. In a quite analogical way we receive concepts of pseudo-convergence, $[n]$-continuity, B-differentiability etc. In these concepts the prefix "pseudo-" means "in a classical sense."

Using the arithmetical hierarchy we shall introduce a hierarchy of constructive analogues of the reals.

For any NN n

a) an $[n]$-constructive real number ($[n]$-CRN) is a RtN or a word of the form $p\diamond^{m+1}q$, where p, q and m are NNs, $m \leq n$, $[\![\langle p\rangle^{[m]}]\!]$ is an $[m]$-SRtNs and $[\![\langle q\rangle^{[m]}]\!]$ is an $[m]$-SNNs being a modulus of fundamentality of $[\![\langle p\rangle^{[m]}]\!]$;

b) an $[n]$-constructive pseudonumber is a RtN or a word of the form $p\diamond^{m+1}$, where p and m are NNs, $m \leq n$ and $[\![\langle p\rangle^{[m]}]\!]$ is a pseudo-fundamental $[m]$-SRtNs ;

c) an $[n]$-pseudonumber $p\diamond^{m+1}$ is called monotone if the $[m]$-SRtNs $[\![\langle p\rangle^{[m]}]\!]$ is monotone ;

d) the set of all $[n]$-CRNs is denoted by $D^{[n]}$, ${}^*D^{[n]} \leftrightharpoons D^{[n]} \cup \{-\infty,+\infty\}$, the set of all $[n]$-pseudonumbers is denoted by $\Pi^{[n]}$, the symbols $x^{[n]}$, $y^{[n]}$, $z^{[n]}$, $v^{[n]}$ and $w^{[n]}$ play the role of variables for $[n]$-CRNs, $\xi^{[n]}$ and $\eta^{[n]}$ play the role of variables for $[n]$-pseudonumbers.

An arithmetical real number (ARN) is a word being an $[n]$-CRN for some NN n. The set of all ARNs is denoted by $\mathcal{A}$. Thus $\mathcal{A} = \bigcup_n D^{[n]}$. Let us mention that $[n]$-CRNs are codes of reals recursive relative to $\emptyset^{(n)}$, and consequently ARNs are codes of reals definable in the elementary theory of NNs. (See Rogers' book.[7]) The symbols X, Y and Z play the role of variables for ARNs.

On the set of all $[n]$-CRNs and $[n]$-pseudonumbers ($0 \leq n$) the relations of equality and of order are defined as predicates in the obvious way. As is known, different operations on $[n]$-algorithms, which are effective relative to $\emptyset^{(n)}$, can be realized effectively as transformations of codes of these $[n]$-algorithms. Naturally, the relativized s-m-n-theorem plays an important role here. For us it is important that, for example, basic algebraic operations on these numbers (i.e. on $\mathcal{A} \cup (\bigcup_n \Pi^{[n]})$) are realized[8] algorithmically and, further, for any $[n]$-sequence $\{x_m\}_m^{[n]}$ of $[n]$-CRNs there is a $[0]$-sequence $\{y_m\}_m^{[0]}$ of $[n]$-CRNs such that $\forall m\, (x_m = y_m)$.

Let us quote several well-known facts on the reals, we have introduced.[8] As for the aritmetical complexity, the set of all $[n]$-CRNs (i.e. $D^{[n]}$) is Π_{n+2}-complete and the set of all $[n]$-pseudonumbers is Π_{n+3}-complete. The set $D^{[0]}$ with the Euclidean metric is a complete separable metric space.[3] The relativization of this assertion holds[8] for $D^{[n]}$. By Shoenfield's theorem on limit computability and by s-m-n-theorem there are $[0]$-algorithms Pseud and Dupl such that

$$\forall n\, x^{[n+1]}\, \xi^{[n]}\, (\mathit{Pseud}(x^{[n+1]}) \in \Pi^{[n]} \;\&\; \mathit{Pseud}(x^{[n+1]}) = x^{[n+1]} \;\&$$

$$\mathit{Dupl}(\xi^{[n]}) \in D^{[n+1]} \quad \& \quad \mathit{Dupl}(\xi^{[n]}) = \xi^{[n]} \;)$$

holds.[8] (Lemma 5.5). This fact shows that we can restrict the using of $[n]$-pseudonumbers to those cases only when it leads to substantial simplifications. For any $[n+1]$-sequence $\{\mathcal{U}_m\}_m^{[n+1]}$ of $[n]$-CRNs there is an $[n]$-sequence $\{v_p\}_p^{[n]}$ of $[n]$-CRNs such that $\forall m \neg\neg\exists p\,(\mathcal{U}_m = v_p)$. We can construct a $[0]$-sequence $\{\vartheta_m\}_m^{[0]}$ of monotone $[n]$-pseudonumbers such that for any word P being an $[n]$-CRN or a monotone $[n]$-pseudonumber there is a NN q such that $P = \vartheta_q$.

The hierarchy of the reals, which we have introduced, is very convenient for using in constructive mathematical analysis. For example, studying measurability or differentiability we find the following lemma very useful.

<u>Lemma 1.1.</u> For any NN n there exists an $[n+2]$-algorithm (resp. an $[n+3]$-algorithm) which transforms any NN m such that $[\![\langle m\rangle^{[n]}]\!]$ is an $[n]$-sequence of $[n]$-CRNs into an element of $^{*}D^{[n+1]}$ (resp. of $^{*}D^{[n+2]}$) being the supremum (resp. limes superior) of the $[n]$-sequence.

On the other hand in some situations we are interested to have some finer hierarchy of arithmetical reals. The well-known hierarchy of T-degrees of sets of NNs is convenient for this purpose.

For any ARN X let $\mathit{Set}(X)$ be an infinite set of NNs A for which there cannot fail to exist an integer j such that X is a pseudo-limit of the A-sequence $\{j+\sum_{k=0}^{t} 2^{-k-1}.F(k)\}_t^A$, where F is the characteristic function of the set A .

For any $[n]$-CRN $x^{[n]}$ we can construct an NN p such that $\langle p\rangle^{[n+1]}$ is the characteristic function of $\mathit{Set}(x^{[n]})$. For any $[n]$-CRN $x^{[n]}$ such that $\neg\exists i k\,(x^{[n]} = i.2^{-k})$ we can even construct an NN q such that $\langle q^{[n]}\rangle$ is the characteristic function of $\mathit{Set}(x^{[n]})$. Thus, for any $[n]$-CRN $x^{[n]}$ the set $\mathit{Set}(x^{[n]})$ cannot fail to be $[n]$-recursive and is $[n+1]$-recursive. On the contrary, for any infinite $[n]$-recursive set B of NNs we can construct an $[n]$-CRN v for which $B = \mathit{Set}(v)$ & $0 < v \leq 1$ holds.

An ARN X is said to be

a) B-computable if there exists a canonically fundamental B-SRtNs which pseudo-converges (i.e., in this case, $[0]$-converges) to X ;
b) weakly B-computable (resp. monotonically weakly B-computable)

if there exists a B-SRtNs (resp. a monotone B-SRtNs) which pseudo--converges (and consequently B'-converges) to X.

Remark 1.2. It is clear that

a) an ARN X cannot fail to be B-computable iff $\mathcal{Sed}(X)$ cannot fail to be B-recursive;

b) an ARN X is $[n]$-computable iff $\exists x^{[n]}(X = x^{[n]})$;

c) an ARN is weakly B-computable iff it is B'-computable;

d) any B-computable ARN is monotonically weakly B-computable;

e) if B is $[n]$-recursive, then for any infinite B-r.e. set C of NNs there is a monotonically weakly B-computable $[n+1]$-CRN v, $0 < v \leq 1$, such that $\mathcal{Sed}(v) = C$.

2. Regular Sets of ARNs

Now we shall devote ourselves to some notions concerning sets of ARNs.

By an $[n]$-segment (resp. $[n]$-interval) we mean any word of the form $x \Delta y$ (resp. $x \nabla y$), where x and y are $[n]$-CRNs and $x < y$ holds. If H is an $[n]$-segment (resp. $[n]$-interval), then $E_\ell(H)$, $E_r(H)$ denote the left and right endpoints, respectively, $|H|$ denotes the length of H, i.e. $E_r(H) - E_\ell(H)$, and $(H)^0$ denotes the interval $E_\ell(H) \nabla E_r(H)$. Any word of the form $a \Delta b$ (resp. $a \nabla b$), where a and b are RtNs, $a < b$, is called a rational segment (resp. rational interval). The relation of membership of ARNs and of $[m]$-pseudonumbers (m arbitrary) to an $[n]$-segment (resp. $[n]$-interval) is defined in a natural way.

Let us suppose to have two fixed enumerations of all rational segments, resp. intervals, given by $[0]$-algorithms $\mathcal{L}$, resp. $\mathcal{L}^0$, where for any NN ℓ $\mathcal{L}^0(\ell)$ is an interval $(\mathcal{L}(\ell))^0$.

For any set of NNs $\mathfrak{M}$ we denote $[\mathfrak{M}] \leftrightharpoons \{X; \neg\neg\exists \ell(\ell \in \mathfrak{M} \,\&\, X \in \mathcal{L}^0(\ell))\}$ and $[\mathfrak{M}]_c \leftrightharpoons \{X; \neg\neg\exists \ell(\ell \in \mathfrak{M} \,\&\, X \in \mathcal{L}(\ell))\}$. Writing about sets of ARNs we shall use H instead of $\{X; X \in H\}$ for any $[n]$-segment (resp. $[n]$-interval) H.

A set of ARNs is said to be regular if it is closed under equality of ARNs. In the sequel writing about sets of ARNs we mean regular ones. We say that a set of ARNs $\mathfrak{M}$ is $[n]$-open (or is of the type $G^{[n]}$) if there is a NN m such that $\mathfrak{M} = [W_m^{[n]}]$. We can also introduce types $G_\delta^{[n]}$, $G_{\delta\sigma}^{[n]}$ etc. Thanks to the relativized s-m-n-theorem we can do it in the following way. E.g. a set of ARNs $\mathcal{P}$ is said to be of the type $G^{[n]}_{\delta\sigma}$ if there exists a NN m such that

$$\mathcal{P} = \bigcup_p \bigcap_q [W^{[n]}_{s_1^2(m,p,q)}] \quad .$$

Remark 2.1. $[0]$-open sets of ARNs can give us some information about A-PRFs (and thus about A-algorithms), where A runs over infinite $[n]$-recursive sets (of NNs) for all NNs n. With help of the s-m-n-theorem we can construct primitive recursive functions f_0 and f_1 such that for any ARN $X \in 0\nabla 1$, $\neg\exists ik(X = i.2^{-k})$, and any NNs m, p, q we have

$$\langle m\rangle^{\mathcal{Sel}(X)}(p) \simeq q \iff X \in [W^{[0]}_{f_0(m,p,q)}] \quad ,$$
$$X \in [W^{[0]}_m] \iff\ !\ \langle f_1(m)\rangle^{\mathcal{Sel}(X)}(0) \quad .$$

Remark 2.2. 1) For any NNs m, n there exists a $[max(m,n)]$--algorithm S such that $!\ S(x^{[n]}\square t) \iff x^{[n]} \in [W^{[m]}_t]$,

$$!\ S(x^{[n]}\square t) \Rightarrow S(x^{[n]}\square t) \in W^{[m]}_t \ \&\ x^{[n]} \in \mathcal{L}^0(S(x^{[n]}\square t)) \ .$$

2) For any NN m and any B-computable ARN X the predicate $X \in [W^{[m]}_t]$ of the variable t is $(\emptyset^{(m)} \oplus B)$-recursively enumerable.

Relativizating results of Lacombe, Zaslavskij, Cejtin[10] we obtain: For any NN n and any RtN $\varepsilon > 0$ there exists an $[n]$-sequence of rational intervals ψ such that $\forall x^{[n]} \exists^{[n]} p(x^{[n]} \in \psi(p))$ and that the sum of the lengths of an arbitrary finite set of these intervals is less than ε. In the sequel we use the following type of $[n]$-coverings of $0\Delta 1$.[8] A $[n]$-sequence Φ of non-overlapping rational segments contained in $0\Delta 1$ is said to be an $[n]$-covering if $E_\ell(\Phi(0)) = 0$, $E_r(\Phi(1)) = 1$ and $\forall x^{[n]}(x^{[n]} \in 0\nabla 1 \Rightarrow \exists^{[n]} pq(E_r(\Phi(p)) = E_\ell(\Phi(q))\ \&\ E_\ell(\Phi(p)) < x^{[n]} < E_r(\Phi(q))))$. An $[n]$-covering Φ is said to be singular if $\neg\neg\exists a \forall p(\sum_{k=0}^{p} |\Phi(k)| \leq a < 1)$. Let us note that there exist both singular and non-singular $[n]$-coverings.

Lemma 2.3. (Cf. Zaslavskij and Cejtin's paper.[10]) Let n be an NN and let Φ be an $[n]$-sequence of $[n]$-segments, let $x^{[n]}\Delta y^{[n]}$ be an $[n]$-segment, and let the series $\sum_k |\Phi(k)|$ $[n]$-converge to an $[n]$-CRN less than $|x^{[n]}\Delta y^{[n]}|$. Then there exists an $[n]$-CRN $w^{[n]}$ such that $w^{[n]} \in x^{[n]}\nabla y^{[n]}\ \&\ \neg\exists k(w^{[n]} \in \Phi(k))$.

Corollary. Let Φ be an $[n]$-covering. Then the $[n]$-SRtNs $\{\sum_{p=0}^{k} |\Phi(p)|\}^{[n]}_k$ is $[n]$-fundamental iff Φ is not singular.

Lemma 2.3 enables us to give reasonable definitions of constructive analogues of the concepts "almost every", "Lebesgue measurable set of reals", and "measure". Naturally, we shall get a hierarchy of concepts. It is a consequence of the systematic use of the theorem of supremum in classical measure theory. Let us note that the $[0]$-level of this hierarchy was introduced in 1969. We shall start with "almost every."

An $[n]$-sequence $\{H_m\}_m^{[n]}$ of non-overlapping $[n]$-segments is termed an $S_6^{[n]}$-set and an $[n]$-CRN z is termed a measure of $\{H_m\}_m^{[n]}$ if the $[n]$-sequence $\{\sum_{m=0}^{k} |H_m|\}_k^{[n]}$ of $[n]$-CRNs pseudo-converges (which is here the same as $[n]$-converges) to z.

If X is an ARN and $\{H_m\}_m^{[n]}$ is an $S_6^{[n]}$-set, then we write $X \in \{H_m\}_m^{[n]}$ for $X \in \hat{\bigcup_m} H_m$.

A property $\mathcal{V}$ of ARNs, resp. $[n]$-CRNs, is said to hold for $[n]$-almost every ARN, resp. $[n]$-CRN (of an $[n]$-segment L) if there exists an $[n]$-sequence $\{\mathcal{Y}^m\}_m^{[n]}$ of $S_6^{[n]}$-sets such that for any NN m the measure of $\mathcal{Y}^m$ is less than 2^{-m} and for any ARN, resp. $[n]$-CRN X (in L) $X \notin \mathcal{Y}^m \Rightarrow \mathcal{V}(X)$ holds. Analogically we can define: A property $\mathcal{V}$ holds for B-almost every ARN.

Now we raise the question of measurability of open sets. (It is useful to compare the following with the definition of $S_6^{[n]}$-sets and their measure.) We shall start with lists of open rational intervals. It is clear that for any NN m the set $[\mathcal{D}_m]$ is effectively measurable, i.e. $[0]$-measurable. There exists a $[0]$-SRtNs μ_0 such that for any NN m the RtN $\mu_0(m)$ is the Lebesgue measure of $[\mathcal{D}_m]$.

Having NNs n and k we know that the Lebesgue measure of the set $[W_k^{[n]}]$ must be the supremum of the non-decreasing $[n]$-SRtNs

$$\{\mu_0(\langle list\rangle_2^{[n]}(k,m))\}_m^{[n]} \quad . \tag{1}$$

If this sequence is not bounded, we say that $[W_k^{[n]}]$ has no finite measure. If there is an $[n]$-CRN z being the supremum of (1), we say that $[W_k^{[n]}]$ is (Lebesgue) $[n]$-measurable and z is its $[n]$-measure. If (1) is quasi-bounded (i.e. cannot fail to be bounded), then it is $[n+1]$-fundamental and we can find an $[n+1]$-CRN y being the supremum of (1); in this case we say that $[W_k^{[n]}]$ is (Lebesgue) $[n+1]$-measurable with the $[n+1]$-measure y. The relativization of the well-known construction of Specker sequence shows that for any NN n there is an $[n]$-open set being a subset of $0\Delta 1$

which is not $[n]$-measurable (but must be $[n+1]$-measurable). It is easy to show that the quasi-union (resp. the intersection) of two $[m]$-measurable $[n]$-open sets is also $[n]$-open and $[m]$-measurable.

In the general case a set $\mathfrak{M}$ of ARNs is said to be (Lebesgue) $[n]$-measurable and an $[n]$-CRN r is said to be the $[n]$-measure of $\mathfrak{M}$ if there exist an $[n]$-SNNs $\{k_p\}_p^{[n]}$ and an $[n]$-sequence $\{\mathcal{G}^p\}_p^{[n]}$ of $S_\sigma^{[m]}$-sets such that for any NN p the measure of $\mathcal{G}^p$ is less than 2^{-p}, and $|\mu_0(k_p) - r| \leq 2^{-p}$ & $\forall X(X \notin \mathcal{G}^p \Rightarrow (X \in [\mathcal{D}_{k_p}] \Leftrightarrow X \in \mathfrak{M}))$.
A set $\mathcal{P}$ of $[n]$-CRNs is said to be (Lebesgue) $[m]$-measurable and an $[m]$-CRN r is said to be the $[m]$-measure of $\mathcal{P}$ if the set of ARNs $\{X; \neg\neg\exists x^{[n]}(x^{[n]} \in \mathcal{P}$ & $X = x^{[n]})\}$ is $[m]$-measurable with the $[m]$-measure r. Let us mention that if $n < m$, then $r = 0$.

Clearly the quasi-union (resp. the intersection) of two $[n]$-measurable sets of ARNs is also $[n]$-measurable.

It is easy to prove the following analogue of a well-known classical theorem.

<u>Theorem 2.4.</u> A set of ARNs $\mathfrak{M}$ is $[n]$-measurable with the $[n]$-measure $x^{[n]}$ iff there exist a NN p and a $[n]$-sequence $\{v_q\}_q^{[n]}$ of $[n]$-CRNs such that

a) for $[n]$-almost every ARN X we have $X \in \mathfrak{M} \Leftrightarrow X \in \bigcap_q [W^{[n]}_{s_1^1(p,q)}]$,

b) for any NN q we have $\mathfrak{M} \subseteq [W^{[n]}_{s_1^1(p,q)}]$, the $[n]$-open set $[W^{[n]}_{s_1^1(p,q)}]$ is $[n]$-measurable with the $[n]$-measure v_q, and $x^{[n]} \leq v_q < x^{[n]} + 2^{-q}$.

We have seen a great importance of sets of the types $G^{[n]}$ and $G_\delta^{[n]}$. Let us remark that any set $\bigcap_p [W^{[n]}_{s_1^1(t,p)}]$ of ARNs, being a subset of a (finitely) measurable set, is certainly $[n+2]$-measurable, and in case that there is an $[n]$-sequence $\{r_p\}_p^{[n]}$ of $[n]$-CRNs such that for any NN p the set $[W^{[n]}_{s_1^1(t,p)}]$ is $[n]$-measurable with the $[n]$-measure r_p, our set is necessarily $[n+1]$-measurable.

Now we shall consider sets of the types $G^{[0]}$ and $G_\delta^{[0]}$ and sets of $[0]$-measure zero and quote important results about them.

<u>Remark 2.5.</u> There exists a NN z such that

a) for any NN k $!\langle z\rangle(k)$, the $[1]$-measure of $[W_{\langle z\rangle(k)}]$ does not exceed 2^{-k}, and $D^{[0]} \subseteq [W_{\langle z\rangle(k+1)}] \subseteq [W_{\langle z\rangle(k)}]$;

b) for any recursive function φ such that [1]-measure of $[W^{[0]}_{\varphi(m)}]$ is not greater than 2^{-m} $(0 \leq m)$, we have $\forall p \exists q ([W^{[0]}_{\varphi(q)}] \subseteq [W_{\langle \gamma \rangle (p)}])$ (cf. Demuth's paper[11]).

Martin-Löf[12] constructed such a [0]-sequence, however, he did not study properties of real numbers from the classes defined below.

We denote $\bigcap_k [W_{\langle \gamma \rangle (k)}]$ by $\mathcal{A}_1$ and $\mathcal{A} \setminus \mathcal{A}_1$ by $\mathcal{A}_2$. $\mathcal{A}_1$ is a [1]-measurable set of ARNs of the type $G^{[0]}_\delta$ whose [1]-measure is zero. Any set of ARNs of [0]-measure zero is contained in $\mathcal{A}_1$, but $\mathcal{A}_1$ contains many [1]-CRNs which are in no set of ARNs of [0]-measure zero (cf. Demuth's paper[13]).

Here we meet "singular properties" of singular [0]-coverings. It is easy to prove the following proposition.

Theorem 2.6. Let a be a RtN, $0 < a < 1$, Φ a [0]-covering, $\forall k (\sum_{p=0}^{k} |\Phi(p)| < a)$, and $\mathfrak{M}$ a [0]-measurable set of ARNs of the [0]-measure less than $1-a$. Then for any NNs k and l we can construct a NN p such that $l \leq p$ and the [0]-measure of $\mathfrak{M} \cap \Phi(p)$ is less than $2^{-k} . |\Phi(p)|$.

Lemma 2.7. (Cf. Demuth's paper.[14]) There exist a NN ν and [1]-recursive functions $\bar{\varphi}$ and $\bar{q}$ such that for any NNs p, t

1) $[W_{\langle \gamma \rangle (p+2)}] \subseteq [W^{[1]}_{s^1_1(\nu,p)}]$, $[W^{[1]}_{s^1_1(\nu,p+1)}] \subseteq [W^{[1]}_{s^1_1(\nu,p)}]$,

the [1]-open set $[W^{[1]}_{s^1_1(\nu,p)}]$ is [1]-measurable with the [1]-measure not greater than 2^{-p}; thus, the set of ARNs $\mathcal{T}_\nu$, where $\mathcal{T}_\nu = \bigcap_q [W^{[1]}_{s^1_1(\nu,q)}]$, is a [1]-measurable set of the type $G^{[1]}_\delta$ with the [1]-measure zero;

2) for any NN l and any ARN X such that $|X| \leq 2^p$ & $X \notin [W^{[1]}_{s^1_1(\nu,p)}]$

a) if $X \in \mathcal{B}(l)$ and $|\mathcal{B}(l)| < 2^{-\bar{\varphi}(t,p)}$, then

$X \in [W^{[0]}_t] \iff \mathcal{B}(l) \cap [\mathcal{D}_{\bar{q}(t,p)}]_\sigma \neq \emptyset \iff \mathcal{B}(l) \subseteq [\mathcal{D}_{\bar{q}(t,p)}]$;

b) if $X \notin [W^{[0]}_t]$, then X is a point of [1]-dispersion of the set $[W^{[0]}_t]$;

3) as a consequence of 2a) and of Remark 2.1, for any ARN $X \notin \mathcal{T}_\nu$ $(\mathcal{S}e\ell(X) \oplus \emptyset') \equiv_T (\mathcal{S}e\ell(X))'$ cannot fail to hold.

Notation In the sequel we shall use the symbols ν and $\mathcal{T}_\nu$ introduced in the previous lemma.

Remark 2.8. (See Demuth's paper.[14]) 1) For each NN t we denote the set $\bigcap_q [W^{[0]}_{s^1_1(t,q)}]$ of ARNs by $\mathcal{P}_t$. As we know, the $\{\mathcal{P}_t\}_t$ is an enumeration of all sets of ARNs of the type $G^{[0]}_\delta$, for any NNs t and q the set $\mathcal{P}_t \cap (-2^q \Delta 2^q)$ is a [2]-measurable set of ARNs, and according to the s-m-n-theorem there exist a recursive function $\hat{\rho}$ and a [2]-recursive function $\bar{\rho}$ such that for any NN p we have

a) $[W^{[2]}_{\hat{\rho}(p)}]$ is a [2]-measurable set of ARNs of the [2]-measure less than 2^{-p} and $\mathcal{T}_q \subseteq [W^{[1]}_{s^1_1(q,p+1)}] \cup D^{[1]} \subseteq [W^{[2]}_{\hat{\rho}(p)}] \subseteq [W^{[2]}_{\hat{\rho}(p-1)}]$;

b) for any NN t and any ARN X, if $|X| \leq 2^p$ & $X \notin [W^{[2]}_{\hat{\rho}(p)}]$, then $X \in \mathcal{P}_t$ is equivalent to $X \in \bigcap_{k=0}^{\bar{\rho}(p,t)} [W^{[0]}_{s^1_1(t,k)}]$.

2) By 1), Lemma 2.7 and Remark 2.1 we get

a) for any NN $n > 1$ there is an $[n]$-algorithm $\mathcal{R}$ applicable to any word of the form $p \square v \square t$, where p and t are NNs and v is an $[n]$-CRN, transforming it into a NN, such that

$$v \notin [W^{[2]}_{\hat{\rho}(p)}] \Rightarrow (v \in \mathcal{P}_t \Longleftrightarrow \mathcal{R}(p \square v \square t) > 0) \quad ;$$

b) for any ARN $X \notin \bigcap_p [W^{[2]}_{\hat{\rho}(p)}]$

$$(\mathcal{Sel}(X) \oplus \emptyset') \equiv_T (\mathcal{Sel}(X))' \quad \& \quad (\mathcal{Sel}(X) \oplus \emptyset'') \equiv_T (\mathcal{Sel}(X))''$$

cannot fail to hold.

Thus, for [1]-almost every [1]-CRN $x^{[1]}$ we have $\neg\neg((\mathcal{Sel}(x^{[1]}))' \equiv_T \emptyset')$ and for [2]-almost every [2]-CRN $x^{[2]}$ we have $\neg\neg((\mathcal{Sel}(x^{[2]}))'' \equiv_T \emptyset'')$.

These results we shall use in our study of differentiability.

3. Everywhere Defined [0]-Constructive Functions of a Real Variable and their Differentiability

Now we come to questions being in the center of our interest here. At first we have to introduce several further notions.

Let us consider constructive analogues of functions of a real variable. An $[m]$-algorithm f is called an $[m;n]$-constructive function of a real variable ($[m;n]$-CFRV) if the following conditions are satisfied:

1) for any $[n]$-CRN r if $!f(r)$, then $f(r)$ is an $[n]$-CRN;

2) $\forall x^{[n]} y^{[n]} (! f(x^{[n]}) \ \& \ x^{[n]} = y^{[n]} \Rightarrow ! f(y^{[n]}) \ \& \ f(x^{[n]}) = f(y^{[n]}))$.

We shall say that f is an everywhere defined $[m;n]$-CFRV if f is an $[m;n]$-CFRV defined at each $[n]$-CRN. In case that $m=n$ we speak about an $[n]$-CFRV.

As a relativization of the well-known result on continuity of $[0]$-CFRVs (Cejtin[15]; Kreisel, Lacombe, Shoenfield) we have the following assertion: If $m \leq n$, then any $[m;n]$-CFRV is $[n]$-continuous at each $[n]$-CRN in its domain. (Cf. Demuth, Kryl and Kučera's paper,[8] Example 5.1.) On the other hand, everywhere defined $[0]$-CFRVs (though continuous) need not be either pseudo-uniformly continuous (i.e. uniformly continuous in the classical sense) nor bounded on $0 \Delta 1$.

As for the arithmetical complexity, the set of (indices of) $[n]$-CFRVs is Π_{n+3}-complete (see Kučera and Kušněr's paper[16] for $n=0$).

For brevity, everywhere defined $[n]$-CFRVs, constant both on $\{x^{[n]}; x^{[n]} \leq 0\}$ and on $\{x^{[n]}; 1 \leq x^{[n]}\}$, are called simply <u>$[n]$-functions</u>. Any monotone $[n]$-function is $[n]$-uniformly continuous.[17] Let us remark, that an $[n]$-function f is pseudo-uniformly continuous (i.e. in this case $[n+1]$-uniformly continuous) iff there exists an everywhere defined $[n+2]$-CFRV g such that $\forall x^{[n]} (f(x^{[n]}) = g(x^{[n]}))$. Consequently for any pseudo-uniformly continuous $[0]$-function f we can construct a $[0]$-algorithm h such that, for any NN n, h is an everywhere defined $[0;n]$-CFRV and $\forall x^{[0]} (f(x^{[0]}) = h(x^{[0]}))$.

By a <u>$[0]$-function of the type $\mathbb{B}$</u> is meant a $[0]$-function which is everywhere defined $[0;n]$-CFRV for each NN n.

Let $\{H_n\}_n^{[0]}$ be a $[0]$-sequence of $[0]$-segments and let F be a $[0]$-function. Then

1) $\mathcal{H}(\{H_n\}_n^{[0]})$ means: $\{H_n\}_n^{[0]}$ is a $[0]$-sequence of non-overlapping $[0]$-segments, $[0]$-sequence $\{|H_n|\}_n^{[0]}$ of $[0]$-CRNs $[0]$-converges to 0, and $\neg \exists n (0 \in (H_n)^0 \vee 1 \in (H_n)^0)$;

2) if $\mathcal{H}(\{H_n\}_n^{[0]})$ holds, then $[F, \{H_n\}_n^{[0]}]$ denotes a $[0]$-function linear on every H_n, satisfying

$$\forall x^{[0]} (\neg \exists n (x^{[0]} \in (H_n)^0) \Rightarrow [F, \{H_n\}_n^{[0]}](x^{[0]}) = F(x^{[0]})) .$$

Let us note that for any $[0]$-covering Φ we have $\mathcal{H}(\Phi)$.

Now we shall introduce principal notions concerning derivatives. We shall deal with bilateral pseudo-derivatives (i.e. derivatives in the classical sense) of everywhere defined $[0]$-CFRVs on the ARNs. But, in general, such $[0]$-CFRVs are defined only on $D^{[0]}$. In corresponding definitions we shall overcome this difficulty using values of $[0]$-CFRVs at RtNs only, which is possible among others thanks to continuity such $[0]$-CFRVs at any $[0]$-CRN.

<u>Definitions</u> (Cf. Demuth's paper.[9]) Let f be an everywhere defined $[0]$-CFRV, $v \Delta w$ a $[0]$-segment, m and p NNs, B a set of NNs and X, Y ARNs. Then 1) $Q(f, v\Delta w)$ denotes $\dfrac{f(w)-f(v)}{w-v}$;

2) a) $D_{cl}(f,X)$ denotes $\forall k \neg\neg \exists m\, \vartheta(f,X,k,m)$ (f is finitely pseudo-differentiable at X), where $\vartheta(f,X,k,m)$ is

$$\forall abcd\, (a<X<b \;\&\; c<X<d \;\&\; b-a<2^{-m} \;\&\; d-c<2^{-m} \Rightarrow |Q(f,a\Delta b)-Q(f,c\Delta d)| \leq 2^{-k});$$

b) $MD(f,X,B,p)$ denotes $\forall k\,(!\langle p\rangle^B(k) \;\&\; \vartheta(f,X,k,\langle p\rangle^B(k)))$

($\langle p\rangle^B$ is a B-recursive function being a modulus of differentiation of f at X);

c) $D^B(f,X)$ denotes $\exists m\, MD(f,X,B,m)$ (f is finitely B-differentiable at X);

3) a) $\overline{D}_{cl}(Y,f,X)$ denotes $\forall k \neg\neg (\exists m \forall ab\, (a<X<b \;\&\; b-a<2^{-m} \Rightarrow Q(f,a\Delta b) < Y + 2^{-k}) \;\&\; \exists cd\, (c<X<d \;\&\; d-c<2^{-k} \;\&\; Q(f,c\Delta d) > Y-2^{-k}))$

(Y is an upper pseudo-derivative of f at X);

b) $\underline{D}_{cl}(Y,f,X)$ is defined analogically;

c) $D_{cl}(Y,f,X)$ denotes $\overline{D}_{cl}(Y,f,X) \;\&\; \underline{D}_{cl}(Y,f,X)$

(Y is a pseudo-derivative of f at X);

d) $D^B(Y,f,X)$ denotes $D_{cl}(Y,f,X) \;\&\; D^B(f,X)$

(Y is a B-derivative of f at X);

4) we suppose that $D_{cl}(-\infty,f,X)$, $D_{cl}(+\infty,f,X)$, $\overline{D}_{cl}(+\infty,f,X)$, $\underline{D}_{cl}(-\infty,f,X)$ are defined analogically;

5) $L_{cl}(Y,f,X)$ denotes $\forall k \neg\neg \exists m \forall a\, (|a-X|<2^{-m} \Rightarrow |f(a)-Y| \leq 2^{-k})$

(Y is a pseudo-limit of f at X);

$L_{cl}(f,X)$ denotes

$$\forall k \neg\neg \exists m \forall ab\, (|a-X|<2^{-m} \;\&\; |b-X|<2^{-m} \Rightarrow |f(a)-f(b)| \leq 2^{-k}).$$

We shall study the following questions:

(i) What is the arithmetical complexity of the problem of pseudo-differentiability of everywhere defined $[0]$-CFRVs at ARNs?

(ii) What is the arithmetical complexity of moduli of differentiation of everywhere defined $[0]$-CFRVs at ARNs?

(iii) Which ARNs are pseudo-derivatives of everywhere defined $[0]$-CFRVs at a given ARN?

Results are quoted from Demuth's papers.[13,14,18,19]

<u>Remark 3.1.</u> For any everywhere defined $[0]$-CFRV f, NN n and any $[n]$-CRN $x^{[n]}$ the predicate $\neg\vartheta(f, x^{[n]}, k, m)$ of variables k, m is $[n]$-recursively enumerable and consequently $[n+1]$-recursive. Further, $\neg\neg\vartheta(f, x^{[n]}, k, m) \Rightarrow \vartheta(f, x^{[n]}, k, m)$ holds. On this account, we have $D_{cl}(f, x^{[n]}) \Rightarrow D^{[n+1]}(f, x^{[n]})$.

If $x^{[n]}$ is in addition B-computable, then $\neg\vartheta(f, x^{[n]}, k, m)$ is B-recursively enumerable and thus B'-recursive, and therefore $D_{cl}(f, x^{[n]}) \Rightarrow D^{B'}(f, x^{[n]})$ holds.

At first we give a few facts concerning pseudo-limits of everywhere defined $[0]$-CFRVs. Let f be such a $[0]$-CFRV.

a) We have $\forall x^{[0]}\ L_{cl}(f(x^{[0]}), f, x^{[0]})$.

b) There exists a NN m_0 such that for any NN k

$$W^{[1]}_{s^1_1(m_0,k)} = \{l; \neg\exists c\, d(c \in \mathcal{L}^0(l)\ \&\ d \in \mathcal{L}^0(l)\ \&\ |f(c) - f(d)| > 2^{-k})\}.$$

Thus, for any ARN X $\quad L_{cl}(f, X) \Leftrightarrow X \in \bigcap_k [W^{[1]}_{s^1_1(m_0,k)}]$ $\quad$ holds,

and according to Remark 2.2 for any NN $n \geq 1$ we have: If $L_{cl}(f, x^{[n]})$ holds, then there exists an $[n]$-CRN $y^{[n]}$ which fulfils $L_{cl}(y^{[n]}, f, x^{[n]})$ and, in addition, if $x^{[n]}$ is B-computable, then $y^{[n]}$ is $(B \oplus \emptyset')$-computable.

c) $\neg(\underline{D}_{cl}(-\infty, f, X) \vee \overline{D}_{cl}(+\infty, f, X)) \Rightarrow L_{cl}(f, X)$ holds for any ARN X; if X is not monotonically weakly $\emptyset$-computable, then $\neg(\underline{D}_{cl}(-\infty, f, X)\ \&\ \overline{D}_{cl}(+\infty, f, X)) \Rightarrow L_{cl}(f, X)$. Thus, if f is finitely pseudo-differentiable at an ARN X, then f has a pseudo-limit at X.

d) There exist a $[1]$-uniformly continuous $[0]$-function g, a $\emptyset'$-recursive set C, a C-computable $[1]$-CRN v, and a $[1]$-CRN w being the pseudo-limit of g at v such that w is not C-computable.

e) Let f be finitely pseudo-differentiable at an ARN X. Then, by c), there exists an ARN Y such that $L_{cl}(Y, f, X)$. The complexity of Y

with respect to arithmetical hierarchy is not greater than that of X (cf. b)). If X is moreover B-computable, then there exists a B-sequence of rational segments $\{a_k \Delta b_k\}_k^B$ such that $\forall k(a_k \leq X < b_k < a_k + 2^{-k})$, a B-sequence $\{Q(f, a_k \Delta b_k)\}_k^B$ of $[0]$-CRNs is pseudo-fundamental, and consequently it is B'-fundamental. Further, the B-sequence $\left\{\frac{f(b_k)-Y}{b_k - X}\right\}_k^B$ of ARNs is quasi-bounded and, on account of this, Y cannot fail to be B-computable.

We have proved the following proposition.

Lemma 3.2. Let f be an everywhere defined $[0]$-CFRV, n a NN, and $x^{[n]}$ an $[n]$-CRN for which $D_{cl}(f, x^{[n]})$ holds. Then there exist an $[n]$-CRN $y^{[n]}$ and an $[n+1]$-CRN $z^{[n+1]}$ such that

i) $L_{cl}(y^{[n]}, f, x^{[n]})$ & $D^{[n+1]}(z^{[n+1]}, f, x^{[n]})$ is fulfilled;

ii) if $x^{[n]}$ is B-computable, then $y^{[n]}$ cannot fail to be B-computable, $z^{[n+1]}$ is weakly B-computable and, thus, B'-computable, and f is finitely B'-differentiable at $x^{[n]}$.

Remark 3.3. Let f be an everywhere defined $[0]$-CFRV. Then there exist NNs m_0 and m_1 such that for any NNs p and q

$W^{[0]}_{s_1^1(m_0, p)} = \{l; |\mathcal{B}(l)| \leq 2^{-p}$ & $Q(f, \mathcal{B}(l)) > 2^p\}$ and

$W^{[0]}_{s_1^2(m_1, p, q)} = \{l; \exists k(\mathcal{B}(l) \subseteq \mathcal{B}(k)$ & $|\mathcal{B}(k)| \leq 2^{-q}$ & $|Q(f, \mathcal{B}(l)) - Q(f, \mathcal{B}(k))| > 2^{-p})\}$.

1) Clearly, $\overline{D}_{cl}(+\infty, f, X) \iff X \in \bigcap_p [W^{[0]}_{s_1^1(m_0, p)}]$ and

$\neg D_{cl}(f, X) \iff X \in \bigcup_p \bigcap_q [W^{[0]}_{s_1^2(m_1, p, q)}]$ hold for any ARN X. According to Remark 2.2 for any NN n the predicate $\overline{D}_{cl}(+\infty, f, x^{[n]})$ of variable $x^{[n]}$ is $[n+2]$-recursive and the predicate $\neg D_{cl}(f, x^{[n]})$ is $[n+2]$-recursively enumerable. Consequently the predicate $D_{cl}(f, x^{[n]})$, which is equivalent to its own double negation, is $[n+3]$-recursive.

2) If f is a non-decreasing $[0]$-function, then for any NN p the $[1]$-measure of the $[0]$-open set $[W^{[0]}_{s_1^1(m_0, p)}]$ is less than $(f(1) - f(0)) \cdot 2^{-p+1}$ and consequently by Remark 2.5 have $\{X; \overline{D}_{cl}(+\infty, f, X)\} \subseteq \mathcal{A}_1$.

So we obtained an estimation of the arithmetical complexity of the predicate $D_{cl}(f, x^{[n]})$ in the general case. But we are interested in the best possible estimations not only for the general case, but also in cases when we restrict ourselves to $[0]$-uniformly or $[1]$-uniformly continuous everywhere defined $[0]$-CFRVs and/or to $[n]$-CRNs belonging to the complement of some $[n]$-measurable set (of $[n]$-CRNs) of small $[n]$-measure. The corresponding results will be formulated in propositions A – D.

It is clear that we can restrict ourselves to $[0]$-functions and to ARNs in $0\nabla 1$ without loss of generality. We shall introduce corresponding notations.

Notations Let f be a $[0]$-function, $\mathfrak{M}$ a set of ARNs, and n a NN.

1) By $Compl(f, \mathfrak{M}, n)$ we denote the least NN m such that there is an $[m]$-algorithm $\mathfrak{K}$ such that for any $[n]$-CRN $x^{[n]}$ in $D^{[n]} \setminus \mathfrak{M}$ $\mathfrak{K}$ is defined at $x^{[n]}$, $\mathfrak{K}(x^{[n]})$ is a NN, and $D_{cl}(f, x^{[n]}) \Leftrightarrow \mathfrak{K}(x^{[n]}) = 0$.

2) By $Compl_0(\mathfrak{M}, n)$ (resp. $Compl_1(\mathfrak{M}, n)$, resp. $Compl_2(\mathfrak{M}, n)$) we denote the greatest of the NNs $Compl(g, \mathfrak{M}, n)$, where g runs through all $[0]$-uniformly continuous (resp. all $[1]$-uniformly continuous, resp. all) $[0]$-functions.

3) For any NN i, $0 \le i \le 1$, and any set B of NNs $Red(i, f, \mathfrak{M}, n, B)$ denotes: for any $[n]$-sequence $\{r_k\}_k^{[n]}$ of $[n]$-CRNs we can construct a $[0]$-sequence $\{y_q\}_q^{[0]}$ of $[n]$-CRNs such that

$$y_q \notin \mathfrak{M} \,\&\, \neg \exists k (y_q = r_k) \,\&\, (D_{cl}(f, y_q) \Leftrightarrow q \in B) \,\&\, (i = 1 \Rightarrow \neg \underline{D}_{cl}(-\infty, f, y_q) \,\&\, \neg \overline{D}_{cl}(+\infty, f, y_q))$$

holds for all NNs q.

We have already proved that $Compl_2(\emptyset, n) \le n+3$ holds for all NNs n. This result cannot be improved by means of $[0]$-measurable sets of ARNs.

Example 3.4. There exists a $[0]$-function g which fulfils the Lipschitz condition and for which $Red(1, g, \mathfrak{M}, n, \setminus \emptyset^{(n+3)})$ holds for all NNs n and for all $[0]$-measurable sets $\mathfrak{M}$ of ARNs with the $[0]$-measure less than 2^{-1}.

The possibility of construction of such a $[0]$-function is given by properties of singular $[0]$-coverings stated in Theorem 2.6.

So we have proved

Theorem A. For any $[0]$-measurable set $\mathfrak{M}$ of ARNs with the $[0]$-measure less than 2^{-1} we have $Compl_0(\mathfrak{M}, n) = n+3$ and consequently $Compl_2(\emptyset, n) = n+3$ for all NNs n.

The last Example has demonstrated that for some $[0]$-function h increasing on $0\Delta 1$ the set $\{X; \neg D_{cl}(h, X)\}$ may be of positive "outer $[0]$-measure." We shall see later that it necessarily has "inner $[0]$-measure" zero (cf. Remark 3.6).

It has turned out that it would be possible to improve our result only by omitting sets of ARNs which are not $[0]$-measurable. In this connection the use of properties of sets of ARNs of the type $G_\delta^{[0]}$ and/or further results on pseudo-differentiability are needed. By a constructive analogue of Vitali's covering theorem (cf. Demuth and Kučera's paper[5]) we can prove the following theorem quoted from Demuth's paper.[20]

Theorem 3.5. For any $[0]$-uniformly continuous $[0]$-function f we can construct a $[0]$-function g increasing on segment $0\Delta 1$ such that for any ARN X in $0\nabla 1$ we have

$$\neg D_{el}(+\infty, g, X) \Rightarrow \neg\neg(\underline{D}_{el}(-\infty, f, X) \,\&\, \overline{D}_{el}(+\infty, f, X) \vee D_{el}(f, X)) .$$

Consequently there exists a set $\mathfrak{M}$ of ARNs of the type $G_\delta^{[0]}$ contained in $0\Delta 1$ for which $\neg D_{el}(+\infty, g, X) \Rightarrow (D_{el}(f, X) \Leftrightarrow X \notin \mathfrak{M})$ (cf. Remark 3.3).

Remark 3.6. For any non-decreasing $[0]$-function g, any NN n, and any $[n]$-measurable set $\mathfrak{M}$ of ARNs contained in $0\Delta 1$ and of positive $[n]$-measure we can construct an $[n]$-CRN $x^{[n]}$ such that $x^{[n]} \in \mathfrak{M} \cap (0\nabla 1)$ and $\neg \overline{D}_{el}(+\infty, g, x^{[n]})$.

Using the last Theorem, Remark 2.2 and Remark 3.3 we get the following proposition.

Corollary 1) a) For any $[0]$-uniformly continuous $[0]$-function f and any NN $n \geq 1$ there exists an $[n+1]$-recursively enumerable predicate R (of variable $x^{[n]}$) such that $x^{[n]} \notin \mathcal{A}_1 \Rightarrow (D_{el}(f, x^{[n]}) \Leftrightarrow R(x^{[n]}))$ and consequently

b) $Compl_0(\mathcal{A}_1, n) \leq n+2$ holds for all NNs n.

2) Any monotone $[0]$-function g is finitely pseudo-differentiable at any ARN in $\mathcal{A}_2$ and consequently at $[1]$-almost every ARN; further, the set $\{x^{[0]}; \neg D_{el}(g, x^{[0]})\}$ of $[0]$-CRNs is of "inner $[0]$-measure" zero and for any $[0]$-measurable set $\mathfrak{M}$ of the $[0]$-measure less than 1 we can construct a $[0]$-CRN r such that $(r \in 0\nabla 1 \setminus \mathfrak{M}) \,\&\, D_{el}(g, r)$ holds.

Example 3.7. There exist a $[0]$-uniformly continuous $[0]$-function g and a $[1]$-uniformly continuous $[0]$-function h such that for any NNs n, t, where $1 \leq n$ and the $[1]$-measure of $[W_t^{[0]}]$ is less than 2^{-1}, $Red(0, g, [W_t^{[0]}] \hat{\cup} \mathcal{A}_1, n, \emptyset^{(n+2)})$ and $Red(1, h, [W_t^{[0]}] \hat{\cup} \mathcal{A}_1, n, \setminus \emptyset^{(n+3)})$ hold.

So we have proved

Theorem B. For any NN $n \geq 1$ and any $[0]$-open set of ARNs $\mathfrak{M}$ with the $[1]$-measure less than 2^{-1} we have $Compl_0(\mathcal{A}_1, n) = Compl_0(\mathcal{A}_1 \hat{\cup} \mathfrak{M}, n) = n+2$ and $Compl_1(\mathcal{A}_1 \hat{\cup} \mathfrak{M}, n) = n+3$.

Lemma 2.7 enables us to get for general $[0]$-functions a result analogical to Theorem 3.5.

Theorem 3.8. Let f be a $[0]$-function. Then for any ARN $X \notin \mathcal{T}_{\mathcal{U}}$ we have $\neg\neg(\underline{D}_{cl}(-\infty, f, X) \& \overline{D}_{cl}(+\infty, f, X) \vee D^{[1]}(f, X))$ and consequently $D_{cl}(f, X) \Leftrightarrow \neg\neg D^{[1]}(f, X)$.

Remark 3.9. According to Lemma 2.7 for any positive NN n there exist an $[n]$-algorithm $\mathcal{E}$ defined on $\{U; \exists p m x^{[n]}(U \mp p \square x^{[n]} \square m)\}$ and an $[n]$-algorithm $\mathcal{J}$ such that $\forall p m x^{[n]}(|x^{[n]}| \leq 2^{p} \& x^{[n]} \notin [W^{[1]}_{s^1_1(\nu, p)}] \Rightarrow$

$$(\mathcal{E}(p \square x^{[n]} \square m) \mp \Lambda \Leftrightarrow x^{[n]} \in [W^{[0]}_{m}]) \& (!\mathcal{J}(p \square x^{[n]} \square m) \Leftrightarrow x^{[n]} \notin \bigcap_{k} [W^{[0]}_{s^1_1(m,k)}])) .$$

Let us notice that for any positive NN n

$x^{[n]} \notin \mathcal{T}_{\mathcal{U}} \Rightarrow \exists^{[n+1]} p\, (x^{[n]} \notin [W^{[1]}_{s^1_1(\nu, p)}])$ holds.

Thus we get the following proposition.

Corollary For any $[0]$-function f and any NN n

$Compl_2(f, \mathcal{T}_{\mathcal{U}}, n) \leq n+1$ holds.

Example 3.10. There exists a $[0]$-uniformly continuous $[0]$-function g such that for any positive NN n, any $[1]$-measurable set $\mathfrak{M}$ of ARNs of the $[1]$-measure less than 2^{-1}, and any $[n]$-measurable set $\mathfrak{N}$ of ARNs of $[n]$-measure zero $Red(0, g, \mathfrak{M} \hat{\cup} \mathcal{T}_{\mathcal{U}}, n, \emptyset^{(n+1)})$ and $Red(0, g, \mathfrak{N} \hat{\cup} \mathcal{T}_{\mathcal{U}}, n, \emptyset^{(n+1)})$ hold.

So we have proved

Theorem C For any positive NN n, any $[1]$-measurable set $\mathfrak{M}$ of ARNs of the $[1]$-measure less than 2^{-1}, any $[n]$-measurable set $\mathfrak{N}$ of ARNs of $[n]$-measure zero, and any NN i, $0 \leq i \leq 2$, we have

$$Compl_i(\mathcal{T}_{\mathcal{U}}, n) = Compl_i(\mathfrak{M} \hat{\cup} \mathcal{T}_{\mathcal{U}}, n) = Compl_i(\mathfrak{N} \hat{\cup} \mathcal{T}_{\mathcal{U}}, n) = n+1 .$$

So we cannot expect to get any improvement of the result by omitting any $[1]$-measurable set of ARNs of small measure. But we have reduced the question of pseudo-differentiability of $[0]$-functions at ARNs of $\mathcal{A} \setminus \mathcal{T}_{\mathcal{U}}$ to the question of belonging such ARNs to special sets of ARNs of the type $G^{[0]}_{\delta}$ which are contained in the rational segment $0 \Delta 1$. So we can use Remark 2.8.

Remark 3.11. We can construct a $[0]$-uniformly continuous $[0]$-function h such that $\neg D_{cl}(h, X) \Leftrightarrow 2^{-1} < X < 1$ holds for all ARNs X. It is easy to show that for any NN n and for any $[n]$-measurable set $\mathfrak{M}$

of ARNs of the $[n]$-measure less than 2^{-1}, $\mathcal{R}ed(0,h,\mathfrak{M},n,\emptyset^{(n)})$ holds.

We have proved.

Theorem D. For any NN $n \geq 2$, for any $[n]$- measurable set $\mathfrak{M}$ of ARNs of the $[n]$- measure less than 2^{-1}, and for any NNs p and i, $0 \leq i \leq 2$, $Compl_i([W^{[2]}_{\xi(p)}],n) = n \leq Compl_i(\mathfrak{M},n)$ holds (cf. Example 3.10).

So, in Theorems A - D we have answered the question (i). Now we shall study the arithmetical complexity of moduli of differentiation, i.e. we shall try to answer the question (ii).

As we have already noted: for any $[0]$-function f, for any NN n, and for any $[n]$-CRN $x^{[n]}$ such that f is finitely pseudo-differentiable at $x^{[n]}$

a) there exists an $[n+1]$-recursive function being a modulus of differentiation of f at $x^{[n]}$, i.e. $D^{[n+1]}(f,x^{[n]})$ holds;

b) if $x^{[n]}$ is, in addition, B-computable, there exists a B'-recursive function being such a modulus and, thus $D^{B'}(f,x^{[n]})$ holds.

Example 3.12. There exists a $[0]$-function f increasing on $0\Delta1$ such that for any NNs $n \geq 1$ and t, where $[W^{[0]}_t]$ is a [1]-measurable set of the [1]-measure less than 2^{-1}, and for any $[n]$-sequence $\{r_k\}^{[n]}_k$ of $[n]$-CRNs there exist an $[n]$-CRN v and an $[n+1]$-CRN w fulfilling

$$v \notin ([W^{[0]}_t] \,\hat{\cup}\, \mathcal{A}_1) \,\&\, \neg \exists k (v = r_k) \,\&\, \mathcal{S}el(v) \equiv_T \emptyset^{(n)} \,\&\, \mathcal{S}el(w) \equiv_T \emptyset^{(n+1)} \,\&\, D_{cl}(w,f,v)$$

and consequently we have $\neg D^{[n]}(f,v) \,\&\, D^{[n+1]}(f,v)$ and for any set B of NNs $\exists p\; MD(f,v,B,p) \Rightarrow \neg\neg(\emptyset^{(n+1)} \leq_T \emptyset^{(n)} \oplus B)$ holds.

Theorem 3.13. Let g be a $[0]$-function increasing on $0\Delta1$. Then there exists a $[0]$-recursive function φ such that for every ARN X and NN p, if $X \notin [W^{[1]}_{s^1_1(\vartheta,p)}]$, then $MD(g,X,\emptyset',\varphi(p))$ holds, i.e. for any NN p, g is uniformly $[1]$-differentiable on the set $\mathcal{A} \setminus [W^{[1]}_{s^1_1(\vartheta,p)}]$ of ARNs (which does not depend on g).

The differentiation of general $[0]$-functions is connected with the differentiation of $[0]$-functions increasing on $0\Delta1$. Indeed, for any $[0]$-function f there exist a $[0]$-sequence of $[0]$-sequences of rational segments $\{\{H^m_n\}^{[0]}_n\}^{[0]}_m$ and two $[0]$-sequences $\{g_{i,m}\}^{[0]}_m$ $(i = 0, 1)$ of $[0]$-functions increasing on $0\Delta1$ such that

a) for any NN m, $\overline{\mathcal{H}}(\{H_n^m\}_n^{[0]})$ and $[f,\{H_n^m\}_n^{[0]}] = g_{0,m} - g_{1,m}$ hold, where the $[0]$-function $g_{1,m}$ is linear on $0\Delta 1$;

b) for any ARN X, if $X \in (\bigcap_m \hat{\bigcup_n} H_n^m) \cap (0\Delta 1)$, then $\underline{D}_{cl}(-\infty, f, X)$.

Let us notice that for any NN m and any $[0]$-CRN $x^{[0]}$ we have

$$|f(x^{[0]}) - [f,\{H_n^m\}_n^{[0]}](x^{[0]})| > 0 \Rightarrow x^{[0]} \in \bigcup_n (H_n^m)^0 .$$

In this context the following result is interesting.

Theorem 3.14. Let h be a $[0]$-function and let $\{H_n\}_n^{[0]}$ be a $[0]$-sequence of rational segments such that $\overline{\mathcal{H}}(\{H_n\}_n^{[0]})$ and $\forall x^{[0]} (0 < |h(x^{[0]})| \Rightarrow x^{[0]} \in \bigcup_n (H_n)^0)$. Then there exist a NN τ and a $[0]$-recursive function $\bar{f}$ of two variables such that for any NNs p and m and ARN X fulfilling $X \notin \hat{\bigcup_n} H_n$ and $X \notin [W_{s_1^1(\vartheta,p)}^{[1]}]$, we have $\underline{D}_{cl}(-\infty, h, X) \Leftrightarrow \overline{D}_{cl}(+\infty, h, X) \Leftrightarrow X \in \bigcap_n [W_{s_1^1(\tau,n)}^{[0]}]$ and $X \notin [W_{s_1^1(\tau,m)}^{[0]}] \Rightarrow MD(h, X, \emptyset', \bar{f}(p,m))$ & $D^{[1]}(0,h,X)$.

According to Theorems 3.13, 3.14, Lemma 2.7 and Remark 3.9 we have proved the following theorem.

Theorem 3.15. For any $[0]$-function f, any NNs p and n, and any $[n]$-CRN $x^{[n]}$, for which $x^{[n]} \notin [W_{s_1^1(\vartheta,p)}^{[1]}]$ & $D_{cl}(f, x^{[n]})$ holds, we have $1 \leq n$ and

a) $\exists^{[n]} q \; MD(f, x^{[n]}, \emptyset', q)$ and consequently $\exists y^{[n]} \; D^{[n]}(y^{[n]}, f, x^{[n]})$;

b) if $x^{[n]}$ is moreover B-computable, then there $(B \oplus \emptyset')$-exists an NN q such that $MD(f, x^{[n]}, \emptyset', q)$.

Using Remark 2.8 instead of Lemma 2.7 (or Remark 3.9) we obtain the following theorem (in which $\hat{\rho}$ is the function introduced in Remark 2.8).

Theorem 3.16. For every $[0]$-function f there exists a $[2]$-recursive function $\hat{e}$ such that for any NN p and ARN X

$X \notin [W_{\hat{\rho}(p)}^{[2]}]$ & $D_{cl}(f,X) \Rightarrow MD(f,X,1,\hat{e}(p))$ holds, i.e. the $[0]$-function f cannot fail to be uniformly $[1]$-differentiable on the set $\{X; X \notin [W_{\hat{\rho}(p)}^{[2]}]$ & $D_{cl}(f,X)\}$ (an index of corresponding modulus [2]-exists) for any NN p .

Example 3.17. There exists a $[0]$-function g fulfilling the Lipschitz condition and such that $\forall X (X \in 0\Delta 1 \Rightarrow D^{[1]}(g,X) \& \neg D^{[0]}(g,X))$. (Cf. Myhill's paper.[21])

Example 3.18. There exist $[0]$-uniformly continuous $[0]$-functions g_1 and g_2 such that

1) for any NN $n \geq 1$ and any $[1]$-measurable set $\mathfrak{M}_1$ of ARNs of the $[1]$-measure less than 2^{-1} we can construct a $[0]$-sequence $\{v_{1,k}\}_k^{[0]}$ of $[n]$-CRNs fulfilling

$$\forall k (v_{i,k} \notin \mathfrak{M}_i \;\&\; D_{cl}(0, g_i, v_{i,k})) \tag{2}$$

with $i = 1$, but there is no $[n-1]$-recursive function $\hat{c}_1$ for which

$$\forall k\, a\, b\ (a < v_{i,k} < b \;\&\; |a\Delta b| < 2^{-\hat{c}_i(k)} \Rightarrow |Q(g_i, a\Delta b)| \leq 1) \tag{3}$$

holds with $i = 1$;

2) $\forall X \neg\neg D^{[0]}(g_2, X)$ and for any NN n and any $[n]$-measurable set $\mathfrak{M}_2$ of ARNs of $[n]$-measure zero we can construct a $[0]$-sequence $\{v_{2,k}\}_k^{[0]}$ of $[n]$-CRNs fulfilling (2) with $i = 2$, but there is no $[n]$-recursive function $\hat{c}_2$ for which (3) holds with $i = 2$.

The results above show how complicated are frontiers between classical existence (i.e. quasi-existence) and $[n]$-existence.

Now we turn to the question (iii).

Notation 1) For any ARN X we denote by $\mathfrak{Der}(X)$ the set of all ARNs Y for which there exists an everywhere defined $[0]$-CFRV f such that $D_{cl}(Y, f, X)$.

2) By $\mathcal{A}_N$ we denote the set of all ARNs X for which there is no everywhere defined $[0]$-CFRV g such that $D_{cl}(+\infty, g, X)$.

Obviously, $\mathfrak{Der}(X)$ contains $D^{[0]}$ and is closed under addition of ARNs and under multiplication by $[0]$-CRNs. We have already learnt that any ARN of $\mathfrak{Der}(X)$ is weakly $\mathfrak{Sel}(X)$-computable, i.e. $(\mathfrak{Sel}(X))'$-computable.

We can again restrict ourselves to ARNs of $0\nabla 1$ ($\mathfrak{Der}(X) = \mathfrak{Der}(X + x^{[0]})$ holds for any ARN X and any $[0]$-CRN $x^{[0]}$) and to $[0]$-functions, even to monotone $[0]$-functions.

Lemma 3.19. For any ARN X, any word V being either $+\infty$ or a positive ARN, and any $[0]$-function f such that $D_{cl}(V, f, X)$ there cannot fail to exist a $[0]$-function g increasing on $0\Delta 1$ and of the type $\mathbb{B}$ fulfilling $\mathcal{D}_{cl}(V,g,X)$ & $\mathcal{Sel}(X) \equiv_T \mathcal{Sel}(g(X))$ & $(L_{cl}(f,X) \Rightarrow L_{cl}(g(X),f,X))$.

The results concerning sets $\mathcal{Der}(X)$ proved till now are summarized in the following propositions. It turned out that decisive for $\mathcal{Der}(X)$ is the answer to the question: is X an element of $\mathcal{A}_{\mathcal{g}}$?

As we have seen, $\mathcal{A}_2 \subseteq \mathcal{A}_{\mathcal{g}}$ (cf. Remark 3.3 and the last Lemma). But there are many ARNs in $\mathcal{A}_1 \cap \mathcal{A}_{\mathcal{g}}$, among them also monotonically weakly $\emptyset$ -computable [1]-CRNs or [1]-CRNs contained in all [0]-coverings.[13] It is easy to show that $X \in \mathcal{A}_1 \cap \mathcal{A}_{\mathcal{g}} \Rightarrow \emptyset'' \leq_T (\mathcal{Sel}(X))'$.

<u>Theorem 3.20</u>. Let X be an ARN in $0\nabla 1$ and let g be a [0]-function increasing on $0\Delta 1$ such that $\mathcal{D}_{cl}(+\infty,g,X)$. Then for any bounded $\mathcal{Sel}(X)$-sequence $\mathcal{S}$ of RtNs there exists a [0]-uniformly continuous [0]-function f such that $\underline{D}_{cl}(Z_0,f,X)$ & $\overline{D}_{cl}(Z_1,f,X)$, where Z_0 and Z_1 are, respectively, the inferior and superior limits of the $\mathcal{Sel}(X)$-sequence $\mathcal{S}$ and consequently, if the $\mathcal{Sel}(X)$-sequence $\mathcal{S}$ is pseudo-fundamental (i.e. if $Z_0 = Z_1$), then $D_{cl}(Z_0,f,X)$.

<u>Corollary</u> For any ARN X in $(0\nabla 1)\setminus\mathcal{A}_{\mathcal{g}}$ the set $\mathcal{Der}(X)$ cannot fail to be the set of all weakly $\mathcal{Sel}(X)$-computable (i.e. $\mathcal{Sel}(X)'$-computable) ARNs.

<u>Theorem 3.21</u>. Let X be an ARN in $(0\nabla 1)\cap\mathcal{A}_{\mathcal{g}}$ and let Y be a monotonically weakly $\mathcal{Sel}(X)$-computable ARN. Then there exists a [0]-uniformly continuous [0]-function f such that $\mathcal{D}_{cl}(Y,f,X)$.

Thus, according to the last two theorems and Remark 1.2 for any ARN X in $0\nabla 1$ and for any infinite $\mathcal{Sel}(X)$-recursively enumerable set C of NNs there cannot fail to exist a [0]-function g increasing on $0\Delta 1$ and an ARN Y such that $\mathcal{D}_{cl}(Y,g,X)$ & $\mathcal{Sel}(Y) = C$ holds.

<u>Example 3.22</u>. There exists a [0]-function g increasing on $0\Delta 1$ such that for any ARN X fulfilling $X \in (0\Delta 1)\cap\mathcal{A}_{\mathcal{g}}$ we have

$$\exists Y\ (\ \mathcal{D}_{cl}(Y,g,X)\ \&\ \mathcal{Sel}(Y) = (\mathcal{Sel}(X))'\ )\ .$$

<u>Theorem 3.23</u>. Let n be a NN and let B be a $[n]$-recursive set of NNs. Then we can construct a [0]-sequence $\{\{d_{p,q}\}_q^B\}_p^{[0]}$ of pseudo-fundamental B-sequences of RtNs such that for the set $\mathcal{Cl}(B)$ of all ARNs being pseudo-limits of B-sequences of this [0]-sequence we have

a) $\mathcal{Cl}(B)$ contains all B-computable and monotonically weakly B-computable ARNs and is closed under arithmetical operations;

b) $\forall XY\ (\mathcal{Sel}(X) \leq_T B\ \&\ X \in \mathcal{A}_{\mathcal{g}}\ \&\ Y \in \mathcal{Der}(X) \Rightarrow Y \in \mathcal{Cl}(B))$ holds.

Let us notice what follows from the definition of $\mathcal{Cl}(B)$: The set $\mathcal{Cl}(B)$ is a "set of B'-measure zero" in the set of all weakly B-computable (i.e. B'-computable) ARNs.

REFERENCES

1. M. Beeson, "Foundations of Constructive Mathematics," Springer-Verlag, Berlin (1985).
2. M. Beeson, Some problems in constructive mathematics, Rend.Sem.Mat. Univers. Politecn.Torino 38:13 (1980).
3. N. A. Šanin, Constructive real numbers and constructive function spaces, in: "Translations of Mathematical Monographs 21," Amer.Math.Soc., Providence, R.I. (1968).
4. B. A. Kušněr, "Lectures on Constructive Mathematical Analysis," Amer. Math.Soc., Providence, R.I. (1984).
5. O. Demuth and A. Kučera, Remarks on constructive mathematical analysis, in: "Logic Colloquium '78," M. Boffa, D. van Dalen, K. McAloon, eds., North-Holland, Amsterdam (1979).
6. N. A. Šanin, On the constructive interpretation of mathematical judgements, in: "Amer.Math.Soc.Transl. (2) 23," Amer.Math.Soc., Providence,R.I. (1963).
7. H. Rogers, "Theory of Recursive Functions and Effective Computability," McGraw-Hill, New York (1967).
8. O. Demuth, R. Kryl, A. Kučera, An application of the theory of functions partial recursive relative to number sets in constructive mathematics (Russian), Acta Univ. Carolinae-Math. et Phys. 19:15 (1978).
9. O. Demuth, Some questions in the theory of constructive functions of a real variable (Russian), Acta Univ. Carolinae-Math. et Phys. 19:61 (1978).
10. I. D. Zaslavskij, G. S. Cejtin, On singular coverings and properties of constructive functions connected with them, in: "Amer. Math Soc. Transl. (2) 98," Amer.Math.Soc., Providence, R.I. (1971).
11. O. Demuth, On some classes of arithmetical real numbers (Russian), Comment.Math.Univ.Carolinae 23:453 (1982).
12. P. Martin-Löf, "Notes on Constructive Mathematics," Almquist and Wiksell Stockholm (1970).
13. O. Demuth, An example of a construction of pseudonumbers by means of recursion theory (to appear).
14. O. Demuth, On arithmetical complexity of differentiation in constructive mathematics (Russian), Comment.Math.Univ.Carolinae 24:301 (1983).
15. G. S. Cejtin, Algorithmic operators in constructive metric spaces, in: "Amer.Math.Soc.Transl. (2) 64," Amer.Math.Soc., Providence,R.I. (1967)
16. A. Kučera, B. A. Kušněr, On the types of recursive isomorphism of some concepts of constructive analysis (Russian), Comment.Math.Univ.Carolinae 19:97 (1978).
17. I. D. Zaslavskij, Some properties of constructive real numbers and constructive functions, in: "Amer.Math.Soc.Transl. (2) 57," Amer.Math. Soc., Providence, R.I. (1966).
18. O. Demuth, Constructive functions of a real variable and reducibilities of sets (to appear).
19. O. Demuth, Derivatives of constructive functions (to appear).
20. O. Demuth, A constructive analogue of Garg's theorem on Dini derivatives (Russian), Comment.Math.Univ.Carolinae 21:457 (1980).
21. J. Myhill, A recursive function defined on a compact interval and having a continuous derivative that is not recursive, Michigan Math.J. 18:97, (1971).

A COMPLETENESS THEOREM FOR HIGHER-ORDER INTUITIONISTIC LOGIC: AN INTUITIONISTIC PROOF

A. G. Dragalin

Debrecen University
Debrecen 4010
Hungary

The Goedel's famous completeness theorem for predicate logic is the foundation of modern model theory in classic logic as well as in intuitionistic one. Moreover, this theorem is intensively used in the theory of mechanical theorem proving [1] and in the theory of logic programming [2]. So there is an insistent aspiration for constructive treating of this theorem. Usual proofs of the completeness theorem are founded on considerations of maximal consistent sets of formulas and are nonconstructive (see, for example, [1] for classical case and [3] for intuitionistic logic). The situation is specially urgent in the intuitionistic case because this logic is intended for effective treating of logical connectives and is used usually in computer science as an instrument for getting a program from a constructive (= intuitionistic) proof of a given formula. A nonconstructive completeness proof in this situation can serve only as a general indication for possible success of a given proof-searching procedure, just as a constructive proof provides precise bounds for complexity of searching.

There is a special interest in investigation of higher-order logics with an impredicative comprehension scheme because these logics are easily treatable in modern programming languages, like PROLOG (see [4]).

The first, not completely successful, attempts to give an intuitionistic completeness proof for intuitionistic logic were given by E. Beth [5]. In particular, he proposed a very important notion of an intuitionistic model (so-called Beth-model notion for the first-order logic). Another well-known model notion for intuitionistic logic was proposed by Kripke [6]. In a report [7] it was shown that utterly intuitionistic proof of the completeness theorem is impossible from the point of view of usual intuitionistic model theory.

Nevertheless, the first correct intuitionistic completeness proof was invented in 1973 by W. Veldman [8]. To avoid the mentioned difficulties, he used modified Kripke models. The modification was in admitting the so-called strange worlds or exploding worlds, i.e., such moments in which every sentence is true. Veldman's theorem has the following form: a modified Kripke (or Beth) model M can be constructed, such that if a sentence A is true in M, then it is deducible in the intuitionistic first-order predicate logic. An analogous proof for a restricted second-order logic was worked out in [9]. However, the distinguished Veldman's model M has continual power, the worlds in this model are intuitionistic free choice sequences, M has a nondiscrete ordering, so it looks rather strange from the point of view of usual constructive reasoning and it seems not appropriate for computational applications.

A countable distinguished model with analogous properties was constructed in [3, Chapter 5] for higher-order intuitionistic logic. But its semantic is abstract algebraic rather than intuitionistically plausible Beth or Kripke-like models.

H. de Swart [10,11] gave a somewhat other form of the completeness theorem. He constructed the whole fan S of modified models, such if a sentence A is true in every model from S, then A is deducible in predicate logic. Every model of S has already a discrete ordering, but the whole family of models is continual. Moreover, the truth-definition of formulas in de Swart meaning has an important disadvantage: if a model from S has at least one strange world, then all worlds of this model turn out to be strange. This fact destroys the monotonocity property in intuitionistic model theory. In this point de Swart's truth-definition distinguishes from the Veldman-like truth-definition.

Essential point in Veldman and de Swart intuitionistic proofs is the using of the intuitionistic fan theorem. This theorem is true for one of specific intuitionistic meaning of free choice sequences, but it is not appropriate for many other directions in constructive mathematics. So we are interested in avoiding the using of this theorem.

Below we give an intuitionistic proof for the completeness theorem in the case of higher-order intuitionistic logic. In fact, for the sake of brevity we consider only second-order logic but a generalization of our proof for the simple type theory (with or without extensionality) is straightforward along the line directed in [12] or in [3, Chapter 5]. We construct a modified Beth-model M with the simple discrete binary tree-like order. In general, this model contains strange worlds, but truth-definition in M is monotone in style of Veldman. In order to avoid a free choice sequences theory we modify a notion of completion in Beth-model and get essentially neutral proof, which is valid from classical as well as from intuitionistic point of view.

Higher-order objects in our model are treated naturally as some families of truth-valued functions. Our families do not exhaust all of such functions so we get a Henkin-style completeness proof for higher-order logic, where in domains of the model not all subsets are used.

We tried to do this paper essentially self-contained; some more details for the case of first-order logic can be found in [13]. An analogous construction for classical higher-order logic is used in [14]; there, our aim was an intuitionistic proof of cut-elimination theorem.

The sign $\leftrightharpoons$ below means "is by definition". By ∇ we mark the beginning of a proof and by $\square$ mark its end. We use logical symbols simultaneously in formal and metamathematical contexts with tne exception of an implication where we use $\Longrightarrow$ for metamathematical contexts and $\supset$ as a sign in formal languages.

1. MODIFIED BETH-MODELS FOR HIGHER-ORDER LOGICS

1.1. Let us describe the language of our logic. It will be second-order logic. For simplicity, we suppose that our language is one-sorted, without equality, without constants and functional symbols. The generalization of the main results for the more complicated situations is rather straightforward and we shall not deal with them.

Three natural numbers 0, 1 and 2 are said to be types. Let us fix for every type τ an infinite set $Var(\tau)$ of variables of the type τ. Variables of the type 0 are considered as variables for elements of an individual domain, variables of type 2 are considered as variables for some subsets of a given individual domain and, at last, variables of type 1 are considered as variables for sentences.

Our language may contain some predicate symbols, such that all argument places of these symbols are of type 0.

1.2. Now we give the inductive definition of expression of type τ. The set of all expressions of type τ we denote as $Exp(\tau)$.

(1) If $x \in Var(\tau)$, then $x \in Exp(\tau)$.

(2) If $t_1,\ldots,t_n \in Exp(0)$ and P is n-place predicate symbol of our language, then $P(t_1,\ldots,t_n) \in Exp(1)$; such expressions are said to be atoms.

(3) If $A,B \in Exp(1)$, $x \in Var(\tau)$, then $(A \wedge B)$, $(A \vee B)$, $(A \supset B)$, $\neg A$, $\forall xA$, $\exists xA$ are elements of $Exp(1)$; the constant $\perp$ ("false") is an element of $Exp(1)$.

(4) If $A \in Exp(1)$ and $x \in Var(0)$, then $\{x|A\} \in Exp(2)$; such expressions are said to be abstracts.

(5) If $t_0 \in Exp(0)$ and $t_2 \in Exp(2)$, then $(t_0 \, \varepsilon \, t_2) \in Exp(1)$.

The definition of expression is finished.

Occurrences of variables into a given expression we divide onto free and bound occurrences by a well-known way: bounding quantifiers in these constructions are $\forall x\ldots$, $\exists x\ldots$, $\{x|\ldots\}$. A variable x is a parameter of a given expression E if it occurs free (at least once) in E. As $A(x \parallel t)$ we denote a result of substitution t instead of all free occurrences x in A, with necessary renaming of bound variables of A avoiding collisions of variables. Here $x \in Var(\tau)$ and $t \in Exp(\tau)$ for the same τ. In doubtless cases $A(t)$ is an abbreviation for $A(x \parallel t)$.

Elements of $Exp(1)$ are said to be formulas (of our language).

1.3. Let Σ be a set ω of all natural numbers or a finite subset of this set. Let us denote by Σ^* the set of all finite sequences of elements Σ, including the empty sequence Λ. By $p*q$ we denote the concatenation p and q, so if $p = \langle i_0,\ldots,i_{m-1}\rangle$ and $q = \langle j_0,\ldots,j_{n-1}\rangle$, then $p*q = \langle i_0,\ldots, i_{m-1},j_0,\ldots,j_{n-1}\rangle$. The number of members of p we denote by ∂p; for example, $\partial\langle i_0,\ldots,i_{m-1}\rangle = m$, $\partial\Lambda = 0$. The one-element sequence is denoted by $\langle i\rangle$. Instead of $p*\langle i\rangle$ we shall write sometimes simply $p*i$.

For $p,q \in \Sigma^*$, q is said to be an extension of p (in symbols $p \leqslant q$), iff p is an initial segment of q, i.e., if $\exists r(p*r = q)$. The strict order relation is introduced by definition: $p < q \leftrightharpoons (p \leqslant q) \wedge \neg (q \leqslant p)$. One-step order relation is defined by $p <\cdot q \leftrightharpoons (\exists i \in \Sigma)(p*\langle i\rangle = q)$.

1.4. A tree is a subset $T \subseteq \Sigma^*$, such that: (i) there is an element $p_0 \in T$ (a root of the tree), such that $(\forall q \in T)(p_0 \leqslant q)$; (ii) $p_0 \leqslant q \leqslant r$, $r \in T \Rightarrow q \in T$; (iii) $(\forall p \in T)(\exists q \in T)(p <\cdot q)$; (iv) T is a decidable subset of Σ^*, i.e.,

$$(\forall p \in \Sigma^*)(p \in T \vee p \notin T).$$

This last condition is important only from intuitionistic point of view. For example, Σ^* itself is a tree.

A function $\alpha : \omega \to T$ is said to be a path in a tree, if $\alpha(0) = p_0$, $\alpha(n) <\cdot \alpha(n+1)$. A path α is said to pass through $p \in T$ iff $\exists n(\alpha(n) = p)$.

1.5. A set $x \subseteq T$ is (order) open iff for all $p,q \in T$, $p \in x, p \leqslant q \Rightarrow q \in x$.

Let $\mathcal{O}$ be the family of all open subsets of T. A set $x \subseteq T$ is said to be complete iff

$$(\forall p \in T)(\forall q(p <\cdot q \Rightarrow q \in x) \Rightarrow p \in x).$$

$\mathfrak{f}$ denotes the family of all complete subsets of T.

Let now $x \subseteq T$ is an arbitrary subset of T. We define a completion of x by the following way:

$$\mathcal{D}x = \bigcap \{y \in \mathfrak{f} \mid x \subseteq y\},$$

i.e., $\mathcal{D}x$ is an intersection of all complete subsets of T containing x.

Evidently, (i) $\mathcal{D}x \in \mathfrak{f}$; (ii) $x \subseteq \mathcal{D}x$; (iii) $(\forall y \in \mathfrak{f})(x \subseteq y \Rightarrow \mathcal{D}x \subseteq y)$.

Remark From intuitionistic point of view it is possible to perceive (i) - (iii) as an independent "generalized inductive" definition of an operator $\mathcal{D}x$ and does not use the original set theoretic definition.

In traditional Beth-model theory (cf., for example [3]), one uses the following completion operator:

$$\mathcal{D}'x \leftrightharpoons \{p \in T \mid \forall \alpha(\exists n(\alpha(n) = p) \Rightarrow (\exists m \geqslant n)(\alpha(m) \in x))\}$$

(" $p \in \mathcal{D}'x$ if every path, passing through p, pass also through some element of x ") rather than our operator $\mathcal{D}x$. It is easy to see, that $x \subseteq \mathcal{D}'x$, $\mathcal{D}'x \in \mathfrak{f}$, hence $\mathcal{D}x \subseteq \mathcal{D}'x$.

Classically it is not difficult to prove also $\mathcal{D}'x \subseteq \mathcal{D}x$, so $\mathcal{D}x = \mathcal{D}'x$. From intuitionistic point of view the using of $\mathcal{D}x$ has some important advantages and, in particular, allows one to avoid employment of the fan theorem.

1.6. Let now $T \subseteq \Sigma^*$ be a tree. For $x,y \subseteq T$ let us define an open implication:

$$x \supset_0 y \leftrightharpoons \{p \in T \mid (\forall q \geqslant p)(q \in x \Rightarrow q \in y)\}.$$

Note further: (i) $x \in \mathcal{O} \Rightarrow \mathcal{D}x \in \mathfrak{f} \cap \mathcal{O}$; (ii) $x \in \mathcal{O}, y \in \mathfrak{f} \Rightarrow (x \supset_0 y) \in \mathfrak{f} \cap \mathcal{O}$; (iii) $x,y \in \mathcal{O} \Rightarrow \mathcal{D}x \cap \mathcal{D}y = \mathcal{D}(x \cap y)$.

We discuss below these properties in a somewhat more general situation. Now we notice only that, from (i) - (iii), follows:

<u>Fact</u> The structure $\langle \mathfrak{f} \cap \mathcal{O}, \subseteq \rangle$ is a complete Heyting algebra. In this algebra: $\mathbb{1} = T$, $\mathbb{0} = \emptyset$; $a \wedge b = a \cap b$, $a \vee b = \mathcal{D}(a \cup b)$; $a \Rightarrow b = (a \supset_0 b)$, $\neg a = (a \supset_0 \emptyset)$. Further, if $Q \subseteq \mathfrak{f} \cap \mathcal{O}$, then $\bigwedge Q = \bigcap Q$, $\bigvee Q = \mathcal{D}(\bigcup Q)$.

Here and below in analogous cases

$$\bigcap Q = \{p \in T \mid (\forall a \in Q)(p \in a)\},$$

so $\bigcap Q = T$, if $Q = \emptyset$.

About Heyting algebras consult, for example [15], where these algebras are named pseudo-Boolean algebras.

1.7. A <u>modified Beth-semimodel</u> is a structure

$$M = \langle T, v, I, V \rangle,$$

where (i) T is a tree (a set of <u>worlds</u> or <u>moments</u> of M); (ii) $v \in \mathfrak{f} \cap \mathcal{O}$ (a <u>zero</u> of M; the world $p \in v$ is said to be <u>strange</u> or <u>exploded</u>); (iii) I is a function defined on the set $\{0,1,2\}$ of types and such that $I(\tau)$ is a non-empty set (a <u>domain of type</u> τ; variables from $Var(\tau)$ are considered as run about $I(\tau)$). Moreover, $(\forall a \in I(1))(v \subseteq a, a \in \mathfrak{f} \cap \mathcal{O})$, and $I(2)$ is a set of functions: if $f \in I(2)$, then $f : I(0) \to I(1)$. Let us define <u>I-valued</u> expression as an expression E' obtained from an expression E of our language by substitution instead of free variables of E to corresponding type elements of $I(\tau)$; therefore, I-valued expression has no free variables, but possibly has elements of $I(0)$, $I(1)$, $I(2)$ as constants. (iv) V is a <u>valuation function</u> of M; namely, V is defined for every I-valued atomic formula P and $v \subseteq V(P) \in \mathfrak{f} \cap \mathcal{O}$.

The definition of a semimodel is finished.

Now we can naturally define a (partial) <u>truth valuation function</u> $\|\cdot\|$. Namely, for a given I-valued expression E of type τ sometimes can be defined an object $\|E\| \in I(\tau)$, by induction on construction of E according to 1.2 as follows:

For an I-valued atom P we put $\|P\| = V(P) \in I(1)$. Further, $\|A \wedge B\| = \|A\| \cap \|B\|$, $\|A \vee B\| = \mathcal{D}(\|A\| \cup \|B\|)$, $\|A \supset B\| = \|A\| \supset_0 \|B\|$, $\|\neg A\| = \|A\| \supset_0 v$, $\|\bot\| = v$, $\|\forall x A(x)\| = \bigcap_{a \in I(\tau)} \|A(a)\|$, $\|\exists x A(x)\| = \mathcal{D}(\bigcup_{a \in I(\tau)} \|A(a)\|)$.

Here every result is considered as defined only in the case, if it belongs to the domain $I(1)$. For example, $\|\exists x A(x)\|$ is defined only in the case, if all $\|A(a)\|$ are defined for all $a \in I(\tau)$, and, moreover, $\mathcal{D}(\bigcup_{a \in I(\tau)} \|A(a)\|) \in I(1)$.

Further, if $\|A(a)\|$ is defined for every $a \in I(0)$ and function f, such that $f(a) = \|A(a)\|$, belongs to $I(2)$, then we put $\|\{x \mid A(x)\}\| = f$. At last, $\|t_0 \varepsilon t_2\| = \|t_2\|(\|t_0\|)$, if $\|t_2\|$ and $\|t_0\|$ are defined.

If t is a constant $t \in I(\tau)$, then $\|t\| = t$.

1.8. A modified Beth-semimodel M is said to be a modified Beth-model if the truth valuation function $\|\cdot\|$ of M is defined for every I-valued expression E.

We say that an I-valued formula A is true in a moment $p \in T$ (in symbols $p \Vdash A$) if $p \in \|A\|$. We say that a formula A is true in a model M (in symbols $M \Vdash A$) iff for every I-valued formula A', obtained from A by substitution, and for every moment p we have $p \Vdash A'$.

Using the fact 1.6, it is a straightforward exercise to prove that for every formula A, deducible in the intuitionistic second-order predicate calculus, and for every modified Beth-model M we have $M \Vdash A$. In particular, quantor axioms and comprehension scheme are valid:

$$\forall zA \supset A(z\|r);\ A(z\|r) \supset \exists zA;\ t \in \{x|A\} \equiv A(x\|t).$$

Here $z \in Var(\tau)$, $r \in Exp(\tau)$, $x \in Var(0)$, $t \in Exp(0)$.

Remark We modify a traditional Beth-model notion in three aspects: (1) in traditional notion $v = \emptyset$, i.e., there are no strange worlds, this modification belongs to Veldman; (2) we use completion operator $\mathcal{D}$, rather than traditional $\mathcal{D}'$; (3) we generalize a traditional model notion for higher-order logic with abstracts.

2. GENERAL SEMANTIC CONSTRUCTIONS

The main task of this and the following point will appear in the paper [13], devoted to the first-order logic but, desiring to be self-contained, we reproduce here main facts with small alterations concerning higher-order features. In this point we consider an arbitrary set T, not necessarily a tree.

2.1. Let us denote by $\mathcal{P}$ the family of all subsets of T. For $a,b \in \mathcal{P}$ let us define a standard implication and standard negation as follows:

$$a \supset_t b \leftrightharpoons \{p \in T | p \in a \Rightarrow p \in b\};\ \neg_t a \leftrightharpoons \{p \in T | p \notin a\}.$$

Fact The structure $\langle \mathcal{P}, \subseteq \rangle$ is a complete Heyting algebra. In this algebra: $\mathbb{1} = T$, $\mathbb{0} = \emptyset$; $a \wedge b = a \cap b$, $a \vee b = a \cup b$; $a \Rightarrow b = (a \supset_t b)$, $\neg a = \neg_t a$. Further, if $Q \subseteq \mathcal{P}$, then $\bigwedge Q = \bigcap Q$, $\bigvee Q = \bigcup Q$.

Note, that classically $\langle \mathcal{P}, \subseteq \rangle$ is even a Boolean algebra and $(a \supset_t b) = (T \setminus a) \cup b$, but intuitionistically we can prove only $(T\setminus a) \cup b \subseteq (a \supset_t b)$.

2.2. Let now $(T,\leqslant)$ be an arbitrary (partially) ordered set. Put

$$p < q \leftrightharpoons (p \leqslant q) \wedge \neg(q \leqslant p),$$
$$x \in \mathcal{O} \leftrightharpoons (x \subseteq T) \wedge (\forall pq \in T)(p \in x \wedge p \leqslant q \Rightarrow q \in x).$$

For $a,b \in \mathcal{P}$ let us define an open implication and open negation as follows: $a \supset_0 b \leftrightharpoons \{p \in T | (\forall q \geqslant p)(q \in a \Rightarrow q \in b)\}$; $\neg_0 a \leftrightharpoons \{p \in T | (\forall q \geqslant p)(q \notin a)\}$.

Fact The structure $\langle \mathcal{O}, \subseteq \rangle$ is a complete Heyting algebra. In this algebra: $\mathbb{1} = T$, $\mathbb{0} = \emptyset$; $a \wedge b = a \cap b$, $a \vee b = a \cup b$; $a \Rightarrow b = (a \supset_o b)$, $\neg a = \neg_0 a$. Further, if $Q \subseteq \mathcal{O}$, then $\bigwedge Q = \bigcap Q$, $\bigvee Q = \bigcup Q$.

2.3. Let $(T,\leqslant)$ again be a partially ordered set.

2.3.1. A completion structure on T is a function $\mathcal{F}$ defined on T and such that for every $p \in T$, $\mathcal{F}(p)$ is a family subset of T. Moreover, for all $p,q \in T$, $a \subseteq T$: $q \in a \in F(p) \Rightarrow p \leqslant q$. A set $x \subseteq T$ is said to be complete (relatively F) if

$$(\forall p \in T)(\forall a \in F(p))(a \subseteq x \Rightarrow p \in x).$$

Let $\mathcal{S}$ denote the family of all complete (relatively F) subsets of T.

2.3.2. Fact $Q \subseteq \mathcal{S} \Rightarrow \bigcap Q \in \mathcal{S}$. Every completion structure F generates a completion operator. Namely, if $x \subseteq T$, then

$$\mathcal{D}x = \bigcap\{b \in \mathcal{S} \mid x \subseteq b\}.$$

2.3.3. Fact For all $a,b \subseteq T$ we have; (i) $a \subseteq \mathcal{D}a$; (ii) $\mathcal{D}a \in \mathcal{S}$; (iii) $a \subseteq b \in \mathcal{S} \Rightarrow \mathcal{D}a \subseteq b$; (iv) $a \subseteq b \Rightarrow \mathcal{D}a \subseteq \mathcal{D}b$; (v) $\mathcal{D}\mathcal{D}a = \mathcal{D}a$; (vi) $a \in \mathcal{S} \Rightarrow \mathcal{D}a = a$; (vii) $\mathcal{D}(a \cup b) = \mathcal{D}(a \cup \mathcal{D}b)$.

2.3.4. Lemma $a \in \mathcal{O}, x \subseteq T \Rightarrow a \cap \mathcal{D}x \subseteq \mathcal{D}(a \cap x)$.

∇ Let $a \in \mathcal{O}$ and $x \subseteq T$. We consider the set $c = (a \supset_t \mathcal{D}(a \cap x))$. Now using 2.3.1 and condition $a \in \mathcal{O}$ it should be checked that $x \subseteq c \in \mathcal{S}$. Therefore (2.3.3) $\mathcal{D}x \subseteq c$, hence, $a \cap \mathcal{D}x \subseteq \mathcal{D}(a \cap x)$. $\square$

2.4. Let $(T,\leqslant)$ be a partially ordered set and $a,b \subseteq T$. Let us define $a \geqslant b \Leftrightarrow (\forall p \in b)(\exists q \in a)(q \leqslant p)$.

2.4.1. A completion structure F on T is said to be ordered iff for all $p.q \in T$:

$$p \leqslant q \Rightarrow (\forall a \in F(p))(\exists b \in F(q))(a \geqslant b).$$

2.4.2. Lemma Let F be an ordered completion structure. Then (i) $x \in \mathcal{O} \Rightarrow \mathcal{D}x \in \mathcal{S} \cap \mathcal{O}$; (ii) $x \in \mathcal{O}, y \in \mathcal{S} \Rightarrow (x \supset_0 y) \in \mathcal{S} \cap \mathcal{O}$; (iii) $x,y \in \mathcal{O} \Rightarrow \mathcal{D}x \cap \mathcal{D}y = \mathcal{D}(x \cap y)$.

∇ (i) Let $x \in \mathcal{O}$. We consider the set:

$$c = \{p \in T \mid (\forall q \geqslant p)(q \in \mathcal{D}x)\}.$$

Now it should be checked that $x \subseteq c \in \mathcal{S}$. In checking $c \in \mathcal{S}$, we use the condition that F is ordered. Hence, $\mathcal{D}x \subseteq c$ and therefore $\mathcal{D}x \in \mathcal{O}$. $\mathcal{D}x \in \mathcal{S}$ follows from 2.3.3.

(ii) The condition $(x \supset_0 y) \in \mathcal{O}$ follows immediately from definition of open implication. Suppose that $x \in \mathcal{O}$, $y \in \mathcal{S}$. For showing $(x \supset_0 y) \in \mathcal{S}$, let us suppose $a \in F(p)$, $a \subseteq (x \supset_0 y)$. It is necessary to prove that $p \in (x \supset_0 y)$. Let us consider $q \geqslant p$, $q \in x$ and conclude that $q \in y$. As F is an ordered structure, for a given $a \in F(p)$ there exists $b \in F(q)$, $a \geqslant b$. We claim $b \subseteq y$. Indeed, let $r \in b$. Because $a \geqslant b$ there exists $s \leqslant r$, $s \in a$. From $a \subseteq (x \supset_0 y) \in \mathcal{O}$ follows $s \in (x \supset_0 y)$ and, hence, $r \in (x \supset_0 y)$. Further, $r \in b \in F(q)$, $q \leqslant r$, so 2.3.1. But $q \in x \in \mathcal{O}$, hence $r \in x$. Thus $r \in x$, $r \in (x \supset_0 y)$, therefore $r \in y$. From $b \subseteq y$, $y \in \mathcal{S}$ and $b \in F(q)$ we conclude $q \in y$.

(iii) Nontrivial is only the inclusion:

$$\mathcal{D}x \cap \mathcal{D}y \subseteq \mathcal{D}(x \cap y).$$

We prove it using (i), (ii) and some evident inclusions in the algebra $\mathcal{O}$ (2.2) as follows:

$$x \cap y \subseteq \mathcal{D}(x \cap y);\ x \subseteq (y \supset_0 \mathcal{D}(x \cap y));$$
$$\mathcal{D}x \subseteq \mathcal{D}(y \supset_0 \mathcal{D}(x \cap y)) = (y \supset_0 \mathcal{D}(x \cap y));$$
$$y \cap \mathcal{D}x \subseteq \mathcal{D}(x \cap y);\ y \subseteq (\mathcal{D}x \supset_0 \mathcal{D}(x \cap y));$$
$$\mathcal{D}y \subseteq \mathcal{D}(\mathcal{D}x \supset_0 \mathcal{D}(x \cap y)) = (\mathcal{D}x \supset_0 \mathcal{D}(x \cap y));$$
$$\mathcal{D}x \cap \mathcal{D}y \subseteq \mathcal{D}(x \cap y). \ \square$$

2.5. Theorem Let F be an ordered completion structure. Then the structure $\langle \mathcal{G} \cap \mathcal{O}, \subseteq \rangle$ is a complete Heyting algebra. In this algebra: $\mathbb{1} = T$, $\mathbb{0} = \mathcal{D}(\emptyset)$, $a \mathbin{\mathbb{\wedge}} b = a \cap b$, $a \mathbin{\mathbb{\vee}} b = \mathcal{D}(a \cup b)$, $a \Rightarrow b = (a \supset_0 b)$, and $\neg a = (a \supset_0 \mathbb{0})$. Further, if $Q \subseteq \mathcal{G} \cap \mathcal{O}$, then $\mathbb{\bigwedge} Q = \bigcap Q$, $\mathbb{\bigvee} Q = \mathcal{D}(\bigcup Q)$.

$\triangledown$ It is a corollary of 2.4.2 and 2.2. $\square$

2.6. Let us call a set $x \subset T$ weak-open (relatively F) if for all $p \in T$, $a \subseteq T$: $a \in F(p)$, $p \in x \Rightarrow a \subseteq x$. The family of all weak-open (relatively F) subsets of T we denote by $\bar{\mathcal{O}}$.

2.6.1. Fact (i) $\mathcal{O} \subseteq \bar{\mathcal{O}}$; (ii) $Q \subseteq \bar{\mathcal{O}} \Rightarrow \bigcap Q, \bigcup Q \in \bar{\mathcal{O}}$.

2.6.2. Lemma $x \in \bar{\mathcal{O}}.\ y \in \mathcal{G} \Rightarrow (x \supset_t y) \in \mathcal{G}$.

$\triangledown$ Let $x \in \bar{\mathcal{O}}$, $y \in \mathcal{G}$. Let us suppose $a \in F(p)$, $a \subseteq (x \supset_t y)$ and conclude $p \in (x \supset_t y)$. Let $p \in x$, then from $x \in \bar{\mathcal{O}}$ it follows $a \subseteq x$. In view of $a \subseteq (x \supset_t y)$, hence $a \subseteq y$. Now from $y \in \mathcal{G}$ it follows $p \in y$. $\square$

2.6.3. A completion structure is said to be monadic iff for every $p \in T$, $F(p)$ is at the most a one-element set. More precisely, F is monadic iff $(\forall ab \in F(p))(a = b)$.

2.6.4. Lemma Let F be a monadic completion structure. Then (i) $x \in \bar{\mathcal{O}} \Rightarrow \mathcal{D}x \in \mathcal{G} \cap \bar{\mathcal{O}}$; (ii) $x, y \in \bar{\mathcal{O}} \Rightarrow \mathcal{D}x \cap \mathcal{D}y = \mathcal{D}(x \cap y)$.

$\triangledown$ (i) Let $x \in \bar{\mathcal{O}}$; it is necessary to show $\mathcal{D}x \in \bar{\mathcal{O}}$ ($\mathcal{D}x \in \mathcal{G}$ follows from 2.3.3). Let us consider the set

$$c = \{p \in \mathcal{D}x \mid (\forall a \in F(p))(a \subseteq \mathcal{D}x)\}.$$

Now it should be checked that $x \subseteq c \in \mathcal{G}$. Suppose, for example, $a \in F(p)$, $a \subseteq c$ and let us show $p \in c$ (proving $c \in \mathcal{G}$). If $b \in F(p)$, then (as F is monadic) $b = a \subseteq c \subseteq \mathcal{D}x$. So $(\forall b \in F(p))(b \subseteq \mathcal{D}x)$. Further, from $b \subseteq \mathcal{D}x \in \mathcal{G}$ it follows $p \in \mathcal{D}x$. Hence, $p \in c$. From $x \subseteq c \in \mathcal{G}$ we get $\mathcal{D}x \subseteq c$ and hence $\mathcal{D}x \in \bar{\mathcal{O}}$.

(ii) A proof is similar to the proof of 2.4.2 (iii). We use (i) 2.6.2 and some inclusions in the algebra $\mathcal{P}$ (2.1) rather than in the algebra $\mathcal{O}$. $\square$

2.7. Let F be a completion structure and $v \subset T$. Let us define a new completion structure:

$$F_0(p) = \{a \subseteq T \mid a \in F(p) \vee (p \in v \wedge a = \emptyset)\}.$$

Let $\mathcal{G}_0$, $\mathcal{D}_0$ are corresponding notions related to F_0.

Fact (i) F_0 is a completion structure on T; (ii) $x \in \mathcal{G}_0 \Leftrightarrow x \in \mathcal{G} \wedge (v \subseteq x)$; (iii) $\mathcal{D}_0 x = \mathcal{D}(x \cup v) = \mathcal{D}(x \cup \mathcal{D}v)$; (iv) if F is an ordered structure and $v \in \mathcal{O}$, then F_0 is also an ordered structure.

3. BRANCHING OF SEQUENTS, SEQUENT TREES

3.1. A sequent is a formal expression of the form $(\Gamma \to \Delta)$, where Γ and Δ are collections of formulas. If S is a sequent $\Gamma \to \Delta$, then let us denote $(S)^0 = \Gamma$ and $(S)^1 = \Delta$. A collection of formulas is by definition a finite (may be empty) set of formulas, in which the repetition of some formulas is admitted. The order of formulas in a collection is not essential. Traditionally we write $\Gamma\Delta$ instead of $\Gamma \cup \Delta$, so $\Gamma\Delta$ and $\Delta\Gamma$ is the same collection. The collection $A\Gamma$ is obtained from Γ by adjoining one copy of A (note that $A\Gamma$ and $AA\Gamma$ are distinct collections).

We consider an intuitionistic sequent calculus IPC2 for our language. This calculus contains cut rule and it is essentially equivalent with the usual second-order intuitionistic predicate calculus without extensionality and with the full impredicative comprehension scheme. For example, one can choose the second-order fragment without extensionality in [12] or in [3, Chapter 5]. The following rules are admissible in IPC2:

$$\frac{A(x\|t)\Gamma \to \Delta}{(t \in \{x|A\})\Gamma \to \Delta} \quad ; \quad \frac{\Gamma \to A(x\|t)}{\Gamma \to (t \in \{x|A\})} \ .$$

The precise formulation of IPC2 is not important for us. In fact, all we need from IPC2 is the fact 3.2.2 below.

A sequent $\Gamma \to \Delta$ is said to be intuitionistic iff Δ is empty or one-element collection. Further, $\Gamma \to \Delta$ is said to be strong deducible (in symbols $\vdash^+$, $\Gamma \to \Delta$) if there exists Δ', $\Delta' \subseteq \Delta$, such that $\Gamma \to \Delta'$ is an intuitionistic sequent and $\Gamma \to \Delta'$ is deducible in IPC2 without cuts.

A sequent is said to be primitive iff it has one of the following forms: $A\Gamma \to \Delta A$ or $\perp\Gamma \to \Delta$, where A is an arbitrary formula. Of course, if S is primitive, then $\vdash^+ S$.

3.2. If S_1, S_2, S_3 are sequents, then let us define two three-place relations:

$$S_1 \prec_0 S_2, S_3 \text{ and } S_1 \prec_1 S_2, S_3$$

(in words: S_1 branches into S_2 and S_3 in reversible manner and, respectively S_1 branches into S_2 and S_3 in nonreversible manner).

To begin with, for an arbitrary sequent S, we define:

$$S \prec_0 S, S.$$

Further, we list all rest cases of both relations $S_1 \prec_i S_2, S_3$. Every case will have a special symbolic name depending on the construction of S_1. The notation $S_1 \prec_i S_2$ is an abbreviation for $S_1 \prec_i S_2, S_2$.

(1) $(\wedge\to)$ $(A \wedge B)\Gamma \to \Delta \prec_0 AB(A \wedge B)\Gamma \to \Delta$;

(2) $(\to\wedge)$ $\Gamma \to \Delta(A \wedge B) \prec_0 \Gamma \to \Delta(A \wedge B)A,\ \Gamma \to \Delta(A \wedge B)B$;

(3) $(\vee\to)$ $(A \vee B)\Gamma \to \Delta \prec_0 A(A \vee B)\Gamma \to \Delta,\ B(A \vee B)\Gamma \to \Delta$;

(4) $(\to\vee)$ $\Gamma \to \Delta(A \vee B) \prec_0 \Gamma \to \Delta(A \vee B)AB$;

(5) $(\supset\to)$ $(A \supset B)\Gamma \to \Delta \prec_0 (A \supset B)\Gamma \to \Delta A,\ B(A \supset B)\Gamma \to \Delta$;

(6) $(\to\supset)\Gamma\to\Delta(A\supset B)\prec_1\Gamma\to\Delta(A\supset B),\ A\Gamma\to B$;

(7) $(\neg\to)\ \neg A\Gamma\to\Delta\prec_0\neg A\Gamma\to\Delta A$;

(8) $(\to\neg)\Gamma\to\Delta\neg A\prec_1\Gamma\to\Delta\neg A,\ A\Gamma\to$;

(9) $(\forall\to)\ \forall xA\Gamma\to\Delta\prec_0 A(x\|t)\ \forall xA\Gamma\to\Delta$;

(10) $(\to\forall)\Gamma\to\Delta\forall xA(x)\prec_1\Gamma\to\Delta\forall xA(x),\ \Gamma\to A(y)$, $(x,y\in \mathrm{Var}(\tau)$, y is not free in $\Gamma)$;

(11) $(\exists\to)\ \exists xA(x)\Gamma\to\Delta\prec_0 A(y)\exists xA(x)\Gamma\to\Delta$, $(x,y\in \mathrm{Var}(\tau)$, y is not free in $\exists xA(x)\Gamma\to\Delta)$;

(12) $(\to\exists)\ \Gamma\to\Delta\exists xA\prec_0\ \Gamma\to\Delta\exists xAA(x\|t)$;

(13) $(\varepsilon\to)(t\ \varepsilon\ \{x|A\})\Gamma\to\Delta\prec_0 A(x\|t)(t\ \varepsilon\ \{x|A\})\Gamma\to\Delta$;

(14) $(\to\varepsilon)\ \Gamma\to\Delta(t\ \varepsilon\{x|A\})\prec_0\ \Gamma\to\Delta(t\ \varepsilon\{x|A\})A(x\|t)$.

In (9), (12), (13), (14) we have $x\in \mathrm{Var}(\tau)$, $t\in \mathrm{Exp}(\tau)$.

As one can see $\prec_1$ relation is used only in cases

$$(\to\supset),\ (\to\neg),\ (\to\forall).$$

3.2.1. <u>Fact</u> (i) If $S_1\prec_i S_2, S_3$, then $(S_1)^0\subseteq(S_2)^0$, $(S_1)^0\subseteq(S_3)^0$, $(S_1)^1\subseteq(S_2)^1$. (ii) If $S_1\prec_0 S_2, S_3$, then $(S_1)^1\subseteq(S_3)^1$.

3.2.2. <u>Fact</u> (i) If $S_1\prec_0 S_2, S_3$ and <u>simultaneously</u> $\vdash^+ S_2$ and $\vdash^+ S_3$, then $\vdash^+ S_1$. (ii) If $S_1\prec_1 S_2, S_3$ and <u>at least</u> $\vdash^+ S_2$ or $\vdash^+ S_3$, then $\vdash^+ S_1$.

3.3. A <u>binary tree</u> T is a tree (see 1.4) such that for every $p\in T$ a set $\{q|p<\cdot q\}$ consists precisely of two elements. For example, $\{0,1\}^*$ is a binary tree.

A <u>sequent tree</u> on a binary tree T is a couple of functions (M,h) defined on T, such that for every $p\in T$, $M(p)$ equals 0 or 1, $h(p)$ is a sequent and, moreover,

$$h(p)\prec_{M(p)} h(p*\langle i_0\rangle),\ h(p*\langle i_1\rangle),$$

where $i_0<i_1$, $p*\langle i_0\rangle$, $p*\langle i_1\rangle\in T$.

For simplicity below we shall write $p*0$, $p*1$ instead of $p*\langle i_0\rangle$, $p*\langle i_1\rangle$, so we shall deal mainly with $T=\{0,1\}^*$; the case of a general binary tree is quite similar.

3.4. A <u>zero</u> of a sequent tree (M,h) is a set:

$$v=\{p\in T|(\exists q\leqslant p)(\vdash^+ h(q))\}.$$

Evidently $v\in\mathcal{O}$.

3.5. Let (M,h) be a sequent tree on a binary tree T. Let us define some completion structures on T. First of all, put

$$F(p)=\{\{p*0,\ p*1\}\}.$$

This structure is ordered (2.4.1) and monadic (2.6.3). The notions, corresponding to $\mathcal{F}$ we denote as $\mathcal{S}$, $\bar{\mathcal{O}}$, $\mathcal{D}$ etc. Note, that $v \subset \mathcal{S} \cap \mathcal{O}$ (3.2.2). Further:

$$\mathcal{F}_0(p) = \{a \subseteq T \mid a \in \mathcal{F}(p) \vee (p \in v \wedge a = \emptyset)\}.$$

This structure is ordered as well (2.7). Corresponding notions are $\mathcal{S}_0$, $\bar{\mathcal{O}}_0$, $\mathcal{D}_0$ etc. According to 2.7:

$$\mathcal{D}_0 x = \mathcal{D}(x \cup v) = \mathcal{D}(x \cup \mathcal{D}v), \quad \mathcal{S}_0 = \{x \in \mathcal{S} \mid v \subseteq x\}.$$

Moreover, in view of 2.5, the structures $\langle \mathcal{S} \cap \mathcal{O}, \subseteq \rangle$ and $\langle \mathcal{S}_0 \cap \mathcal{O}, \subseteq \rangle$ are complete Heyting algebras. Zero of $\mathcal{S}_0 \cap \mathcal{O}$ is precisely a zero v of the tree (M,h).

At last, let us define two further completion structures:

$$\mathcal{F}_1(p) = \begin{cases} \{\{p*0, p*1\}\}, & \text{if } M(p) = 0; \\ \{\{p*0\}\}, & \text{if } M(p) = 1. \end{cases}$$

$$\mathcal{F}_2(p) = \begin{cases} \{\{p*0, p*1\}\}, & \text{if } M(p) = 0; \\ \{\{p*0\}, \{p*1\}\}, & \text{if } M(p) = 1. \end{cases}$$

Respectively, arise notions $\mathcal{S}_1$, $\mathcal{S}_2$, $\mathcal{D}_2$, $\bar{\mathcal{O}}_2$ etc.

The structure $\mathcal{F}_1$ is monadic so (2.6.4):

$$x, y \in \bar{\mathcal{O}}_1 \Rightarrow \mathcal{D}_1 x \cap \mathcal{D}_1 y = \mathcal{D}_1(x \cap y).$$

$\mathcal{F}_2$ is not ordered, not monadic, so the useful relation for it will be mainly 2.3.4: $a \in \mathcal{O},\ x \subseteq T \Rightarrow a \cap \mathcal{D}_2 x \subseteq \mathcal{D}_2(a \cap x)$.

3.5.1. Lemma (i) $v \in \mathcal{O}$, $v \in \bar{\mathcal{O}}_1$, $v \in \mathcal{S}$, $v \in \mathcal{S}_i$ for i = 0,1, and 2; (ii) $\mathcal{S}_2 \subseteq \mathcal{S}_1 \subseteq \mathcal{S}$; (iii) $\mathcal{D}x \subseteq \mathcal{D}_1 x \subseteq \mathcal{D}_2 x$.

▽ (i) Use 3.2.2. □

3.6. Let (M,h) be a sequent tree on a binary tree T. For every formula A we define two sets L(A) and R(A):

$$L(A) = \{p \in T \mid A \in (h(p))^0\}; \quad R(A) = \{p \in T \mid A \in (h(p))^1\}.$$

3.6.1. Lemma $L(A) \in \mathcal{O}$, $R(A) \in \bar{\mathcal{O}}_1$.

▽ Use 3.2.1. □

3.6.2. Lemma (i) $L(\perp) \subseteq v$; (ii) $L(A) \cap R(A) \subseteq v$.

▽ A primitive sequent is strongly deducible. □

4. SYSTEMATIC SEQUENT TREES

4.1. A sequent tree (M,h) on a binary tree T is said to be systematic if the following conditions are fulfilled:

(1) $L(A \wedge B) \subseteq \mathcal{D}(L(A) \cap L(B))$;

(2) $R(A \wedge B) \subseteq \mathcal{D}_1(R(A) \cap R(B))$;

(3) $L(A \wedge B) \subseteq \mathcal{D}(L(A) \cup L(B))$;

(4) $R(A \wedge B) \subseteq \mathcal{D}_1(R(A) \cap R(B))$;

(5) $L(A \supset B) \subseteq \mathcal{D}(R(A) \cup L(B))$;

(6) $R(A \supset B) \subseteq \mathcal{D}_2(L(A) \cap R(B))$;

(7) $L(\neg A) \subseteq \mathcal{D}(R(A))$;

(8) $R(\neg A) \subseteq \mathcal{D}_2(L(A))$;

(9) $L(\forall x A) \subseteq \mathcal{D}(L(A(x\|t)))$, $x \in Var(\tau)$, $t \in Exp(\tau)$;

(10) $R(\forall x A(x)) \subseteq \mathcal{D}_2(\bigcup_{y \in Var(\tau)} R(A(y)))$;

(11) $L(\exists x A(x)) \subseteq \mathcal{D}(\bigcup_{y \in Var(\tau)} L(A(y)))$;

(12) $R(\exists x A) \subseteq \mathcal{D}_1(R(A(x\|t)))$, $x \in Var(\tau)$, $t \in Exp(\tau)$;

(13) $L(t\ \varepsilon\{x|A\}) \subseteq \mathcal{D}(L(A(x\|t)))$, $x \in Var(0)$, $t \in Exp(0)$;

(14) $R(t\ \varepsilon\{x|A\}) \subseteq \mathcal{D}_1(R(A(x\|t)))$, $x \in Var(0)$, $t \in Exp(0)$.

4.2. Theorem Let S be a sequent and T be a binary tree. Then a systematic sequent tree (M,h) on T can be constructed, such that $h(p_0) = S$. Here p_0 is a root of T.

∇ We define M(p) and h(p) by induction on ∂p. If $\partial p = 0$, i.e., $p = p_0$, then put $h(p_0) = S$. Let us suppose now $\partial p > 0$, h(p) is already defined, and M(p'), h(p') are defined for every $p' \in T$, $\partial p' < \partial p$. Note that M(p) is not defined yet.

In this situation we define M(p) and h(p*0), h(p*1). Let us represent ∂p in the form

$$\partial p = 2^{m_0} \cdot 3^{m_1} \cdot 5^{m_2} \cdot m_3,$$

where 2,3,5 do not divide m_3.

The moment p is said to be expressive iff: (i) m_0 equals 0 or 1; and (ii) m_1 is a (Goedel) number of some formula A, such that A is nonatomic and A differs from $\perp$; moreover, if $m_0 = 0$, then A occurs in $(h(p))^0$, and if $m_0 = 1$, then A occurs in $(h(p))^1$; (iii) if $m_0 = 0$ and A begins from $\forall x$, where $x \in Var(\tau)$, then m_2 is a (Goedel) number of some expression $t \in Exp(\tau)$ and if $m_0 = 1$ and A begins from $\exists x$ with $x \in Var(\tau)$, then m_2 is a number of some $t \in Exp(\tau)$.

Now, if p is not expressive, we put M(p) = 0 and

$$h(p*0) = h(p*1) = h(p).$$

If p is expressive, we define

$$M(p) = i,\ h(p*0) = S_2,\ h(p*1) = S_3$$

in such a way that $h(p) \prec_i S_2, S_3$ with a given formula A with number m_1, corresponding to cases (1) - (14) in 3.2. Moreover, in the cases (9) $(\forall \rightarrow)$ and (12) $(\rightarrow \exists)$, we use a given expression t with number m_2. In (10) $(\rightarrow \forall)$ (11) $(\exists \rightarrow)$ we simply chose as y a first variable not free in a worked out sequent.

The definition of functions M, h is finished. It is clear that (M,h) is a sequent tree on T.

For checking the systematic conditions 4.1, we need the following:

<u>Lemma</u> (i) Let $n \in \omega$ be a natural number, $p \in L(A)$ and $u = \{q | q \geqslant p, q \in L(A), \partial q = \partial p + n\}$, then $p \in \mathcal{D}u$; (ii) Let $n \in \omega$, $p \in R(A)$ and also $u = \{q | q \geqslant p, q \in R(A), \partial q = \partial p + n\}$, then $p \in \mathcal{D}_1 u$ and $p \in \mathcal{D}_2 u$.

∇ Induction on n. Let us consider, for example, (ii) and operator $\mathcal{D}_1$. If $n = 0$, then $u = \{p\}$, $p \in u$ and, hence, $p \in \mathcal{D}_1 u$.

Let $n > 0$ and for $i = 0,1$, define

$$u_i = \{q \geqslant p{*}i | q \in R(A), \partial q = \partial(p{*}i) + (n - 1)\}.$$

(1) If $M(p) = 0$, then $p{*}0 \in R(A)$ and $p{*}1 \in R(A)$ (3.2.1), so on inductive supposition $p{*}i \in \mathcal{D}_1 u_i$. But $u_i \subseteq u$, therefore $p{*}i \in \mathcal{D}_1 u$. Let us consider $a = \{p{*}0, p{*}1\}$. In our case $a \in \mathcal{F}_1(p)$, $a \subseteq \mathcal{D}_1 u$ (note $M(p) = 0$), hence $p \in \mathcal{D}_1 u$.

(2) If $M(p) = 1$, then $p{*}0 \in R(A)$, so on inductive supposition $p{*}0 \in \mathcal{D}_1 u_0$. But $u_0 \subseteq u$, so $p{*}0 \in \mathcal{D}_1 u$. If $a = \{p{*}0\}$, then $a \in \mathcal{F}_1(p)$, $a \subseteq \mathcal{D}_1 u$ (note $M(p) = 1$), hence $p \in \mathcal{D}_1 u$. □

Now let us check several of the conditions (1) - (14) in 4.1.

(5) $L(A \supset B) \subseteq \mathcal{D}(R(A) \cup L(B))$. Let $p \in L(A \supset B)$. Let us consider a natural number $m = 2^0 \cdot 3^{m_1} \cdot 5^{m_2} \cdot m_3$, $m > \partial p$, such that m_1 is a (Goedel) number of $(A \supset B)$. Let $u = \{q \geqslant p | \partial q = m, q \in L(A \supset B)\}$.

According to the Lemna above, $p \in \mathcal{D}u$. In this situation every $q \in u$ is expressive and in accordance with the construction of (M,h), $M(q) = 0$ and $u \subseteq \mathcal{D}(R(A) \cup L(B))$ (cf. 3.2,5 ($\supset\rightarrow$)).

Hence $p \in \mathcal{D}u \subseteq \mathcal{D}(R(A) \cup L(B))$.

(6) $R(A \supset B) \subseteq \mathcal{D}_2(L(A) \cap R(B))$. Let $p \in R(A \supset B)$. Let us consider a number $m = 2^1 \cdot 3^{m_1} \cdot 5^{m_2} \cdot m_3$ such that m_1 is a Goedel number of $(A \supset B)$. Let $m > \partial p$, $u = \{q \geqslant p | \partial q = m, q \in R(A \supset B)\}$.

Then according to the Lemma, $p \in \mathcal{D}_2 u$. In this situation every $q \in u$ is expressive and in accordance with (M,h), $M(q) = 1$ and the branching (3.2,6($\rightarrow\supset$)) is used. Hence, $u \subseteq \mathcal{D}_2(L(A) \cap R(B))$ and $p \in \mathcal{D}_2 u \subseteq \mathcal{D}_2(L(A) \cap R(B))$.

(10) $R(\forall x A(x)) \subseteq \mathcal{D}_2(\bigcup_{y \in Var(\tau)} R(A(y)))$. Let $p \in R(\forall x A(x))$. Let us consider a number $m = 2^1 \cdot 3^{m_1} \cdot 5^{m_2} \cdot m_3$, $m > \partial p$, where m_1 is a (Goedel) number of $\forall x A(x)$. Let

$$u = \{q \geqslant p | \partial q = m, q \in R(\forall x A(x))\}.$$

Then $p \in \mathcal{D}_2 u$. In this situation every $q \in u$ is expressive, $M(q) = 1$, and the branching (3.2,10($\rightarrow\forall$)) is used, so for every $q \in u$ there exists a variable $y \in Var(\tau)$, such that $q{*}1 \in R(A(y))$. Hence

$$u \subseteq \mathcal{D}_2(\bigcup_{y \in Var(\tau)} R(A(y))),$$

and

$$p \in \mathcal{D}_2 u \subseteq \mathcal{D}_2(\bigcup_{y \in Var(\tau)} R(A(y))).$$

(12) $R(\exists xA) \subseteq \mathcal{D}_1(R(A(x\|t)))$. Let us fix some $t \in Exp(\tau)$ and also $p \in R(\exists xA)$. Let us consider a natural $m = 2^1 \cdot 3^{m_1} \cdot 5^{m_2} \cdot m_3$, where $m > \partial p$, m_1 is a Goedel number of formula $\exists xA$ and m_2 is a number of t. Let, further, $u = \{q \geqslant p \mid \partial q = m,\ q \in R(\exists xA)\}$.

According to the Lemma above, $p \in \mathcal{D}_1 u$. In this situation all $q \in u$ are expressive, $M(q) = 0$ and the branching (3.2,12) ($\rightarrow \exists$) is used with the fixed expression t. Therefore, for all $q \in u$ we have $q{*}0, q{*}1 \in R(A(x\|t))$, so $u \subseteq \mathcal{D}_1(R(A(x\|t)))$, hence $p \in \mathcal{D}_1 u \subseteq \mathcal{D}_1(R(A(x\|t)))$. □

5. BETH-MODEL ASSOCIATED WITH A SEQUENT-TREE

5.1. Let (M,h) be a systematic sequent tree. For every type τ, we define a domain $I(\tau)$ and some relation $a \approx t$, where $a \in I(\tau)$ and $t \in Exp(\tau)$. Namely:

For $\tau = 0$. $I(0) = Exp(0)$, $a \approx t \leftrightharpoons (a \text{ is } t)$.

For $\tau = 1$. If $a \in \mathfrak{b}_0 \cap \mathcal{O}$ (cf. 3.5) and $t \in Exp(1)$, we define $a \approx t \leftrightharpoons (L(t) \subseteq a) \wedge (R(t) \cap a \subseteq v)$. Further, $a \in I(1) \leftrightharpoons (a \in \mathfrak{b}_0 \cap \mathcal{O}) \wedge (\exists t \in Exp(1))\ (a \approx t)$.

For $\tau = 2$. Let a be a function $a : I(0) \rightarrow \mathfrak{b}_0 \cap \mathcal{O}$ and $t \in Exp(2)$. We define $a \approx t$ if for every $a_0 \in I(0)$ and $t_0 \in Exp(0)$ from $a_0 \approx t_0$ follows $a(a_0) \approx (t_0\ \varepsilon\ t)$. Further, $a \in I(2) \leftrightharpoons (a : I(0) \rightarrow \mathfrak{b}_0 \cap \mathcal{O}) \wedge (\exists t \in Exp(2))\ (a \approx t)$.

5.1.1. <u>Lemma</u> For every $t \in Exp(\tau)$ there exists $a \in I(\tau)$, $a \approx t$.

∇ For $\tau = 0$, we put simply a is t.

For $\tau = 1$, we put $a = \mathcal{D}(v \cup L(t))$. Let us prove $R(t) \cap a \subseteq v$. We have $R(t) \cap L(t) \subseteq v$ (3.6.2), hence $R(t) \cap (v \cup L(t)) \subseteq v$. Further, $R(t)$, $v \cup L(t) \in \bar{\mathcal{O}}_1$ (3.6.1 and 3.5.1), so using 2.6.4 we conclude that $\mathcal{D}_1(R(t)) \cap \mathcal{D}_1(v \cup L(t)) \subseteq v$. From 3.5.1 (iii) we have $R(t) \cap \mathcal{D}(v \cup L(t)) \subseteq v$.

For $\tau = 2$ we define $a : I(0) \rightarrow \mathfrak{b}_0 \cap \mathcal{O}$ such that for every $a_0 \in I(0)$, $a(a_0) = \mathcal{D}(\bigcup \{v \cup L(t_0\ \varepsilon\ t)\ t_0 \in Exp(0),\ a_0 \approx t_0\})$. For proving $a \approx t$ we consider $a_0 \approx t_0 \in Exp(0)$, i.e., a_0 is t_0. Then $a(a_0) = \mathcal{D}(v \cup L(t_0\ \varepsilon\ t))$ and $a(a_0) \approx (t_0\ \varepsilon\ t)$ can be proved as in the previous case $\tau = 1$. □

5.2. With every sequent tree (M,h) on T, we associate some modified Beth-semimodel

$$M = \langle T, v, I, V \rangle$$

(1.7) in the following way: (1) a zero v of M is a zero of the sequent tree (3.4); (2) a domain function I of the semimodel is defined in 5.1;

(3) for every atomic I-valued formula P we put $V(P) = \mathcal{D}(v \cup L(P))$. Note, that $V(P) \in \mathcal{S}_0 \cap \mathcal{O}$; moreover, $V(P) \approx P$ (see proof of 5.1.1 for the case $\tau = 1$).

In fact, if (M,h) is a <u>systematic</u> sequent tree, then M is a (modified) <u>Beth-model</u>. We get this result from the following key theorem.

5.3 <u>Theorem</u> Let (M,h) be a systematic sequent tree on T. Let $t(x_1,\ldots,x_n) \in Exp(\tau)$, where $x_1,\ldots,x_n$ is the list of all parameters of t, $x_i \in Var(\tau_i)$. Let $t_1,\ldots,t_n$ be a list of expressions $t_i \in Exp(\tau_i)$, and $a_1,\ldots,a_n$ be a list of elements $a_i \in I(\tau_i)$.

Let us denote $t' = t(t_1,\ldots,t_n)$ and $t'' = t(a_1,\ldots,a_n)$. Now let us suppose that $a_i \approx t_i$ for all $i = 1,\ldots,n$. Then $\|t''\|$ is defined and also $\|t''\| \approx t'$.

∇ By straightforward induction on the building of t (1.2), we consider several representative cases of this induction.

1. If t is a variable x_i, then the result follows immediately from conditions of the theorem.

2. If t is an atom, then $t'' = t'$, $\|t''\| = V(t'')$ and $V(t'') \approx t'$ follows from the definition 5.2 (3).

3. Let, for example, t be $(A \supset B)$. On inductive supposition $\|A''\|$ and $\|B''\|$ are defined and, moreover, $\|A''\| \approx A'$, $\|B''\| \approx B'$. So $\|A''\|$ and $\|B''\|$ are elements of I(1). We claim that $(\|A''\| \supset_0 \|B''\|) \approx (A' \supset B')$, and therefore $\|A'' \supset B''\| = (\|A''\| \supset_0 \|B''\|)$ is defined and is an element of I(1).

(i) $L(A' \supset B') \subseteq (\|A''\| \supset_0 \|B''\|)$. Indeed, from $\|A''\| \approx A'$ and $\|B''\| \approx B'$ we conclude $R(A') \cap \|A''\| \subseteq v$ and $L(B') \subseteq \|B''\|$. Hence, $R(A') \cap \|A''\| \subseteq \|B''\|$ and $L(B') \cap \|A''\| \subseteq \|B''\|$, so

$$(R(A') \cup L(B')) \cap \|A''\| \subseteq \|B''\|.$$

Further, $\mathcal{D}((R(A') \cup L(B')) \cap \|A''\|) \subseteq \mathcal{D}(\|B''\|) = \|B''\|$. As $\|A''\| \in \mathcal{O}$, using 2.3.4, we have

$$\mathcal{D}(R(A') \cup L(B')) \cap \|A''\| \subseteq \|B''\|.$$

Now using the systematic condition 4.1 (5), we get

$$L(A' \supset B') \cap \|A''\| \subseteq \|B''\|.$$

Now acting in Heyting algebra $\mathcal{O}$ (2.5, 3.5):

$$L(A' \supset B') \subseteq (\|A''\| \supset_0 \|B''\|).$$

(ii) $R(A' \supset B') \cap (\|A''\| \supset_0 \|B''\|) \subseteq v$. Indeed, from $\|A''\| \approx A'$ and $\|B''\| \approx B'$, we conclude $R(B') \cap \|B''\| \subseteq v$, $L(A') \subseteq \|A''\|$. In Heyting algebra $\mathcal{S}_0 \cap \mathcal{O}$, moreover $\|A''\| \cap (\|A''\| \supset_0 \|B''\|) \subseteq \|B''\|$, so $\|A''\| \cap R(b') \cap (\|A''\| \supset_0 \|B''\|) \subseteq v$ and, further, $L(A') \cap R(B') \cap (\|A''\| \supset_0 \|B''\|) \subseteq v$.

Hence, $\mathcal{D}_2(L(A') \cap R(B') \cap (\|A''\| \supset_0 \|B''\|)) \subseteq \mathcal{D}_2 v$. But $v \in \mathcal{S}_2$ (3.5.1) and $(\|A''\| \supset_0 \|B''\|) \in \mathcal{O}$, so using 2.3.4, we get $\mathcal{D}_2(L(A') \cap R(B')) \cap (\|A''\| \supset_0 \|B''\|) \subseteq v$.

Now using the condition 4.1 (6), we get

$$R(A' \supset B') \cap (\|A''\| \supset_0 \|B''\|) \subseteq v.$$

4. Let, for example, t be $\exists xA$, where $x \in Var(\tau)$. On inductive supposition for every $b \in I(\tau)$, $r \in Exp(\tau)$, $b \approx r$, $\|A''(x\|b)\|$ is defined and $\|A''(x\|b)\| \approx A'(x\|r)$, so $\|A''(x\|b)\|$ is an element of I(1). According to the definition of $I(\tau)$, for every $b \in I(\tau)$ there exists $r \in Exp(\tau)$, $b \approx r$; hence $\|A''(x\|b)\| \in I(1)$ for every $b \in I(\tau)$. Let us define

$$c = \mathcal{D}(\bigcup_{b \in I(\tau)} \|A''(x\|b)\|).$$

We claim that $c \approx \exists xA'$, therefore $\|\exists xA''\| = c$ is defined and is an element of I(1).

(i) $L(\exists xA') \subseteq c$. Let us fix an arbitrary $y \in Var(\tau)$ and find for this y an element $b \in I(\tau)$, $b \approx y$ (5.1.1). From inductive supposition it follows $L(A'(y)) \subseteq \|A''(x\|b)\|$, so also $L(A'(y)) \subseteq c$. Hence, $\bigcup_{y \in Var(\tau)} L(A'(y)) \subseteq c$ and therefore

$$\mathcal{D}(\bigcup_{y \in Var(\tau)} L(A'(y))) \subseteq \mathcal{D}c = c.$$

Using the condition 4.1 (11), we get the result.

(ii) $R(\exists xA') \cap c \subseteq v$. Indeed, let us fix an arbitrary $b \in I(\tau)$. For this b we can find $r \in Exp(\tau)$, $b \approx r$. From inductive supposition $R(A' \times (x\|r)) \cap \|A''(x\|b)\| \subseteq v$. Further (3.5.1), $\mathcal{D}_1(R(A'(x\|r)) \cap \|A''(x\|b)\|) \subseteq \mathcal{D}_1 v = v$. Using 2.3.4 we get $\mathcal{D}_1(R(A'(x\|r))) \cap \|A''(x\|b)\| \subseteq v$.

Moreover, $R(\exists xA') \subseteq \mathcal{D}_1(R(A'(x\|r)))$ (4.1 (12)), so $R(\exists xA') \cap \|A''(x\|b)\| \subseteq v$. Let us take a join on the left for all $b \in I(\tau)$: $R(\exists xA') \cap \bigcup_{b \in I(\tau)} \times \|A''(x\|b)\| \subseteq v$. Note that $R(\exists xA') \in \bar{\mathcal{O}}_1$ (3.6.1), $\bigcup_{b \in I(\tau)} \|A''(x\|b)\| \in \bar{\mathcal{O}}_1$ (2.6.1), so using 2.6.4 $\mathcal{D}_1(R(\exists xA'(x)) \cap \mathcal{D}_1(\bigcup_{b \in I(\tau)} \|A''(x\|b)\|) \subseteq v$. Now from 3.5.1 we conclude $R(\exists xA') \cap \mathcal{D}(\bigcup_{b \in I(\tau)} \|A''(x\|b)\|) \subseteq v$.

5. Let t be $\{x|A\}$. On inductive supposition for every $a_0 \in I(0)$, $t_0 \in Exp(0)$, $a_0 \approx t_0$, we have $\|A''(x\|a_0)\| \approx A'(x\|t_o)$, so the function $f : I(0) \to I(1)$ is defined, $f(a_0) = \|A''(x\|a_0)\|$. For proving $f \approx \{x|A'\}$ (5.1) let us consider $a_0 \in I(0)$, $t_0 \in Exp(0)$, $a_0 \approx t_0$ and show that $f(a_0) \approx (t_0 \,\varepsilon\, \{x|A'\})$.

(i) $L(t_0 \,\varepsilon\, \{x|A'\}) \subseteq \|A''(x\|a_0)\|$. From inductive supposition then $L(A'(x\|t_0)) \subseteq \|A''(x\|a_0)\|$. Further, $\mathcal{D}(L(t_0 \,\varepsilon\, \{x|A'\})) \subseteq \|A''(x\|a_0)\|$, and using 4.1 (13) we get the result.

(ii) $R(t_0 \,\varepsilon\, \{x|A'\}) \cap \|A''(x\|a_0)\| \subseteq v$. We have $r(A'(x\|t_0)) \cap \|A'' \times (x\|a_0)\| \subseteq v$. Using 3.5.1, 3.6.1 and 2.6.4, we get $\mathcal{D}_1(R(A'(x\|t_0))) \cap \mathcal{D}_1 \times (\|A''(x\|a_0)\|) \subseteq \mathcal{D}_1 v = v$. By 4.1 (14): $R(t_0 \,\varepsilon\, \{x|A'\}) \cap \|A''(x\|a_0)\| \subseteq v$. □

5.4. Corollary Let (M,h) is a systematic sequent tree on T and M is an associated modified Beth-model.

Let a sequent $A_1 \ldots A_n \to B$ be true in M. Then

$$L(A_1) \cap \ldots \cap L(A_n) \cap R(B) \subseteq v.$$

∇ Let $t(x_1,\ldots,x_m) \in Exp(\tau)$, where $x_1,\ldots,x_m$ is the list of all parameters of t, $x_i \in Var(\tau_i)$. Let us choose $a_i \approx x_i$ according to 5.1.1. Now let us define $\|t\| = \|t(a_1,\ldots,a_m)\|$.

Now if $A_1 \ldots A_n \to B$ is true in M, then $\|A_1 \ldots A_n \to B\| = T$, which is equivalent to $\|A_1\| \cap \ldots \cap \|A_n\| \subseteq \|B\|$. But in view of 5.3 $L(A_i) \subseteq \|A_i\|$ and $\|B\| \cap R(B) \subseteq v$. Hence $L(A_1) \cap \ldots \cap L(A_n) \cap R(B) \subseteq v$. □

6. MAIN RESULTS

6.1. Theorem For every binary tree T and every intuitionistic sequent S a modified Beth-model M_S on T can be constructed, such that if S is true in M_S, then $\vdash^+ S$.

∇ Let S be $A_1 \ldots A_n \to B$. For a given S and T we construct a systematic sequent tree (M,h), such that $h(p_0) = S$ where p_0 is a root of T (4.2). Let M_S be an associate with (M,h) model. Then (5.4):

$$p_0 \in L(A_1) \cap \ldots \cap L(A_n) \cap R(B) \subseteq v.$$

Hence, $p_0 \in v$, and by definition of zero (3.4) we get $\vdash^+ h(p_0)$. □

6.2. Corollary (Cut-elimination theorem, model-theoretic proof.) If an intuitionistic sequent S is deducible in intuitionistic second order sequent calculus IPC2, then it is deducible also without cuts.

∇ If S is deducible, then it is true in every Beth-model and, in particular, in M_S (6.1). □

6.3. Corollary For every formula A and binary tree T a modified Beth-model M_A on T can be constructed, such that if A is true in M_A, then A is deducible in the intuitionistic second order predicate calculus.

∇ Use 6.1 with S to be $\to A$. □

6.4. Remark In fact we can get some strengthening of 6.3 in the following form: a modified Beth-model M on a binary tree T can be constructed, such that for every formula A, if A is true in M, then A is deducible in the intuitionistic second order predicate calculus. For this result one can use some more sophisticated version of the theorem 4.2 or "glue" the model M sought from models M_A for all formulas A (see a first order case in [13, 5.5]). For higher order logics details are rather cumbersome so we put them aside for some other place.

REFERENCES

1. Chin-Liang Chang and R. Char-Tung Lee, "Symbolic Logic and Mechanical Theorem Proving", Academic Press (1973).
2. K. R. Apt and M. H. van Emden, Contributions to the theory of logic programming, J. Assoc. Comput. Mach., 29, 3:841-862 (1982).
3. A. G. Dragalin, Mathematical intuitionism. Introduction to proof theory (in Russian), Nauka, Moscow (1979).

4. K. L. Clark, An introduction to logic programming, in: "Introductory Readings in Expert Systems", D. Michie, ed., Gordon & Breach Sci. Publ., pp. 93-112 (1982).
5. E. W. Beth, "The Foundations of Mathematics", North- Holland, Amsterdam (1959).
6. S. A. Kripke, Semantical analysis of intuitionistic logic I, in: "Formal Systems and Recursive Functions", North-Holland, Amsterdam, pp. 92-129 (1965).
7. V. H. Dyson and G. Kreisel, Analysis of Beth's semantic construction of intuitionistic logic, Stanford Report (1961).
8. W. Veldman, An intuitionistic completeness theorem for intuitionistic predicate logic, J. Symbolic Logic, 41, No. 1:159-166 (1976).
9. G. K. Lopez-Escobar and W. Veldman, Intuitionistic completeness of a restricted second order logic, Lect. Notes Math., 500;198-232 (1975).
10. H. de Swart, Another intuitionistic completeness proof, J. Symbolic Logic, 41, No. 3:644-662 (1976).
11. H. de Swart, First steps in intuitionistic model theory, J. Symbolic Logic, 43, No. 1;3-12 (1978).
12. Takahashi Moto-o, Cut-elimination theorem and Brouwerian-valued models for intuitionistic type theory, Comment Math. Univ. St. Pauli, XIX-I:55-72 (1970).
13. A. G. Dragalin, "A Completeness Theorem for Intuitionistic Predicate Logic. An Intuitionistic Proof", Publicationes Mathematicae Debrecen, to appear.
14. A. G. Dragalin, "Cut-elimination Theorem for Higher-Order Classical Logic. An Intuitionistic Proof", this volume.
15. H. Rasiowa and R. Sikorski, "The Mathematics of Metamathematics", second ed., Warszawa (1968).

REASONING IN TREES

Herman Ruge Jervell

University of Oslo
Inst. of Informatics
Box 1080 Blindern
0316 Oslo 3, Norway

1. THE TURING SWAMP

Most theories of computation start with the following:

a set of states S;

the initial states I - a subset of S;

the terminal states τ - a subset of S;

a set of moves M - mapping S in S.

This is, so to say, the kernel. There may be various superstructures built on the top of it. But in most cases we would define a calculation - or an execution - as a sequence of moves starting with an initial state and going on until a terminal is reached. If we do not know more about the calculation than this, we are in the middle of the Turing swamp. There is very little we can say about such calculations. We have no way of analyzing the quantifier-combinations involved in going on until. We need to know more about the calculation to do that.

Part of the problem with the Turing swamp is that it is so easy to get stuck in it. We may make a nice theory of computation using the unanalyzed going on until. The theory is of very little help when we come to concrete calculations.

2. GÖDEL'S ABSTRACT IDEAS

The way out of the Turing swamp is to use abstract ideas. In his Dialectica-paper Gödel talked about them [8]. The point is that we usually know more about the calculations than come out in the usual theories of computations. We may know:

a proof that the calculation terminates;

some property that is invariant under the local moves in the calculation;

Dedicated to Kurt Gödel 1906-1978.

some interpretation of the calculation;

that something is finite;

that something has a model;

...

The abstract properties involve quantifier combinations like going on until. To give an example: in general we cannot decide whether something is finite or not. There is a quantifier combination in to be finite. But it can be used as assumption in arguments - and we may go from the abstract idea to be finite to the abstract idea going on until.

In all formal arguments there are assumptions involving abstract ideas. (For example, the idea of a finite string.) The goal in logic is, of course, not to eliminate these abstract ideas. We cannot do without them. But the goal is to drag them out into daylight and make them both visible and useful.

Here I will show that some useful abstract ideas can be formulated in terms of trees. They will be useful for both reasoning in trees and reasoning about trees. And, of course, we know that trees come up all the time in logic and in computer science.

3. PREDICATE LOGIC

Let us start with some of the reasoning in trees implicit in predicate logic. Say we have a Gentzen-style sequential calculus [4]. A common way of proving the completeness theorem for classical predicate calculus goes like this:

start with a formula F;

with F at the bottom node try to systematically construct a tree $T(F)$ with formulae at the nodes;

$T(F)$ has three types of nodes - terminal nodes, nodes with no branching and nodes with binary branching;

if $T(F)$ is well-founded, then we get a derivation in sequential calculus of F;

if there is an infinite branch in $T(F)$, we can from that branch construct a term-model M falsifying F;

if we have a model N falsifying F, we can from the model N construct an infinite branch in $T(F)$.

The construction is of course standard - but there are a few things which are not so often noted. First some obvious facts about the construction:

it is decidable whether some node is terminal or not;

for a non-terminal node it is at most a binary branching;

for a non-terminal node it is decidable what kind of branching we have;

the term-model M is not assumed to be total;

the falsifying model N is not assumed to be total.

The construction can readily be used as a base for mechanizing predicate logic. The way the logicians usually give the construction there is a lot of wasteful copying - but that can to a large extent be eliminated. One should also have a strategy for which formulae in the sequents should be analyzed first. More important, the construction gives - to my mind - the best way of formulating the two major problems in mechanizing predicate logic:

1. Herbrand universe: which terms in the Herbrand universe are superfluous in the construction? Or, if one likes, which terms do we need to analyze existential quantifiers with respect to?

2. Using cuts: which cuts should be allowed?

The first problem has been taken up by Hao Wang [15]. Let us look a little at the second. It is not hard to give examples of arguments where cuts are necessary to make them short. One could argue that the first essential use of cut was by Archimedes in his short note on the Sandreckoner [1]. In less than a page he shows how to talk about numbers which are larger than the number of sand-grains in the universe. If he had done that in a cut-free way he could very well have ended up with a description which took more room than the sand-grains he was counting. From the work of Gentzen, it is clear that a derivation of length d with cut-formulae of complexity < n can be transformed into a cut-free derivation of length less than:

$$\left.2^{2^{\cdot^{\cdot^{2^d}}}}\right\} = n$$

and that this estimate cannot be essentially reduced.

The resolution method of Prawitz [12] and Robinson [13] involves the cut-rule, but only such cuts as can be derived from a unification procedure. This unification gives an answer to which cuts should be allowed, but one can give examples which show that it is too restrictive. There are arguments proposing that one should have used more cuts than those given by resolution.

The importance of this comes from the following observations which are in Gentzen's work [4]:

any formal (or half-formal) argument in logic can be transformed step for step into a formal derivation in the system of natural deduction;

there is no loss of efficiency in the transformation;

the derivation in natural deduction can be transformed into a derivation in sequential calculus using cuts, and with no loss of efficiency.

The outcome of this is that we do not lose efficiency as long as we use cuts liberally. The use of cuts can be eliminated with an enormous loss in efficiency. On the other hand, the liberal use of cuts is a hinder to mechanizing logic. With liberal use of cuts we have a tendency to end up with a British Museum variant of proof-procedure - search through all possible proofs until you find one which does the job.

4. LOGIC PROGRAMMING

Logic Programming has become a catch-word and with the advent of Prolog, it may seem to be not far away from being realized. In Clocksin-Mellish's introduction to Prolog [2] we read:

"In the last few sections we have seen how Prolog is based on the idea of a theorem prover. As a result of this, we can see that our programs are rather like our hypothesis about the world, and our questions are rather like theorems that we would like to have proved. So programming in Prolog is not so much like telling the computer what to do and when, but rather like telling it what is true and asking it to try and draw conclusions."

The authors are of course aware that this way of thinking is not realized in Prolog. After the first few programs you soon become aware that you have to take into account a number of side effects. The execution depends heavily on the way you give "the hypothesis about the world" and you have to know a lot about the backtracking mechanism to make programs of some complexity. But let us now leave these objections aside and look at "logic programming" from a more theoretical point of view. We get a system of "logic programming" as soon as we have a mechanical theorem prover for predicate logic (and also a way of getting values out of proved existential formulae).

Given some hypothesis H of the world and some question F that we want to ask, formulated as formulae in predicate logic. The problem is whether H logically implies F. Starting with H and F the theorem prover constructs a tree over the formulae as before. We call the tree $T(H,F)$. There are two ways of reasoning in this tree:

we may concentrate on what is true in the world given the hypothesis H and ask whether F also had to be true;

we may concentrate on the possible derivations of F from the hypothesis H.

Gödel's completeness theorem tells us that these things are equivalent. But that is not the whole story. If we look at these ways of reasoning from the view of abstract ideas, a lot more should be said.

The models used in "logic programming" tend to be partial. In mathematics it is major step to go from a prime ideal to a maximal ideal. We need some extra argument to do it. In the same way it is a major step to go from a partial model to a total model. It is no accident that our theorem prover - using the construction of $T(H,F)$ - gives partial models when the construction fails to give a derivation of F from H. A theorem prover which gives total models under failure cannot be as efficient. The theorem prover we get from the Henkin proof of the completeness theorem is not better than the British Museum variant - first enumerate all proofs and then test them one by one until we find one that fits.

In the formulations of "logic programming" we assume that there is a vast difference between it and ordinary imperative programming. This is of course not so. The following argument comes from Turing's major paper [14]. Consider an execution according to a program P. Turing showed that using simple description of the possible states the machine can be in and the moves given by the program, we can construct in a straightforward way a formula $\tau(P)$ such that the following is equivalent:

the execution according to P terminates;

$\tau(P)$ is derivable in predicate logic.

Turing used the equivalence to show that the Entscheidungsproblem were equivalent to the Halting problem. We can use it to show that there is not such a vast difference between imperative programming and logic programming. (We would then not only be interested in termination of the programs but also the result after termination. Turing's argument is readily extended to give formulae expressing such things.)

In Turing's argument there is also information about efficiency. The execution of P can be translated step for step into a derivation of $\tau(P)$, and the other way around (being a little liberal with respect to the cuts allowed in the derivation). So we do not lose any efficiency in translating the one to the other. (As a side remark, there are connections between deterministic calculations and Horn-formulae.)

The remarks here are formulated in terms of "imperative programming". The same remarks can of course be made for "functional programming".

Do we gain anything at all by logic programming? We have shown that it is straightforward to translate an imperative program P into a formula $\tau(P)$ such that derivations of the formula do the same job as execution of the program. But this is only translating it into the syntactic part of the completeness theorem. "Logic programming" is concerned with the semantic part. What is involved in the equivalence in the completeness theorem? Let us see which parts are straightforward:

we can in an elementary recursive way construct the tree $T(F)$ given the formula F;

given a falsifying model N of F we can in a straightforward way construct an infinite branch in $T(F)$.

But the problems come with the following:

from the non-existence of a derivation there is no straightforward way to get an infinite branch in the tree.

The argument for this uses:

<u>Theorem 1</u>

Weak König's lemma. Given a tree T with the following properties:

there is an elementary recursive construction of the nodes of the tree;

for each node we can decide whether it is a terminal node or not;

for each non-terminal node there is an upper bound on the number of successor nodes;

for each non-terminal node we know the exact number of successor nodes;

there are infinitely many nodes.

We can thus conclude that there is an infinite branch.

One of the assumptions is more abstract than the others. The first four refer to algorithms for constructing or proving. The last "there are infinitely many nodes" can be interpreted in many ways. If we understand the quantifier combination there in a particular way, this way can then also be used to give an understanding of the quantifier combination in the conclusion "there is an <u>infinite</u> branch".

The weak König's lemma is implicit in "logic programming" and is part of the reason why it is supposed to be so much easier to think in terms of models than in terms of executions of programs (or derivations of formulae). The following "formula" makes a little sense:

Logic Programming

‖

Imperative Programming + Weak König's Lemma

or we could say that in "logic programming" we talk about the possible branches in the trees, while in "imperative programming" we talk about the nodes. In "logic programming" we get the abstract argument involved in weak König's lemma for free, but we can do exactly the same arguments in imperative programming if we add the weak König's lemma as an extra ingredient.

The "logic" in "logic programming" is here used as a synonym for predicate logic. In an uninteresting way we can do all our programs in logic programming. But then we are in the middle of the Turing swamp. The interesting things come with the abstract arguments and the abstract ideas which we use to understand the programs. The point about "logic programming" was that it had built into it a use of the completeness theorem of predicate calculus. This is, using reasonable assumptions, equivalent to weak König's lemma. After the work of Friedmann et al. [3] this is well understood.

We can also see why the Henkin proof of the completeness theorem is different from the above. The Henkin proof gives total models. "Logic programming" with total models corresponds to the following stronger principle:

<u>Theorem 2</u>

Strong König's lemma. Given a tree T with the following properties:

there is an elementary recursive construction of the nodes in the tree;

for each node we can decide whether it is a terminal node or not;

for each non-terminal node there is an upper bound on the number of successor nodes;

there are infinitely many nodes.

We can thus conclude that there is an infinite branch.

The only difference is that we do not know the exact number of successor nodes to a non-terminal node. We only know an upper bound to the number of successor nodes. Friedmann [3] has given a number of statements from ordinary mathematics equivalent to one of the two versions of the König's lemma. The importance of such work is to make clear the abstract ideas involved in various mathematical arguments. We have seen that it can be used to see what "logic programming" could be about. There is a moral here:

In all our formal reasoning we use some abstract ideas. It could be the idea of a formula (as a <u>finite</u> string of symbols) or other ideas. These abstract ideas are also used in general arguments like the König's lemmas. The abstractness is shown by the absence of calculatory content. Nevertheless, we need them to govern our calculations. The crucial assumption in the König's lemma is "there are infinitely many nodes". If we have a nice analysis of the quantifier combination in "there are infinitely many", then this analysis can be transferred to a similar analysis of the quantifier combination in "there is an infinite branch". If we have some analysis of the assumption with calculatory content, we get also calculatory content in the conclusion.

We cannot in advance enumerate all possible abstract ideas. To get abstract ideas usually involves creative work and this cannot be mechanized. A critique of "logic programming" is that it has connected itself too much with one abstract argument - the one giving weak König's lemma.

If we have an argument for termination, or correctness, of calculation involving the weak König's lemma, then it is possible to transform this into a problem in "logic programming" with no use of the weak König's lemma. This is an advance. But if we need the strong König's lemma, then we are in exactly the same situation as in "imperative programming".

5. β-COMPLETENESS

Friedmann [3] has given principles beyond the two König's lemmas. These principles fit well into the headline "reasoning in trees". The next principle for him is the kind of bar-rule used to analyze formal theories of hyper-arithmetical functions. After that he comes to a principle connected with Π^1_1-completeness. This principle is of course connected with the idea that a tree with natural number branching should be well founded.

In the remainder of this paper we will show how to go beyond that. The main person here is of course Jean-Yves Girard [5,6,7].

The problem of β-completeness came up in works of Mostowski [11]. He was interested in logics where the notion of well-ordering was absolute. Girard translated this into the framework of many sorted predicate calculus where we have one particular sort with predefined meaning [6]:

a sort W of elements from a well-ordering;

a linear ordering $\prec$ defined on W.

This logic is called β-logic. Models in β-logic should always have W as well-ordered by $\prec$. Below we will give a completeness theorem for this logic. To analyze validity here we need of course some fairly complicated abstract ideas. They must be of logical complexity Π^1_2 - as is not hard to see.

To get started we assume that in our language we have a fixed list of variables of type W:

$$o_0, o_1, o_2, o_3, \ldots$$

Then we go through the same construction as in predicate logic. We start with a formula F not containing any free variables (nor for simplicity any constants of type W). Then we go through the construction of the tree of formulae above F minding that the free variables of type W introduced should all be from the fixed list. We end up with the tree $T(F)$. In most cases this tree is an infinite tree which locally looks like a derivation.

So far we are not going beyond ordinary predicate calculus. Now we introduce ordinals.

<u>Definition 1</u>

Let α be an ordinal. We let α^* be the set of all finite sequences of ordinals $< \alpha$. The elements of α^* are often written

$$\sigma = \langle \sigma_0, \sigma_1, \ldots, \sigma_{n-1} \rangle .$$

Definition 2

A sequent (in sequential calculus) involving ordinals and less than ($\prec$) is secured if the subsequent, of the formulae which are atomic, is true.

Definition 3

Given $\langle \sigma_0, \sigma_1, \ldots, \sigma_{n-1} \rangle$ from α^*. We define

$$T(F)/\langle \sigma_0, \sigma_1, \ldots, \sigma_{n-1} \rangle$$

as the result of textual substitution of σ_i for o_i in $T(F)$ where $i = 0, 1, \ldots, n - 1$.

The result of such a textual substitution may not any longer look like a derivation (problems with the o's used as eigenparameters), but we do not care.

Definition 4

Let σ be from α^*. We say that σ is secured relative to $T(F)$ if in all branches in $T(F)/\sigma$ there is a secured sequent at some node.

Definition 5

Let alpha be an ordinal. The α-tree based on F, written $TREE(F)[\alpha]$, is defined as the tree of all not secured sequences from α^*.

After this rather heavy going it is time for an example. Consider the formula, τI, expressing the principle of transfinite induction over our ordering $\prec$. We then first construct the tree of formulae (or sequents) with τI at the bottom node. This tree, $T(\tau I)$, is infinite because the formula is not derivable in predicate logic. A sequence $\langle \sigma_0, \sigma_1, \ldots, {}_{n-1} \rangle$ from α^* is not secured there, if it is strictly descending

$$\sigma_0 \succ \sigma_1 \succ \ldots \succ \sigma_{n-1}$$

and $TREE(\tau I)[\alpha]$ consists of all strictly descending sequences from α^*.

Lemma 1

All initial segments of elements from $TREE(F)[\alpha]$ are themselves elements. So we naturally have a tree structure.

Definition 6

We write $\sigma \sim \tau$ where σ and τ are from α^* if they are order isomorphic.

Lemma 2 (Homogeneity)

For $\sigma \sim \tau$ from α^*: if one of them is in $TREE(F)[\alpha]$, then so is the other.

Any finite sequence is order isomorphic to a sequence of natural numbers. This gives:

Definition 7

Let T be a set of sequences of ordinals and β an ordinal. We define the extension (or restriction) of T with β as:

$T[\beta] == \{\sigma \text{ in } \beta^* |$ for some τ in T we have $\sigma \sim \tau\}$.

Lemma 3

For all α: TREE(F)$[\alpha]$ = TREE(F)$[\omega][\alpha]$.

We must connect our big trees with logic.

Definition 8

For each α we define the auxiliary logic where the universal quantifiers over the type W are taken to be in conjunction over all ordinals $< \alpha$, and existential quantifiers as disjunctions over all ordinals $< \alpha$. Derivability in this logic is denoted by the sign: $\vdash_\alpha$. For alpha $\geqslant \omega$ these logics are infinitary. It is not hard to see that:

It is not hard to see:

Lemma 4

For any formula F: if TREE(F)$[\alpha]$ is well founded, then $\vdash_\alpha F$.

Lemma 5

The following is equivalent for a formula F:

F is valid in our β-logic;

for all α: TREE(F)$[\alpha]$ is well founded.

Remember now that the point about abstract ideas is not to eliminate but to make them visible and useful. We need some complicated abstract ideas to explain validity in β-logic. They must be of logical complexity Π^1_2. Friedmann used well foundedness in trees with natural number branching to explain concepts of complexity Π^1_1 [3]. Here we introduce a new notion, strongly well founded.

Definition 9

Let T be a tree made of sequences of natural numbers. Then:

T is a homogeneous if $T = T[\omega]$;

T is strongly well founded if for all α: $T(\alpha)$ is well founded.

We then get:

Theorem 3 (β-Completeness)

For a formula F the following is equivalent:

F is valid in β-logic;

TREE(F)$[\omega]$ is strongly well founded.

How can we use the abstract idea "strongly well founded"? It is used in much the same way as we use other abstract ideas. There are many uses of "finiteness" even if it is not a decidable concept. We know already one interesting strongly well founded tree. Above we introduced the tree connected with "transfinite induction". It consisted of all strictly descending sequences of ordinals (less than some α). This is obviously strongly well founded.

There is by now a number of results connected with these concepts. There is, for example, a concept of composition between strongly well founded trees. And there is a powerful recursion principle. But for these we must refer to the literature [5,9].

6. BEYOND β-LOGIC

The argument behind the proof of the β-completeness theorem is quite general. In fact we only needed the following fact about ordinals:

A Subset of Ordinals is Isomorphic to an Ordinal.

So if we now start with some class C of structures which we want to be absolute in our logic. We assume that we have a particular sort which are always interpreted as one of the structures. There is one basic assumption:

The Class C of Structures is Closed
Undertaking Substructures.

With this assumption the argument above can be carried through, and we get a completeness theorem for C-logic. The concepts "homogeneous" and "strongly well founded" carry nicely over into this new framework. This is done in [7,10]. There it was used to get what is called Π^1_n-completeness but that is just one particular use of this idea.

This paper is called "reasoning in trees". We have shown that trees come up in a number of arguments. The abstract ideas used are connected with properties of trees. But there is more to it than that. We have shown that a number of complicated logical situations can be analyzed into two parts:

the construction of a tree T in an elementary way;

T has some abstract property.

There are a number of candidates for such abstract properties:

T is finite;

T is well founded;

T is strongly well founded.

and the trees may have different types of branching etc.

Gödel emphasized in his work that one should not be afraid of abstract properties. For one thing they had to be used in any case. But more important, they can help us in understanding new logical situations.

REFERENCES

1. Archimedes. "The Sandreckoner".
2. W. F. Clocksin and C. S. Mellish, "Programming in Prolog", Springer Verlag (1981).
3. H. Friedmann, S. Simpson and S. Smith, Countable algebras and set existence axioms, Annals of Pure & App. Logic (1983).
4. G. Gentzen, Untersuchungen über das logische Schliessen, Mathematische Zeitschrift, 39:176-210, 405-431 (1935).
5. J-Y. Girard, Π^1_2-logic, Part I: Dilators, Annals of Math. Logic, vol. 21:75-219 (1981).
6. J-Y. Girard, The Ω-Rule. Proceedings International Congress of Mathematicians, Warszawa (1983).

7. J-Y. Girard and J. P. Ressayre, Elements de logique Π^1_n. Proceedings of the AMS Symposium on Recursion Theory, Cornell (1982).
8. K. Gödel, Ueber eine bisher noch nicht benützte Erweiterung des finiten Standpunktes, Dialectica, 12:280-287 (1958).
9. H. R. Jervell, Introducing Π^1_2-logic. Proceedings from a Symposium in Oslo. To appear in Springer Lecture Notes.
10. H. R. Jervell, Π^1_n-completeness. Proceedings from a Symposium in Oslo. To appear in Springer Lecture Notes.
11. A. Mostowski, Formal systems of analysis based on an infinistic rule, in: "Infinitistic Methods", Warszawa (1960).
12. D. Prawitz, Mekanisk bevisföring i predikatkalkylen, Uppsats för seminariet i teoretisk filosofi (mimeographed), Stockholm (1957).
13. J. A. Robinson, Theorem-Proving on the Computer, J. Assoc. for Computing Machinery, Vol. 10, No. 2, April (1963).
14. A. M. Turing, On Computable Numbers, with an Application to the Entscheidungsproblem. Proceedings of the London Mathematical Society, Ser. 2, Vol. 42:230-265 (1936-1937); Vol. 43:544-546 (1937).
15. H. Wang, Proving theorems by pattern recognition: I, Communications of the Assoc. Computing Machinery, Vol. 3, No. 4, April (1960). ("H. Wang" should be "W. Hao".)

NON-DETERMINISTIC PROGRAM SCHEMATA AND THEIR RELATION TO DYNAMIC LOGIC

V. A. Nepomniaschy and N. V. Shilov

Computing Center
Siberian Division of USSR Academy of Sciences
Novosibirsk 630090, USSR

Program schemata [1,2] theory is a branch of theoretical computer science which deals with properties of programs with non-interpreted functional and predicate symbols. Because of decidability of main properties (equivalence, halting and so on), a special place in this theory belongs to so-called Janov schemes [1], e.g., a complete system of equivalent transformations for Janov schemes in Algol-like syntax was developed by A. P. Ershov [2].

Some authors investigated non-deterministic analogs of deterministic program schemes; so-called R-schemes [3], so-called non-deterministic Janov schemes [4] and so on.

But those works used no Algol-like syntax, which is accepted, and the results have not such a complete form as a result in the theory of deterministic schemes.

At the same time, non-deterministic schemes provide the foundation for Dynamic Logic (DL) [5] and its variants [6,7]. Many recent papers have been devoted to the complexity of the decidability of those variants. For example, exponential up and lower estimations of complexity of Propositional DL (PDL) [7] and Deterministic PDL (DPDL) [6] are well-known. PDL with the predicate of looping (PDL^+) is decidable for the time exp(N) [8].

In this work non-deterministic Janov schemes (njas) are defined in Algol-like syntax; the complete system of equivalent transformations for njas is constructed and decidability for the time $O(N^N)$ of the so-called relative halting (generalization of halting) is proved.

Then we propose an algorithm of PDL and PDL^+ "embedding" into a class of non-deterministic Janov schemes. This embedding transforms valid formulas of PDL and PDL^+ into valid formulas of DPDL and $DPDL^+$ respectively, which, in turn, are "translated" into relatively halting schemes. Thus it follows that PDL and DPDL are decidable for the time $O(N^N)$ but PDL^+ and $DPDL^+$ for the time exp(N log N).

Some of these results have been published [9].

1. NON-DETERMINISTIC JANOV SCHEMES

Let begin, end, crash, loop, ... be labels; p, q, r, ... predicate symbols; f, g, h, ... functional symbols. Let L, L^+ and L^- be finite sets of labels. If f is a functional symbol then the expression "f goto L" is called "assignment operator"; if p is a predicate symbol then "if p then goto L^+ else goto L^-" is called a "conditional operator". If op is an operator and ℓ is a label which differs from end, crash and loop, then "ℓ : op" is called an "operator with mark ℓ"; ℓ is said to be the marker (label) of the op operator. Label end, crash and loop are marking empty expression.

A non-deterministic Janov scheme (njas) is a finite set of operators with different labels.

Interpretation I consists of non-empty domain (set of states) D_I and of the interpretation $(\ )_I$ of predicate and functional symbols by the subsets of D_I and operations over D_I respectively.

Let I be an interpretation, $s \in D_I$, S be a njas and ℓ a label. The tree of computations T(S, ℓ; I, s) is defined inductively.

If ℓ does not mark any operator in S then T(S, ℓ, I, s) is indefinite.

If ℓ = loop then the root of T(S, ℓ, I, s) is marked by the pair (ℓ, s) and its successor is the root of a similar tree T(S, ℓ, I, s).

If ℓ = end or crash then tree T(S, ℓ, I, s) consists of the only node (ℓ, s).

If ℓ is the label of assignment "f goto $\ell_1,\dots,\ell_n$" then the root of T(S, ℓ, I, s) is the node (ℓ, s); its successors are roots of T(S, ℓ_1, I, s'),...,T(S, ℓ_n, I, s') where $s' = (f)_I(s)$.

If ℓ is the label of conditional operator "if p then goto L^+ else goto L^-" then let $\ell_1,\dots,\ell_n$ be equal to L^+ if s $(p)_I$ and to L^- in the other case. Then the root of T(S, ℓ, I, s) is the pair (ℓ, s); its successors are the roots of the trees T(S, ℓ_1, I, s),...,T(S, ℓ_n, I, s). The tree T(S, begin, I, s) is denoted by T(S, I, s). Let us define val(S, I, s) as $\{s' \mid$ (end, s') is a leaf of T(S, I, s)$\}$.

We shall write $s \models_I$ loop(S) (S is njas) if there exists an infinite way through T(S, I, s). Finally, we define val^+(S, I, s) as val(S, I, s) $\cup$ {loop}, if $s \models_I$ loop(S) and as val(S, I, s) in the other case.

Two njas S and S' are said to be equivalent (denotation: $S \simeq S'$), if for any interpretation I and arbitrary state s $val^+(S, I, s) = val^+(S', I, s)$. For example, in [10] the so-called guarded commands "if $p_1 \to S_1 \| \dots \| p_n \to S_n$ if" and "do $p_1 \to S_1 \| \dots \| p_n \to S_n$ od" are defined. Without precise definition of their semantics we accept that the first construction is equivalent to the following njas (in slightly abuse syntax): {(begin: goto $\ell_1,\dots,\ell_n$); (ℓ_1: if p_1 then S_1 else goto begin), ...,(ℓ_n: if p_n then S_n else goto begin)}. The second command can be written in terms of njas too.

Njas S halts in the class of interpretations K if for each I from K and arbitrary s from D_I, T(S, I, s) has a finite way from root to leaf.

For example, let k be an arbitrary non-negative interger and R be a set of predicate symbols. We denote by K(R, k) the class of interpretations such that $I \in K(R, k)$ if for an arbitrary s from D_I and r from R, any term t which consists of k functional symbols, $((t)_I(s))$, does not belong to $(r)_I$.

The scheme halts with respect to R ("is relatively halting", when R is meant), if the scheme is halting in the class $K(R) = \bigcup_k K(R, k)$.

In the sequel we shall use slightly abuse syntax similarly to the above example with guarded commands. The following constructions will be used too. Let S and S' be the schemes. Then (S; S') is constructed from S and S' by the replacement of the label end in S by a new label ℓ and the label begin in S', by ℓ; $(S \cup S')$ is constructed from S and S' by the replacement of the label begin in S by a new label ℓ, begin in S' by a new label ℓ' and the addition of the new operator (begin: goto ℓ, ℓ'). The scheme (S^*) is constructed from S by the replacement of the label begin in S by a new label ℓ, and operator end by the (begin: goto end, ℓ).

2. EQUIVALENCE OF NON-DETERMINISTIC JANOV SCHEMES

Let F be a scheme with the only so-called "entrance label" ℓ and so-called "exit labels" L. Such schemes (with entrance and exit labels) will be called "fragments". Let F' be another fragment with the same entrance label ℓ and the same exit labels L. F' is said to be equivalent to F $(F \sim F')$ if for any interpretation I and an arbitrary s $\{(\ell', s') \mid (\ell', s')$ is a leaf of $T(F', \ell, I, s)\} = \{(\ell'', s'') \mid (\ell'', s'')$ is a leaf of $T(F, \ell, I, s)\}$ and $s \models_I loop(S)$ if $s \models_I loop(S')$. Here S and S' are constructed from F and F' by the replacement of ℓ by begin.

Proposition 1

Let njas S' be constructed from the scheme S by the replacement of fragment F in S by the equivalent fragment F'. Then S' is equivalent to S.

Paper [1] contains complete (in the obvious sense) systems A1-A8 of transformations for the so-called simple deterministic Janov schemes. Let us augment these systems by new rules A9-A12.

Rule A9 is given in Figure 1 in graphic form. Here ℓ is an entrance label, $L_1 \cup \dots \cup L_5$ is a set of exit labels, α, β, γ are logical values, p and q are predicate symbols, $\overset{\alpha}{\to}$ is a notation for one arc with logical label α, $\overset{\beta}{\to}$ is a notation for a finite set of arcs which are marked by the logic value β.

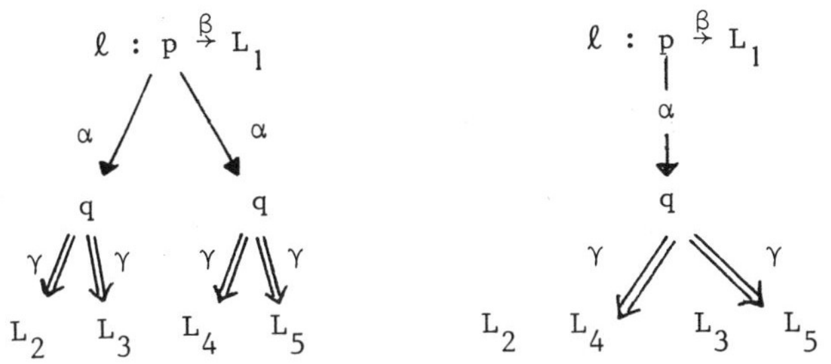

Figure 1. Rule A9

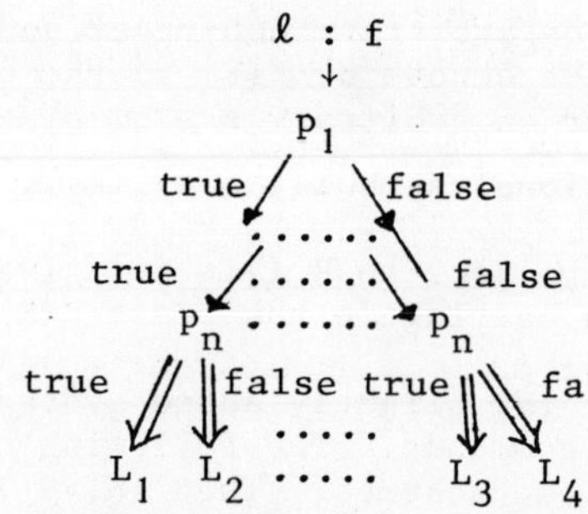

Figure 2. Canonical Fragment

Let us denote by Ja(n) the class of non-deterministic Janov schemes which use predicate symbols $p_1,\ldots,p_n$ only. Canonical fragment of class Ja(n) is displayed in Figure 2. Scheme S (from class Ja(n)) is said to be canonical if the following three conditions hold: (1) S is constructed from canonical fragments of class Ja(n) and labels end, crash, loop; (2) each operator is reachable from operator begin; and (3) operator end is reachable from each operator which differs from crash.

Proposition 2

Let S be Janov scheme of class Ja(n) with M assignments. There exists canonical equivalent scheme C(S) which can be constructed for the time $O(M(2n)^n + M^M)$ and which halts with respect to R if S halts with respect to R.

This proposition is proved similarly to deterministic case [1,2]. We shall omit such standard proofs in the sequel.

The definition of $val^+(S, I, s)$ does not use leaves of T(S, I, s) which are marked by pairs of the form (crash,...). Hence, the following rule A10 preserves equivalence: for each set of labels L, expression "goto L" in scheme S can be replaced by the expression "goto L $\cup$ {crash}" or by expression "goto L \ {crash}".

Scheme S is said to be simple if each functional symbol from S participates in one assignment only.

Proposition 3

System A1-A10 is complete in the class of simple schemes.

Interpretation I, whose domain is the set of all terms such that for any term t and any functional symbol f $(f)_I(t) = f(t)$, is called free interpretation. Let λ denote the empty term.

Proposition 4

(1) Two schemes S and S' are equivalent if for each free interpretation I $val^+(S, I, \lambda) = val^+(S', I, \lambda)$.

(2) Let R be a set of predicate symbols, S be the scheme. S is total with respect to R if for arbitrary free interpretation I from K(R), tree T(S, I, λ) has finite way.

Let S be a canonical scheme from Ja(n) with M assignments; labels ℓ and ℓ' are marking different assignments with one and the same functional symbol; let h be equal to $(M \cdot n + 1)$. Let I be free interpretation. Let us denote by $v(S, \ell, I)$ set $\{t \mid$ there exists ℓ'' (which differs from crash) such that (ℓ'', t) is node of $T(S, \ell, I, \lambda)$ at depth $h\}$. $v(S', \ell', I)$ is defined similarly. If for each free I, $v(S, \ell, I) = v(S, \ell', I)$ then we write $\ell \sim \ell'$ (they are said to be equivalent).

Rule A11 If $\ell \sim \ell'$ then S is equivalent to S' which is constructed from S by the replacement of ℓ by ℓ' everywhere.

In automata theory terms A11 "sticks" together equivalent states. Correctness of this rule follows from the sequel. For any free interpretation I, arbitrary non-negative k and term t the following holds: there exists some ℓ'' (which differs from crash) such that (ℓ'', t) is a node in $T(S, \ell, I, \lambda)$ at depth k if there exists some ℓ''' (which differs from crash) such that (ℓ''', t) is marking node in $T(S, \ell', I, \ell)$ at depth k.

Let S be a canonical scheme from Ja(n), L_1 a set of labels of assignments from S, L_2 a set of labels of assignments with functional symbol f. Let V be the set of all mappings from $\{p_1,\ldots,p_n\}$ into {true, false}, and v a member of V. We shall write $L_1 < f, v > L_2$ in S if $L_2 = \{\ell_2 \mid$ for some ℓ_1 from L_1 there exists canonical fragment F with root ℓ_1 such that (ℓ_2, λ) is leaf of $T(F, \ell_1, v, \lambda)\}$.

Rule A12 Let S be a canonical scheme from class Ja(n). If S' is constructed from canonical fragments of class Ja(n), labels of S' are sets of labels from S, and for any sets of labels from S - L_1 and L_2, arbitrary f and v $L_1 < f, v > L_2$ in S if $\{L_1\} < f, v > \{L_2\}$ in S', then S is equivalent to S'.

Rule A12 corresponds to determinization of a finite automaton. Its correctness follows from assertion: for arbitrary free interpretation I, any term t and any non-negative k $(\{\ell \mid (\ell, t)$ is marking node in $T(S, I, \lambda)$ at depth $k\}, t)$ is marking node in $T(S', I, \lambda)$ at depth k.

Theorem 1

System A1-A12 is complete in the class of non-deterministic Janov schemes.

3. DECIDABILITY AND COMPLEXITY OF HALTING PROBLEM

Let S be a canonical scheme from Ja(n), V be the set of all mappings from $\{p_1,\ldots,p_n\}$ into {true, false}, p(S) be the set of all subsets of set of labels from S, L, L', L", $L_1,\ldots,L_k$ be elements of p(S). Let us denote $(p(S) \times V)$ by D.

Let I be a free interpreation of S, t be a term, and v be an element of V. I is consistent with v on t (denotation: $T(t) = v$) if for any p, $v(p) =$ true if $t \in (p)_I$. Sequence $(L_1, v_1),\ldots,(L_k, v_k)$ of D elements is said to be consistent with I on t if there exists sequence of terms that $t_1 = t,\ldots,t_k$ and sequence of functional symbols $f_1,\ldots,f_k$ such that for each i $(1 \leqslant i \leqslant k)$: (1) $t_i = f_t(t_{i-1})$; (2) $I(t_{i-1}) = v_{i-1}$ and (3) $L_{i-1} < f_i, v_i > L_i$.

Let D_0 be a subset of D, $\bar{D}_1$ a subset of D_0. Now we are going to define the pair of recursive algorithms $U(D_0, \bar{D}_1)$ and $\bar{U}(D_0, \bar{D}_1)$ (U and $\bar{U}$ when D_0 and $\bar{D}_1$ are meant). U and $\bar{U}$ construct a pair of the subsets of D_0 with the same names U and $\bar{U}$:

Step 1. $D_1 := D_0 \setminus \bar{D}_1$; $i := 1$;

Step 2. $D_{i+1} := D_i \setminus \{(L, v) \mid$ for any v' there exists L' such that for some f $L < f, v > L'$ and $(L', v') \in \bar{D}_i\}$;

Step 3. $\bar{D}_{i+1} := \bar{D}_i \cup (D_{i+1} \setminus D_i)$;

Step 4. If $D_i \neq D_{i+1}$ then $i := i + 1$and goes to step 3, else $U := = D_i$, $\bar{U} := \bar{D}_i$ and for each $j \geqslant i$, $D_j := D_i$, $\bar{D}_j := \bar{D}_i$.

For details see [9].

Proposition 5

Let S be canonical njas, $\bar{D}_1 \subseteq D_0 \subseteq D$. Then for all j the following is equivalent:

(1) $(L_1, v_1) \in \bar{D}_j$;

(2) For any free interpretation I which is consistent with v_1 on λ there exists sequence $(L_1, v_1), \ldots, (L_j, v_j)$ which is consistent with I on λ, finishes on element of $\bar{D}_1$.

Proof Induction on j. Basis follows from definitions. Inductive step, we have: $(L_1, v_1) \in \bar{D}_{j+1}$ if for any v_2 there exist L_2 and f_2 such that $(L_2, v_2) \in \bar{D}_j$ and $L_1 < f_2, v_1 > L_2$. It is equivalent to the following: for arbitrary tree J, which is consistent with v_2 on λ, there exists the sequence $(L_2, v_2), \ldots, (L_{j+1}, v_{j+1})$, which is consistent with J on λ, such that $(L_{j+1}, v_{j+1}) \in \bar{D}_1$ and $L_1 < f_2, v_1 > L_2$.

Let us define new interpretation I in this manner: $I(\lambda) = v$ and for arbitrary term t, $I(t(f)) = J(t)$. Hence, the last assertion from the above is equivalent to the following: for arbitrary free interpretation I, which is consistent with v_1 on λ, there exists sequence $(L_1, v_1), \ldots, \times$ $\times (L_{j+1}, v_{j+1})$ which is consistent with I on λ such that $(L_{j+1}, v_{j+1}) \in \bar{D}_1$. The proof is over.

Proposition 6

Let S be scheme from Ja(n) with M assignments. Then for arbitrary D_0 and $\bar{D}_1$, U and $\bar{U}$ can be constructed along the time $O(4^{M+n})$.

Proof D consists of 2^{M+n} elements; hence, steps 2 and 3 are repeated not more than 2^{M+n} times. Complexity of each repetition of steps 2 and 3 is $O(2^{M+n})$. Hence, the complexity of algorithm for U and $\bar{U}$ is $O(4^{M+n})$.

Theorem 2

The halting problem (in the class of all interpretations) for the non-deterministic Janov scheme with M assignments from class Ja(n) is decidable for the time $O((2n)^n M^M)$.

<u>Proof</u> As it follows from Proposition 2, arbitrary scheme S halts if canonical scheme C(S) halts. As it follows from Proposition 5, if $D_0 = D$ and $\bar{D}_1 = \{(\underline{end}, v), (\underline{crash}, v) | v \in V\}$ the following is equivalent:

(1) For some v, ($\{\underline{begin}\}$, v) $\in$ U;

(2) For some v, some interpretation I such that $I(\lambda) = v$ there is no sequence of D_0 elements which is consistent with I on λ and which is finishing in $\bar{D}_1$.

If S contains M assignments and belongs to Ja(n) then scheme C(S) is constructed along the time $O((2n)^n + M^M)$, and it contains not more than M assignments and belongs to Ja(n). We have: $O((2n)^n + M^M) + O(4^{M+n}) \leqslant$ $\leqslant O((2n)^n M^M)$.

Let R be an arbitrary finite set of predicate symbols.

<u>Proposition 7</u>

Canonical scheme from Ja(n) with M assignments halts with respect to R if it halts in class $K(R, 2^M)$.

<u>Proof</u> Let us suppose, on the contrary, that S is total in $K(R, 2^M)$ but for some free J from K(R), T(S, J, λ) has not finite way. For any such J tree T(S, J, λ) can be divided into two parts: $T^-(J) = \{(\ell, t) | (\ell, t)$ is node from T(S, J, λ) such that for all r from R, $t \notin (r)_J\}$ and $T^+(J) =$ $= T(S, J, \lambda) \setminus T^-(J)$.

Depth of $T^+(J)$ is always dividable by (n + 1), because S is canonical scheme from Ja(n). If depth of $T^+(J)$ is k(n + 1) then it is possible to consider J as interpretation from K(R, k).

Let us take the minimal k such that for some J from K(R, k), tree T(S, J, λ) has not finite way. As it follows from our assumption, there exists such k and it is greater than 2^M. As it follows from the choice of k, there exists J from K(R, k) such that T(S, J, λ) has not finite way and the depth of $T^+(J)$ is k(n + 1). Hence, there exists h and h' such that $h < h' \leqslant 2^M$ and L(h) (the set of all labels of assignments in $T^+(J)$ at depth h(n + 1)) is equal to L(h') (the set of all labels of all assignments in $T^+(J)$ at depth h'(n + 1)). For each functional symbol f let us choose term t_f such that for some ℓ from L, (ℓ, t_f) is a node in $T^+(J)$ at depth h'(n + 1). Now we are ready to define the new free interpretation I as follows: for any term t whose length is less than h, I(t) = J(t); for each term t of length h such that for some ℓ (ℓ, t) is a node in $T^+(J)$ at depth h(n + 1) and ℓ is the labels of assignment with symbol f for arbitrary term t', $I(t'(t)) = J(t'(t_f))$; for all other terms t and the symbols of the predicate p, $t \notin (p)_I$.

It is easy to check that $T^+(I) \subseteq T^+(J)$, $T^-(J) \subseteq T^-(I)$ but $I \in K(R, k - 1)$. There is contradiction with the choice of k and J. Hence, S halts with respect to R.

Let V(R) be the set {v|v ∈ v but for all r from R v(r) = false}, D_0 be (p(S) × V(R)) and $\bar{D}_1$ be {(end, v), (crash, v)|v ∈ V(R)}. Let U be equal to $U(D_0, \bar{D}_1)$, $\bar{U}$ equal to $\bar{U}(D_0, \bar{D}_1)$.

Let us define $\bar{w}_0$ as $\bar{U}$, $\bar{w}_1$ as {(L, v)| for all v' from V(R) there exists L' such that for some f, L<f, v>L' and (L', v') ∈ $\bar{w}_0$}. Similarly, for all i 1 we define $\bar{w}_{i+1}$ as {(L, v)| for all v' from V there exists L' such that for some f, L<f, v>L' and (L', v') ∈ $\bar{w}_i$}.

Theorem 3

Let S be a scheme from Ja(n) with M assignments, R be a set of predicate symbols. Then the halting problem with respect to R is decidable for time $O((2n)^{n_M M})$.

Proof As it follows from Propositions 2 and 7, scheme S with M assignments from Ja(n) is total with respect to R, if C(S) halts in class K(R, 2^M). Hence, S halts with respect to R if for all i, $0 \leqslant i \leqslant 2^M$, and ({begin}, v) ∈ w_i. Other details are similar to Theorem 2.

4. PROPOSITIONAL DYNAMIC LOGIC WITH LOOPING IS DECIDABLE

Let A be a propositional formula. Let us denote by (A?) the following scheme: {(begin: if A then goto end else goto loop), (end:), (loop:)}.

Njas which is written in syntax ;, ∪, * and ? will be called "program" and will be denoted by Greek letters. The program of the form (A?) is called "test".

If α is a program, I is an interpretation, s and s' are states and (end, s') is a leaf of T(α, I, s), then we shall write $s<\alpha>_I s'$.

The set of formulas of Deterministic Propositional Dynamic Logic with simple tests (DPDL(0)) [6] can be defined as follows. If p is a predicate symbol then (p) is formula; if A and B are formulas then disjunction (A V B) and negation (¬A) are formulas. If α is a program and A is a formula, then (<α>A) is formula too.

DPDL(0) with looping (DPDL$^+$(0)) is the enlargement of DPDL(0) by permission of formulas of the form loop(α).

Let I be an interpretation, s and s' be states. Let us define the relation $\models_I$ by induction on the structure of formulas: $s \models_I (p)$ if $s \in (p)_I$; $s \models_I (A \wedge B)$ if $s \models_I A$ and $s \models_I B$; $s \models_I (<\alpha>A)$ if for some s such that $s<\alpha>_I s'$, $s' \models_I A$; $s \models_I (\models A)$ if it is not the case $s \models_I A$. Formula A is said to be valid in the state s under the interpretation I if $s \models_I A$.

Formula A is said to be valid in the interpretation I ($\models_I A$) if for each s from D_I $s \models_I A$. Formula A is said to be valid ($\models A$) if for arbitrary I, $\models_I A$.

We accept the following abbreviations: (A ∧ B) (conjunction) for ¬((¬A) ∨ (¬B)), ([α]A) for ¬(<α>(¬A)), (end(α)) for ¬(loop(α)). Notation DPDL$^{(+)}$(0) will refer to DPDL(0) and DPDL$^+$(0) respectively.

Formula is said to be positive if the negation "$\neg$" in this formula s applying to predicate symbols only.

:oposition 8

Each formula of $DPDL^{(+)}(0)$ is equivalent to some positive formula of PDL$^{(+)}(0)$, which can be constructed for the linear time. Hence, without oss of generality, until the end of this article we can deal with positive ormulas of $DPDL^{(+)}(0)$.

Now we are going to describe the recursive procedure S from [9]; S translates" positive formulas of DPDL(0) and programs with positive tests .nto the class of njas. We use a slightly abuse notation for schemes:

1. Let A be a propositional formula. Then S(A) = {(begin: if A then goto end else goto loop), (loop:), (end:)}.
2. $S(A \vee B) = (S(A) \cup S(B))$.
3. $S(A \wedge B)$ = {begin: if p then S(A) else S(B)}. Here p is a new symbol.
4. $S(\langle\alpha\rangle A) = (\alpha; S(A))$.
5. $S([\alpha]A) = (S(\alpha); S(A))$.
6. $S(\alpha; \beta) = (S(\alpha); S(\beta))$.
7. $S(\alpha \cup \beta)$ = {begin: if p then $S(\alpha)$ else $S(\beta)$}. Here p is a new symbol.
8. $S(\alpha^*) = (((r?); S(\alpha))^*; (\neg r?))$. Here r is a new symbol.
9. S(A?) = {(begin: if A then goto end else goto crash); (end:),(crash:)}.
10. S(f) = f.

Let us denote the set of all new symbols r by R(A).

The following result is proved in [9].

Proposition 9

Let A be any positive formula, α be a program with positive test from DPDL(0), I be an interpretation, s and s' be states. Then

(1) $s \models_I A$ if for all J from K(R(A)), which coincides with I on the symbol from A, T(S(A), J, s) has finite way.

(2) $s \langle\alpha\rangle_I s'$ if for some J from K(R(A)), which coincides with I on symbols from α, $s \langle S(\alpha)\rangle_J s'$ and $S(\alpha)$ halts in the class of all such J.

Let us complete the algorithm S by the new rules for loop(α) and end(α) (we use slightly abuse but clear notations):

11. S (loop(α)) is constructed from α by the replacement of each assignment f by the program (if r then f else goto crash), where r is a new symbol.
12. S (end(α)) is constructed from α by the replacement of each expression of the form "goto $\ell_1, \ldots, \ell_n$" by the program (if p_1 then goto ℓ_1 else (if p_2 then goto ℓ_2 else (...(if p_{n-1} then goto ℓ_{n-1} else goto ℓ_n)....))). Here $p_1, \ldots, p_{n-1}$ are new symbols for each $\ell_1, \ldots, \ell_n$.

We denote by R(A) the set of all new symbols r (as stated above).

Proposition 10

Let α be a program of DPDL(0), I be an interpretation and s be a sta[te]. Then:

(1) $s \models_I$ loop(α) if for an arbitrary J from K({r}) which coincides with I on the symbols from α, tree T(S(loop(α)), J, s) has a finite way.

(2) $s \models_I$ end(α) if for an arbitrary J which coincides with I on the symbols from α, tree T(S(end(α)), J, s) has a finite way.

Proof (1) $s \models_I$ loop(α) if an arbitrary non-negative k tree T(α, I, has way which length is at least k. It is equivalent to the following: fo[r] an arbitrary $k \geqslant 0$ and an arbitrary J from K({r}, k), which coincides with I on common symbols, T(S,(loop(α)), J, s) has a finite way. (2) S (end(α)) can coincide as "determinization" of α with the help of new predicate symbols. Hence, $s \models_I$ end(α) if all ways through T(α, J, s) are finite. It is equivalent to what follows: for an arbitrary H which coincides with I on common symbols, T(S(end(α)), J, s) has the finite way.

Proposition 11

For an arbitrary formula of DPDL(0) ($DPDL^+$(0)) for the linear (square for $DPDL^+$(0)) time, one can construct the scheme and the set of predicate symbols R, such that scheme halts with respect to R if the formula is valid

The proof immediately follows from Propositions 9 and 10.

We denote by $DPDL^{(+)}$ [7] the enlargement of $DPDL^{(+)}$(0) which permits programs of the form (A?) (tests) for arbitrary formulas, not for propositional formulas only. Semantics is defined as follows: $s \langle A? \rangle_I s'$ if $s = s'$ and $s \models_I A$.

Proposition 12

Let A be the formula of $DPDL^{(+)}$. One can construct for square time on linear space formula of $DPDL^{(+)}$(0) which is valid if A is valid.

Proof. Let H be the set of all functional symbols from A. We denote by Y the following program $(\bigcup_{f \in H} f)^*$. Let p be a new symbol, B formula of $DPDL^{(+)}$(0) such that (B?) is a sub-program of A. If A' denotes formula which is constructed from A by the replacement of B in A by p, then ([Y]($p \equiv B$)) $\supset$ A') is valid if A is valid. It is constructed along linear time, on the space $|A|$ + const($|A|$ is length of A) and number of complex tests in it is less than in A.

Repetition of this procedure proves the Proposition.

Theorem 4

For arbitrary formula of $DPDL^{(+)}$ one can construct for the square time and linear (square for $DPDL^+$) space a non-deterministic Janov scheme and set of predicate symbols R such that the scheme halts with respect to R if formula is valid.

Propositional Dynamic Logic (with Looping) PDL (PDL^+) has the same syntax as DPDL ($DPDL^+$ respectively) but the semantics of each assignment is binary relation. For details see [6-9].

Proposition 13

For an arbitrary formula of PDL (PDL^+) it is possible to construct for the linear time formula of DPDL ($DPDL^+$) which is valid if initial formula is valid.

Proof Let us replace all assignments by programs ((f^*); g) where f and g are new symbols for each assignment. It proves the Proposition.

Theorem 5

For an arbitrary formula of PDL (PDL^+) it is possible to construct for the square time and the linear (square for PDL^+) space non-deterministic Janov scheme and R a set of the predicate symbols such that the scheme halts with respect to R if the initial formula is valid.

Theorem 6

(1) PDL, DPDL and DPDL(0) are decidable for the time $O(N^N)$. (2) PDL^+, $DPDL^+$ and $DPDL^+(0)$ are decidable for the time $\exp(N^2 \log N)$.

REFERENCES

1. V. E. Kotov, "Introduction to Program Schemata Theory", Science, Novosibirsk (1978) (in Russian).
2. A. P. Ershov, "Introduction to Theoretical Computer Science", Science, Moscow (1977) (in Russian).
3. R. I. Podlovchenco, R-schemes and their equivalence, in: "Problems of Cybernetics", v. 27, Science, Moscow (1973) (in Russian).
4. J. Engelfriet, "Simple Program Schemes and Formal Languages", Lecture Notes in Comput. Sci., v. 20, Springer-Verlag (1974).
5. D. Harel, "First-order Dynamic Logic", Lecture Notes in Comput. Sci., v. 68, Springer-Verlag (1978).
6. M. Ben-Ari, J. Y. Halpern and A. Pnueli, Deterministic propostional dynamic logic: finite models, complexity and completeness, J. Comput. System. Sci., 25:3 (1982).
7. R. S. Street, Propositional dynamic logic of looping and converse in elementary decidable, information and control, 54:2 (1982).
8. M. Y. Vordi, The taming of converse: reasoning about two-way computation, in: "Logic of Programs", R. Parikh, ed., Lecture Notes in Comput. Sci., 193, Springer-Verlag (1985).
9. V. A. Nepomniaschy and N. V. Shilov, Non-deterministic program schemata and their relation to dynamic logic, in: "Translation and Transformations of Programs", I. V. Pottosin, ed., Computer Center of Siberian Division of USSR Academy of Science, Novosibirsk (1984) (in Russian).
10. E. W. Dÿkstra, "A Discipline of Programming", Prentice-Hall (1976).

THE PARALLEL EVALUATION OF FUNCTIONAL PROGRAMS

Alberto Pettorossi
IASI-CNR
Viale Manzoni 30
00185 Roma (Italy)

Andrzej Skowron
Institute of Mathematics
Warsaw University
PKiN IX p. 907
00-901 Warsaw (Poland)

ABSTRACT

We address the correctness problem of parallel implementation of functional programs. Those functional programs are evaluated by a set of concurrent agents communicating with each other and cooperating together while the computations progress. New communications among agents are introduced to improve the performance, because properties or facts about functions to be computed are exploited. In particular we show that those communications may avoid redundant computations of intermediate results. We provide the logical theories for proving correctness of implementation of functional programs together with facts about those programs.

1. INTRODUCTION

Functional languages have been advocated as a formalism for expressing algorithms which allow to overcome the limitations imposed by the von Neumann computer architecture [1]. Applicative expressions in fact, can be evaluated in a parallel way by a set of concurrent computing agents, because there is a unique value associated with any subexpression, independently from the context in which it occurs (by the referential transparency property). That context independence property allows us to compute the various subexpressions in a parallel way by assigning each of them to an individual computing agent.

In order to fix our ideas and to introduce our Hope-like notations [3], let us consider the following simple program for computing binomial coefficients:

$$\text{BO:}\quad \begin{cases} \text{bin}(n,0)=1 \\ \text{bin}(n,n)=1 \\ \text{bin}(n,m)=0 \quad \underline{\text{if}}\ n<m \\ \text{bin}(n+1,m+1) = \text{bin}(n,m)+\text{bin}(n,m+1) \quad \underline{\text{if}}\ \ n>m \end{cases}$$

The computation of the function bin(n,m) can be informally described as a rewriting of sets of agents. (The formal definition of a generic computation will be given later). For the time being, an <u>agent</u> can be thought of as a pair consisting of a string, i.e., the name of the agent, and an expression, i.e., the expression which the agent has to evaluate.

We may assume that initially there is the agent ε::bin(n,m), which can be rewritten into the set of agents: {ε:: .0+.1, 0::bin(n-1,m-1), 1::bin(n-1,m)} where $n>m\geq 1$. In the rewriting we see that: i) new agents are generated by the recursive calls of the program equations, ii) the left and right sons of the initial agent ε have names 0 and 1 respectively, and iii) .k denotes in any given expression the result of the computation of the agent whose name is k for k=0 or 1.

We assume that the convention for giving names to the agents is the following: the initial agent has name ε, and if the agent x has k+1 sons, i.e., it uses for the rewriting a recursive equation with k+1 recursive calls, the names of its sons are x0, x1,..., xk, respectively, according to the left-to-right order of those recursive calls.

For our binomial function, in the second step of the computation the agents 0 and 1 generated by the agent ε, can be rewritten independently from each other, so that the computation of bin(n,m) can be more precisely viewed as a nondeterministic and parallel rewriting of sets of agents into new sets of agents (see figure 1). The nondeterminism consists in choosing one (or more) agents to be rewritten, and parallelism consists in rewriting all agents which have been chosen, according to the recursive equations of the program.

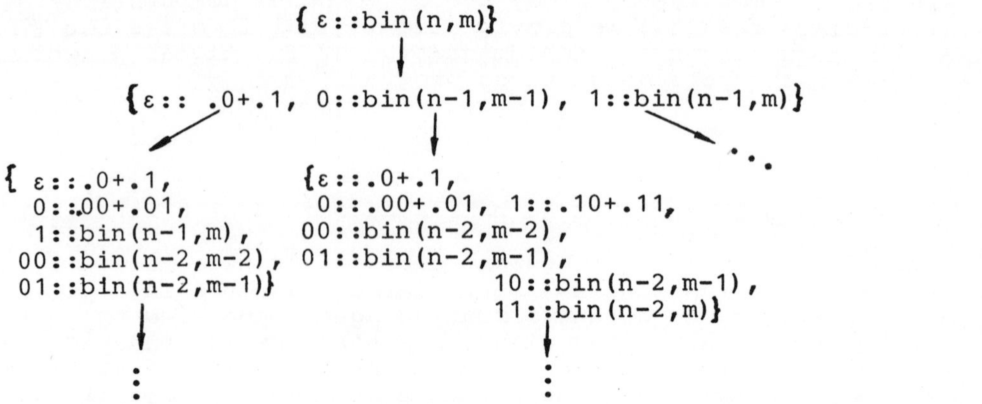

figure 1. Possible rewritings of the parallel computation of bin(n,m) when $n>m\geq 2$.

Notice that, according to the naming convention, each set of agents can be viewed as a tree-structured set, and therefore we may say that the computation progresses as a <u>nondeterministic and parallel rewritings of trees of agents</u>.

We propose a method based on the discovery or knowledge of some "facts" concerning the functions to be evaluated. Those facts will be implemented as suitable communications among computing agents and the occurrence of those communications will realize the expected efficiency improvements.

Suppose, for instance, that we know the following fact about the binomial coefficients function: if $n>m\geq 2$ then the left son of the right son of bin(n,m) is equal to the right son of the left son of bin(n,m). In the language of facts which we will formally introduce later, we write:

$$\mathrm{bin}(n,m)]01 = \mathrm{bin}(n,m)]10$$

(see figure 2).

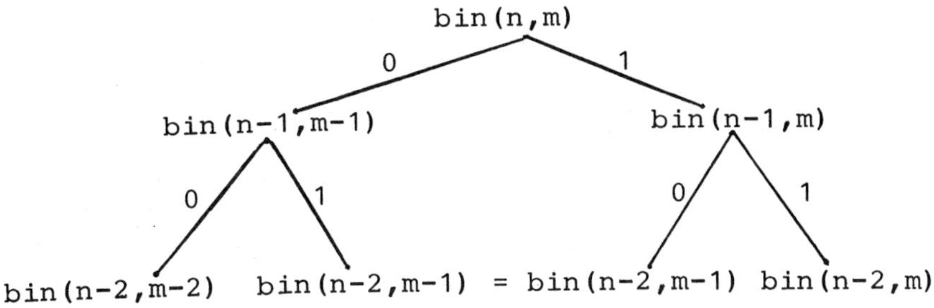

figure 2. A fact for the binomial coefficients function: bin(n,m)]01=bin(n,m)]10 where $n>m\geq 2$.

In the rewriting of the tree-structured set of agents we may then avoid the expansion of the agent 10 and allow the expansion of the agent 01 only. Once the value of bin(n-2,m-1) has been obtained, the agent 01 may send it to the agent 10, and save its computation. We will see in the sequel how that communication takes place and how it can be implemented by associating a local memory with each agent.

Avoiding the expansion of the agent 10 reduces the number of agents we need for performing the required computation of bin(n,m) and in most cases that saving is crucial, because in practice the number of available computing agents cannot be considered as unbounded.

In the Section 2 we will present a method for implementing facts about recursive functions and we will see how they can be translated into communications among agents.

In the subsequent Section we will give a more formal presentation of the ideas introduced in Section 1 and 2. Next we will present the logic [6] in which we can express and prove the properties of parallel programs. Finally we apply that logic to show the correctness of the parallel implementation of functional programs considered together with facts about functions defined by those programs.

2. THE GENERAL SCENARIO

We consider the functional programs written as a set of recursive equations in a language LO (to be defined later), and we assume that the programmer discovers or knows some useful facts about the functions to be computed. Those facts, written in the language of facts LF, are first submitted to a calculus which may or may not accept them. The role of the calculus is twofold: i) it makes sure that the facts are correct with respect to the given programs, and ii) it validates the efficiency improvements they determine, once implemented

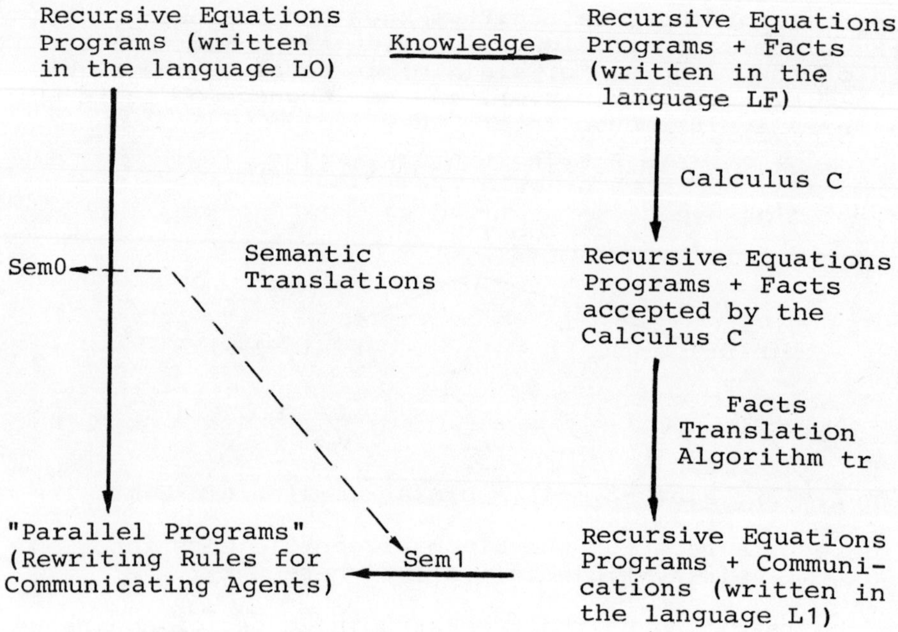

figure 3. The general scenario for improving the parallel execution of functional programs.

as communications among agents. Those two aspects of the calculus are very important, because otherwise we may derive, as we will see, erroneous or inefficient programs.

Facts which are accepted by the calculus are then translated by a translation algorithm, which produces a new set of recursive equations, written in a language L1, where it is possible to denote communications among function calls.

The general scenario of our approach is the one depicted in figure 3.

The scenario we presented can be considered as an extension of the ideas of the transformation system of Burstall and Darlington [2]. However in our approach the knowledge the programmer has about the functions to be computed, is expressed in a language LF of facts (not as "eureka steps" [2]). Moreover, the presence of the calculus which deals with the facts and ensures their correctness and their usefulness for improving efficiency, avoids the problems inherent to the Burstall-Darlington approach, where it may be that during the transformation process correctness is not preserved [5] and efficiency is not improved.

Notice that we explicitly deal with functional programs executed in a parallel way, so that the facts discovered by the programmer may refer to the behaviour of the associated sets of computing agents, but we allow only facts which can be written in a high level language LF. In this language it is possible to describe properties of sequences of expressions generated by the standard rewriting technique, so that the programmer is not involved in all the peculiarities connected with the computing agents behaviour. The system can accept facts if they are "true" facts and "efficiency improving" facts. Accepted facts about functional programs are used for increasing the efficiency of their parallel execution.

The semantics Sem0 and Sem1 are identified as 2 translations. Given a program P in L0 (or L1) Sem0 (or Sem1) produces a corresponding parallel program P'. Indeed, Sem0 (or Sem1) defines a set of agents for evaluating the functions defined in P, and a set of rewriting rules which specify the concurrent behaviour of those agents.

The meaning of P' is given by a transition relation which defines for any given finite set of computing agents, i.e., a configuration, all possible future configurations. That transition relation is able to capture the nondeterministic and parallel execution of the program P' by constructing a tree of configurations as indicated in figure 1.

We will show that the translation Sem0 from a program P to P' (or the translation Sem1•tr of P with some additional facts to P') can be done automatically for some classes of programs and facts.

The system is able to improve efficiency by automatically introducing suitable communications among computing agents. Those communications are derived on the basis of the given facts only.

It will be shown that the introduction of the intermediate language L1 in which functional programs with communications are written, allows us to have a simple calculus for checking the correctness of the communications among computing agents. We present a theorem about correctness of that calculus. The theorem assures that the implementation of given facts is correct if those facts are accepted by the calculus.

The language of parallel programs should be considered as a lower level language, which is understood by the system which realizes the parallel execution of functional programs via a set of computing agents.

3. PARALLEL PROGRAMS

In this section we introduce the notion of parallel programs [6-8].

AExp, MExp, Exp are sets of terms constructed in the standard way from constants, variables and function symbols. (The sets of constants, variables and function symbols are fixed in our considerations.) The elements of the sets AExp, MExp, Exp are called agent-name expressions, message expressions and expressions, respectively. If T is a set of terms then by CT we denote the subset of T s.t. $t \in CT$ iff t is a term without variables.

An <u>agent expression</u> is a triple

$$\langle agn, msg \rangle :: e$$

where $agn \in AExp$, $msg \in MExp$, $e \in Exp$.

A *computing agent* (or simply an *agent*) is an agent expression without variables.

Configurations are finite sets of agents. By CON we denote the set of all configurations.

A *rule* (or *recursive equation*) is an expression of the form:

$$lh \Leftarrow rh \ \underline{if} \ cond$$

where lh,rh are finite sets of agent expressions and cond is a boolean expression.

By r(x1,...,xk) we denote the rule r in which x1,...,xk are the only variable occurrences. If a1,...,ak are constants (of suitable types) then by r(a1,...,ak) (or $\underline{r}$ or $\underline{lh} \Leftarrow \underline{rh}$ $\underline{if}$ $\underline{cond}$) we denote a concrete instance of r which we derive by substituting a1,...,ak for x1,...,xk in r.

A *parallel program* is any non-empty finite set of rules.

Let c,c'∈CON and let $\underline{r}$ be a concrete instance of the rule r and let us assume that:

$c \xrightarrow{\underline{r}} c'$ holds iff $\underline{cond}$ is true and $\underline{lh} \subset c$ and $c' = (c-\underline{lh}) \cup \underline{rh}$.

$\xrightarrow{\underline{r}}$ is called the transition relation of $\underline{r}$.

The transition relation corresponding to a given sequence s=$\underline{r1}$,...,$\underline{rk}$ of instances of rules is defined as the composition of the transition relations:

$$\xrightarrow{\underline{r1}}, \ldots, \xrightarrow{\underline{rk}} \text{ and it is denoted by } \xrightarrow{s}.$$

The transition relation of a program P (denoted by $\xrightarrow{P}$) is defined as follows:

$c \xrightarrow{P} c'$ holds iff there is a non-empty finite sequence s of instances of rules in P (derived by the same substitution) s.t. for an arbitrary permutation s' of s we have:

$$c \xrightarrow{s'} c'.$$

As an example, let us consider the parallel program BO' for computing the binomial coefficients. BO' is a set of rewriting rules which implements BO.

```
       {<x,E>::bin(n+1,m+1)}⇐{<x,E>::.x0+.x1,<x0,E>::bin(n,m),
                                          <x1,E>::bin(n,m+1)}
                                                       if n>m
       {<x,E>::.x0+e, <x0,E>::n}⇐{<x,E>::n+e}
BO' :   {<x,E>::e+.x1, <x1,E>::n}⇐{<x,E>::e+n}
       {<x,E>::bin(n,0)}⇐{<x,E>::1}
       {<x,E>::bin(n,n)}⇐{<x,E>::1}
       {<x,E>::bin(n,m)}⇐{<x,E>::0}  if n<m
       {<x,E>::n+m}⇐{<x,E>::k}  where k=n+m
       {<x,E>::n-m}⇐{<x,E>::k}  where k=n-m
```

where E is the empty message. The messages will play significant role in others examples, here they do not.

We see the above rules in action in the following computation of bin(3,2), where we write x::exp instead of <x,E>::exp and by → we denote the transition relation of BO'.

{ε::bin(3,2)} → {ε::.0+.1,0::bin(2,1),1::bin(2,2)} →
{ε::.0+.1,0::.00+.01,00::bin(1,0),01::bin(1,1),1::1} →
{ε::.0+ 1,0::.00+.01,00::1,01::1}→{ε::.0+1,0::1+1}→
{ε::.0+1,0::2}→{ε::2+1}→{ε::3}.

4. THE TRANSLATION FUNCTIONS Sem0 AND Sem1

This section is a technical section where we formally define the language LO in which preliminary program versions are written, the language LF in which facts as equality of terms are written, and the language L1 in which recursive programs and facts are translated. We also present methods for translating programs in LO (or L1) into parallel programs.

We will define a <calculus,fact-translation> pair for which facts accepted by the calculus can be correctly translated into communications among recursive calls. A correct fact-translation tr is the one which makes the diagram of fig. 3 commute.

We will give an explanation of the various notions by discussing the problem of computing the binomial coefficients.

The language LO in which the functional program for computing binomial coefficients is written is a Hope-like language [3]. An <u>expression</u> e of LO is defined by

e::= n|x|g(e,...)|f(e,...)

where n∈Constants, x∈Variables, g∈Basic-Functions and f∈Recursive-Functions.

A <u>program</u> P in LO is a set of recursive equations each of which is of the form:

f(e,...)=n (base case)

or

f(e,...)=e' with f occuring in e' (recursive case).

For LO expressions we assume a parallel and distributed evaluation using computing agents [7]. The translation Sem0 of functional programs in LO into parallel programs is defined by the following list of rules:

1. <u>Generation of sons</u>

f(e0,...,ek)=g(...,f_0(e,...),...,f_p(e',...),...) produces:
{<x,E>::f(e0,...,ek)} ⇐ {<x,E>::g(...,.x0,...,.xp,...),
<x0,E>::f(e,...),...,<xp,E>::f(e',...)}

2. <u>Base configurations</u>

f(e0,...,ek)=n produces:
{<x,E>::f(e0,...,ek)} ⇐ {<x,E>::n}

3. Values to fathers

{<x,E>::g(...,xj,...),<xj,E>::n} ⇐ {<x,E>::g(...,n,...)}

4. Basic functions evaluation

{<x,E>::g(n1,...)} ⇐ {<x,E>::m}   if m=g(n1,...)

Initial agent. For computing f(n1,...) the initial configuration {<ε,E>::f(n1,...)} is generated.

E is the empty message. It will play a significant role in Sem1, here it does not.

One can check that our parallel program BO' is the result of the above translation applied to BO.

In order to produce an improved version of the program BO we may use our system by supplying it with "facts", i.e., properties of the computation of bin. We will restrict our attention to facts which can be expressed as equalities of agent expressions. They can be discovered by symbolic evaluation.

The syntax of the language LF of facts is as follows:

e::=...(as in LO)| e]s with s∈{0,1,...,k}*
fact::=f(e,...)]s1=f(e,...)]s2.

A fact is checked by a Calculus and, once accepted, a Translation algorithm produces from it an improved version of the given program.

The Calculus for checking facts is given by the rewriting rules, which we will present for any program PO in L1 with one recursive case only. Let PO be:

$$\begin{cases} f(e,\dots)=n1,\dots,f(e',\dots)=nk \text{ (base cases)} \\ f(e,\dots)=e' \text{ (recursive case).} \end{cases}$$

The rules for the Calculus are:

1. e]ε ↦ e
2. n]s ↦ stop if s≠ε
3. x]s ↦ stop if s≠ε
4. g(e0,...,ek)]js ↦ if 0≤j≤k then ej]s else stop
5. f(e0,...ek)]s ↦ if f(e0,...,ek)=e is an instance of the recursive case then e]s else stop.

We write:

e]s = e']s'

iff it is possible to reduce e]s and e']s' to the same expression (or one of them to stop) after applying to them a finite number of times the rules 1-5 and the rules of the Basic Functions algebra.

For instance, in the program BO we have for n>m and m≥2:

bin(n,m)]01 ↦ bin(n-1,m-1)]1 ↦ bin(n-2,m-1)
bin(n,m)]10 ↦ bin(n-1,m)]0 ↦ bin(n-2,m-1), and bin(n,m)]01= bin(n,m)]10.

The program produced by the translation is written in the following language L1. The syntax of L1 is like that of L0, with the following additions:

e::=...|e(s comm l) with $s\in\{0,1,...,k\}^*$, l∈Locations.

The r.h.s. of an equation can be of the form:

e or e decl l.

The parallel programs which we will get after the translation of programs in L1 have a special form of messages.

Messages are either E, i.e. the empty message or em≃L, where:

i) em is the empty elementary message Φ or a constant elementary message n∈Constants,

ii) L is a set of names of son agents which may read or write the associated elementary message. L is represented as the list [s1,...,sn], where each $si\in\{0,1,...,k\}^*$.

For instance, if the left son of an agent has to read/write the message of its father, that message will initially be: Φ≃[0].

Now we are ready to define the translation of the set of programs in L1 into the set of parallel programs.

A program P in L1 is translated as a program in L0 (i.e., rules 2,4 and the initial agent rule for Sem1 are like those for Sem0) with the following additions and changes:

1'. Generation of sons with communications

$f(e0,...,ek)=g(...,f_0(e,...),...,f_j(e',...)(s\ comm\ l),...)\ \underline{decl}\ l$

produces:

{<x,E>::f(e0,...,ek)} ⇐ { <x,Φ≃L>::g(...,.x0,...,.xj,...),
<x0,E>::f(e,...),...,<xj,E>::f(e',...),...}

where L={js|s comm l occurring in the j-th recursive call}.

3. Values to fathers

{<x,m>::g(...,.xj,...), <xj,m'>::n} ⇐
{<x,m>::g(...,n,...),<xj,m'>::n}.

Notice that contrary to the previous case, after reporting to its father a son agent remains present. It may need to communicate its value to a message location.

5. Writing a message location

{<x,Φ≃L>::e,<xs,m>::n} ⇐ { <x,n≃L-{s}>::e,<xs,m>::n} if s∈L

Notice that after writing a location a son agent remains present because afterwards it may have to return its value to its father.

6. Reading a message location ("reading communications")

$$\{\langle x,n\simeq L\rangle ::e,\langle xs,m\rangle ::e1\} \Leftarrow \{\langle x,n\simeq L-\{s\}\rangle ::e,\langle xs,m\rangle ::n\} \text{ \underline{if} } s\in L.$$

The Calculus and the Translation algorithm enjoy the necessary properties with respect to correctness and efficiency of the derived programs, as the results presented latter show. First we will formulate the formal language in which we can express and prove properties of parallel programs [7]. That language is used to prove the correctness of the derived programs.

5. THE KERNEL LANGUAGE

In this Section we present a logical framework for defining for any given set $\mathbb{P}$ of programs a modal-temporal theory in which we can prove properties of behaviour of programs in $\mathbb{P}$.

By $\mathbb{E}$ we denote the set of all rules occuring in programs from $\mathbb{P}$ together with two additional rules

$$\Leftarrow \{\langle x,m\rangle ::e\}, \quad \{\langle x,m\rangle ::e\} \Leftarrow.$$

By $\underline{\mathbb{E}}$ we denote the set of all concrete instances of rules from $\mathbb{E}$.

The alphabet of our kernel language consists of:

1. propositional variables $A(x,m,e)$ for $x\in CAExp$, $m\in CMExp$ and $e\in CExp$;
2. propositional connectives: implication ($\Rightarrow$), negation ($\sim$) and conjunction for countable sets of formulas ($\bigwedge$);
3. temporal operators: "always in the future" (G) and "for a cut in the future" (F);
4. modal operators: $\Box_{\underline{r}}$ for $\underline{r}\in \underline{\mathbb{E}}$;
5. auxiliary symbols: (,).

The set $\pi(\mathbb{P})$ of formulas over a set of programs $\mathbb{P}$ is defined as the least set such that:

a. all variables are in $\pi(\mathbb{P})$;
b. if $\alpha,\beta\in\pi(\mathbb{P})$, $P\in\mathbb{P}$ and $r\in\mathbb{E}$ then $\alpha\Rightarrow\beta$, $\sim\alpha$, $\Box_r\alpha$, $F(P)\alpha$, $G(P)\alpha$ are in $\pi(\mathbb{P})$;
c. if Φ is a countable subset of $\pi(\mathbb{P})$ then $\bigwedge\Phi$ is in $\pi(\mathbb{P})$.

Usual abbreviations:

$\alpha\&\beta = \sim(\alpha\Rightarrow(\sim\beta))$; $\alpha\vee\beta = ((\sim\alpha)\Rightarrow\beta)$;
$g(P)\alpha = \sim G(P)\sim\alpha$; $f(P)\alpha = \sim F(P)\sim\alpha$;
$\sim\Phi = \{\sim\alpha \mid \alpha\in\Phi\}$; $\bigvee\Phi = \sim\bigwedge\sim\Phi$; $\langle\rangle_{\underline{r}}\alpha = \sim\Box_{\underline{r}}\sim\alpha$.

If $\underline{w} = \underline{r1}\ldots\underline{rk}$ then $\Box_{\underline{w}}$ denotes $\Box_{\underline{r1}}\ldots\Box_{\underline{rk}}$.

Let us consider a non-empty set U, $u\in U$ and $R\subset U\times U$. We denote by:

$U^*(U^+)$ the set of all (non-empty) sequences of elements of U,

U^∞ the set of all infinite sequences of elements of U,

R^* the transitive and reflexive closure of R,

R^+ the transitive closure of R,

$$\text{MAX-PATH-FROM}(R,u) = \{(u,u_1,\ldots,u_k,\ldots)\in U^\infty \mid uRu_1\ldots u_{k-1}Ru_k\ldots\} \cup \{(u,u_1,\ldots,u_k)\in U^+ \mid uRu_1\ldots \ldots u_{k-1}Ru_k \& \sim(\exists v)(u_kRv)\},$$

$$\text{FUTURE}(R,u) = \{v\in U \mid uR^+v\}.$$

If $X \subset U^+ \cup U^\infty$ then $E(X) = \{u \mid u$ occurs in a sequence $x\in X\}$.

<u>Models for $\pi(\mathbf{P})$</u>

A Kripke-model for $\pi(\mathbf{P})$ is the following triple:

$$M = (U, \{R_{\underline{r}}\}_{\underline{r}\in\underline{\mathbf{E}}}, Val)$$

where U is an arbitrary non-empty set (called the set of states of M); $R_{\underline{r}}$ is a partial function from U into U for every $\underline{r}\in\underline{\mathbf{E}}$; Val is a function from the set of propositional variables V into the power set of U (called the valuation of M).

<u>Satisfiability relation</u> : $\models$

Given a model M, and a state u of the model and a formula α in $\pi(\mathbf{P})$ we define the satisfiability relation $\models$ by structural induction on the formula α:

1. $M,u\models A(x,m,e)$ iff $u\in Val(A(x,m,e))$;
2. $M,u\models \sim\alpha$ iff non$(M,u\models\alpha)$;
3. $M,u\models\alpha\Rightarrow\beta$ iff either non$(M,u\models\alpha)$ or $(M,u\models\beta)$;
4. $M,u\models\bigwedge\Phi$ iff for every $\alpha\in\Phi$ $M,u\models\alpha$;
5. $M,u\models G(P)\alpha$ iff for every $v\in\text{FUTURE}(R_P,u)$ $M,v\models\alpha$

   where R_P is the union of all $R_{\underline{r}}$ for all r in P;
6. $M,u\models F(P)\alpha$ iff for every $p\in\text{MAX-PATH-FROM}(R_P,u)$

   there exists $v\in E(p)$ s.t. $M,v\models\alpha$;
7. $M,u\models \Box_{\underline{r}}\alpha$ iff for every $v\in U$ $uR_{\underline{r}}v$ implies $M,v\models\alpha$.

We say that $\alpha\in\pi(\mathbf{P})$ is true in the model M ($M\models\alpha$) iff for every $u\in U$ $M,u\models\alpha$ and α is true (is a tautology) iff α is true in every model for $\pi(\mathbf{P})$.

A theory is any non-empty subset of $\pi(\mathbf{P})$. If $\Gamma\subset\pi(\mathbf{P})$, then by $\Gamma\models\alpha$ we denote the fact that for every model M

if $M\models\Gamma$ (i.e. for every $\beta\in\Gamma$ $M\models\beta$), then $M\models\alpha$.

6. THEORIES OF PROGRAMS

In this section we restrict our considerations to a class of Kripke models which are (in some sense) isomorphic with the model defined by transition relations of concrete instances of rules from a given set of programs $\mathbf{P}$ and with the valuation s.t. for any propositional variable $A(x,m,e)$

$$M,u\models A(x,m,e)$$

has intended meaning "in a state u there exists an agent with the name x, message m and expression e".

For any set of programs $\mathbb{P}$ we present a list $\Gamma(\mathbb{P})$ of specific axioms s.t. if $M \models \Gamma(\mathbb{P})$, then M has mentioned above properties.

First we introduce some notations. Let

$c = \{<x1,m1>::e1,\ldots,<xk,mk>::ek\} \in CON$ and $y \in CAExp$.

$Agnof(c) = \{x1,\ldots,xk\}$.

By [c], (c), Exist(y), [c,~y] we denote the following formulas:

$$\bigwedge_{i=1}^{k} A(xi,mi,ei);\ [c] \& \bigwedge_{y \notin Agnof(c)} \sim Exist(y);$$

$$\bigvee_{m \in CMExp} \bigvee_{e \in CExp} A(y,m,e);\ [c] \& \sim Exist(y).$$

We assume [ ]=true for the empty configuration.

List of specific axioms $\Gamma(\mathbb{P})$

1. $\bigvee_{c \in CON} (c);$

2. For every $\underline{r} \in \mathbb{E}$ if $\underline{r}$ is of the form $\underline{lh} \Leftarrow \underline{rh}$ $\underline{if}$ $\underline{cond}$ and $\underline{cond}$ is true:

   r.1 $[\underline{lh}] \Rightarrow <>_{\underline{r}} [\underline{rh}];$

   r.2 $[\underline{lh} \cup \{<y,m>::e\}] \Rightarrow <>_{\underline{r}} A(y,m,e)$

   for $y \notin Agnof(\underline{lh}) \cup Agnof(\underline{rh});$

   r.3 $[\underline{lh}] \Rightarrow <>_{\underline{r}} \sim Exist(y)$ for $y \in Agnof(\underline{lh}) - Agnof(\underline{rh});$

   r.4 $[\underline{lh},\sim y] \Rightarrow <>_{\underline{r}} \sim Exist(y)$ for $y \notin Agnof(\underline{lh}) \cup Agnof(\underline{rh});$

   r.5 $\sim[\underline{lh}] \Rightarrow \Box_{\underline{r}} ff;$

   r.6 $<>_{\underline{r}} \alpha \Rightarrow \Box_{\underline{r}} \alpha$ for $\alpha \in \pi(\mathbb{P});$

3. $\bigvee_{n} \bigwedge_{\underline{w} \in \underline{P}^n} \Box_{\underline{w}} ff$ for $P \in \mathbb{P}$.

Axiom 1 says: In arbitrary state only finite number of agents exists. Axioms r.1-r.6 characterize the partial operation $R_{\underline{r}}$ in our models e.g. Axiom r.5 states that if it is not possible to match the left hand side of the rule $\underline{r}$ (with $\underline{cond}$ = true) in a given state u, then there is no v s.t. $uR_{\underline{r}}v$. Axiom r.1 says that, in opposite case, such a state exists and a configuration $\underline{rh}$ is a "part" of that state. To simplify our considerations we assume also that all programs in $\mathbb{P}$ define total functions (Axiom 3).

By $\pi_o(\mathbb{P})$ we denote the set of all formulas in $\pi(\mathbb{P})$ without temporal operators. The following fact establishes the relationship between $\pi(\mathbb{P})$ and $\pi_o(\mathbb{P})$.

Theorem 1 There exists a function $Tr:\pi(\mathbb{P}) \rightarrow \pi_o(\mathbb{P})$ with the following property:

for every model M and a state u of M if $M \models \Gamma(\mathbb{P})$ then $M,u \models \alpha \Leftrightarrow Tr(\alpha)$, where $\alpha \in \pi(\mathbb{P})$.

Proof. It is enough to put: $Tr(\alpha)=\alpha$ for $\alpha \in \pi_o(P)$,

$Tr(G(P)\alpha) = \bigwedge_{\underline{w} \in \underline{P}^+} \Box_{\underline{w}} Tr(\alpha)$ and $Tr(F(P)\alpha) = B(P, Tr(\alpha))$,

where $B(P,\alpha) = \sim \bigvee_{\underline{w} \in \underline{P}^+} [\langle\rangle_{\underline{w}} tt \,\&\, \bigwedge_{\underline{r} \in \underline{P}} \Box_{\underline{wr}} ff \,\&\, \bigwedge_{0 \leq i \leq length(w)} \langle\rangle_{Pref_i(w)} \sim\alpha]$

where $Pref_i(w)$ denotes the prefix of w of the length i.

□

Let A be the least fragment [9] of the infinitary propositional modal logic containing $Tr(\pi(\mathbb{P}))$.

The logical axioms of A are all instances of the following schemas of formulas [9]:

1. all propositional tautologies in A;
2. $\bigwedge \Phi \Rightarrow \alpha$ for $\alpha \in \Phi$ and $\bigwedge \Phi \in A$;
3. $\Box_{\underline{r}}(\alpha \Rightarrow \beta) \Rightarrow (\Box_{\underline{r}}\alpha \Rightarrow \Box_{\underline{r}}\beta)$ for $\underline{r} \in \underline{\mathbf{E}}$ and $\alpha, \beta \in A$;
4. $\bigwedge \{ \Box_{\underline{r}}\alpha \mid \alpha \in \Phi \} \Rightarrow \Box_{\underline{r}} \bigwedge \Phi$ for $\underline{r} \in \underline{\mathbf{E}}$ and $\bigwedge \Phi \in A$.

As rules of inferrence we take:

Modus Ponens: $\dfrac{\alpha, \alpha \Rightarrow \beta}{\beta}$; Necessitation: (for $\underline{r} \in \underline{\mathbf{E}}$) $\dfrac{\alpha}{\Box_{\underline{r}}\alpha}$;

Infinite Conjunction: $\dfrac{\alpha \Rightarrow \beta, \text{ for all } \beta \in \Phi}{\alpha \Rightarrow \bigwedge \Phi}$

where $\alpha, \beta, \bigwedge \Phi \in A$.

The notion of the proof we define in the standard way and

$$\Gamma(\mathbb{P}) \vdash_A \alpha$$

means that there exists a proof of α from $\Gamma(\mathbb{P})$ in the fragment A of infinitary propositional modal logic.

Theorem 2 For every $\alpha \in A$ $\Gamma(\mathbb{P}) \models \alpha$ iff $\Gamma(\mathbb{P}) \vdash_A \alpha$.

Proof Follows from the main result in [9].

Corollary For every $\alpha \in \pi(\mathbb{P})$

$$\Gamma(\mathbb{P}) \models \alpha \quad \text{iff} \quad \Gamma(\mathbb{P}) \vdash_A Tr(\alpha).$$

Proof Follows from Theorem 1 and Theorem 2.

□

For a given set of programs $\mathbb{P}$ we define the canonical model $M_{\mathbb{P}}$ as follows:

$$M_{\mathbb{P}} = (CON, \{R_{\underline{r}}\}_{\underline{r} \in \underline{\mathbf{E}}}, Val_{\mathbb{P}})$$

where

CON is the set of all configurations;

$\underline{\mathbf{E}}$ is the set of all instances of rules from $\mathbf{E}$;

$R_{\underline{r}} = \xrightarrow{\underline{r}}$;

$Val_{\mathbb{P}}(A(x,m,e)) = \{ c \in CON \mid \langle x,m \rangle ::e \in c \}$.

Theorem 3 Let $\mathbb{P}$ be a set of recursive equations programs and let M be a model for $\Gamma(\mathbb{P})$ (i.e. $M \models \Gamma(\mathbb{P})$) and let ≡ be the following equivalence relation in the set of states of M:

$$u \equiv u' \text{ iff } (M,u \models A(x,m,e) \text{ iff } M,u' \models A(x,m,e) \text{ for every } x,m,e).$$

Then $M/\equiv \; \simeq M_{\mathbb{P}}$ (i.e. $M/\equiv$ and $M_{\mathbb{P}}$ are isomorphic).

Proof. Let $M \models \Gamma(\mathbb{P})$. An arbitrary state u of M defines a configuration

$$c_u = \{<x,m>::e \mid M,u \models A(x,m,e)\}.$$

The isomorphism of $M/\equiv$ and $M_{\mathbb{P}}$ is defined by the function f s.t.

$$f([u]_{\equiv}) = c_u \quad \text{for every state u of M.}$$

□

7. CORRECTNESS OF FUNCTIONAL PROGRAMS IMPLEMENTATION

We are ready to formulate the correctness theorem which states that to prove the correctness of implementation of given facts it is enough to check if those facts are accepted by our very simple calculus.

Let $\mathbb{P}O$ be the class of all programs in LO with 1 recursive definition only and let $\mathbb{P}1$ be the class of all programs in L1. We consider the class TR of translations from $\mathbb{P}O$ into $\mathbb{P}1$ such that if tr∈TR then tr adds s comm l and decl l annotations only.

Correctness Theorem for Communications If for every program PO∈$\mathbb{P}$O and s comm l, s' comm l occurring in tr(PO) in the recursive call at position j and j' respectively, we have in our Calculus:

$$f(\ldots)]js = f(\ldots)]j's'$$

then tr is correct, i.e. for every PO∈$\mathbb{P}$O the parallel programs Sem0(PO) and Sem1(tr(PO)) compute the same function.

Sketch of the proof It is enough to apply Theorem 3 and to prove for every r in tr(PO) and configurations c and c'

$$M_{\{PO,tr(PO)\}} \models (c) \;\&\; \langle\rangle_{\underline{r}}(c') \Rightarrow ((h(c)) \Rightarrow g(PO)(h(c')))$$

where h is an erasing messages homomorphism.

□

Remark Sem1(tr(PO)) is more efficient than Sem0(PO) if some reading communications take place.

Let us consider the following program B1 in L1:

$$B1: \begin{cases} bin(n,0)=1 \\ bin(n,n)=1 \\ bin(n,m)=0 \quad \underline{if}\ n<m \\ bin(n+1,m+1)=\overline{bin}(n,m)(1\ \underline{comm}\ 1) + bin(n,m+1)(0\ \underline{comm}\ 1)\overline{\underline{decl}\ 1} \\ \qquad \underline{if}\ n>m\geq 0 \end{cases}$$

The program B1 is equivalent to BO, i.e., the parallel programs Sem0(BO) and Sem1(B1) compute the same function bin. This follows from the Theorem above and from the fact:

$$bin(n,m)]01 = bin(n,m)]10$$

accepted by the Calculus when $n>m\geq 1$.

The following parallel program Sem1(B1) is the result of translating B1:

```
{<x,E>::bin(n+1,m+1)} ⇐{<x,Φ≃[01,10]>::.x0+.x1,<x0,E>::bin(n,m),
                         <x1,E>::bin(n,m+1)}  if n>m≥0
{<x,mes>::bin(n,0)} ⇐{<x,E>::1}
{<x,mes>::bin(n,n)} ⇐{<x,E>::1}

{<x,mes>::bin(n,m)} ⇐{<x,E>::0}    if n<m
{<x,mes>::n±m}⇐{<x,E>::k}          if k=n±m
{<x,mes>::e+.x1,<x1,mes1>::n}⇐{<x,mes>::e+n,<x1,mes1>::n}
{<x,mes>::.x0+e,<x0,mes1>::n}⇐{<x,mes>::n+e,<x0,mes1>::n}
{<x,Φ≃[01,10]>::e,  <x01,mes>::n}⇐{<x,n≃[10]>::e,<x01,mes>::n}
{<x,Φ≃[01,10]>::e,  <x10,mes>::n}⇐{<x,n≃[01]>::e,<x10,mes>::n}
{<x,n≃[01]>::e,   <x01,mes>::e1}⇐{<x,n≃[ ]>::e,<x01,mes>::n}
{<x,n≃[10]>::e,   <x10,mes>::e1}⇐{<x,n≃[ ]>::e,<x10,mes>::n}.
```

Let us consider an example of computation of the program Sem1(B1) with the initial configuration {<ε,E>::bin(5,3)} :

```
{<ε,E>::bin(5,3)}→

{<ε,Φ≃[01,10]::.0+.1,
 <0,E>::bin(4,2),      <1,E>::bin(4,3)}→ ...→

          {<ε,Φ≃[01,10]>::.0+.1
 <0,Φ≃[01,10]>::3+.01            <1,Φ≃[01,10]>::.10+1
        ... <01,Φ≃[01,10]>::3    <10,E>::bin(3,2)} ...

 A
 →        {<ε,3≃[10]>::.0+.1
 <0,Φ≃[01,10]>::3+.01             <1,Φ≃[01,10]>::.10+1
        ... <01,Φ≃[01,10]>::3     <10,E>::bin(3,2)} ...
 B
 →        {<ε,3≃[ ]>::.0+.1
 <0,Φ≃[01,10]>::3+3              <1,Φ≃[01,10]>::.10+1
                                    <10,E>::3}
...
```

In the transition A the value computed by the agent 01 is communicated to the agent ε and in B this value is communicated to the agent 10 (by the agent ε). That communication avoids the computation of bin(3,2) by the agent 10.

Let us consider now a program FIB in LO:

$$\text{FIB:}\ \begin{cases} \text{fib}(0)=1 \\ \text{fib}(1)=1 \\ \text{fib}(n+2)=\text{fib}(n+1)+\text{fib}(n) \end{cases}$$

and a program FIB1 derived from FIB and the fact

$$\text{fib}(n)]01 = \text{fib}(n)]10 \quad (n\geq 2)$$

$$\text{FIB1:}\begin{cases}\text{fib}(0)=1\\ \text{fib}(1)=1\\ \text{fib}(n+2)=(\text{fib}(n+1)(0\ \underline{\text{comm}}\ l)+\text{fib}(n)W(\varepsilon\ \underline{\text{comm}}\ l))\underline{\text{decl}}\ l\end{cases}$$

where we use the annotation W to tell the system that the newly generated computing agent with the task fib(n) is forced to wait until the corresponding value appears in the location l.

The reader can easily modify our language L1 and the fact-translation algorithm for coping with the new kind of annotations.

Fact. Let FIB1' be the result of application of that new translation algorithm to FIB1. The parallel program FIB1' uses a linear number of computing agents (with respect to n) in the computation with the initial configuration

$$\{<\varepsilon,E>::\text{fib}(n)\}.$$

The length of this computation is a linear function of n. □

In the <calculus,fact-translation> pair we use, it is possible to express facts as equality of terms, so that repeated evaluations of common subexpressions are avoided. The calculus is based on the unfolding rule [5] and the symbolic evaluation technique. The associated translation realizes communications among concurrent agents and improves the efficiency of their computation.

8. CONCLUSIONS

We have presented some basic idea for the construction of a system for automatic implementation of functional programs in the set of parallel programs. The system uses a calculus for checking the correctness of supplied "factual knowledge" (or facts) [8] about the functions to be computed. It than translates those facts into suitable communications among computing agents so that the derived computations are more efficient and may satisfy given complexity constraints [8]. The main goal of our project is to build such a system for different classes of problems.

REFERENCES

[1] J. Backus, "Can Programming be Liberated from the von-Neumann Style? A Functional Style and Its Algebra of Programs", J.A.C.M. Vol. 21, No.8 (1978) pp. 613-641.

[2] R.M. Burstall, J. Darlington, "A transformation system for developing recursive programs", J.A.C.M. Vol. 24, No. 1 (1977) pp. 44-67.

[3] R.M. Burstall, D.B. MacQueen, D.T. Sannella, "HOPE: An Experimental Applicative Language", Proc. LISP Conf. Standord University (1980).

[4] J. Darlington, M. Reeve, "Alice: A Multiprocessor Reduction Machine for Parallel Evaluation of Applicative Languages", Proc. ACM/MIT Conference on Functional Programming Languages and Computer Architecture (1981).

[5] L. Kott, "About transformation system: A theoretical study", 3ème Colloque International sur la Programmation, Dunod, Paris (1978).

[6] A. Pettorossi, A. Skowron, "Theories for Verifying Communicating Agents Behaviour in Recursive Equations Programs", Proc. 20-th Annual Conf. on Information Sciences and Systems, Princeton (1986).
[7] A. Pettorossi, A. Skowron, "Using Facts for Improving the Parallel Execution of Functional Programs", Proc. 15-th Conf. on Parallel Processing, St. Charles, Illinois (1986).
[8] A. Pettorossi, A. Skowron, "Factual Knowledge for Developing Concurrent Programs", Proc. AAAI-86, Philadelphia (1986).
[9] S. Radev, "Infinitary Modal Logic and Programming Languages", Ph. D. Thesis, Warsaw University 1980.
[10] W.L. Scherlis, "Expression Procedures and Program Derivation", Ph.D. Thesis, Stanford Univ. Computer Science Report STAN-CS-80-818 (1980).
[11] D.A. Turner, "Functional Programs as Executable Specifications" in "Mathematical Logic and Programming Languages" C.A.R. Hoare, J.C. Shepherdson eds., Prentice Hall (1984).
[12] N. Wirth, "Program Development by Stepwise Refinement", C.A.C.M. Vol. 14 (1971) pp. 221-227.

LOGIC APPROXIMATING SEQUENCES OF SETS *

Helena Rasiowa

Institute of Mathematics
University of Warsaw
PKiN, 00-901 Warsaw, Poland

Introduction

In various studies concerning computer science or artificial intelligence, in which approximation tools could be applied, there appears a need of gradual approximating descending set sequences $X = (X_m)$ (e.g. of documents, objects, points) formed of elements satisfying some stronger and stronger conditions. Gradual approximations (both: lower and upper ones) are determined by a descending sequence $(\tilde{=}_j)$ of equivalence relations, going to be established progressively. Approximations of grade j+1 are better than those of grade j. Approximations determined by $\tilde{=}_\omega$, which is the intersection of $\tilde{=}_j$ for $j < \omega$, are the most precise.

In order to deal with a gradual approximating of descending sequences of sets (in particular relations) with equivalence relations determining finer and finer equivalence

* This is a modified and - in a sense - expounded version of the paper by Rasiowa 1986, prepared for 16th ISMVL, Blacksburg, VA, USA

classes, a multiple-valued (ω^+-valued) approximation logic MAPL of first order is offered.
A semantics is inspired by set-theoretical representations of Post algebras of order ω^+ (Rasiowa 1985)-different from those of Maksimova and Vakarelov 1972-and on ideas presented in papers by Rasiowa and Skowron 1985, 1986. They are connected with rough sets approach (Pawlak 1982).
Formalized systems of that logic are also formulated and the completeness theorem is proved.

1. Generalized Post Algebras of Order ω^+ and Their Representations by Sequences of Sets

In order to make the paper self-contained the notion of a generalized Post algebra of order ω^+ will be reminded and fundamental properties shortly summarized*). For the sake of brievity we shall in this paper write sometimes GP--algebra instead of Post algebra of order ω^+.

The simplest and very important example of that notion is offered by a linear GP-algebra to be denoted by $\underline{P}_\omega$. It is formed of a chain of order ω^+

$$\bot = e_0 \leq e_1 \leq \dots \leq e_\omega = \top \tag{1}$$

considered as a linear pseudo-Boolean algebra with respect to binary operations $\cup$, $\cap$, $\Rightarrow$ and one unary operation $\neg$ defined as follows:

$$e_i \cup e_j = e_{max(i,j)}, \quad e_i \cap e_j = e_{min(i,j)} \tag{2}$$

$$e_i \Rightarrow e_j = \begin{cases} e_\omega & \text{if } i \leq j \\ e_0 & \text{otherwise} \end{cases} \tag{3}$$

---

*) For more details see Rasiowa 1973

$$\neg e_i = e_i \Rightarrow e_o \; . \tag{4}$$

Moreover the following unary operations d_m , $m \in N$ - where N will always denote the set of all positive integers - are adopted:

$$d_m e_i = \begin{cases} e_\omega & \text{if } m \leq i \\ e_o & \text{otherwise} \end{cases} \; . \tag{5}$$

Thus $\underline{P}_\omega$ is the following algebra

$$\underline{P}_\omega = (P_\omega, \top, \cup, \cap, \Rightarrow, \neg, (d_m)_{m \in N}, (e_i)_{0 \leq i \leq \omega})$$

where $P_\omega = \{e_o, \ldots, e_\omega\}$ and the operations in $\underline{P}_\omega$ are defined by means of (2) - (5). Notice that $(\{\top, \bot\}, \cup, \cap, \Rightarrow, \neg)$ is the two-element Boolean algebra. The role of $\underline{P}_\omega$ in the class of all GP-algebras is analogous to that of the two-element Boolean algebra in the class of all Boolean algebras. $\underline{P}_\omega$ has been taken as a semantic basis for ω^+-valued predicate logic (Rasiowa 1973) and will be adopted as a basis to define an algebraic semantics for MAPL.

The class of all GP-algebras can be characterized by the following system of axioms (see Rasiowa 1973):

$(P, \top, \cup, \cap, \Rightarrow, \neg)$ is a pseudo-Boolean algebra with unit element $\top$ and zero element $\bot = \neg \top$ $\quad (p_o)$

$$d_m(a \cup b) = d_m a \cup d_m b \tag{p_1}$$

$$d_m(a \cap b) = d_m a \cap d_m b \tag{p_2}$$

$$d_m(a \Rightarrow b) = (d_1 a \Rightarrow d_1 b) \cap \ldots \cap (d_m a \Rightarrow d_m b) \tag{p_3}$$

$$d_m(\neg a) = \neg d_1 a \tag{p_4}$$

$$d_m d_k a = d_k a \tag{p_5}$$

$$d_m e_i = \begin{cases} \top & \text{if } m \leq i \\ \bot & \text{if } m > i \end{cases}, \quad 0 \leq i \leq \omega \tag{p_6}$$

$$d_1 a \cup \neg d_1 a = \top \qquad (p_7)$$

$$d_{m+1} a \leq d_m a \qquad (p_8)$$

$$e_\omega = \top \qquad (p_9)$$

$$a = \bigcup_{1 \leq m < \omega} (d_m a \cap e_m) , \qquad (p_{10})$$

where $\bigcup$ denotes l.u.b in the lattice under consideration.

Recall the following basic properties of GP-algebras. In any GP-algebra:

$$\text{if } a \leq b, \text{ then } d_m a \leq d_m b \qquad (6)$$

$$a = b \text{ iff } d_m a = d_m b \text{ for all } m \in N \qquad (7)$$

$$a \Rightarrow d_m b = d_1 a \Rightarrow d_m b \ , \quad m \in N \qquad (8)$$

if $\underline{P}$ is non-degenerate, then

$$i \neq j \text{ implies } e_i \neq e_j \ , \quad \text{for } 0 \leq i,j \leq \omega . \qquad (9)$$

1.1. The set $B_{\underline{P}} = \{d_m a: m \in N \text{ and } a \in P\}$ coincides with the set of all complemented elements in $(P, \cup, \cap)$ and $\underline{B}_{\underline{P}} = (B_{\underline{P}}, \top, \cup, \cap, \Rightarrow, \neg)$ is a Boolean algebra that is said to correspond to $\underline{P}$.

Let (Q) be a set of infinite joins and meets in $\underline{P}$:

$$\begin{aligned} a_s &= \bigcup_{t \in T_s} a_{st} , \quad s \in S \\ b_u &= \bigcap_{t \in T_u} b_{ut} , \quad u \in U . \end{aligned} \qquad (Q)$$

1.2. (Rasiowa 1973) Equalities (Q) are equivalent with the following ones

$$\begin{aligned} d_m a_s &= \bigcup_{t \in T_s} d_m a_{st} , \quad s \in S \\ d_m b_u &= \bigcap_{t \in T_u} d_m b_{ut} , \quad u \in U . \end{aligned} \qquad (DQ)$$

1.3. (Rasiowa 1973) Let ∇_o be any prime filter in $\underline{B}_{\underline{P}}$ preserving infinite joins and meets (DQ). Let

$$\nabla = \{a \in P\colon d_m a \in \nabla_0 \ \text{ for all } \ m \in N\} \ . \tag{10}$$

Then ∇ is a prime filter in $\underline{P}$ preserving infinite joins and meets (Q) and (DQ) and $\nabla \cap B_{\underline{P}} = \nabla_0$. Moreover

$$a \in \nabla \quad \text{iff} \quad d_m a \in \nabla \quad \text{for all} \quad m \in N \tag{11}$$

$$b_u \in \nabla \quad \text{iff} \quad b_{ut} \in \nabla \quad \text{for all} \quad u \in U \tag{12}$$

$\underline{P}/\nabla$ is isomorphic to the linear GP-algebra $\underline{P}_\omega$ and for any $|a| \in P/\nabla$,

$$|a| = T \quad \text{iff} \quad a \in \nabla , \tag{13}$$

a condition: if $a \in \nabla$, then $d_m b \in \nabla$ implies that

$$a \Rightarrow d_m b \in \nabla , \quad m \in N . \tag{14}$$

All GP-algebras can be obtained up to isomorphisms applying the following method.

Let $U \neq \emptyset$ be an arbitrary space and let $\underline{B}(U) = (B(U), U, \cup, \cap, \Rightarrow, -)$ be a field of subsets of U. Let DS(B(U)) be the set of all descending sequences $Y = (Y_1, Y_2, \ldots)$, $Y_1 \supset Y_2 \supset \ldots$, of subsets in B(U). Define Post operations on DS(B(U)) thus

$$T = e_\omega = (U, U, \ldots)$$

$$Y \cup Z = (Y_1 \cup Z_1, Y_2 \cup Z_2, \ldots)$$

$$Y \cap Z = (Y_1 \cap Z_1, Y_2 \cap Z_2, \ldots)$$

$$Y \Rightarrow Z = (-Y_1 \cup Z_1, (-Y_1 \cup Z_1) \cap (-Y_2 \cup Z_2), \ldots)$$

$$\neg Y = (-Y_1, -Y_1, \ldots)$$

$$d_m Y = (Y_m, Y_m, \ldots), \quad \text{for} \quad 1 \leq m < \omega ,$$

$$e_i = (\underbrace{U, \ldots, U}_{i\text{-times}}, \emptyset, \ldots), \quad 0 \leq i < \omega .$$

Then

$$\underline{DS}(B(U)) = (DS(B(U)), T, \cup, \cap, \Rightarrow, \neg, (d_m)_{m \in N}, (e_i)_{0 \leq i \leq \omega})$$

and all its subalgebras are GP-algebras constructed of descending sequences of sets in B(U).

Observe, that taking as a given space U^n, $n \in N$, we obtain GP-algebras formed of descending sequences of n-ary relations.

Clearly, the analogous method can be applied to Boolean algebras, leading to obtain GP-algebras formed of decreasing sequences of elements in a Boolean algebra.

The following representation theorem is fundamental in this paper.

1.4. For every non-degenerate GP-algebra $\underline{P}$ there is a monomorphism h from $\underline{P}$ into a full GP-algebra $\underline{DS}(B(U))$ of all descending sequences of sets in a field of sets B(U). Moreover, it can be assumed that for any enumerable set (Q) of infinite joins and meets in $\underline{P}$, h preserves infinite joins and meets in (Q) and (DQ), i.e. maps on set-theoretical operations.

Sketch of a proof Let U be the set of all prime filters in $\underline{B}_P$ preserving infinite joins and meets in (DQ). Put $h_o(d_m a) = \{\nabla_o \in U: d_m a \in \nabla_o\}$. Then h_o is an isomorphism from $\underline{B}_P$ onto the field
$B(U) = \{h_o(d_m a): a \in P \text{ and } m \in N\}$.
Let $h(a) = (h_o(d_m a))_{m \in N} \in DS(B(U))$. It can be proved that h is a monomorphism from $\underline{P}$ into $\underline{DS}(B(U))$ and preserves infinite joins and meets in (Q) and (DQ).

2. Generalized Post Fields

Another method of constructing GP-algebras taking as a starting point a field B(U), $U \neq \emptyset$, consists in transforming any descending sequence $(X_m)_{m \in N}$ of sets in B(U) into a union of subsets of disjoint spaces. More exactly, given B(U) assign a GP-space

$$U_P = (\{U_m, g_m\}_{m \in N}, B(U), B(U_P)) \tag{1}$$

connected with B(U), defined thus:

$$U_P = \bigcup_{m \in N} U_m \tag{2}$$

is the union of disjoint spaces U_m, $m \in N$ any U_m being said to be m-th stratus of U_P,

$$g_m : U_m \longrightarrow U \text{ are bijections onto } U, \tag{3}$$

$$B(U_P) = \left\{ \bigcup_{m \in N} g_m^{-1}(X) : X \in B(U) \right\}. \tag{4}$$

The family $B(U_P)$ is a field of subsets of U_P.

Let us set

$$E_0 = \emptyset, \ E_1 = U_1, \ldots, E_m = E_1 \cup \ldots \cup E_m, \ldots, E_\omega = U_P. \tag{5}$$

Let $P(U_P)$ be the family of all subsets Y of U_P of the form

$$Y = \bigcup_{m \in N} g_m^{-1}(X_m), \text{ where } X_1 \supset X_2 \supset \ldots, \text{ and } X_n \in B(U). \tag{6}$$

Sets in $P(U_P)$ are called stratified sets.

Let us define operator D_m on $P(U_P)$ as follows:

$$D_m Y = \bigcup_{k \in N} g_k^{-1}(X_m), \text{ for } Y \text{ defined by (6).} \tag{7}$$

$$\text{Clearly } D_m Y \in B(U_P) \text{ and } D_1 Y \supset D_2 Y \supset \ldots \tag{8}$$

Moreover, by a calculation it may be proved that

$$Y = \bigcup_{m \in N} (D_m Y \cap E_m) \tag{9}$$

and that this monotonic representation is unique.

So there is one-to-one correspondence between

$$X = (X_m)_{m \in N} \quad \text{in} \quad \underline{DS}(B(U)) \quad \text{and} \quad Y = \bigcup_{m \in N} (D_m Y \cap E_m) \; .$$

Family $P(U_P)$ is a set lattice. Adopt operations $\Rightarrow$, $\neg$ defined as follows

$$Y \Rightarrow Z = \bigcup_{m \in N} ((-D_1 Y \cup D_1 Z) \cap \ldots \cap (-D_m Y \cup D_m Z) \cap E_m) \qquad (10)$$

$$\neg Y = Y \Rightarrow \emptyset \; .$$

Then $\underline{P}(U_P) = (P(U_P), U_P, \cup, \cap, \Rightarrow, \neg, (D_m)_{m \in N}, (E_i)_{0 \leq i \leq \omega})$ is a GP-algebra to be said to be a full GP-field over U_P . Subalgebras of $\underline{P}(U_P)$ are also GP-algebras, to be called GP-fields over U_P .

2.1. (Rasiowa 1985) For every non-degenerate GP-algebra P there is a monomorphism h from $\underline{P}$ into a full GP-field $\underline{P}(U_P)$ of subsets of U_P . Moreover it can be assumed that for any enumerable set (Q) of infinite joins and meets in $\underline{P}$, h preserves infinite joins and meets in (Q) and (DQ), i.e. maps on set-theoretical operations.

3. GP-Functions vs. Descending Sequences of Sets and Stratified Sets

Let $U \neq \emptyset$ be any space, $B(U)$ – the field of all subsets of U and let $Fn(U)$ be the set of all functions $f : U \rightarrow P_\omega$ to be called GP-functions over U . To every $f \in Fn(U)$ assign Boolean functions $d_m f$, $m \in N$, which according to (p_6) in Sec.1 are defined thus:

$$d_m f(u) = \begin{cases} \top & \text{if} \quad f(u) \geq e_m \\ \bot & \text{if} \quad f(u) < e_m \end{cases} \qquad (1)$$

By (p_{10}) in Sec.1,

$$f = \bigcup_{m\in N} (d_m f \cap e_m) . \tag{2}$$

Let us set

$$S(d_m f) = \{u \in U: d_m(u) = \top\} = \{u \in U: f(u) \geq e_m\} . \tag{3}$$

Obviously $d_m f$ are characteristic functions of $S(d_m f)$, $m \in N$. By (p_8) in Sec.1 $S(d_m f) \supset S(d_{m+1} f)$, $m \in N$.

Thus $(S(d_m f))_{m\in N}$ (4)

is a descending sequence of subsets of U. It is easy to prove that a mapping t assigning to each $f \in Fn(u)$ in form (2) a sequence (4) is one-to-one and maps $Fn(U)$ onto $DS(B(U))$.

On the other hand a mapping s assigning to every $f \in Fn(U)$ a stratified set $Y_f = \bigcup_{m\in N} (D_m Y_f \cap E_m)$, where $D_m Y_f = \bigcup_{k\in N} g_k^{-1}(S(d_m f))$ is also one-to-one and maps $Fn(U)$ on the full GP-field $\underline{P}(U_P)$ of stratified sets over a GP--space connected with $B(U)$.

4. Gradual Approximation Spaces and Approximation Operators

Let $EQ(U)$, $U \neq \emptyset$, be the set of all functions $eq: U^2 \rightarrow P$ satisfying the following conditions

$$eq(u\ ,u\ ) = e_\omega \quad , \quad eq(u_1,u_2) = eq(u_2,u_1) \tag{1}$$

$$eq(u_1,u_2) \cap eq(u_2,u_3) \leq eq(u_1,u_3) \ , \tag{2}$$

where $\cap$ is the meet operation in $\underline{P}$.

Notice that $d_m f: U^2 \longrightarrow \{\top, \bot\}$, for $m \in N$, belong to $EQ(U)$ provided $f \in EQ(U)$ and $d_{m+1} f(u) \leq d_m f(u)$ for $u \in U$.

4.1. Every $eq \in EQ(U)$ determines equivalence relations $\cong_j$, $j \in N$, and $\cong_\omega$ on U as follows:

$$u_1 \cong_j u_2 \text{ iff } d_j eq(u_1,u_2) = T, \text{ for } j \in N \tag{3}$$

$$u_1 \cong_\omega u_2 \text{ iff } (d_j eq(u_1,u_2) = T \text{ for all } j \in N). \tag{4}$$

Relations $\cong_j$ for $1 \leq j \leq \omega$ can be extended on U^n, $n \in N$ by adopting

$$(u_1,\ldots,u_n) \cong_j (u_1',\ldots,u_n') \text{ iff } (u_i \cong_j u_i' \text{ for all } i = 1,\ldots,n). \tag{5}$$

Moreover,

$$\cong_{j+1} \subset \cong_j, \text{ for } j \in N \tag{6}$$

$$\cong_\omega = \bigcap_{j<\omega} \cong_j. \tag{7}$$

An approximation space of order ω^+, determined by $eq \in EQ(U)$, is a system $\underline{A} = (U,(\cong_j)_{1 \leq j \leq \omega})$, where equivalence relations $\cong_j$, $1 \leq j \leq \omega$ are defined by means of (3), (4). This approximation space will also be denoted by $\underline{A} = (U,eq)$.

Equivalence relations $\cong_j$, $1 \leq j \leq \omega$,are said to be of grade j. They determine interior operations I_j, and closure operations C_j, for $1 \leq j \leq \omega$ in U^n, $n \in N$ by adopting the family of unions of equivalence classes with respect to $\cong_j$ in $U^n \times U^n$ and the empty set as a basis for the open sets in U^n.

Thus for every $X \subset U^n$, $u \in U^n$ and $1 \leq j \leq \omega$:

$$u \in I_j X \text{ iff for each } u' \in U^n, \text{ if } u \cong_j u' \text{ then } u' \in X \tag{8}$$

$$u \in C_j X \text{ iff there is } u' \in U^n \text{ such that } u \cong_j u' \text{ and } u' \in X. \tag{9}$$

Obviously, $I_j X = -C_j - X$. (10)

For every $X \subset U^n$, $I_j X$ and $C_j X$ for $1 \leq j \leq \omega$, are adopted as a lower approximation of X (of grade j) and as an upper approximation of X (of grade j), respectively.

Now, consider:

any function $f = \bigcup_{m \in N} (d_m f \cap e_m)$ in $Fn(U^n)$, (11)

$$tf = (Sd_m f)_{m \in N} \text{ in } \underline{DS}(B(U^n)) \tag{12}$$

$$sf = Y_f = \bigcup_{m \in N} (D_m Y_f \cap E_m) \quad , \tag{13}$$

where $D_m Y_f = \bigcup_{k \in N} g_k^{-1} S(d_m f)$ for $m \in N$ and mappings t, f were defined in Sec. 3.

It is quite natural to adopt

$$I_j (Sd_m f)_{m \in N} \overset{df}{=} (I_j Sd_m f)_{m \in N} \ , \quad 1 \leq j \leq \omega \tag{14}$$

$$C_j (Sd_m f)_{m \in N} \overset{df}{=} (C_j Sd_m f)_{m \in N} \quad , 1 \leq j \leq \omega \ . \tag{15}$$

Definitions of operations I_j, C_j, $1 \leq j \leq \omega$ on Boolean functions $d_m f : U^n \longrightarrow \{\top, \bot\}$ correspond to (8) and (9):

$$I_j d_m f(u) = \begin{cases} \top & \text{if for each } u' \in U^n, \text{ if } u \cong_j u' \text{ then } d_m f(u') = \top \\ \bot & \text{otherwise} \end{cases} \tag{16}$$

$$C_j d_m f(u) = \begin{cases} \top & \text{if there is } u' \in U^n \text{ such that } u \cong_j u' \text{ and } d_m f(u') = \top \\ \bot & \text{otherwise} \end{cases} \tag{17}$$

Notice that $I_j d_m f \geqq I_j d_{m+1} f$ and $C_j d_m f \geqq C_j d_{m+1} f$, $m \in N$.

Let us set

$$I_j f \overset{df}{=} \bigcup_{m \in N} (I_j d_m f \cap e_m) \ , \quad C_j f \overset{df}{=} \bigcup_{m \in N} (C_j d_m f \cap e_m) \ . \tag{18}$$

By the uniqueness of representations of $f \in Fn(U^n)$, we obtain

$$d_m I_j f = I_j d_m f \quad \text{and} \quad d_m C_j f = C_j d_m f \ . \tag{19}$$

Observe that by the definitions (14), (15), (16), (17), (18)

$$t(I_j f) = I_j(tf) \quad \text{and} \quad t(C_j f) = C_j(tf), \ 1 \leq j \leq \omega \ . \tag{20}$$

Now, let us put

$$I_j D_m Y_f = \bigcup_{k \in N} g_k^{-1} \ I_j S(d_m f) \tag{21}$$

$$C_j D_m Y_f = \bigcup_{k \in N} g_k^{-1} \, C_j S(d_m f), \quad 1 \leq j \leq \omega \tag{22}$$

and define analogously to (18)

$$I_j Y_f \overset{df}{=} \bigcup_{m \in N} (I_j D_m Y_f \cap E_m), \quad C_j Y_f \overset{df}{=} \bigcup_{m \in N} (C_j D_m Y_f \cap E_m). \tag{23}$$

This yields $D_m I_j Y_f = I_j D_m Y_f$ and $D_m C_j Y_f = C_j D_m Y_f$ by the uniqueness of monotonic representations of stratified sets, and

$$s(I_j f) = I_j(sf), \quad s(C_j f) = C_j(sf), \quad 1 \leq j \leq \omega . \tag{24}$$

5. Formalized Systems of MAPL

Consider countable first order predicate languages $\underline{L} = (\text{Alph}, F)$. Assume that Var is the set of free individual variables in the alphabet Alph of $\underline{L}$. The set $F_{at} \subset F$ of atomic formulas consists of propositional constants $\underline{e}_i$, $0 \leq i \leq \omega$ and of formulas $\underline{e}(x,y)$, $p_i(x_1,\ldots,x_{n(i)})$ for any $x,y,x_1,\ldots,x_{n(i)}$ in Var and $i=1,\ldots,q$, where $\underline{e}$ is an equivalence predicate and p_i for $i=1,\ldots,q$ are n(i)-ary predicates. Other formulas are built of atomic ones by means of propositional connectives $\cup$, $\cap$, $\Rightarrow$, $\neg$, d_m for $m \in N$, I_j and C_j for $1 \leq j \leq \omega$, quantifiers $\forall$, $\exists$ and parentheses. Instead of $(A \Rightarrow B) \cap (B \Rightarrow A)$ we write $A \Leftrightarrow B$. For each formula $A \in F$, Var A will denote the set of all free individual variables in A.

An algebraic semantics for $\underline{L}$ is defined by means of realizations in an arbitrary gradual approximation space $\underline{A} = (U, eq)$ (see Sec.4) and valuations in U, i.e. functions $v : \text{Var} \longrightarrow U$. If $\underline{A}$ is established, then Val will denote the set of all valuations in U. It is worth mentioning that algebraic semantics determines two equivalent ones viz. replacing GP-functions by means of descending sequences of sets or stratified sets.

A realization R in $\underline{A}$ assigns to predicate $\underline{e}$ the function $eq \in EQ(U)$ in $\underline{A}$, to each $\underline{e}_i$ for $0 \leq i \leq \omega$, the element $e_i \in P_\omega$ and to each p_i for $i=1,\ldots,q$, a GP-function $f_i : U^{n(i)} \longrightarrow P_\omega$. Propositional connectives $\cup, \cap, \Rightarrow, \neg, d_m$ for $m \in N$ are realized as corresponding operations in the GP-algebra $\underline{P}_\omega$, approximation operators I_j, C_j for $1 \leq j \leq \omega$ as corresponding operators on GP-functions in $\underline{A} = (U, eq)$ (see (16),(17),(18) in Sec.4) and quantifiers $\forall$, $\exists$ as infinite meets and joins in $\underline{P}_\omega$ (see Rasiowa 1973).

For any $A \in F$, a realization of A by valuation v will be denoted by $A_R(v)$. If Var A $= \{x_1,\ldots,x_n\}$, then A_R is conceived to be a GP-function in $Fn(U^n)$. Thus it follows from our assumptions and (16), (17) in Sec.4, that

$$I_j d_m A_R(v) = \begin{cases} T & \text{if for all } u_1,\ldots,u_n \in U, \text{ if } v(x_i) \cong_j u_i \\ & \text{for } i=1,\ldots,n \text{ then } d_m A_R(v')=T, \text{ where} \\ & v'(y)=v(y) \text{ for } y \neq x_1,\ldots,x_n \text{ and} \\ & v'(x_i) = u_i \text{ for } i=1,\ldots,n \\ \bot & \text{in the opposite case} \end{cases} \quad (1)$$

$$I_j d_m A_R(v) = d_m A_R(v) \quad \text{if} \quad \text{Var } A = \emptyset \quad (1')$$

$$C_j d_m A_R(v) = \begin{cases} T & \text{if there are } u_1,\ldots,u_n \in U \text{ such that} \\ & v(x_i) \cong_j u_i \text{ for } i=1,\ldots,n \text{ and} \\ & d_m A_R(v')=T, \text{ where } v'(y)=v(y) \text{ for} \\ & y \neq x_1,\ldots,x_n \text{ and } v'(x_i)=u_i \text{ for } i=1,\ldots,n \\ \bot & \text{in the opposite case} \end{cases} \quad (2)$$

$$C_j d_m A_R(v) = d_m A_R(v) \quad \text{if} \quad \text{Var } A = \emptyset \ . \quad (2')$$

Adopt as logical axioms in $\underline{L} = (\text{Alph}, F)$: all formulas in F which are substitutions of axioms of the intuitionistic propositional calculus and moreover the following axiom schemes for ω^+-valued logic, being counterparts of axioms (p_1) - (p_{10}) in Sec.1.

(ax 1) $d_m(A \cup B) \Leftrightarrow (d_m A \cup d_m B)$,

(ax 2) $d_m(A \cap B) \Leftrightarrow (d_m A \cap d_m B)$,

(ax 3) $d_m(A \Rightarrow B) \Leftrightarrow (d_1 A \Rightarrow d_1 B) \cap \ldots \cap (d_m A \Rightarrow d_m B)$,

(ax 4) $d_m \neg A \Leftrightarrow \neg d_1 A$,

(ax 5) $d_m d_k A \Leftrightarrow d_k A$,

(ax 6) $d_m \underline{e}_i$ for $m \leq i$, $\neg d_m \underline{e}_i$ for $i > m$, $0 \leq i \leq \omega$,

(ax 7) $d_{m+1} A \Rightarrow d_m A$,

(ax 8) $\underline{e}_\omega$

(ax 9) $d_1 A \cup \neg d_1 A$

(ax10) $(d_m A \cap \underline{e}_m) \Rightarrow A$

where $m,k \in N$, $0 \leq i \leq \omega$,

axioms for equivalence predicate $\underline{e}$

(ax11) $\underline{e}(x,x)$, (ax12) $\underline{e}(x,y) \Leftrightarrow \underline{e}(y,x)$

(ax13) $(\underline{e}(x,y) \cap \underline{e}(y,z)) \Rightarrow \underline{e}(x,z)$,

and the following axiom schemes for approximation operators I_j, C_j, $1 \leq j \leq \omega$, $n,m \in N$

(ax14) $I_j A \Leftrightarrow A$ if $\mathrm{Var}\, A = \emptyset$

(ax15) $\forall y_1 \ldots \forall y_n((\underline{e}(x_1,y_1) \cap \ldots \cap \underline{e}(x_n,y_n)) \Rightarrow d_m A(y_1,\ldots,y_n)) \Leftrightarrow I_\omega d_m A(x_1,\ldots,x_n)$, where $\mathrm{Var}\, A = \{x_1,\ldots,x_n\}$

(ax16) $\forall y_1 \ldots \forall y_n((d_j \underline{e}(x_1,y_1) \cap \ldots \cap d_j \underline{e}(x_n,y_n)) \Rightarrow d_m A(y_1,\ldots,y_n)) \Leftrightarrow I_j d_m A(x_1,\ldots,x_n)$, where $\mathrm{Var}\, A = \{x_1,\ldots,x_n\}$ and $1 \leq j \leq \omega$

(ax17) $d_m I_j A \Leftrightarrow I_j d_m A$, (ax18) $d_m C_j A \Leftrightarrow C_j d_m A$,

(ax19) $C_j d_m A \Leftrightarrow \neg I_j \neg d_m A$.

Rules of inference for intuitionistic predicate calculi

are adopted and moreover the following ones:

$(r_1)\quad \frac{A}{d_m A}$ for $m \in N$, $\quad (r_2)\quad \frac{\{d_m A\}_{m\in N}}{A}$

For any set $\Sigma \subset F$, $\Sigma \vdash A$ denotes that A is provable from Σ.

5.1. For any $\Sigma \subset F$ and $A,B \in F$:

(i) if $\Sigma \vdash A \Rightarrow B$, then $\Sigma \vdash I_j A \Rightarrow I_j B$ and $\Sigma \vdash C_j A \Rightarrow C_j B$, $1 \leq j \leq \omega$

(ii) if $\Sigma \vdash A$, then $\Sigma \vdash I_j A$

(iii) $\vdash I_j A \Rightarrow A$, $1 \leq j \leq \omega$

(iv) $\vdash \underline{e}(x_1,y_1) \cap \ldots \cap \underline{e}(x_n,y_n) \Rightarrow (I_\omega d_m A(x_1,\ldots,x_n) \Rightarrow d_m A(y_1,\ldots,y_n))$

(v) $\vdash d_j\underline{e}(x_1,y_1) \cap \ldots \cap d_j\underline{e}(x_n,y_n) \Rightarrow (I_j d_m A(x_1,\ldots,x_n) \Rightarrow d_m A(y_1,\ldots,y_n))$ for $1 \leq j < \omega$.

We shall sketch an algebraic proof of the following

<u>Completeness Theorem</u> A formula A_o in F is provable from a set $\Sigma \subset F$ if and only if for every realization R in a gradual approximation space $\underline{A} = (U,eq)$, which is a model of Σ , R is a model of A_o.

Soundness, i.e. implication: if $\Sigma \vdash A_o$, then $\Sigma \models A_o$, is easy to prove by verifying logical axioms and rules of inference.

Given $\Sigma \subset F$, consider a relation $\equiv$ over F, defined by adopting $A \equiv B$ iff $\Sigma \vdash A \Rightarrow B$ and $\Sigma \vdash B \Rightarrow A$. It follows from intuitionistic axioms, (ax 3), rule (r_1) and 5.1 (i), that $\equiv$ is a congruence on the algebra of formulas. Let

$\underline{F/\equiv} = (F/\equiv\ , |\underline{e}_\omega|, \cup, \cap, \Rightarrow, \neg, (d_m)_{m\in N}, (\ \underline{e}_i\ ), (I_j), (C_j))$

where $0 \leq i \leq \omega$, $0 \leq j \leq \omega$, be the quotient algebra consisting of equivalence classes $|A|$ of $\equiv$ for $A \in F$. Let $\underline{P}$ be the reduct of $\underline{F}/\equiv$ obtained by elimination of I_j and C_j for $1 \leq j \leq \omega$. $\underline{P}$ is a GP-algebra in which

$$|A| = T \text{ iff } \Sigma \vdash A, \ |A| \leq |B| \text{ iff } \Sigma \vdash A \Rightarrow B, \ |\underline{e}_i| = e_i \tag{3}$$

$$|A| = \bigcup_{m \in N} (|d_m A| \cap |\underline{e}_m|) \text{ for each } A \in F \tag{4}$$

$$|\forall y A(y)| = \bigcap_{x \in Var} |A(y/x)|, \quad |\exists y A(y)| = \bigcup_{x \in Var} |A(y/x)| \tag{Q}$$

$$|d_m \forall y A(y)| = \bigcap_{x \in Var} |d_m A(y/x)|, \quad |d_m \exists y A(y)| = \bigcup_{x \in Var} |d_m A(y/x)| \tag{DQ}$$

for x not being in a scope of any quantifier in $A(y)$.

Assume non $\Sigma \vdash A_o$. Then by (3), $|A_o| \neq T$. Hence by (7) in Sec.1, there is $m \in N$ such that $|d_m A_o| \neq T$. On account of lemma in paper by Rasiowa and Sikorski 1950 there is a prime filter ∇_o in the Boolean algebra $\underline{B}_{\underline{P}}$ (of complemented elements in $\underline{P}$) which preserves infinite joins and infinite meets in (DQ) and such that

$$|d_m A_o| \notin \nabla_o. \tag{5}$$

By 1.3 in Sec.1,

$$\nabla = \{|A| : |d_m A| \in \nabla_o \quad \text{for all } m \in N\}$$

is a prime filter in $\underline{P}$, preserving infinite joins and meets in (Q) and (DQ) and such that

$$|A_o| \notin \nabla. \tag{6}$$

Moreover, the quotient algebra $\underline{P}/\nabla$ is isomorphic to $\underline{P}_\omega$. We have for every $A \in F$,

$$\|A\| = T \text{ iff } |A| \in \nabla . \tag{7}$$

Hence and (6),

$$\|A_o\| \neq T . \tag{8}$$

Let us set $\underline{A}$ = (Var, eq), where

$$eq(x,y) \overset{df}{=} \|\underline{e}(x,y)\| \quad \text{for } x,y \in Var . \tag{9}$$

Axioms (ax11), (ax12), (ax13) and (3) yield that eq $\in$ EQ(Var). Thus $\underline{A}$ is a gradual approximation space with equivalence relations $\cong_j$, $1 \leq j \leq \omega$, defined as follows

$$x_1 \cong_j x_2, \text{ for } 1 \leq j < \omega \text{ , iff } d_j \|\underline{e}(x_1,x_2)\| = T$$
$$\text{iff } d_j |\underline{e}(x_1,x_2)| \in \nabla \tag{10}$$

$$x_1 \cong_\omega x_2 \text{ iff } d_j \|\underline{e}(x_1,x_2)\| = T \text{ for all } 1 \leq j < \omega \text{ iff}$$
$$d_j |\underline{e}(x_1,x_2)| \in \nabla \text{ for all}$$
$$1 \leq j < \omega . \tag{11}$$

Now we are going to define a canonical realization R in the canonical approximation space $\underline{A}$ = (Var, eq).

Let us set

$$\underline{e}_R = eq \text{ and } p_i(x_1,\ldots,x_n)_R = \|p_i(x_1,\ldots,x_n)\| . \tag{12}$$

<u>Lemma</u> For each A $\in$ F and valuation v : Var $\longrightarrow$ Var

$$A_R(v) = \|vA\| , \tag{13}$$

where vA is a formula obtained from A by the simultaneous replacement of each free occurrance of individual variable x by v(x).

The proof is by an inductive argument with respect to the length of a formula A. For atomic formulas (13) follows from (12), (9) and (3). In the case of formulas $B \cup D$, $B \cap D$, $B \Rightarrow D$,

$\neg B$, $d_m B$ for $m \in N$, $\forall x\, A(x)$, $\exists\, xA(x)$, proofs are standard on applying the definition of realization of formulas, (Q), (DQ) the fact that in $\underline{P}/\equiv$ all infinite joins and meets in (Q) and (DQ) are preserved.

Now assume Var A = $\{x_1,\dots,x_n\}$ and

$$d_m A(x_1,\dots,x_n)_R(v) = \|v d_m A(x_1,\dots,x_n)\| \text{ for } v \in Val,\ m \in N \quad (14)$$

Claim:

$$I_j d_m A(x_1,\dots,x_n)_R(v) = \|v I_j d_m A(x_1,\dots,x_n)\|\ ,\ 1 \leq j \leq \omega\ . \quad (15)$$

It follows from (1) (ax17), (ax 6), that both sides of (15) can only adopt values $\top$, $\bot$. Thus to prove (15) it is sufficient to demonstrate, that the left-hand side adopts value $\top$ iff the right-hand side adopts the value $\top$.

Assume $\|v I_j d_m A(x_1,\dots,x_n)\| = \top$. By (7) $|v I_j d_m A(x_1,\dots,x_n)| \in \nabla$. Suppose $v(x_i) \cong_j y_i$, $i=1,\dots,n$. Then by (10), $d_j |\underline{e}(v(x_i),y_i)| \in \nabla$ for $i=1,\dots,n$. Hence and from 5.1 (v) it follows that $|v' d_m A(x_1,\dots,x_n)| =$ $= |d_m A(y_1,\dots,y_n)| \in \nabla$ by valuation v', such that $v'(x_i) = y_i$, $i=1,\dots,n$ and $v'(z) = v(z)$ for $z \neq x_1,\dots,x_n$. By (14) this implies that $d_m A(x_1,\dots,x_n)_R(v') = \top$. Hence by (1) it follows that $I_j d_m A(x_1,\dots,x_n)_R(v) = \top$.

In a similar way, applying (7), (11), 5.1 (iv) and (1) it can be proved that the assumption $\|v I_\omega d_m A(x_1,\dots,x_n)\| = \top$ implies $I_\omega d_m A(x_1,\dots,x_n)_R(v) = \top$.

Now assume $I_j d_m A(x_1,\dots,x_n)_R(v) = \top$. This implies by (1) that: for all $y_1,\dots,y_n \in$ Var, if $v(x_i) \cong_j y_i$ for $i=1,\dots,n$, then $d_m A(x_1,\dots,x_n)_R(v') = \top$, where $v'(z)=v(z)$ for $z \neq x_1,\dots,x_n$ and $v'(x_i)=y_i$ for $i=1,\dots,n$. Hence, by (14), and (7),

$|v' d_m A(x_1,\dots,x_n)| \in \nabla$ provided $|d_j \underline{e}(v(x_i),y_i)| \in \nabla$ for

$$i = 1,\ldots,n. \tag{16}$$

On applying 1.3 (14) in Sec.1, and definition of v' we get

$|d_j\underline{e}(v(x_1),y_1)| \cap \ldots \cap |d_j\underline{e}(v(x_n),y_n)| \Rightarrow | d_m A(y_1,\ldots,y_n)| \in \nabla$

for all $y_1,\ldots,y_n \in$ Var. This by 1.3 (12) and (DQ) yields

$$|\forall y_1 \ldots \forall y_n (d_j\underline{e}(v(x_1),y_1) \cap \ldots \cap d_j\underline{e}(v(x_n),y_n) \Rightarrow d_m A(y_1,\ldots,y_n)| \in \nabla .$$

Hence, by (ax16)

$|vI_j d_m A(x_1,\ldots,x_n)| \in \nabla$, i.e. $\| vI_j d_m A(x_1,\ldots x_n)\| = \top$.

On applying an analogous argument and (14), (ax15) it is easy to prove that assumption $I_\omega d_m A(x_1,\ldots,x_n)_R(v) = \top$ implies that $\| vI_\omega d_m A(x_1,\ldots,x_n)\| = \top$. This completes the proof of (15).

In the case of Var A = $\emptyset$, $I_j d_m A_R(v) = d_m A_R(v) = \| vd_m A \| = \| vI_j d_m A \|$ by (1'), (14), (ax14).

The inductive step for $C_j d_m A$ is based on (1), (2), the inductive step for $I_j d_m A$ just proved and (ax19).

Now assume that A is any formula and for all $m \in N$ and $v \in$ Val , $1 \leq j \leq \omega$, $I_j d_m A_R(v) = \| vI_j d_m A \|$ and $C_j d_m A_R(v) = \| vC_j d_m A \|$. (17)

By (18) in Sec.4, (ax17), (4) we obtain

$$I_j A_R(v) = \bigcup_{m \in N} (I_j d_m A_R(v) \cap \| \underline{e}_m \|) = \bigcup_{m \in N} (\| vI_j d_m A \| \cap \| \underline{e}_m \|) =$$

$$= \bigcup_{m \in N} (\| vd_m I_j A \cap \underline{e}_m \|) = \| vI_j A \| . \tag{18}$$

Analogously on applying (ax18) instead of (ax17) we get

$$C_j A_R(v) = \| vC_j A \| . \tag{19}$$

This completes the proof of Lemma.

It follows from Lemma that R is a model of Σ .

But by the identical valuation v_o, such that $v_o(x) = x$ for $x \in$ Var, $A_{oR}(v_o) = \| A_o \| \neq \top$ by (8).

This completes the proof of the Completeness Theorem.

References

Creswell M.J. and Hughes G.E., 1980, An introduction to modal logic. London: Metheuen and Co. Ltd.

Epstein G., The lattice theory of Post algebras, 1960, Trans. Amer. Math. Soc. 95, 63-74

Hunter G.M. and Steiglitz K., 1979, Operations on images using quadtrees, IEEE Trans.Pattern Recog. Mach. Int., 145-153

Maksimova L. and Vakarelov D., 1974, Representation Theorems for Generalized Post Algebras of Order ω^+, Bull. Pol. Acad. Sci., Ser. Math. Astron. Phys. 22, 757-764

Marek W. and Rasiowa H., 1986, Approximating Sets with Equivalence Relations, Proc. Int. Symp. on Methodologies for Intelligent Systems, October 23-25, Knoxville - Tennessee, USA, to appear

Pawlak Z., 1982, Rough Sets, Int. Journ. of Comp. and Information Science 11(5), 341-356

Rasiowa H., 1973, On generalized Post algebras of order ω^+ and ω^+-valued predicate calculi, Bul. Ac. Pol. Sci., Ser. Math. Astr. Phys. 21 No 3, 209-219

Rasiowa H., 1985, Topological Representations of Post Algebras of Order ω^+ and Open Theories Based on ω^+-Valued Post Logic, Studia Logica 44 No 4, 353-368

Rasiowa H., 1986, Rough Concepts and ω^+-Valued Logic, Proc. 16th Int. Symp. Multiple-Valued Logic, Blacksburg, VA, USA, Computer Society Press, 282-288

Rasiowa H. and Skowron A., 1985, Rough Concepts Logic, in: Computation Theory, ed. A.Skowron, LNCS 208, 288-297

Rasiowa H. and Skowron A., 1986, Approximation Logic, in: Mathematical Methods of Specification and Synthesis of Software Systems'85, ed. W.Bibel and K.P.Jantke, Mathematical Research 31, Akademie Verlag, Berlin, 123-139

Ullman J.D., 1983, Principles of Database Systems, Computer Science Press, Rockville MD

INTUITIONISTIC FORMAL SPACES - A FIRST COMMUNICATION

Giovanni Sambin

Dipartimento di Matematica, Universitá di Siena
Via del Capitano 15, 53100 Siena, Italy

The notion of formal space was introduced by Fourman and Grayson [FG] only a few years ago, but it is only a recent though important step of a long story whose roots involve such names as Brouwer and Stone and whose development is due to mathematicians from different fields, mainly algebraic geometry, category theory and logic.

I am not going to tell this story (but see for instance [J] and [G]). For our purposes here, it is enough to say that the main idea is to reverse the traditional conceptual order of definitions in topology and define points as particular filters of neighbourhoods, rather than opens as particular sets of points (I adjust the language to this point of view, by considering open also as a noun). This explains why our topic is sometimes called pointless topology; formal or abstract topology, or just topology tout court would be preferable.

The basic notion is that of locale which, roughly speaking, is a lattice satisfying all those properties of opens in a topological space which are expressible without mention of points. Thus locales are complete lattices, where $\wedge$ and $\vee$ correspond to finite intersection and arbitrary union, satisfying the law of infinite distributivity $a \wedge \bigvee\{b_i : i \in I\} = \bigvee\{a \wedge b_i : i \in I\}$. One could also look at locales as the solution of x : topological spaces = boolean algebras : Stone spaces.

In [FG] a method is given to construct a locale from an entailment relation and the result is called a formal space. We will show here that, with minor modifications, their method is general enough to yield all locales. Our basic notion is that of covering relation, which, besides entailment, is strictly related to what is known in the literature under a wide variety of names: coverage, J-operator, congruence, ...

This is not a mere technical device. In fact, from the intuitionistic point of view, which is here taken seriously and thus is not reduced to putting an asterisk where the axiom of choice is used, the notion of covering permits a study of topology which avoids such problematic notions as the powerset of a given set or quantification over subsets. I here try to show how this is possible, that is begin to develop formal topology in the framework of Martin-Löf's intuitionistic type theory (or constructive set theory).

One of the fundamental aspects of Martin-Löf's type theory is the distinction between sets (or (data) types) and categories (or logical types). However, while a formal treatment of sets has by now been developed and reached maturity (see for instance [ITT]), a similar work for categories is still in progress. The lack of texts with rules to handle categories is particularly felt here, where we deal with the category of opens, among others (but see [SI] for the basic logical types of propositions and truths). I thus have to ask the reader to rely on a pragmatic principle, which is based on my understanding of Martin-Löf's views by direct talking: all what we are going to do informally, can also be done formally in Martin-Löf's foundational theory, once it will appear in complete form. Only a few additions to [ITT] are necessary here, and we give them in the preliminaries below.

During my visit to Stockholm in Spring 1984, following Martin-Löf's proposal I began to work with him on the topics of this paper; one of his aims then was to give a general form to the connection he had discovered between denotational and operational semantics for programming languages (cf. [SG] and the course given in Udine at the Course on Computation Theory in September 1984). Since then, we have gradually changed basic definitions, also under the influence of the literature, mainly [FG]. The first public outcome was the 'short course' I gave at the VIII Incontro di Logica Matematica in Siena in January 1985. The present formulation is due to Martin-Löf and was presented by him in a few seminars in Stockholm in May 1985.

This paper contains little beyond definitions and a few meaningful examples, including Scott's domains. Much work remains to be done, which, together with time, will certainly cause modifications on what is presented here. It should be clear however that even the present fragment would not exist without the work by Per Martin-Löf; most ideas, definitions and examples have been suggested by him and discussed together (but of course responsibility here is only mine). I am glad to thank him for teaching me so much. I also thank Isa Bossi and Silvio Valentini.

Preliminaries

The following simple though radical additions to [ITT] will be sufficient here. A subset U of a set S is a unary propositional function with argument ranging over S, shortly U(a) prop ($a \in S$); we write as usual $U \subseteq S$, and $a \in U$ for U(a) (note that, contrary to [ITT], $a \in U$ is a proposition here). We also often write {a: U(a)} for U, and {f(a): U(a)} for $(\exists a \in S)(I(S,x,f(a))\ \&\ U(a))$. A relation between A and B, where A, B are either sets or categories, is just a propositional function with two arguments, one in A and one in B. Recall that propositions, and hence a fortiori subsets of any set, form a category which is not a set. For A, B propositions, we here write $A \leq B$ for A true $\rightarrow$ B true and thus $A \leq B$ iff $A \rightarrow B = T$, where $T \equiv$ any true proposition and $(A = B) \equiv (A \leq B\ \&\ B \leq A)$. Finally, recall that any set is given together with an equality relation between elements; such equality is denoted simply by = here since it should always be clear to which set it refers.

1. Formal topologies

The classical notion of topological space is not suitable, as it stands, for an intuitionistic treatment, mainly because opens generally form a proper category and coverings are always defined pointwise. To bring opens into our framework, we build them up from basic neighbourhoods, which are supposed to form a set, by means of an abstract covering relation; and we define points in terms of the algebra of opens ('pointless topology'), and thus ultimately as particular filters of neighbourhoods.

Thus the usual notion of topological space corresponds here to two notions, formal topology and formal space. We must begin with the former. The hints above should justify the following definition (but see also section 2 below for some intuitive motivations):

Definition 1.1 A formal topology *A* consists of:

1. a formal base, namely a set S_A with a binary operation $\wedge_A$ and a distinguished element Δ_A, such that $S_A, \wedge_A, \Delta_A$ form a semilattice with one;
2. a covering relation, that is a relation $a \leq_A U$ prop ($a \in S$, $U \subseteq S$) which for arbitrary $a, b \in S_A$, $U, V \subseteq S_A$ satisfies

reflexivity $$\frac{a \in U}{a \leq_A U}$$

transitivity $$\frac{a \leq_A U \quad U \leq_A V}{a \leq_A V} \qquad \text{where } U \leq_A V \equiv (\forall a \in U)(a \leq_A V)$$

$\wedge$-left $$\frac{a \leq_A U}{a \wedge_A b \leq_A U} \qquad \frac{b \leq_A U}{a \wedge_A b \leq_A U}$$

$\wedge$-right $$\frac{a \leq_A U \quad a \leq_A V}{a \leq_A U \wedge_A V} \qquad \text{where } U \wedge_A V \equiv \{b \wedge_A c:\ b \in U, c \in V\}$$

3. a consistency predicate, that is a property $Pos_A(a)$ prop $(a \in S)$ which for arbitrary $a \in S_A$, $U \subseteq S_A$ satisfies

monotonicity $$\frac{Pos_A(a) \quad a \leq_A U}{Pos_A(U)}$$ where $Pos_A(U) \equiv (\exists b \in U)Pos_A(b)$

ex falso quodlibet $$\frac{\neg Pos_A(a)}{a \leq_A \varnothing}$$

openness $$\frac{a \leq_A U}{a \leq_A U^+}$$ where $U^+ \equiv \{b \in U : Pos_A(b)\}$

Elements a, b, c, ... of S_A are called formal basic neighbourhoods, or simply neighbourhoods; $a \leq_A U$ is read U covers a, or U is a cover of a, and $Pos_A(a)$ is read a is positive, or consistent. We read also $U \leq_A V$ as V covers U; it is then natural to say that two subsets are equal if they cover each other:

$$(U =_A V) \equiv (U \leq_A V \;\&\; V \leq_A U)$$

It is immediate to see that also for subsets $\leq_A$ is reflexive and transitive, and hence that $=_A$ is an equivalence relation.

For convenience, we will omit the subscript *A*, except when it is necessary to avoid confusion; in particular, it is needed to distinguish $=_A$ above from extensional equality of subsets, denoted by = as usual.

Definition 1.2 For any formal topology *A*, Open(*A*) is the category *P*(S) of subsets of S, with equality $=_A$. The objects of Open(*A*) are called formal opens of *A*.

In other words, an open in *A* is the equivalence class under $=_A$ of a subset of S.

It is well known that opens of a topological space form a complete Heyting algebra (see for instance [J], p. 39). Our next aim is to see that this is true also for formal opens. However, it is easier if we first have a little stock of derived rules at our disposal.

From reflexivity we have $U \subseteq V \rightarrow U \leq V$, and hence

weakening $$\frac{a \leq U \quad U \subseteq V}{a \leq V}$$

Applying weakening and substitution rules of [ITT], we have

substitution $$\frac{a \leq U \quad U = V}{a \leq V} \qquad \frac{a = b \quad b \leq V}{a \leq V}$$

A fortiori, we obtain that $=_A$ respects $\leq$; that is, writing $a \leq b$ for $a \leq \{b\}$, and hence $a =_A b$ for $\{a\} =_A \{b\}$

$$\frac{a =_A b \quad b \leq U \quad U =_A V}{a \leq V}$$

Since $a \wedge \Delta = a$, by $\wedge$-left applied to $\Delta \leq \Delta$ we have $a \leq \Delta$ and hence

$$U \leq \Delta$$

Applying first $\wedge$-left and then $\wedge$-right, we obtain

stability $$\frac{a \leq U \quad b \leq V}{a \wedge b \leq U \wedge V}$$

which, since $b \le b$, gives in particular

$$\text{localisation} \quad \frac{a \le U}{a \wedge b \le U \wedge b} \qquad \text{where } U \wedge b \equiv \{a \wedge b : a \in U\}$$

This completes the list of derived rules we need.

We have already seen that $\le$ is a preorder on $P(S)$. So, to see that it is a partial order on Open(A), it is enough to check that $\le$ is respected by $=_A$, namely

$$(U \le V') \;\&\; (U =_A U') \;\&\; (V =_A V') \;\rightarrow\; U' \le V'$$

which is obvious by transitivity.

Next we show that $\wedge$ gives to Open(A) the structure of a semilattice. To begin with, we must check that $=_A$ respects $\wedge$, namely

$$(U =_A U') \;\&\; (V =_A V') \;\rightarrow\; U \wedge V =_A U' \wedge V'$$

(Proof. If $a \in U \wedge V$, then there exist $b \in U$, $c \in V$ such that $a = b \wedge c$. Since $U \le U'$, $V \le V'$, we have $b \le U'$, $c \le V'$ and hence $b \wedge c \le U' \wedge V'$ by stability. This proves $U \wedge V \le U' \wedge V'$, and the converse holds by symmetry.) Then it is enough to show that $\wedge$ gives the infimum, or meet, with respect to $\le$, namely

$$U \wedge V \le U, \quad U \wedge V \le V$$

$$W \le U \;\&\; W \le V \;\rightarrow\; W \le U \wedge V$$

which are obvious by $\wedge$-left and $\wedge$-right respectively.

Since $U \le \Delta$ and obviously $\varnothing \le U$ for any U, $\varnothing$ and $\{\Delta\}$ are zero and one of Open(A), respectively.

The last step is to define arbitrary suprema, or joins. For any family of subsets $(U_i)_{i \in I}$ we put

$$\bigvee_{i \in I} U_i \equiv \bigcup_{i \in I} U_i$$

(in section 3 we try to give a reason for this definition). Again, we first have to check that $=_A$ respects $\bigvee$, that is

$$(\forall i \in I)(U_i =_A U_i') \;\rightarrow\; \bigvee_{i \in I} U_i =_A \bigvee_{i \in I} U_i'$$

This is not difficult, by using weakening. Now we easily see (the first by weakening, the second by intuitionistic logic) that indeed $\bigvee$ gives joins:

$$(\forall i \in I)(U_i \le \bigvee_{i \in I} U_i)$$

$$(\forall i \in I)(U_i \le V) \;\rightarrow\; \bigvee_{i \in I} U_i \le V$$

Finally, the definition of $\wedge$ tells us that $V \wedge \bigcup_{i \in I} U_i = \bigcup_{i \in I} (V \wedge U_i)$, and hence a fortiori infinite distributivity holds:

$$V \wedge \bigvee_{i \in I} U_i =_A \bigvee_{i \in I} (V \wedge U_i)$$

We follow [FG] here and use the word frame for complete lattices with infinite distributivity; so we have shown above that Open(A) is a frame. It is well known that in any frame the operation of implication is definable, that is we can make a complete Heyting algebra out of any frame. In our case, such definition reduces to:

$$U \rightarrow_A V \equiv \{a \in S : U \wedge a \le V\}$$

We leave it to the reader to show, first of all, that $\rightarrow_A$ is well defined (i.e. it is respected by $=_A$), and that it satisfies the usual characterization of implication, namely

$$W \le U \rightarrow_A V \quad \text{iff} \quad W \wedge U \le V$$

We thus have completed the proof of

Proposition 1.3 Open(A), $\wedge_A$, $\vee_A$, $\varnothing$, Δ_A, $\rightarrow_A$ form a complete Heyting algebra

Proving that any complete Heyting algebra, or better any frame, is obtainable as above, is now an easy task. We say that a frame H is based on a set D_0 if H is the closure of D_0 under finite meets and arbitrary joins. The closure of D_0 under finite meets, including the top element of H, forms a set D, which is a semilattice with one. We define a covering on D putting

$$a \leq_K U \equiv a \leq_H \vee U$$

Then the assignment $U \rightarrow \vee U$ is an isomorphism of Open(A) onto H, because by infinite distributivity $\wedge$ is preserved and any element of H can be written as join of elements in D.

From the classical point of view, any frame is trivially set-based and hence representable as above, which means that our approach is not restrictive. On the other hand, it is our claim that the notion of formal topology itself is the intuitionistic counterpart of classical frames, and what follows should justify it.

We now turn to the consistency predicate. Recalling that $Pos(U) \equiv (\exists b \in U)Pos(b)$, we can easily extend monotonicity to subsets

$$\frac{Pos(U) \quad U \leq V}{Pos(V)}$$

In particular $=_A$ respects Pos, that is

$$Pos(U) \;\&\; (U =_A V) \rightarrow Pos(V)$$

so that we may speak of consistency for opens. The empty open is not consistent,

$$\neg Pos(\varnothing)$$

since trivially $\neg(\exists a \in \varnothing)Pos(a)$. Therefore also

$$\frac{U \leq \varnothing}{\neg Pos(U)}$$

since Pos(U) together with $U \leq \varnothing$ gives $Pos(\varnothing)$ by monotonicity. But by ex falso quodlibet also the converse holds

$$\frac{\neg Pos(U)}{U \leq \varnothing}$$

and hence

$$\neg Pos(U) \leftrightarrow U =_A \varnothing$$

that is, $\varnothing$ is the only inconsistent open.

We intend Pos(U) to be a positive way to assert that U is not empty (classically, from what above we would have $Pos(U) \leftrightarrow U \neq_A \varnothing$). A similar predicate is also introduced in [FG], p. 113, but its definition requires a quantification over subsets, which is not accepted here.

Openness is necessary to express the fact that only consistent opens contribute to covers. From the classical point of view, openness has little meaning, since one can easily prove that:

Proposition 1.4 Any decidable property satisfying monotonicity and ex falso quodlibet, also satisfies openness, and hence is a consistency predicate.

Finally note that openness is equivalent to the equation $U =_A U^+$.

2. Example. Concrete spaces

Consider the following naive definition of topological spaces:

Definition 2.1 A concrete space $\mathcal{M}$ consists of:
1. a set M of concrete points m, n, ...
2. a set S of indices a, b, ..., with a binary operation $\wedge$ and a distinguished element Δ
3. a neighbourhood relation, that is a relation N(m,a) prop ($m \in M, a \in S$) which for arbitrary $m \in M$, $a, b \in S$ satisfies:

$$N(m, \Delta) = T, \quad N(m, a \wedge b) = N(m,a) \;\&\; N(m,b)$$

Of course, a is meant to be an index for the subset $N_a \equiv \{m \in M: N(m,a)\}$, called a concrete neighbourhood. The family $(N_a)_{a \in S}$ is thus a concrete base for a concrete topology on M, since, by the assumptions in 3., $N_\Delta = M$ and $N_{a \wedge b} = N_a \cap N_b$. Concrete opens are then defined, as usual, to be unions of concrete neighbourhoods: for every $U \subseteq S$, $N_U \equiv \{m \in M: (\exists a \in U)N(m,a)\}$ is the concrete open which is the union of the family $(N_a)_{a \in S}$. We might equivalently define concrete opens to be those subsets P of M for which the usual condition $(\forall m \in P)(\exists a \in S)(N_a \subseteq P \;\&\; m \in N_a)$ holds.

The trouble is that many interesting spaces can not be presented in this way, because the points we want to consider do not form a set or the covering relation is not defined pointwise. This is why we introduce the notion of formal topology, which now can be seen as the abstract result of the following concrete actions on concrete spaces:
1. add to S, $\wedge$, Δ a relation $a \leq U$, which holds iff $(N_b)_{b \in U}$ covers N_a and a predicate Pos(a) which holds iff $(\exists m \in M)(m \in N_a)$;
2. write down all the properties of $\wedge$, Δ, $\leq$ and Pos which can be expressed without mentioning concrete points;
3. get rid of concrete points, and hence also of the neighbourhood relation.

The content of this intuitive explanation is formally expressed as follows: if, for any concrete space $\mathcal{M}$, we put

$$S_{\mathcal{A}(\mathcal{M})} \equiv S, \quad \wedge_{\mathcal{A}(\mathcal{M})} \equiv \wedge, \quad \Delta_{\mathcal{A}(\mathcal{M})} \equiv \Delta$$

$$a \leq_{\mathcal{A}(\mathcal{M})} U \equiv (\forall m \in M)(N(m,a) \rightarrow (\exists b \in U)N(m,b)) \qquad Pos_{\mathcal{A}(\mathcal{M})}(a) \equiv (\exists m \in M)(N(m,a)$$

we obtain a formal topology $\mathcal{A}(\mathcal{M})$ with base S.

Let us look at a concrete example of concrete space. Contrary to propositions, the formulae of a fixed formal language L, say that of predicate logic, form a set. In particular, let S be the set of formulae A(x) with at most x free. Given a structure for L based on the set M, we put $N(m,A) \equiv A(m)$ is true. Then M, S, N form a concrete space, whose concrete neighbourhoods are the L-definable subsets of M, and of course two formal neighbourhoods are extensionally equal iff they define the same subset of M.

3. Coverings as closure operators

Some reader may be annoyed by the fact that we give formal opens of a formal topology $\mathcal{A}$ only up to $=_{\mathcal{A}}$-equivalence classes. If so, the following alternative approach can be taken.

With any formal topology $\mathcal{A}$, we associate an operator $Cl_{\mathcal{A}}$ acting on subsets of S, by putting

$$Cl_{\mathcal{A}}(U) \equiv \{a \in S: a \leq_{\mathcal{A}} U\}$$

To minimize subscripts, we often write $\mathcal{A}(U)$ for $Cl_{\mathcal{A}}(U)$. Note that by definition $a \in \mathcal{A}(U)$ iff $a \leq U$. Therefore

$$U \leq V \;\text{ iff }\; U \subseteq \mathcal{A}(V)$$

Using this equivalence, one easily shows that:

Proposition 3.1 For any formal topology A, the following hold:

1. $U \subseteq A(U)$
2. $U \subseteq V \rightarrow A(U) \subseteq A(V)$
3. $A(A(U)) = A(U)$

That is, Cl_A is a closure operator on $P(S)$. Again by the above equivalence, we have

$$U \leq V \text{ iff } A(U) \subseteq A(V)$$

and hence

$$U = V \text{ iff } A(U) = A(V)$$

In particular, also

$$U =_A A(U)$$

Let us say that U is A-closed, or saturated, when $U = A(U)$; then each $=_A$-equivalence class is represented by one and only one A-closed subset, which is the greatest in the class. That is, A : Open(A) → Sat(A), where Sat(A) is the category of saturated subsets, is a bijection.

It is well known that, when A is a closure operator, A-closed subsets form a complete lattice, with meets given by intersection and joins defined by $\vee_{i \in I} A(U_i) \equiv A(\cup_{i \in I} A(U_i))$. With standard calculations, we see that $A(\cup_{i \in I} U_i) = \vee_{i \in I} A(U_i)$, which at the same time means that the function A preserves joins and justifies the very definition of joins as unions which we gave for opens.

Since $A(\Delta) = S$, A preserves one, and since $A(\varnothing)$ is the least saturated subset, A also preserves zero. Thus to show that A is a frame homomorphism, we only have to prove that A preserves meets, i.e.

4. $A(U \wedge V) = A(U) \cap A(V)$

The inclusion from left to right follows immediately from $U \wedge V \leq U$, $U \wedge V \leq V$, while the converse inclusion holds by $\wedge$-right.

Finally, note that $U \rightarrow_A V$ is saturated for any $U, V \subseteq S$ (we leave this as an exercise). This means both that $\rightarrow_A$ is a good implication also in Sat(A), and that $A(U \rightarrow_A V) = A(U) \rightarrow_A A(V)$, that is A preserves $\rightarrow_A$. This completes the proof of:

Proposition 3.2 Sat(A) is a complete Heyting algebra isomorphic to Open(A).

We will in the sequel confuse the two algebras; as hinted above, in Sat(A) equality is extensional.

We have shown that for any formal topology A, Cl_A is a closure operator which preserves meets, i.e. is an operator on $P(S)$ satisfying 1.-4. An interesting fact is that the converse also holds, in the following sense. For every operator $C : P(S) \rightarrow P(S)$, we define a relation $\leq_C$ by putting

$$a \leq_C U \equiv a \in C(U)$$

Then it is not difficult to reverse what we did, and show that when C is a closure operator $\leq_C$ is reflexive and transitive, and moreover that, when C satisfies 4., $\wedge$-rules hold for $\leq_C$. Since the correspondence between C and $\leq_C$ is obviously biunivocal, we have

Proposition 3.3 There is an isomorphism between covering relations and meet-preserving closure operators.

The notion of J-operator (see e.g. [J], p. 48) is strictly related to that of meet-preserving closure operator. In fact, one can show that J-operators on Open(A) coincide with those meet-preserving closure operators B which satisfy $A(U) \subseteq B(U)$ for any $U \subseteq S$ (cf. definition 4.4 below).

4. Points and spaces

We have introduced bases, opens and coverings up to now, but still miss points. To grasp the following definition, it may help to look at it as the result of describing properties of concrete points without mentioning them.

Definition 4.1 A formal point on the formal topology A is a subset $\alpha(a)$ prop ($a \in S$) of S which for arbitrary a, b $\in$ S, U $\subseteq$ S satisfies:
1. $\alpha(a \wedge b) = \alpha(a) \;\&\; \alpha(b)$, $\alpha(\Delta) = T$
2. $a \leq_A U \rightarrow \alpha(a) \leq \alpha(U)$ where $\alpha(U) \equiv (\exists b \in U)\alpha(b)$
3. $\alpha(a) \rightarrow \mathrm{Pos}(a)$

The definition above is simply the translation into our approach of the usual definition of points as completely prime filters over a frame (see e.g. [J], p. 41). To see this, let us first say that a propositional function F(U) prop ($U \subseteq S$) is a filter on Open(A) if $F(U \wedge V) = F(U) \;\&\; F(V)$, $F(\Delta) = T$ and $U \leq V \rightarrow F(U) \leq F(V)$. We then say that the filter F is completely prime if $F(\cup_{i \in I} U_i) = (\exists i \in I)F(U_i)$ and consistent if $F(U) \rightarrow \mathrm{Pos}(U)$ (note that classically consistency is expressed by requiring $\neg F(\emptyset)$). Now, given a point α we put as above $\alpha(U) \equiv (\exists a \in U)\alpha(a)$ and then can easily check that $\alpha(U)$ prop ($U \subseteq S$) is a completely prime consistent filter on Open(A). Conversely, given a completely prime consistent filter F we put $\alpha_F(a) \equiv F(\{a\})$ and check (exercise) that α_F is a point. Since the correspondence is obviously biunivocal, we have

Proposition 4.2 There is an isomorphism between points on A and completely prime consistent filters on Open(A).

We denote by Pt(A) the category of points on A. Writing $\varphi(\alpha,a)$ for $\alpha(a)$, we immediately see that $\varphi(\alpha,a)$ prop ($\alpha \in$ Pt(A), $a \in S$) is a neighbourhood relation, since by the definition of point

$$\varphi(\alpha, \Delta) = T\,,\quad \varphi(\alpha,a \wedge b) = \varphi(\alpha,a) \;\&\; \varphi(\alpha,b)$$

We don't have a concrete space, however, for the simple reason that Pt(A) is in general not a set. Still we say that a is a formal neighbourhood of α if $\varphi(\alpha,a)$, that is $\alpha(a)$, is true. More generally, we can define formal topological notions on Pt(A) in terms of A (for instance, see the definition of subspaces below and of continuos functions in section 6). This justifies the following

Definition 4.3 For any formal topology A, we call Pt(A) the formal space induced by A.

Of course, a reader who doesn't distinguish sets from categories will consider Pt(A) as an ordinary topological space, with the base given by the family $\varphi(a) \equiv \{\alpha \in \mathrm{Pt}(A): \varphi(\alpha,a)\}$ for $a \in S$ and where of course the covering relation is just inclusion. Then the assertion that $\varphi(U) \equiv \cup\{\varphi a: a \in U\}$ covers φa in Pt(A) iff U covers a in A, or equivalently $(\forall\alpha \in \mathrm{Pt}(a))(\alpha(a) \rightarrow \alpha(U)) \rightarrow a \leq U$, is far from being trivial, and actually is often a strong existence principle (see for instance end of sections 7 and 9). Here we take the other way round, and define $\varphi a \subseteq \varphi(U)$ to mean that $a \leq U$ (which is possible only because in constructive set theory no other meaning is given to quantification over the category of points). For similar reasons, we say that Pt(A) is proper when Pos(Δ) is true.

Given any two formal topologies A, B on the same base S, we say that A is finer than B (and B coarser than A) if every open in B is also an open in A, and every neighbourhood which is consistent in B is also consistent in A. To put it in symbols, $A(U) \subseteq B(U)$ and $\mathrm{Pos}_B(U) \rightarrow \mathrm{Pos}_A(U)$ for any $U \subseteq S$. It is immediate to see that when A is finer than B any point on B is also a point on A. This justifies

Definition 4.4 If A is finer than B, then Pt(B) is called a subspace of Pt(A).

Two kinds of spaces deserve specific attention, and hence specific names.

Definition 4.5 If the covering relation $\leq$ of a formal topology A satisfies 1. (or 2.) below, then A is called a Stone (or Scott, resp.) topology, and Pt(A) a Stone (or Scott, resp.) space:
1. $a \leq U \rightarrow a \leq U_0$ for some finite $U_0 \subseteq U$
2. $\mathrm{Pos}(a) \;\&\; a \leq U \rightarrow a \leq b$ for some $b \in U$

The names I have chosen are justified below (after my lecture in Druzhba, Y. Ershov has kindly brought to my attention f-spaces, which he introduced independently of Scott and which should correspond exactly to what I here call Scott spaces; thus it seems that the name Scott here is justified not only by the results in section 8, but also by gaps in my knowledge, which I am not able to fill in now).

A perspicuous characterization of Stone and Scott topologies is easily obtained through closure operators. Recall that a closure operator C is called algebraic if $C(U) = \cup\{C(U_0): U_0$ is a finite subset of $U\}$ holds for any $U \subseteq S$. Here we also say that C is irreducible if for any consistent $U \subseteq S$, $C(U) = \cup\{C(a): a \in U\}$ (where of course $C(a) \equiv C(\{a\})$). Then, working out definitions and using the equality $A(U) = A(U^+)$, we obtain the following characterizations:

Proposition 4.6 For any formal topology A,
1. A is Stone iff Cl_A is algebraic;
2. A is Scott iff Cl_A is irreducible.

Points on Stone and Scott topologies also have a neat characterization. As usual, let us say that a subset $\alpha(a)$ prop $(a \in S)$ is a consistent filter on $S, \wedge, \leq_A$ if it satisfies 1.1, 1.3 above and $a \leq_A b \rightarrow \alpha(a) \leq \alpha(b)$; we also say that a filter α is prime if it satisfies $a \leq_A \{b_1, \cdots, b_n\} \rightarrow \alpha(a) \leq \alpha(b_1) \vee \cdots \vee \alpha(b_n)$. Then, by a little more than working out definitions, we obtain

Proposition 4.7 For any formal topology A,
1. if A is Stone, Pt(A) is the category of consistent prime filters;
2. if A is Scott, Pt(A) is the category of all consistent filters.

It is clear that any Stone topology is compact, in the sense that from $\Delta \leq U$ we have $\Delta \leq U_0$ for some finite U_0. Given an arbitrary topology A, we can get its Stone compactification $C(A)$ quite easily, namely by declaring $a \leq_{C(A)} U$ to hold iff $a \leq_A U_0$ holds for some finite $U_0 \subseteq U$, while keeping the same consistency predicate. It is routine to check that we indeed obtain a topology, which is Stone by definition, and that any other Stone topology finer than A is also finer than $C(A)$. We can similarly obtain a Scott topology $S(A)$ out of A by requiring, if Pos(a) is true, $a \leq_{S(A)} U$ to hold iff $(\exists b \in U)(a \leq_A b)$. We thus have

Proposition 4.8 For any formal topology A, the topology $C(A)$ (or $S(A)$) defined above is the coarsest Stone (or Scott, resp.) topology finer than A.

5. Some examples

We present here some, hopefully instructive, examples of formal topologies and formal spaces. In some of them we have used notions defined in successive sections.

a. The real numbers

Let Q be the set of rational numbers, S the set of pairs (p,q) where p, q are either rational or $\pm\infty$. We define Δ to be $(-\infty,+\infty)$ and $\wedge$ by $(p,q) \wedge (p',q') \equiv (\max\{p,p'\},\min\{q,q'\})$; then S, $\wedge$, Δ form a semilattice. We can define on S a covering relation essentially as in [J], p.123, and a positivity predicate by putting $Pos((p,q)) \equiv (p < q)$, where $<$ is the order of Q with top and least elements $\pm\infty$. The result will be a formal topology, whose corresponding formal space is the space of real numbers (see [J], pp. 123 - 125 for details).

b. Compactifications of N

The set N of natural numbers is constructed as usual by means of the rules

$$0 \in N \qquad\qquad \frac{n \in N}{s(n) \in N}$$

Basing on them, we introduce the set S of formal neighbourhoods by the production rules

$$\underline{0} \in S \qquad\qquad \Delta \in S$$

$$\frac{a \in S}{s(a) \in S} \qquad\qquad \frac{a \in S \quad b \in S}{a \wedge b \in S}$$

$+$ S, $\wedge$, Δ form a semilattice with one

The idea is that a formal neighbourhood gives a piece of information: Δ gives no information, s(a) is met by successors of those objects which meet a, $\underline{0}$ is met only by 0, etc. We introduce a relation Approx(n,a) prop (n $\in$ N,a $\in$ S) (read: the information a applies to n, or a approximates n) by the rules

$$\text{Approx}(0, \underline{0}) \qquad\qquad \text{Approx}(n,\Delta)$$

$$\frac{\text{Approx}(n,a)}{\text{Approx}(s(n),s(a))} \qquad\qquad \frac{\text{Approx}(n,a) \quad \text{Approx}(n,b)}{\text{Approx}(n,a \wedge b)}$$

Since Approx is clearly a neighbourhood relation, N, S, Approx give a concrete space *N*. Then we have also a formal topology *A*(*N*), whose covering relation and consistency predicate are defined, as in section 2, by

$$a \leq U \equiv (\forall n \in N)(\text{Approx}(n,a) \rightarrow (\exists b \in U)(\text{Approx}(n,b))$$

$$\text{Pos}(a) \equiv (\exists n \in N)\text{Approx}(n,a)$$

The consistent formal neighbourhoods and their ordering in *A*(*N*) are indicated in the picture:

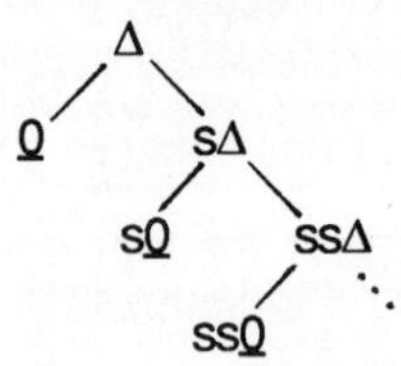

However, *A*(*N*) doesn't say much: since $\Delta \leq \{\underline{0},s\underline{0},ss\underline{0}, ...\}$, points on *A*(*N*) are just finite branches in the picture, and thus we get nothing but a copy of N.

The situation is quite different if we compactify *A*(*N*). In the Stone compactification, Δ is not covered by $\{\underline{0},s\underline{0},ss\underline{0}, ...\}$, and therefore, beside all finite branches, the infinite branch $\omega \equiv \{\Delta,s\Delta,ss\Delta, ...\}$ is also a point. The idea behind ω is that we can never exclude that it is a natural number; we thus may call it a non standard natural number. In the Scott compactification, a path from any node up to Δ in the picture is also a point; we may call it a partial number.

It is possible to obtain the three formal spaces above without any reference to N; we here only give a hint, and leave the details as an exercise. Think of formal neighbourhoods as pieces of information, as suggested above. Then the rules for S can be integrated with $s(a \wedge b) = sa \wedge sb$; also, we declare $\underline{0}$, Δ to be possible, a to be equipossible with s(a), $\underline{0} \wedge s\Delta$ to be impossible. Defining $a \leq b$ as a is more informative than b, that is $a \leq b \equiv (a \wedge b = a)$, we obtain for possible pieces of information the same picture as above. Now, according to how we handle disjunction of information, we will have three possible coverings. If we accept the infinite disjunction $\Delta \leq \{\underline{0},s\underline{0},ss\underline{0}, ...\}$, we obtain the topology *A*(*N*) above, and hence a copy of N. If we only accept finite disjunctions, we obtain the Stone topology above (more precisely, we assume $\Delta \leq \{\underline{0},s\Delta\}$ and close under the rule $a \leq U \rightarrow s(a) \leq \{s(b): b \in U\}$, beside the rules for topologies; of course, $\Delta \leq \{\underline{0},s\Delta\}$ is interpreted as saying that being 0 or a successor amounts to no information). Finally, in the Scott topology no disjunction is allowed, i.e. $a \leq U$ means that U already contains one piece of information covering a.

c. The topology of ideals

Assume that L is a set with operations +, $\cdot$ and constants 0, 1 such that L, +, $\cdot$, 0, 1 form a distributive lattice with zero and one. We write $\leq_L$ for the partial order defined as usual by $a \leq_L b \equiv (a \cdot b = a)$. For any $U \subseteq L$, we define $I(U)$ to be the ideal of L generated by U, that is we put

$$I(U) \equiv \{a \in L: (\exists b_1,...,b_n \in U)(a \leq_L b_1 + \cdots + b_n)\}$$

It is obvious, and well known, that $I: P(L) \to P(L)$ is an algebraic closure operator. So to show that I induces a covering $\leq_I$ it is enough to show that $I(U \cdot V) = I(U) \cdot I(V)$, which is easily done using distributivity.

Now assume we also have a consistency predicate Pos (the easiest case is when equality in L is decidable, so that we can put $\text{Pos}(a) \equiv (a \neq 0)$). Then L, $\cdot$, 1, $\leq_I$, Pos give a formal topology $I(L)$, called the topology of ideals of L, which is Stone since I is algebraic. Note that $\text{Sat}(I(L))$ is just the well known complete Heyting algebra of ideals of L. We will show in section 7 that all Stone topologies are obtained in this way (up to isomorphism).

d. The subsets of the one-element set

Let 1 be a set with just one element, say 0 (N_1 of [ITT] is such a set). 1 is trivially a semilattice with one, and hence it can be taken as a formal base for a topology. We define a covering relation $\leq$ by putting

$$0 \leq U \equiv U(0) \equiv 0 \in U$$

for any subset U of 1, and we declare 0 to be consistent. We then trivially have a topology, whose algebra of opens is just $P(1)$ with extensional equality, because $U \leq V \equiv U \subseteq V$. Since 0 is the only consistent neighbourhood, {0} is the only point of $P(1)$.

Since subsets of 1 are propositional functions with an argument which can only be 0, $P(1)$ is isomorphic to the category of propositions (using a tiny bit of Martin-Löf's theory of expressions, we can see that the isomorphism maps a subset U of 1 into U(0), which is a proposition, and conversely a proposition A into its abstraction (x)A, which is a unary propositional function; the claim then follows because $U = (x)(U(0))$ in $P(1)$, while $A \equiv ((x)A)(0)$ in the category of propositions). As a corollary, we obtain that for any formal topology A, Pt(A) is isomorphic to the category of morphisms from A into $P(1)$ (cf. [FG], p. 122).

e. Free topologies

Let S be any set, Fin(S) the set of finite subsets of S. For any d, e $\in$ Fin(S), we put $d \wedge e \equiv d \cup e$ (and hence $d \leq e \equiv d \supseteq e$) and $\Delta \equiv \varnothing$. Then Fin(S) becomes a semilattice with one, and actually the semilattice freely generated by S (see for instance [J], p. 27). We now define Pos so that Pos(d) holds for any d $\in$ Fin(S) and put

$$d \leq U \equiv (\exists e \in U)(d \leq e)$$

Then $\leq$ is the least covering relation on Fin(S), and the resulting topology $F(S)$ is called the free topology on S. In fact, given any formal topology A and a surjective map $f_0: S \to S_0$, where S_0 is a subset of S_A generating S_A as a semilattice, there is a unique full morphism from $F(S)$ to A which extends f.

If we define a continuous function $f^*: \text{Pt}(A) \to \text{Pt}(B)$ to be injective precisely when f is full, then Pt(F(S)) can be called a universal space, since all formal spaces which are small enough can be embedded in it. Since F(S) is Scott by definition, Pt(F(S)) then plays the role of a universal Scott domain.

6. Morphisms of topologies and continuos functions

The usual definition says that a function between two topological spaces is continuous if its inverse is a homomorphism between the frames of opens. Since here opens are more basic than points, reversing the order we will obtain a continuous function as the adjoint (that is, a sort of generalized inverse) of a homomorphism between the frames of opens, which in its turn can be described in terms of the underlying formal topologies. We are thus led to:

Definition 6.1 Let A, B be arbitrary formal topologies. A function f: $S_A \to \mathcal{P}(S_B)$ is called a morphism of A into B, written f: $A \to B$, if for any a, b $\in$ S, U $\subseteq$ S:

1. $f(a \wedge b) =_B fa \wedge fb$, $f(\Delta_A) = \Delta_B$
2. $a \leq_A U \to fa \leq_B f(U)$ where $f(U) \equiv \cup\{fb: b \in U\}$
3. $Pos_B(fa) \to Pos_A(a)$

In other words, f is a function which preserves all the strucure of A. Two little facts, however, should be pointed out: first, we can not require the base S_A always to be mapped into the base S_B, which explains why values of f are subsets of S_B, i.e. opens; second, consistency is preserved backwards, which is essential when defining f* below.

It is easy to see that for a morphism f: $A \to B$ properties 1. and 2. can be extended to opens, that is $f(U \wedge V) =_B f(U) \wedge f(V)$ and $U \leq_A V \to f(U) \leq_B f(V)$ hold. This means that the function f^0: Open(A) $\to$ Open(B), where $f^0U \equiv f(U)$, is well defined and that it preserves the order and finite meets. By its definition, f^0 also preserves arbitrary joins. So f^0 is a homomorphism between frames.

Conversely, given an arbitrary homomorphism h: Open(A) $\to$ Open(B), consider its trace h_0 on the base S_A, which is defined by $h_0(a) \equiv h(\{a\})$ for $a \in S_A$. Then h_0 is obviously a morphism of topologies and $h_0{}^0 = h$ because h preserves infinite joins. So we have:

Proposition 6.2 Any morphism of topologies f: $A \to B$ induces, as described above, a frame homomorphism f^0: Open(A) $\to$ Open(B), and all frame homomorphisms are obtained in this way.

A morphism f: $A \to B$ is called faithful if $a \leq_A U \leftrightarrow fa \leq_B f(U)$ and full if for any $V \subseteq S_B$ there exists $U \subseteq S_A$ such that $f(U) =_B V$. A morphism which is both faithful and full is called an equivalence of topologies, since it is not difficult to prove that

Proposition 6.3 A morphism f: $A \to B$ is faithful iff f^0: Open(A) $\to$ Open(B) is injective and full iff f^0 is surjective.

Given a morphism f: $A \to B$ and a point β of B, we define a propositional function $f^*\beta$ by putting

$$(f^*\beta)(a) \equiv \beta(fa) \quad \text{for } a \in S_A$$

It is easy to check that $f^*\beta$ is a point on A (here the fact that f preserves consistency backwards is essential to show that $f^*\beta$ is consistent). Now assume that a is a formal neighbourhood of $f^*\beta$, i.e. $(f^*\beta)(a)$ true. Then $\beta(fa)$ is true, and hence $(\exists b \in fa)\beta(b)$, i.e. there is a neighbourhood b of β such that $b \leq_B fa$ (which is our way of expressing the classical statement $(\forall \alpha \in Pt(B))(\alpha(b) \to (f^*\alpha)(a))$, that is $f^*(\varphi b) \subseteq \varphi a$). This justifies the following:

Definition 6.4 If f: $A \to B$ is a morphism, then f^*: Pt(B) $\to$ Pt(A) is called a continuous function.

It is not difficult to check that formal topologies and morphisms form a category in MacLane's sense, and that Pt(-): $A \to$ Pt(A), f $\to$ f* is a contravariant functor. Hence our category of formal spaces corresponds to the category of locales (see e.g. [J], p. 39). A natural question is then how much of the theory of locales can be transferred into our framework; this indeed is a good project, but yet for another paper.

Some reader may not like the definition of morphisms above, since they are functions with subsets as values. Then, instead of f, (s)he may consider the relation F(a,b) prop ($a \in S_A, b \in S_B$) defined by $F(a,b) \equiv b \in fa$, and find those properties of F which correspond to 1.-3. above. Then (s)he will find that the job is easier if f is assumed to be saturated, that is if fa is B-closed for each $a \in S_A$ (note that f and its saturation $f_B(a) \equiv B(fa)$ induce the same frame homomorphism). The result is that f is a saturated morphism iff F satisfies

1. $F(a \wedge b,c) = F(a,c) \,\&\, F(b,c)$, $F(\Delta_A, \Delta_B)$
2. $a \leq_A U \,\&\, F(a,b) \to b \leq_B \{c \in S_B: (\exists a \in U)F(a,c)\}$
3. $Pos_B(b) \,\&\, F(a,b) \to Pos_A(a)$
4. $(\forall b \in V)F(a,b) \,\&\, (b \leq_B V) \to F(a,b)$

So, since clearly any relation F(a,b) prop ($a \in S_A; b \in S_B$) induces a function $fa \equiv \{b \in S_B: F(a,b)\}$ and the correspondence is biunivocal, there is an isomorphism between saturated morphisms and relations satisfying 1.-4.

Conditions 2. and 4. are not very perspicuous. When *A*, *B* are Scott, however, they acquire a much simpler form, namely

2'. $a \leq_A a'$ & $F(a,b) \rightarrow F(a',b)$
4'. $F(a,b)$ & $b' \leq_B b \rightarrow F(a,b')$

Apart from minor notational details, 1.,2'.,3.,4'. form the very definition of approximable mappings in [S2].

7. Compact opens and Stone representation

The usual notion of compactness is easily expressed in pointless words:

Definition 7.1 An open U in the formal topology *A* is called compact if, for arbitrary $V \subseteq S$, $U \leq V \rightarrow U \leq V_0$ for some finite $V_0 \subseteq V$.

First of all, note that this is a good definition; hence compact opens of *A* are just those elements of Open(*A*) which are compact (or algebraic, or finite) according to the usual definition in complete lattices (see e.g. [J], p. 63). Note that finite subsets of S need not be compact (the standard example here is that of trees, in section 9). On the other hand we do have: if U is compact, then $U =_A U_0$ for some finite $U_0 \subseteq U$. Using this, it is easy to see that compact opens are closed under finite unions, that is form a join-subsemilattice of Open(*A*).

By the definitions, *A* is Stone iff every basic neighbourhood is compact. Thus, because of closure under finite unions, we have:

Proposition 7.2 A formal topology *A* is Stone iff every finite subset of S is compact.

The following results are our version of Stone's representation theorems (see e.g. [J], pp. 64-66).

Proposition 7.3 Any Stone topology is equivalent to the topology of ideals on a distributive lattice.

Proof. The idea is to induce on Fin(S), the set of finite subsets of S, the structure of *A*. That is, for d, e $\in$ Fin(S), U $\subseteq$ Fin(S), define $d \leq_I U \equiv d \leq_A \cup U$, $d \wedge e \equiv \{a \wedge b: a \in d, b \in c\}$, $d \vee e \equiv d \cup e$. In particular, Fin(S) with equality $=_I$ is a distributive lattice and the name is well chosen, that is $d \leq_I U$ iff d is in the ideal of Fin(S) generated by U, because d is compact. It is now obvious that the identity mapping i: Fin(S) $\rightarrow$ $P(S_A)$ is an equivalence of topologies.

Proposition 7.4 Every distributive lattice L is isomorphic to the sublattice of compact opens of the Stone topology *I*(L).

Proof. Since *I*(L) is Stone, U is compact in *I*(L) iff $U =_I \{a_1, \ldots, a_n\}$ for some $a_1, \ldots, a_n \in L$, and hence $U =_I a_1 + \cdots + a_n$. So the identity mapping is the isomorphism we want.

It may be instructive to see how, admitting classical principles, Stone's representation of a distributive lattice L would be derived from the above result. Recall that classically Pt(*I*(L)) is a topological space with base $(\varphi a)_{a \in L}$. Then φ is a function from Open(*I*(L)) onto the opens of Pt(*I*(L)), since any open is of the form $\varphi(I) \equiv \cup\{\varphi a: a \in I\}$ for some ideal I of L. To show that φ is injective, it is enough and necessary to have

$$\varphi a \subseteq \varphi(I) \rightarrow a \in I$$

for any ideal I. By definition, $\varphi a \subseteq \varphi(I)$ means, classically, that $(\forall \alpha \in Pt(I(L)))(a \in \alpha \rightarrow (\exists b \in I)(b \in \alpha))$; so by classical logic the condition above is equivalent to

$$a \notin I \rightarrow (\exists \alpha \in Pt(I(L)))(a \in \alpha \ \& \ I \cap \alpha = \emptyset)$$

which, by proposition 4.6, is exactly a formulation of the prime filter theorem. So the prime filter theorem is equivalent to φ being an isomorphism. Since obviously U is a compact subset in $\mathcal{I}$(L) iff φ(U) is compact in the traditional sense, proposition 4 above gives Stone's theorem: assuming the prime filter theorem, L is isomorphic to the lattice of compact opens in Pt($\mathcal{I}$(L)).

8. Scott spaces and Scott domains

Various presentations of the so called Scott domains have been given by D. Scott himself. In the so called axiomatic presentation, a Scott domain is a structure D, $\perp$, $\leq$ which is a complete partial order (i.e. $\leq$ is a partial order with bottom $\perp$, in which every directed subset $E \subseteq D$ has a supremum $\vee E$) which is algebraic (i.e. for every element d of D, algebraic elements below d form a directed subset whose supremum is d) and any two algebraic elements which are majorized in D have a supremum $d \vee e$. The connection with our approach is immediate:

Proposition 8.1 Any Scott space Pt(*A*) is a Scott domain.

In fact, by proposition 4.7, points of *A* are just filters on S. So the order $\alpha \leq \beta \equiv (\forall a \in S)(\alpha(a) \rightarrow \beta(a)) \equiv \alpha \subseteq \beta$ is a complete partial order on Pt(*A*), with the filter generated by Δ as bottom, which is algebraic since algebraic elements are exactly principal filters. The last condition is then taken care of by the consistency predicate: if the filters generated by a, b $\in$ S are majorized by α in Pt(*A*), then $\alpha(a \wedge b)$ holds, hence Pos($a \wedge b$) and the filter generated by $a \wedge b$ is consistent.

On the other hand, let D be a Scott domain and S the set of its algebraic elements. Then any element of D may be identified with the directed set $E_d \equiv \{a \in S: a \leq d\}$, since $d = \vee E_d$. But then we may think of E_d as a filter on the dual of S, that is the structure with order $\geq$ and meet $\vee$. In more recent presentations, Scott has introduced neighbourhood systems [S1] and information systems [S2], which can be seen as a way to axiomatize the structure of S above. In particular, information systems are provided with a consistency predicate (which intuitively asserts a to be consistent with b when the meet of a and b exists), and Scott domains are then given as the collection of consistent filters over them. By the above proposition, Scott topologies play a similar role. We can see it more clearly as follows.

Given an arbitrary formal topology *A*, consider the quotient of the base S under equality $=_A$ between formal neighbourhoods (recall that $a \leq b \equiv a \leq \{b\}$ and $(a =_A b) \equiv (a \leq b \;\&\; b \leq a)$). Since $\wedge$ respects $=_A$, we obtain a semilattice with one. Note that $a \wedge b =_A b$ iff $a \leq b$. Moreover, the properties of the consistency predicate Pos of *A* imply that

$$\frac{\mathrm{Pos}(a) \quad a \leq b}{\mathrm{Pos}(b)} \qquad\qquad \frac{\neg\mathrm{Pos}(a)}{a \leq b}$$

We put all this into a definition, and hence also a little result:

Definition 8.2 Any semilattice with one and with a predicate Pos satisfying the above two conditions is called a Scott semilattice.

Proposition 8.3 Any formal topology *A* traces on the base S a Scott semilattice, denoted by *S*(*A*).

The idea is that algebraic elements of a Scott domain give rise to a Scott semilattice; we now work it out in our approach. Given a Scott semilattice *S* consisting of S, $\wedge$, Δ, Pos, by the topology generated by *S* we mean the topology obtained by taking $a \leq_{A(S)} b$ as axiom whenever $a \leq b$ holds in *S*, and closing up under all the rules for topologies. It is easy to show, by induction on the rules, that

$$\mathrm{Pos}(a) \;\&\; (a \leq_{A(S)} U) \rightarrow (\exists b \in U)(a \leq b)$$

(the assumption Pos(a) is used for the inductive steps for the rules of openness and ex falso quodlibet). As a particular case, we have

$$\mathrm{Pos}(a) \;\&\; (a \leq_{A(S)} b) \rightarrow a \leq b$$

Since trivially $a \leq b \rightarrow a \leq_{A(S)} b$, the first implication above means that $A(S)$ is a Scott topology, while the second shows that the partially ordered subset of consistent elements of S and of $S(A(S))$ coincide. Hence, in particular, by proposition 4.7 the Scott space Pt($A(S)$) is just the space of consistent filters of S (note that what we called filters on S in section 4 are actually standard filters on the quotient of S by $=_A$).

On the other hand, given any Scott topology A, the topology $A(S(A))$, generated by the trace $S(A)$ of A on S, is also a Scott topology; by what above, the Scott space Pt($A(S(A))$) coincides with the space of filters on $S(A)$ and hence, by proposition 4.7, with Pt(A). Summing up:

Proposition 8.4 For any Scott semilattice S, the space of consistent filters on S is a Scott space, and any Scott space is obtainable in this way.

I would have liked a stronger result, namely that any Scott semilattice S coincides with $S(A(S))$, but I am not able to show that $a \leq_{A(S)} b \rightarrow a \leq b$. In any case, propositions 8.4 and 8.1 should justify our proposal of taking Scott semilattices (may be with a shorter name) as the basic structures on which the theory of Scott domains can be built up. Surely the connections established above open the way to much further work.

9. Trees and bars

One of the motivations to the study of formal spaces is to give an interpretation, in the foundational framework adopted, of Brouwer's notion of choice sequence. The main idea is that choice sequences on a given tree should be points of a suitable formal topology based on that tree. However, this is easier said than done. In fact, I was able to materialize that idea only at the cost of modifying the definition of formal topology itself, namely by requiring $a \leq U^+$ to be derivable from $a \leq U$ only when Pos(a) holds (weak openness). Whether this is a good reason to adopt this modification from the start, I still do not know. In any case, I believe the topic to be interesting enough to present here, though briefly and informally, the present state of work.

A tree is here given by two families of sets

$$A(n) \text{ set } (n \in N)$$

$$B(n,x) \text{ set } (n \in N, x \in A(n))$$

which satisfy

$$A(n+1) = (\Sigma x \in A(n))B(n,x)$$

Intuitively, we begin with A(0), which usually contains only one element Δ, called the root. Then we have a set $B(0, \Delta)$ of choices on how to proceed. Once $b \in B(0, \Delta)$ is chosen, we form $a_1 \equiv (\Delta,b) \in A(1)$. Then we have a set $B(1,a_1)$ of choices, etc. So A(n) is the set of elements which are obtained with n successive choices and $S \equiv (\Sigma x \in N)A(n)$ is the set of all nodes. May be a better definition can be found, but at least the one above has the advantage of lying entirely within the framework of [ITT]. Of course, common examples of trees (like the complete tree over a given set T, which is often {0,1} or N) fall under it (take $B(n,x) = T$, $A(0) = \{\Delta\}$ and put $A(n+1) \equiv A(n) \times T$).

We define the ordering on S in which Δ is the top element by putting $(n,a) \leq (m,b)$ when $n \geq m$ and b is obtained from a by right projection applied n-m times. We often write a for (n,a), and say that a is a node of length n.

We now want to put a topology on a given tree S. To this aim, we first close S under an operation $\wedge$ of formal meet which respects the order of nodes; that is, we impose $\wedge$ to satisfy $a \leq b \leftrightarrow a \wedge b = a$, beside idempotency, commutativity and associativity. What about the 'formal' nodes $a \wedge b$ obtained when a, b are incomparable? The consistency predicate solves this problem: we declare them to be inconsistent (and hence they will not appear in a drawing).

For every node a, we define immediate successors of a to be those nodes which are reached from a with just one choice. More formally, the immediate cover of $(n,a) \in S$ is the set $C(a) \equiv \{(n+1,(a,b)) : b \in B(n,a)\}$ and any element of C(a) is called an immediate successor of a. Now assume we have a predicate Pos satisfying

$$\begin{array}{l} Pos(a) \;\&\; a \le b \rightarrow Pos(b) \\ Pos(a) \rightarrow Pos(C(a)) \\ Pos(a \wedge b) \rightarrow a \le b \vee b \le a \end{array}$$

(this means of course that even 'standard' nodes may be inconsistent, i.e. sterilized in Brouwer's terminology). We then define a covering relation $\le_C$ by assuming

$$a \le_C C(a)^+ \quad \text{for any } a \in S$$

and closing under the rules for coverings. We want to show that this indeed defines a formal topology C, called the inductive topology (a name borrowed from [FG]) on the tree S with consistency predicate Pos.

First of all, note that when $a \le b$ holds, then $a \wedge b = a$ and hence $a \le_C b$ follows from $b \le_C b$ by $\wedge$-left; so, even if the converse $a \le_C b \rightarrow a \le b$ is in general false (for instance when b is the only immediate successor of a), we will drop the subscript C.

The rule of ex falso quodlibet is easily seen to be derivable: if $\neg Pos(a)$, then also $\neg Pos(C(a))$ by the assumptions on Pos, and hence $C(a)^+ = \varnothing$, so that $a \le \varnothing$ is simply an axiom.

Any derivation of $a \le U$, where Pos(a) is true, can be reduced to a sort of canonical form as follows. First, it is easy to show that the rule of $\wedge$-right is equivalent to the rule of localisation introduced in section 1; thus we eliminate $\wedge$-right in favour of localisation. The advantage is that applications of localisation can be lifted over all other rules (for example,

$$\dfrac{\dfrac{a \le U}{a \wedge b \le U}}{a \wedge b \wedge c \le U} \qquad \text{reduces to} \qquad \dfrac{\dfrac{a \le U}{a \wedge c \le U \wedge c}}{a \wedge c \wedge b \le U \wedge c}$$

and other rules are treated similarly). Thus we can reduce the derivation to one in which localisation is applied only to axioms, that is to obtain covers of the form $b \wedge c \le C(b)^+ \wedge c$ (note that any number of consecutive applications of localisation can be reduced to one).

Now the derivation of $a \le U$ can be reduced to a derivation of $a \le U^+$ in which only consistent nodes occurr. To see how this is possible, imagine to apply weak openness to the conclusion of the derivation, and then lift it as much as possible (an example of reduction is

$$\dfrac{\dfrac{b \le V \quad V \le W}{b \le W}}{b \le V} \qquad \text{reduces to} \qquad \dfrac{\dfrac{b \le V}{b \le V^+} \quad \dfrac{V^+ \le W}{V^+ \le W^+}}{b \le W^+}$$

where it is better to think of $V \le W$ as a free-variable derivation of $c \le W$ from the assumption $c \in V$, so that $V^+ \le W$ is obtained simply by restricting to the assumption $c \in V^+$; the cases for reflexivity and $\wedge$-left are even simpler). The result is that only consistent nodes appear, at least all the way up to any conclusion $b \wedge c \le C(b)^+ \wedge c$ of localisation. Then in particular $Pos(b \wedge c)$ is true, and hence $b \le c \vee c \le b$. So we can substitute localisation either by nothing (when $b \le c$, $b \wedge c = b$ and $C(b)^+ \wedge c = C(b)^+$) or by reflexivity (when $c < b$ $b \wedge c = c$ and $C(b)^+ \wedge c = \{c\}$). Note that the result is a real derivation of $a \le U^+$ in which on the three rules of transitivity, $\wedge$-left and reflexivity are applied.

Now it is easy to prove closure under monotonicity and weak openness by induction on such three rules (the assumption Pos(a) is used to show that $a \le U^+$ whenever $a \in U$). This completes the proof of the fact that C is a formal topology.

The definition of inductive covering is a precise formulation of the following informal but clear notion: U is an intuitive cover of a if starting from a we fall into U after a finite number of arbitrary choices. Certainly the axioms and rules for $\le$ fulfill such interpretation; the claim is that they embrace it completely. Incidentally, note that the above informal notion is equivalent to that of wellfounded tree, i.e. a cover of Δ.

Similarly, the definition of point is the precise counterpart of the informal notion of branch (or arrow, or infinitely proceeding sequence). By condition 2 of definition 4.1, a point α which meets a node a, in the sense that $\alpha(a)$ is true, also meets any cover U of a, in the sense that $\alpha(U)$. In particular, $\alpha(a) \rightarrow (\exists b \in C(a))\alpha(b)$ holds for any node a. Moreover, whenever $b, c \in C(a)$ and $\alpha(b)$, $\alpha(c)$ are both true, then $\alpha(b \wedge c)$ and hence $Pos(b \wedge c)$, which can only hold if $b = c$, because b, c have equal length. So α satisfies

(1) $\alpha(a) \rightarrow (\exists! b \in C(a))\alpha(b)$

which, together with $\alpha(\Delta)$ and $\alpha(a) \rightarrow Pos(a)$, tells that α is a branch. The converse holds if we agree that any branch satisfies (1) and contains only consistent nodes, beside the root. In fact, from $\alpha(a) \rightarrow (\exists b \in C(a))(\alpha(b)$, which takes care of the axioms, condition 2 is proved by induction on the rules, while condition 1 follows by the uniqueness of the immediate successor granted by (1). Of course, the question remains open whether (1) is inherent our intuitive notion of branch, in particular since it asserts the constructive existence of the immediate successor (one could be satisfied with $\alpha(a) \rightarrow \neg\neg(\exists b \in C(a))(\alpha(b))$. This is why I have not used the word choice sequence.

Usually the order of definitions is the opposite (two good sources on such matters are [FIM] and [MG]): one assumes to know what a branch is so well to make sense of quantification over branches and then defines wellfoundedness, or equivalently covers, by saying that U is a bar of a if

(2) $(\forall\alpha)(\alpha(a) \rightarrow \alpha(U))$

holds, where α ranges over branches. But the notion of bar so defined is difficult to be used, unless one assumes also that, under some weak restrictions, bars can be defined inductively (principle of bar induction). Inductive bars are defined by the rules (read U i.b. a as U is an inductive bar of a):

$$\eta \; \frac{a \in U}{U \text{ i.b. } a} \qquad \mathcal{F} \; \frac{\begin{matrix}(b \in C(a)) \\ | \\ U \text{ i.b. } b\end{matrix}}{U \text{ i.b. } a}$$

A subset U is monotone if $a \in U \;\&\; b \leq a \rightarrow b \in U$. Then monotone bar induction is the statement

$$BI_M \quad U \text{ monotone } \& \; (\forall\alpha)(\alpha(a) \rightarrow \alpha(U)) \rightarrow U \text{ i.b. } a$$

One of Brouwer's arguments for BI_M is based on the assumption (Brouwer's Dogma) that any fully analysed proof of (2) only makes use of $\eta, \mathcal{F}, \zeta$-inferences, where a ζ-inference allows to obtain U bars b from $b \in C(a)$ and U bars a. It is easy to see that, for a consistent node a, a fully analysed proof of U bars a exists iff $a \leq U$ (hint: first reduce the derivation of $a \leq U$ to one in which transitivity is applied only with an axiom as left premiss, and note that then η, ζ,$\mathcal{F}$-rules correspond exactly to our reflexivity, $\wedge$-left and transitivity+axioms, respectively). Hence, assuming BD, BI_M is proved simply by lifting $\wedge$-left over transitivity (obvious) and then noting that any application of $\wedge$-left under reflexivity can be eliminated if U is monotone. Moreover, the fact that any derivation of $a \leq U$ can be reduced to a fully analysed proof in Brouwer's sense, is a little argument in favour of BD.

Brouwer himself gives also another, shorter argument for BI_M: thought through intuitionistically, a bar is nothing else than an inductive bar. We agree with him (and hence also with Kleene, see [FIM], end of p. 50). However, our foundational framework allows to go a little step further, and simply define the meaning of $(\forall\alpha)(\alpha(a) \rightarrow \alpha(U))$ to be $a \leq U$. The content of BI_M then becomes part of the definition of universal quantification over points, and the above claim by Brouwer is substituted by: thought through intuitionistically, an intuitive cover is nothing else than an inductive cover.

Finally, it is now easy to keep a promise made in section 4. Conceiving $Pt(\mathcal{C})$ in the classical way, the function φ from opens of $\mathcal{C}$ onto opens of $Pt(\mathcal{C})$ is an isomorphism iff for any monotone subset U and any consistent node a, $\varphi(a) \subseteq \varphi(U) \rightarrow a \leq U$; but $\varphi(a) \subseteq \varphi(U)$ is just (2), and we have shown above that $a \leq U$ is equivalent, when U is monotone and a is consistent, to U i.b. a. Thus φ is an isomorphism iff BI_M holds.

References

[FG] M. Fourman - R. Grayson, Formal spaces, in: The L.E.J. Brouwer Centenary Symposium, eds. A. S. Troelstra and D. van Dalen, North Holland 1982, pp. 107-122

[G] J. W. Gray, Fragments of the history of sheaf theory, in: Applications of Sheaves, eds. M. Fourman, C. Mulvey and D. Scott, Springer 1979, pp. 1-79

[J] P. Johnstone, Stone spaces, Cambridge U. P. 1982

[FIM] S. C. Kleene - R. E. Vesley, The foundations of intuitionistic mathematics, North Holland 1965

[ITT] P. Martin-Löf, Intuitionistic type theory, Bibliopolis, Napoli 1984

[SI] P. Martin-Löf, On the meanings of logical constants and the justifications of logical laws, in: Atti degli Incontri di Logica Matematica v. 2, eds. C. Bernardi and P. Pagli, Dip. di Matem., Univ. di Siena, Siena 1985, pp. 203-281

[SG] P. Martin-Löf, Unifying Scott's theory of domains for denotational semantics and intuitionistic type theory, in: Logica e Filosofia della Scienza, oggi. San Gimignano, 7-11 Dicembre 1983, v.1, CLUEB, Bologna 1986, pp. 79-83

[MG] E. Martino - P. Giaretta, Brouwer, Dummett and the bar theorem, in: Atti del Convegno Nazionale di Logica, Montecatini Terme, 1-5 ottobre 1979, Bibliopolis, Napoli 1981, pp. 541-558

[S1] D. Scott, Lectures on a mathematical theory of computation, Oxford Univ. Computing Lab.,Techn. Monograph PRG-19, 1981

[S2] D. Scott, Domains for denotational semantics, in: Automata, Languages and Programming, eds. M. Nielsen and E. M. Schmidt, Springer 1982, pp. 577-613

ON THE LOGIC OF SMALL CHANGES IN THEORIES II

Krister Segerberg

Department of Philosophy
University of Auckland
Auckland
New Zealand

1. INTRODUCTION

The idea that it might be worth investigating the logic of theory change, in the sense of the present paper, is due, independently to Carlos Alchourrón and David Makinson on the one hand and to Peter Gärdenfors on the other. Their joint paper [1] is the most substantial contribution to this area to date. The idea is that if T is a theory (which we think of as a set of formulae) and A is a formula, then T + A is a new theory, the theory which results if A is "added" to T. What makes this notion problematic - and interesting - is that this operation is supposed to be defined even when A is inconsistent with T.

The work of Alchourrón, Gärdenfors and Makinson (from now on, AGM) is axiomatic. They regard the following postulates as reasonable (the numbering is theirs):

(+2) $A \in T + A$.
(+3) $T + A = T/A$, if $\neg A \notin T$.
(+4) $T + A$ is consistent, if $\{A\}$ is consistent.
(+5) $T + A = T + B$, if A and B are logically equivalent.
(+7) $T + A \wedge B \subseteq (T + A)/B$.
(+8) $(T + A)/B \subseteq T + A \wedge B$, if $\neg B \notin T + A$.

They list postulates for a more or less dual operation - of subtraction as well, but there is no need to include them as subtraction is definable in terms of addition:

$$T - A = T \cap (T + \neg A).$$

Although Gärdenfors has considered the question of semantics for theory change, for example in [2], it is an open problem whether the meta-theory of [1] has one. This problem was addressed in Part I of this paper [4] even though the result was inconclusive. The line of approach taken there was this. There is no assumption in [1] that the object language in which the changing theories are formulated contains any non-Boolean operators but neither is there anything to preclude this possibility. On the assumption that the object language contains a conditional operator $\sqsupset$ ("if it were the case that - then it would be case that"), obeying some underlying conditional logic, two modellings were proposed in [4]. One was the *plain* modelling, according to which

$$T + A = \begin{cases} T/A, \text{ if } \neg A \notin T, \\ T \oplus A, \text{ if } \neg A \in T. \end{cases}$$

(Here $T/A = \{B : A \supset B \in T\}$; while $T \oplus A = \{B : A \sqsupset B \in T\}$.) The other, the *improved* modelling, specified that

$$T + A = \begin{cases} T/A, \text{ if } \neg A \notin T, \\ T \oplus A, \text{ if } \neg A \in T \text{ but } A \sqsupset \bot \notin T, \\ /A, \text{ if } \neg A \in T \text{ and } A \sqsupset \bot \in T. \end{cases}$$

(Here $/A = \{B : A \supset B \in L\}$, where L is the underlying conditional logic.) If there is a semantics for the underlying conditional logic, then either modelling at once induces a semantics for the theory of theory change. It was noted in [4] that if the underlying logic is David Lewis's VC, then the AGM postulates are verified in both modellings with some notable exceptions: the plain modelling verified (+4) and (+8) only in the presence of extra assumptions, and while the improved modelling verified (+4), it required the same assumption as the plain modelling in order to verify (+8). Not noted in [4] but easy to check is that if one takes as the underlying conditional logic any traditional modal logic extending VC, where thus $\sqsupset$ is definable by the condition $A \sqsupset B = \Box(A \supset B)$, then the improved modelling delivers all the AGM postulates. It would be interesting to know whether any other conditional logic over the improved modelling does. An affirmative answer to this question would of course throw some doubt on the interest of the AGM postulates. However, the author is inclined to conjecture that the answer is negative.

The AGM postulates are the first of their kind, and they will serve as a standard against which to compare future metatheories. However, modellings of the kind introduced in [4] seem interesting in their own right, even if they do not fit the AGM theory perfectly. So little is yet known in this new field that there is ample room for experimentation. In this paper we shall examine and try to improve on the improved modelling. The result will be a third modelling, called the *de luxe* modelling. As it turns out, it too does not quite fit the AGM standard, but for some applications it is preferable to either of the modellings in [4].

2. INFORMAL CONSIDERATIONS

Recall that in [3] Lewis defines $\Diamond A = \neg(A \sqsupset \bot)$ and $\Box A = (\neg A) \sqsupset \bot$; Lewis reads $\Diamond A$ as "possibly, A" or "it is entertainable that A" and $\Box A$ as "necessarily A" or "it would be the case, no matter what, that A". The central question in the theory of theory change is what happens when you add a formula A to a theory T when $\neg A$ is already in T, and this difficulty is of course aggravated if $\Box\neg A$ is also in T. If both $\neg A$ and $\Box\neg A$ are in T, then the plain modelling (over VC) defines T + A as the inconsistent theory, which is quite drastic. What the improved modelling does is also drastic, for it gives up all of T and defines T + A as /A. If the plain modelling may be said to respond to this particular situation by giving up the game, the improved modelling may be said to do it by going back to square one and starting the game all over. Either approach seems unnecessarily defeatist.

The situation warrants a closer analysis. To begin with we shall consider a special case, namely, the addition of a formula A to a maximal consistent theory x. There are two cases.

Case 1: $A \in x$. In this case there is no need to do anything: the definition of x + A = /A is intuitively right.

Case 2: $\neg A \in x$. Here there are two cases.

Subcase 2.1: $\Diamond A \in x$. In this case some A-world is accessible from x. Thus the set $\{B : A \sqsupset B \in x\}$ is not empty, and so it seems intuitively quite reasonable that this set should be identified with x + A. This, by the way, is in agreement with both the plain and the improved modelling.

Subcase 2.2: $\Box\neg A \in x$. Then $A \sqsupset \bot \in x$, and so $\{B : A \sqsupset B \in x\}$ is the inconsistent theory. Is it a reasonable candidate for x + A? Yes, said the plain modelling but was then faced with the failure of the AGM postulate (+4). No, said the improved modelling and went on to define x + A = /A. One reason why the latter definition is unsatisfactory is that it does not make x + A depend on x. It seems that the improved modelling gives up too easily, for the game can be played a good deal longer. That $A \sqsupset \bot \in x$ can perhaps be taken to mean that it is part of the theory x that one cannot meaningfully add A to it: the negation of A is, to some extent, "entrenched" in x. However, even a view to the effect that A cannot be consistently added to x could itself be subject to review. Accordingly there are two cases to consider.

Subsubcase 2.2.1: $\Diamond^2 A \in x$. As we saw, it is not possible to add A directly to x (for $x \oplus A$ is inconsistent). But it *is* possible to add $\Diamond A$ to x and obtain $x \oplus \Diamond A$ without incurring inconsistency; and it *is* possible to add A to $x \oplus \Diamond A$ and thus obtain $x \oplus \Diamond A \oplus A$ (association to the left obviates the need for parentheses). The latter is a theory which has been reached in the attempt to add A to x, and it seems intuitively natural to define it as x + A. To be certain, the construction was carried out in two steps rather than one, but that does not seem to constitute an objection to the theory.

Subsubcase 2.2.2: $\Box^2\neg A \in x$. Thus $(\Diamond A) \sqsupset \bot \in x$. Now we can no longer add $\Diamond A$ to x as we did in the former subsubcase, for $x \oplus \Diamond A$ is now inconsistent. In this case the negation of A is still more "entrenched" in the theory x. However, even a view to the effect that A cannot be consistently added to x could itself be subject to revision; so again there are two cases, etc.

It is clear that this line of argument can be continued indefinitely.

The process described in the preceding few paragraphs can be surveyed in Fig 1. Note that we are still left with the problem of deciding what should happen at the end of the infinite branch: how should we define x + A if, for all n, $\Box^n\neg A \in x$ - if the negation of A is "infinitely deeply entrenched", as it were?

3. FORMAL DEVELOPMENT

Against this informal background let us now proceed to develop a formal theory reflecting the intuitions delineated. We assume a propositional language containing a conditional operator (the *would*-operator) as well as the usual Boolean operators. In addition we use $>$, $\Box$ and $\Diamond$ as abbreviations: $>$ (the *might*-operator) is defined by the condition $A > B = \neg(A \sqsupset \neg B)$, and $\Box A$ and $\Diamond A$ were defined above. A *theory* is any set of formulae that contains all tautologies and is closed under modus ponens (that is, if it contains A and $A \supset B$, then it contains B). We will use A, B, C, D to denote formulae, T to denote theories, x to

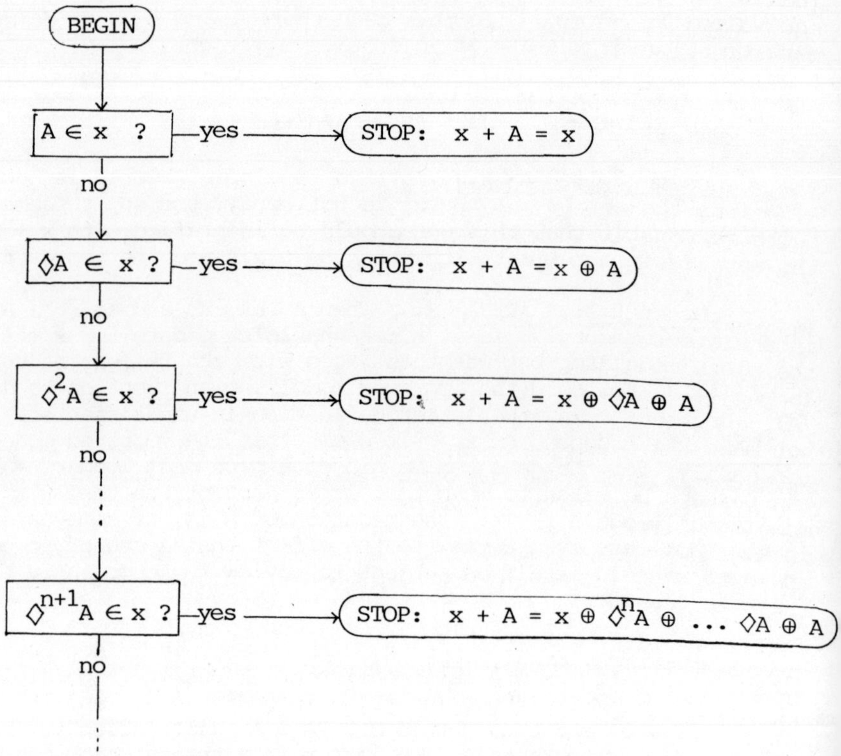

Fig. 1

denote maximal consistent sets (over some specified logic), and m, n, p to denote natural numbers 0, 1, 2, ... With respect to a given logic we write $A_0, \ldots, A_{n-1} \vdash B$ if the formula $(A_0 \wedge \ldots \wedge A_{n-1}) \supset B$ is a thesis of the logic; thus $\vdash A$ if and only if A is a thesis of the logic. Furthermore, we say that two formulae A and B are *logically equivalent*, writing $A \dashv\vdash B$, if both $A \vdash B$ and $B \vdash A$. In order to reduce the number of parentheses we shall assume that unary operators bind more strongly than others and among the binary operators the Boolean ones bind more strongly than the others.

David Lewis's logic VC can be represented as the smallest set of formulae to satisfy the following conditions:

(TF) $\vdash A$, if A is a tautology,
(MP) $A, A \supset B \vdash B$,
(K1) $A \sqsupset B \wedge C \dashv\vdash (A \sqsupset B) \wedge (A \sqsupset C)$,
(K2) $\vdash A \sqsupset T$,
(#1) $\vdash A \sqsupset A$,
(#2) $A > B \vdash \Diamond B$,
(#3) $A, B \vdash A \sqsupset B$,
(#4) $A, B \vdash A > B$,
(#5) $A \wedge B \sqsupset C \vdash A \sqsupset B \supset C$,
(#6) $A > B, A \sqsupset B \supset C \vdash A \wedge B \sqsupset C$;

moreover, it satisfies the following closure conditions:

(EA) if $A \dashv\vdash B$, then $A \sqsupset C \dashv\vdash B \sqsupset C$,
(EC) if $B \dashv\vdash C$, then $A \sqsupset B \dashv\vdash A \sqsupset C$.

Thus VC is closed under substitution and replacement of provable equivalents; we shall often refer to these properties without explicit mention.

We introduce the following definitions:

$$A \sqsupset^0 B = A \supset B,$$
$$A \sqsupset^{n+1} B = \Diamond^n A \sqsupset (A \sqsupset^n B);$$
$$A >^0 B = A \wedge B,$$
$$A >^{n+1} B = \Diamond^n A > (A >^n B).$$

We note the following auxiliary result:

Lemma 3.1. For all n, $\Diamond^n A \sqsupset (A \sqsupset^n B)$ and $\Diamond A \sqsupset^n (A \sqsupset B)$ are equivalent in VC.

Proof. By induction on n. If n = 0, then LHS is $A \sqsupset (A \wedge B)$, which in VC reduces to $A \sqsupset B$. On the other hand, RHS is $\Diamond A \supset (A \sqsupset B)$. Now, $\Box \neg A \supset (A \sqsupset B)$ is a thesis of VC, as is $\Diamond A \vee \Box \neg A$. Hence RHS, too, reduces to $A \sqsupset B$.

Suppose the result has been proved for n = m, for any m. If n = m + 1, then LHS is $\Diamond^{m+1} A \sqsupset (A \sqsupset^{m+1} B)$, which by the definition of $\sqsupset^{m+1}$ logically equivalent to $\Diamond^{m+1} A \sqsupset ((\Diamond^m A \sqsupset (A \sqsupset^m B))$. By the induction hypothesis, this is logically equivalent to $\Diamond^{m+1} A \sqsupset (\Diamond A \sqsupset^m (A \sqsupset B))$, which by the definition of $\sqsupset^{m+1}$ again is logically equivalent to $\Diamond A \sqsupset^{m+1} (A \sqsupset B)$, the RHS. QED

We define the *degree* of a theory T with respect to a formula A - in symbols, deg(T, A) - as the smallest number n such that $\Box^n \neg A \notin T$, if such a number exists, and as ω otherwise. Notice that while only consistent theories can have finite degrees, a theory may have an infinite degree without being inconsistent. The operation $\oplus$ is defined as in [4] and mentioned above. We introduce the following notation:

$$x \oplus^0 A = x,$$
$$x \oplus^{n+1} A = x \oplus^n \Diamond A \oplus A.$$

(Again we assume association to the left.) Note that under this convention $x \oplus^1 A = x \oplus A$. In general,

$$x \oplus^{n+1} A = x \oplus \Diamond^n A \oplus \ldots \oplus \Diamond A \oplus A.$$

We define

$$x + A = \begin{cases} x \oplus^n A, & \text{if } \deg(x, A) = n, \\ /A, & \text{if } \deg(x, A) = \omega. \end{cases}$$

This finally puts us in a position to define T + A for theories T in general - our *de luxe* modelling:

$$T + A = \begin{cases} \bigcap\{x + A : T \subseteq x\}, & \text{if T is consistent,} \\ /A, & \text{if T is inconsistent.} \end{cases}$$

Lemma 3.2. If deg(x, A) = n, then $x + A = \{B : A \sqsupset^n B \in x\}$.

Proof. By induction on n. If n = 0, then the assumption that deg(x, A) = 0 implies that $\neg A \notin x$, and so $A \in x$ by maximality. Hence LHS and RHS are both equal to x.

Assume that the lemma has been proved for n. Suppose that deg(x, A) = n + 1. Then, for any formula B,

$$B \in x + A \text{ iff } A \sqsupset B \in x \oplus^{n} \Diamond A \quad \text{(by the definition of } \oplus^{n+1}\text{)}$$
$$\text{iff } \Diamond A \sqsupset^{n} (A \sqsupset B) \in x \quad \text{(by the induction hypothesis)}$$
$$\text{iff } \Diamond^{n} A \sqsupset (A \sqsupset^{n} B) \in x \quad \text{(by Lemma 3.1)}$$
$$\text{iff } A \sqsupset^{n+1} B \text{ (by the definition of } \sqsupset^{n+1}\text{).} \qquad \text{QED}$$

The following fact is important.

Theorem 3.3. For all theories T and formulae A,

$$T + A = \begin{cases} \{B : A \sqsupset^{n} B \in T\}, & \text{if deg } (T, A) = n, \\ /A, \text{ if } \deg(T, A) = \omega. & \end{cases}$$

Proof. Suppose that deg(T, A) = n. Then, for all formulae B, we have

$$B \in T + A \text{ iff, for all } x \text{ such that } T \subseteq x,\ B \in x + A$$
$$\text{iff, for all } x \text{ such that } T \subseteq x,\ A \sqsupset^{n} B \in x \text{ (by Lemma 3.2)}$$
$$\text{iff } A \sqsupset^{n} B \in T \text{ (by Lindenbaum's Lemma).} \qquad \text{QED}$$

The following result shows that the de luxe modelling agrees with AGM on how to add to a theory a formulae consistent with it:

Corollary 3.4. If $\neg A \in T$, then T + A = T/A.

Proof. Suppose that $\neg A \in T$. Then deg(T, A) = 0. Hence, by the theorem, T + A = $\{B : A \supset B \in T\}$, which is the same as saying that T + A = T/A. QED

4. THE DE LUXE MODELLING OVER VC^{∞}

The de luxe modelling over VC is not so interesting; for our purposes we need a slightly stronger logic which apparently has not been described before. For any formula A, let us write $(A)^n$ for the result of replacing in A each occurrence of $\sqsupset$ in A by an occurrence of $\sqsupset^n$. It can be shown that $(K1)^n$, $(K2)^n$, $(\#1)^n$ $(\#2)^n$, $(\#3)^n$, $(\#4)^n$ and $(\#5)^n$ are all derivable in VC, for every n, but that if $n > 1$ there are instances of type $(\#6)^n$ which are not. Let us write VC^{∞} for the smallest set of formulae containing all of VC as well as all instances of $(\#6)^n$, for all n, and which is closed under conditions (MP), (EA) and (EC). The author intends to offer a detailed study of this and some related logics in a separate article.

Let us now compare our new modelling with the AGM standard:

Theorem 4.1. All the AGM postulates are satisfied in the de luxe modelling over VC^{∞} except for the following qualification: (+8) holds for all theories T and formulae A provided that if deg(T, A) is finite, then $A >^{p} B$ T, where p = deg(T, A).

Proof. We give the cases for (+7) and (+8). Note that they reduce to the single condition

$$\text{if } \neg B \in T + A, \text{ then } T + A/B = T + A \wedge B.$$

Thus it will be enough to show that, given the proviso stated in the theorem,

$T + A/B = T + A \wedge B.$

Let us use the notation LHS for $T + A/B$ and RHS for $T + A \wedge B$. There are two cases.

Case 1: T is consistent. There are two subcases.

Subcase 1.1: deg(T, A) is finite. Let p be the degree of T with respect to A. Then, by Theorem 3.3, $T + A = \{D : A \sqsupset^p D \in T\}$. Consequently, for all formulae C,

$$C \in LHS \text{ iff } A \sqsupset^p B \supset C \in T,$$

$$C \in RHS \text{ iff } A \wedge B \sqsupset^p C \in T.$$

But VC^∞ contains all instances of $(\#5)^p$ and $(\#6)^p$. By our special proviso, $A >^p B \in T$. Therefore LHS = RHS.

Subcase 1.2: deg (T, A) is infinite. In this case $\Box^n \neg A \in T$, for all n. It follows that $\Box^n \neg(A \wedge B) \in T$, for all n, so deg(T, A) is also infinite. Therefore LHS = /A/B and RHS = $/A \wedge B$.

Case 2: T is inconsistent. Then again LHS = /A/B and RHS = $/A \wedge B$.

QED

It is worth pointing out that when deg(T, A) is finite, then our proviso is stronger than that of AGM. For suppose deg(T, A) = p, for some p. If $\neg B \in T + A$, then by Theorem 3.3 $A \sqsupset^p \neg B \in T$, which contradicts our proviso; hence $\neg B \notin T + A$. On the other hand we have already seen that we have no need for a proviso when deg(T, A) is infinite.

REFERENCES

[1] Alchourrón, Carlos E., Gärdenfors, Peter and Makinson, David, "On the logic of theory change", *The journal of symbolic logic* 50:510-530 (1985).

[2] Gärdenfors, Peter, "Conditionals and changes of belief", *Acta philosophica fennica* 30:133-153 (1978).

[3] Lewis, David K., *Counterfactuals*, Harvard University Press, Cambridge, Mass., 1973.

[4] Segerberg, Krister, "On the logic of small changes in theories I", *Acta philosophica fennica*, to appear.

ON 'LOGICAL RELATIONS' IN PROGRAM SEMANTICS

Boris A. Trakhtenbrot

MIT Lab. for Computer Science
Tel-Aviv University*

1. INTRODUCTION

The simplest way to think about logical relations and invariance is to start with permutations on a 'ground' set D; a permutation m_1 is obviously lifted to a permutation m_2 on functionals F from D into D by

$$m_2F = m_1 \circ F \circ (m_1)^{-1}$$

and this process can be performed for all higher types as well. Then a functional G (of some type) is invariant with respect to a class M of permutations iff mG = G for all m in M.

Clearly, a permutation m can be considered as a (binary) relation; namely, each x is related to mx. Permutations are a particular case of logical relations introduced in G. Plotkin's seminal [6], where invariance of functionals with respect to certain logical relations was proven to be a necessary (and sometimes sufficient) condition for their λ-definability. In [6] and [10] even more general classes of relations were considered and their impact on λ-definability was investigated (see [4,10] for detailed bibliography on logical relations and the contributions of other authors in the area).

According to [10], "the story of the [simple] typed λ-calculus is the story of a certain class of hereditarily defined relations called 'logical relations'". On the other hand, the role of the typed λ-calculus in clarifying the essence of programming concepts is well-known. Hence by transitivity one can expect interesting applications of logical relations as a conceptual and technical tool in the theory of programming. This is indeed the case, in particular when the concern is about properties of (typed) programming languages which intuitively amount to representation independence [3,7,4]. Among the theorems and conjectures in this area several have the following format:

If two models M and N are related in a certain way then the meaning of any program T in M is related to its meaning in N in the same way.

---

* This work was supported in part by NSF Grant No. A511190-DCR and by ONR Grant No. N00014-83-K-0125.

Here 'related in a certain way' means that suitable logical relations hold.

Another way to express phenomena like this is to characterize the invariance properties of functionals which provide the meaning (in the style of denotational semantics) of programming constructs.

The appropriate use of logical relations is illuminating in various semantical considerations of programming languages. Sometimes it may even result in a considerable simplification of the techniques to be used in the area. A recent impressive illustration is in [1] where a few simple facts about logical relations allowed avoidance of such heavy techniques (previously used) as generalized domain theory (with limits of transfinite chains) and nonstandard power domain constructions. Well, it is hard to expect such miracles to occur very often. Intricate semantical problems may still require tricky guesses or cumbersome techniques even when logical relations appear from the very beginning as the most suitable tool for the investigation. This is for example the case in [4], where representation independence issues for the second order λ-calculus are considered.

On the other hand some restricted form of logical relations and invariance arguments appeared implicitly in different areas even independently of the theory of logical relations. For example already in the late fifties invariance arguments were used by S. Yablonski and A. Kuznetsov (both in Moscow; exact references unavailable) to prove nondefinability results about functions in multivalued logics. Another example is [11] where the concern was about computability, sequentiality and invariance of functionals on the one hand, and language constructs which are able to express them on the other hand. At that time there was enough consensus about what computability means, less about sequentiality and invariance. In [11] invariance was characterized mainly via permutations. Only after [6] we realized that the right way to deal with invariance is through more general classes of logical relations.

In this paper we want to illustrate the impact of logical relations revising some work we were involved earlier ([11,2]), presenting a simple version of results on observational equivalence from [4] and formulating some open questions. More concretely, we address the following topics:

1. Schematological abstractness characterizes languages which use a first order signature Σ. It amounts to representation independence from the concrete Σ-algebra in which the signature is interpreted. The formalization is in terms of rank 2 functionals that are invariant with respect to specific logical relations (partial isomorphisms).

2. Observational equivalence of two models with respect to a given programming language may sound as reminiscent of the elementary equivalence of two models with respect to the first order predicate-calculus. This notion may be adapted to characterize the behavioral equivalence of two user defined implementations of an abstract data type. Ultimately it reflects the possibility of defining an appropriate logical relation between the models under consideration [7,3].

3. Invariance with respect to locations. The concern is about specific languages with imperative features and side effects we call Algol-like [12,2]. The crucial point in the semantical considerations there is to capture the intuition that some locations of the memory are not relevant for the meaning of a given programming construct. Again, the formal definition uses specific logical relations to make things precise and valuable as a guide for reasoning about programs.

The paper is organized as follows:

Section 2 covers some provisos and auxiliary material about functional frames, models and language construct. The common notations and terminology concerning the syntax and semantics of simple typed λ-calculus are assumed to be known (see, for example, [2]). More details are given about the subclass of completely partial ordered (CPO)-models we call here FLAT's; these models are consistently complete and ω-algebraic. In [8] and [11] sequentiality and invariance were investigated for functionals in FLAT using the infinitary languages of strategies and generalized recursive schemes. Here as a language tool we rely rather on Böhm-trees, a natural generalization of λ-terms, which seems quite adequate for this purpose. Since in this paper we do not care particularly about computability, arbitrary infinite Böhm-trees are considered rather than finite λ-terms augmented with fixed-point operators.

Section 3 presents a short survey on logical relations and their fundamental properties. First, λ-models are considered and the fundamental theorem of logical relations is formulated in both versions: for the model M itself [6] and for the 'starred' model M^* with indeterminates [10]. Since we want to allow recursion, CPO-models are of special interest; therefore, the fundamental theorems have to be adapted, namely inclusive relations should be considered instead of arbitrary logical relations. Another impact is that there is not more reason to confine with λ-terms; infinite Böhm-trees work as well. These adaptations are usually straightforward, but sometimes the attempt to prove or to disprove analogies with the 'classical' cases run into difficulties we could not overcome. Some open technical questions of this sort are formulated in this section.

The last three sections respectively cover the three topics listed above.

Proofs are omitted, except Section 5 about observational equivalence and some hints in other places. The point about observational equivalence is that it exhibits a typical situation when usual logical relations do not provide the expected characterization and the more powerful (and unfortunately less intuitive) class of Statman's S-relations enters into the play. In [4] such a characterization is given even for the more general case of second order λ-calculus, but just because of this generality the main idea is hidden under the heavy techniques.

In some cases we remain deliberately a bit vague in formulations; this is the price to be paid in order to avoid the cumbersome notational and definitional machinery of formal semantics.

2. TYPES, FRAMES, BÖHM-TREES

Type Expressions

Ground types are $o_1,\dots,o_k$. If α and β are type expressions so is $\alpha \to \beta$. Each type expression α may be represented as $\alpha_1 \to (\alpha_2 \to \dots(\alpha_m \to o)\dots)$ with o a ground type. The brackets may be dropped; for readability use also the notation $\alpha_1,\dots,\alpha_m \to o$.

Ranks

The rank of a ground type is 0; the rank of $\alpha \to \beta$ is 1 + max{rank α, rank β}.

Functional Type Frame

Frame (for short) is a mapping from type expressions to sets such that $\emptyset \neq D^{\alpha\to\beta} \subseteq D^{\alpha} \to D^{\beta}$ = the set of all mappings from D^{α} into D^{β}. The elements of D^{α} are called functionals of type α. For simplicity we confine always when no misunderstandings can arise with one ground type o. Some frames of special interest are:

1. P_D - the full frame over a given ground set D, i.e., all possible mappings are included in the frame.

2. CPO-frames M: the M^{α} are complete partial orders and only continuous mappings are considered. In particular a full continuous frame is a CPO with all continuous mappings included on each inductive step.

3. FLAT frames are full continuous frames over a flat ground domain in which $x \subseteq y$ if $x = y$ or $x = \bot$. Think about the ground domain of FLAT as $\{\bot, 0, 1, \ldots\}$, though in fact we could consider ground domains of arbitrary cardinality.

The following features of FLAT should be mentioned:

a) Consistent completeness: each subset which has an upper bound in the frame has an lub, i.e., a least upper bound there. Use the infix $\cup$ for the lub of two ground elements.

b) Though $\cup$ does not belong to FLAT because it is partial, it is continuous everywhere it exists.

c) The following rank 1 functionals exist in FLAT; note that their definition is 'polymorphic' - it does not matter what the ground domain really is:

The conditionals if (sequential) and IF (parallel) both of type o,o,o,o → o, where

$$\begin{aligned} \text{if}\,xyuv = \; & v \quad \text{if } \bot \neq x \neq y \neq \bot, \\ & u \quad \text{if } \bot \neq x = y \neq \bot, \\ & \bot \quad \text{otherwise.} \end{aligned}$$

IF behaves as if except that IFxyuu = u.

The 'voting' function $\Gamma_{n,m}$ of n arguments ($n < 2m$) such that a value $\neq \bot$ is returned if at least m of the arguments have just this value.

All the frames considered above are models for the λ-calculus.

Given a set C of typed constants $\lambda(C)$ is the notation for the language of typed λ-terms which may contain these constants. CPO models provide semantics in the case when C consists of the constants $Y^{(\alpha\to\alpha)\to\alpha}$ interpreted as the respective fixed-point operators and of $\bot$, interpreted as the bottom of ground type. More generally, they provide semantics for the infinitary language of typed Böhm-trees Δ. The nodes of the (in general infinite) tree Δ are labeled and the labels have the format

$$\lambda\vec{x}.y$$

where $\lambda\vec{x}$ is a prefix of abstractors and y is a variable or a constant (the constant $\bot$ is always allowed); the prefix may happen to be empty. More formally

$$\Delta = \lambda\vec{x}.y\Delta_0\Delta_1 \ldots \Delta_m$$

where the Δ_i are themselves Böhm-trees. It is also assumed that in each tree there is only a finite set of unbound (i.e., free or constant) identifiers, though some of these identifiers may have infinite many occurrences.

We omit the formulation of the typing rules, which essentially require the following: to each subtree of Δ (its root is an arbitrary node in Δ) a type should be assigned in a way that is consistent with the types of the atoms.

Long $\beta\eta$-normal form $\Delta =_{def} \lambda\vec{x}.y\Delta_0\Delta_1 \ldots$ where $y\Delta_0\Delta_1 \ldots$ has ground type, and the Δ_i are long $\beta\eta$-normal forms.

<u>Partial Ordering Among the Trees of a Given Type</u>

Notation: for $\alpha = (\beta,\ldots,\gamma \to o)$ consider

$$\bot_\alpha =_{def} \lambda u^\beta \ldots \lambda v^\gamma.\bot$$

Now the ordering $\subseteq$ is defined by: (1) $\Delta \subseteq \Delta$; (2) $\bot_\alpha \subseteq \Delta$ for each Δ of type α; and (3) $\Delta_2' \subseteq \Delta_2$ implies $\lambda\vec{x}.y\Delta_1'\Delta_2' \ldots \subseteq \lambda\vec{x}.y\Delta_1\Delta_2 \ldots$.

Claim: for each type this ordering is a CPO which is consistently complete, i.e., each set of Böhm-trees that has an upper bound has at least one.

<u>Semantics</u>. $[\![\Delta]\!]_M\rho$ - the meaning of tree Δ in model M with respect to environment ρ - is defined as the <u>lub</u> of $[\![\delta]\!]_M\rho$ for finite trees (λ-terms) that approximate Δ, i.e., $\delta \subseteq \Delta$.

<u>Facts</u>

1. Each λ-term t with fix-point operators is equivalent (semantically) to a tree Δ such that the free variables of Δ are among the free variables of t and the constants of Δ are among the constants of t but the fix-point operators do not occur in Δ.

2. Each tree Δ is equivalent to a long $\beta\eta$-normal form Δ'.

3. Application rule: if the types of trees Δ_1,Δ_2 match then there exists a tree Δ which is equivalent to their application.

4. There are algorithms that perform the transformations above.

For a given set of typed constants m,n,...,p let $\Lambda(m,n,\ldots,p)$ be the class of Böhm trees in which these constants are allowed. Given a CPO M and an interpretation of the constants, a tree $\Delta \in \Lambda(m,n,\ldots,p)$ defines an element f in M; we say that f is Λ definable from m,n,...,p, $\Lambda(n,\ldots,p)$ definable from m, etc. Use also "reducible to" instead of "definable from", suggesting the usual terminology of reducibility, of reducibility degrees, etc.

For FLAT, one step more may be done, allowing in the trees the constant $\cup$. In general such a tree is meaningless because $\cup$ is partial, but it may happen to have a meaning. By abuse of notation we use $\Lambda(\ldots)$ for both the trees and their meanings.

<u>Notation</u>: $F \in \lim\Lambda$ iff F is the limit of an ascending chain of functionals in Λ. Clearly,

$$\Lambda \subseteq \lim\Lambda \subseteq \Lambda(\cup).$$

Question 1 Which of the inclusions above is strong, if any?

Assume the ground domain of FLAT is $\{\bot, 0, 1, \ldots\}$ and the constants if, suc, 0 with the self-explanatory interpretations. Then:

Fact [6,8]: all functionals are expressible through trees in Λ(if, suc, 0, $\cup$).

Note, that in [11] and [8] it is argued (using other notations and terminology) that Λ(if, suc, 0) coincides with the class SEQUENTIAL of functionals in FLAT to be recognized as sequential, whereas those functionals which essentially need the $\cup$ are the parallel ones. In this sense IF is indeed a parallel function and Λ(if) is a proper subclass of SEQUENTIAL.

3. LOGICAL RELATIONS

Notations and Terminology

M is a model of the typed λ-calculus. M^α is the domain associated with the type α (hence $M^{\beta\to\gamma} \subseteq (M^\beta \to M^\gamma)$). $[\![t]\!]^\rho_M$ is the meaning of the (typed λ-term) t in the model M with respect to the environment ρ.

Given n models $M_1, \ldots, M_n$, an n-ary logical relation $R \subseteq M_1 \times \ldots \times M_n$ is a set of relations R^α (one for each type α)

$$R^\alpha \subseteq {M_1}^\alpha \times \ldots \times {M_n}^\alpha$$

satisfying the condition:

$$R^{\beta\to\gamma}(d_1, \ldots, d_n) \Leftrightarrow \forall e_1, \ldots, e_n [R\ (e_1, \ldots, e_n) \supset R^\gamma(d_1 e_1, \ldots, d_n e_n)].$$

Theorem 1 (the fundamental theorem of logical relations [6]). Given arbitrary λ-term t, models $M_1, \ldots, M_n$, environments $\rho_1, \ldots, \rho_n$ and a logical relation $R \subseteq M_1 \times \ldots \times M_n$. Assume the conditions: (1) $R([\![c]\!]_{M_1}, \ldots, [\![c]\!]_{M_n})$ for each constant c; (2) $R(\rho_1 z, \ldots, \rho_n z)$ for each free variable z. Then

$$R([\![t]\!]^{\rho_1}_{M_1}, \ldots, [\![t]\!]^{\rho_n}_{M_n})$$

holds.

Consider a logical relation $R \subseteq M \times \ldots \times M$.

Definitions: (1) f is invariant with respect to R (f respects R, R preserves f) $=_{def} R(f, \ldots, f)$; (2) g inherits the logical relations that preserve $\{f_i\}$ $=_{def} \forall R(\forall f_i (R \text{ preserves } f_i) \supset R \text{ preserves } g)$.

Corollaries from Theorem 1: (1) the meaning of a closed λ-term is invariant with respect to all logical relations; (2) g is λ-definable from $\{f_i\}$ implies that g inherits the logical relations of $\{f_i\}$.

Notes

1. Clearly, a logical relation is fully defined as soon as it is defined for the ground types.

2. Permutations of the ground domains (particularly the identity permutation) induce binary logical relations. For binary relations we shall use mainly the infix notation.

3. Unary logical relations may be considered as well and we shall make use of them. Assume that the unary relation R holds for the functional F; then it obviously follows from the definitions that F is also invariant with respect to R.

4. In general if functionals F, G are invariant with respect to R so is their application.

Example

Consider the logical relations C_k (k = 2,3,4,...) over FLAT, which on the ground domain behave as follows: $C_k(x_1,\ldots,x_k)$ is true iff at least one of the arguments equals $\perp$.

See that if respects all the C_k, IF does not respect C_3, for k > 2 $\Gamma_{k+1,k}$ respects C_k but $\Gamma_{k,k-1}$ does not. It is easy to see that if is λ-definable from IF and also that $\Gamma_{k+1,k}$ is λ-definable from $\Gamma_{k,k-1}$. Hence introducing in a self-explanatory way the notion of λ-reducibility and λ degrees, we discover an infinite antichain of degrees among the rank 1 functions $\Gamma_{k+1,k}$.

Note

A similar way of justifying nondefinability was used in fact in the investigation of completeness criteria for multivalued logics (Yablonski, Kuznetsov).

It is not known whether Theorem 1 or its corollaries are reversible. Here is a particular case when it happens:

Theorem 2 (Plotkin [6]). Consider the full frame P_D with infinite ground set D; then for each functional f of rank $\leq$ 2 f respects all logical relations iff f is λ-definable.

This is a sample of what may be called a characterization theorem in which four parameters are involved: (1) a model M; (2) a language L; (3) a class R of relations (logical relations in the sense above or some generalizations); and (4) type restrictions on the functionals.

The format of such theorems is: for each functional f in M with the intended type restrictions f is definable in L iff f is invariant with respect to all relations in R.

Now we are going to consider the approach developed by R. Statman which provides a very general characterization theorem.

Given a model M, consider for each type α a countable set Ind_α of indeterminates (fresh variables). The 'starred' model M_α^* is defined as a suitable factorization (see below) of the set of all formal applicative and typed terms upon elements of M and indeterminates. Suppose ϕ,ψ are such terms and consider the semantics of each of them as a function of environments ρ which assign to the indeterminates values in M. It may happen that under these conditions ϕ and ψ induce the same function; in this case they are identified as elements of M^*. If ϕ does not contain indeterminates then it is obviously identified with an element of M.

Now consider a logical relation R on M^*; its restriction on M may not be a logical relation on M, but one can show [10] that every logical relation on M is the restriction of some logical relation on M^*.

Terminology: S-relation on M = <def> the restriction on M of a logical relation on M^*. Similarly, S-relations on $M_1 \times M_2 \times \ldots$ are defined. Simplicity for confining below with binary S-relations.

Theorem 3 Fundamental Theorem of S-Relations:

1. Let R be an S-relation between M_1, M_2 and τ an arbitrary closed λ-term. Then $[\![\tau]\!]_{M_1} R [\![\tau]\!]_{M_2}$.

2. More generally, if τ is λ-definable from $f_1,\ldots,f_k$ and $[\![f_i]\!]_{M_1} R [\![f_i]\!]_{M_2}$ then $[\![\tau]\!]_{M_1} R [\![\tau]\!]_{M_2}$.

This theorem generalizes the fundamental theorem of logical relations.

Theorem 4 (Statman) Characterization theorem for arbitrary λ-model M: f is λ-definable iff it respects all S-relations.

In fact the more general fact holds: f is λ-definable from $\{f_i\}$ iff it inherits all S-relations of the $\{f_i\}$.

Inclusive Relations, Lub-Relations

When CPOs and in particular FLATs are considered it is natural to confine with those logical relations (S-relations) which are appropriately connected to the partial ordering under consideration. A logical relation R is inclusive if it satisfies:

1. R preserves $\bot$.

2. R is closed under limits.

Inclusive S-relations $=_{def}$ restrictions of inclusive logical relations between the "starred" CPO-models.

Notes

1. Condition 2 holds if it holds for the ground domains.

2. The fixed point operators respect all inclusive logical relations (and S-relations).

3. The logical relation induced by $Rzy =_{def} x \subseteq y$ for the ground domain is inclusive and "f respects R" means "f is monotonic". Hence all elements in a CPO model are invariant with respect to this R.

For FLAT it is clear that a logical relation is inclusive if it preserves $\bot$. In particular every permutation in the ground domain which preserves $\bot$ induces an inclusive relation. Obviously the logical relations C_k are inclusive as well. Now Theorems 1 and 3 can be adapted to CPO and Böhm trees:

Claim 1

If a functional F is in Λ then it respects all inclusive logical relations and even more, F respects all inclusive S-relations. In general if F is in $\Lambda(m,n,\ldots)$ then it inherits all inclusive relations respected by $\{m,n,\ldots\}$.

The proofs are immediate, taking into account that the meaning of a tree is the limit of the approximating λ-terms.

In particular the fix-points operators respect all inclusive logical relations.

But what about the characterization theorems?

Claim 2

Assume for a given CPO M, that $\Lambda = \lim\Lambda$; then for each F in M we have, F is in Λ iff F respects all inclusive S-relations.

Note that the proof of this claim makes essential use of the Fact 3 about the semantics of Böhm trees (see Section 2).

As far as FLAT is concerned one can consider the more restrictive class of Lub-relations which contains only relations that are closed under Lubs.

Claim 3

For each F in FLAT F is in $\Lambda(\cup)$ iff F respects all lub-S-relations.

We would expect for FLAT the analogy of Theorem 2.

Question 2. Is it true for rank 2 functionals in FLAT that F is in Λ iff it respects all inclusive logical relations?

Use the notation Inherit(m,n,...) for the class of functionals which inherit all logical relations preserved by {m,n,...}. Clearly, Λ(if) $\subseteq$ Inherits(if).

Question 3. Does the inverse claim hold for rank 2 functionals? Recall that the functionals in Λ(if) are a part of the sequential ones in the sense explained at the end of Section 2.

4. SCHEMATOLOGICAL FUNCTIONALS

In (first-order) schematology one considers program schemes which in addition to the 'core constructs' of the language use also an uninterpreted signature Σ consisting of functional and predicate symbols. Below we concentrate on the case when the signature consists only of functional and individual symbols (no predicates allowed). Now abstracting from the symtax of the language L but explicitly referring to its first-order signature Σ (which implicitly includes the symbol $\perp$), the semantics is expected to be a total mapping which sends each Σ-algebra over a carrier D into some element of D. The specific element $\perp \in D$ is assumed as usually to avoid partial mappings . The following (minimal) requirements reflect the expected representation independence properties of the language and its semantics:

1. Respecting isomorphisms. If two algebras A_1, A_1 are isomorphic then the returned values should correspond to each other in the isomorphism.

2. The Herbrand property. For the algebra A with carrier D, let $D_H \subseteq D$ be the Herbrand subdomain of D: it consists of those elements of A which are representable as applicative terms upon the signature. Let A_H be the Herbrand subalgebra of A upon the carrier D_H. Then A and A_H return the same value.

Note: other requirements may concern the computability of the mapping under consideration.

In the sequel, we confine to the case when the carrier D is fixed once for over and namely is a flat domain. Then conditions 1 - 2 amount to the restriction to some specific class of functionals of rank 2 in the model FLAT: call them schematological functionals.

Characterization of Schematological Functionals F via Logical Relations

Claim 1

Condition 1 $\Leftrightarrow$ invariance of F with respect to permutation.

With each subdomain $M \subseteq D$, associate the binary logical relation R (particular case of partial isomorphisms, see 2.2.2) induced by

$$xRy =_{def} x = y \in M$$

Note that a subdomain is assumed to include $\perp$. In this case, we have the direct definition of the Herbrand property for a rank 2 functional F: $\alpha_1,\ldots,\alpha_k \to o$. Namely, F has the Herbrand property $\Leftrightarrow$ for arbitrary $\vec{f} = f_1,\ldots,f_k$ and $\vec{f}' = f_1',\ldots,f_k'$ if $\vec{f}'$ equals $\vec{f}$ on the subdomain generated by $\vec{f}$ then $F\vec{f} = F\vec{f}'$.

Claim 2

F has the Herbrand property iff F is invariant with respect to all subdomain relations.

In the sequel we use 'partial isomorphisms' for permutations and subdomain relations.

Examples

The following "core" languages being equipped with first order signatures indeed associate with each program Π a rank 2 functional which respects all partial isomorphisms.

1. The typed λ-calculus with fix-point operators and the conditional if. All logical relations inherited from if are respected. Among them - all partial isomorphisms.

2. Add to the language above arbitrary rank 1 functions f which respect permutations. (In fact it suffices to add the function $\Gamma_{3,2}$ to which as it may be shown each such f is λ(if) reducible.)

3. Allow in the language Böhm trees, i.e., consider $\Lambda(\underline{\text{if}},\Gamma_{3,2})$.

But in all these cases there still exist rank 2 functionals which are invariant with respect to all partial isomorphisms (and are even computable) but are not expressible in the language. In order to cover all these functionals, consider the more powerful language $\Lambda(\underline{\text{if}},\cup)$.

Theorem 1 (Characterization of schematological functionals). A rank 2 functional in FLAT is $\Lambda(\underline{\text{if}},\cup)$-definable iff it respects all partial isomorphisms.

Note. The 'only if' part holds for all ranks and not just for rank 2. This is because if and $\cup$ respect partial isomorphisms. The claim about $\cup$ needs an explanation, because it is a partial function whereas logical relations and invariance were defined earlier only for total functions.

In this particular case what we have in mind is the following: given a partial isomorphism R and ground elements a,b,c,d such that aRc, bRd hold then $(a \cup b)R(c \cup d)$ holds assuming that the lubs exist. It is easy to show that Claim 1 from Section 2 still holds for Böhm trees with the constant $\cup$.

For the 'if' part the restriction to rank 2 is essential in order to assure the existence of the defining tree. In this case the tree may be even chosen with no nestings of $\cup$ in the scope of functional symbols.

The parallel rank 1 functions IF, $\Gamma_{k+1,k}$ are schematological. So is the parallel existential quantifier defined as follows:

$$
fgab: = \begin{cases} a & \text{if } f\bot = a, \\ b & \text{if } f(g(g(\ldots(ga)\ldots))) \text{ happens to equal } b, \\ \bot & \text{otherwise.} \end{cases}
$$

Sazonov [8] and Plotkin [5] proved (in some other terms) that each functional in FLAT is Λ-definable from IF, , suc, 0. We would like to manage without the noninvariant suc and 0.

In particular:

Question 4 Does there exist a finite set S of functionals in Λ(if,$\cup$) such that every schematological functional f (of rank 2) is Λ-reducible to S?.

Clearly, both Λ(if) and Inherit(if) are subsets of [Λ(if,$\cup$)] $\cap$ SEQUENTIAL.

Question 5 What of these inclusions is strong, if any? (at least for rank 2 functionals).

A more general task is to describe the structure of degrees among the schematological functionals with respect to 'sequential' reducibility, where f reduces sequentially to g means f is in Λ(if,g).

5. OBSERVATIONAL EQUIVALENCE

We want to consider a language L with several ground types. For simplicity assume that there are only two ground types: the "observable" type int and the other one data. Consider also two models of the λ-calculus M_1, M_2 upon these types, which have in common the domain D_{int} but perhaps different domains D_{data}. L is assumed to be the typed λ-calculus with the ground types above and with a set Σ of constants - the signature of the language. But note, that unlike the previous Section 4 the symbols in Σ are not necessarily of rank 1 or 0. For example, constants to be interpreted as fix-point operators might be allowed. Let Σ_1, Σ_2 be the interpretations of the (symbols in the) signature in the two models above. As programs in L consider only closed terms of type int.

$$M_1 \cong_{obsL} M_2 =_{def} \text{ for each program } \tau \in L$$

$$[\![\tau]\!]_{M_1} = [\![\tau]\!]_{M_2}$$

Note: it may happen that the value $[\![\tau]\!]$ do not exhaust the semantical ground domain D_{int}, but only some "observational" part $D^{obs}_{int} \subseteq D_{int}$.

The task: criteria or observational equivalence.

Theorem 1 Assume there exists a logical relation $R \subseteq M_1 \times M_2$ such that:

1. R_{int} is the identity on D^{obs}_{int}.

2. $\Sigma_1 R \Sigma_2$, i.e., R holds for each pair of interpretations for signature symbols.

Then $M_1 \cong_{obs} M_2$.

Proof Obvious consequence of two facts: (1) the fundamental theorem 1 of logical relations; (2) for the observable type, validity of R means identity.

Is Theorem 1 reversible? Here is a particular case when it is.

Theorem 2 Assume the signature Σ is first order (rank ≤ 1). Then $M_1 \cong_{obs} M_2$ implies the existence of an R as above.

Proof Define R on the ground domains

$$\xi R \eta =_{def} \exists \text{term in } L\{\xi = [\![\text{term}]\!]_{M_1} \wedge \eta = [\![\text{term}]\!]_{M_2}\}.$$

Now obviously for this logical relation R, there holds $\Sigma_1 R \Sigma_2$ (since Σ is first order). Hence condition (2) of Theorem 1 holds. It remains to check that condition (1) holds as well, namely that R^{int} is the identity on D^{obs}_{int}.

Assume $\xi \in D^{obs}_{int}$; then there exists a term τ such that $[\![\tau]\!]_{M_2} = \xi$. Since $M_1 =_{obs} M_2$, it must be also $[\![\tau]\!]_{M_1} = \xi$ and therefore $\xi R \xi$ holds by the definition of R. On the other hand, for $\xi \neq \eta, \xi, \eta \in D^{obs}_{int}$, there cannot be such a term.

Example

Consider the language L with the ground types real, complex and with the first order signature:

mod: comp → real
sum: complex × complex → complex
mult: complex × complex → complex
imag: complex → real

The models M_1, M_2 are assumed to have the same ground domains D_{real}, $D_{complex} = D_{real} \times D_{real}$ and the interpretations of the signature are assumed to deal respectively with the cartesian and polar representations of the complex numbers. Now it is easy to check that in this case there exists a logical relation R with respect to which the interpretations of the signature behave properly (as required in the theorem). Namely, define for ground type complex: $d_1 R d_2$ iff there exists a complex number with d_1, d_2 as its cartesian and polar representations, respectively. Assuming that R is the identity for type real and lifting to higher types we get the suitable logical relation.

It is an easy exercise to generalize the theorems above to the case when in addition to the first order signature the language allows fixed point operators. (Hint: rely on inclusive logical relations.) If the signature Σ contains higher rank constants the part of the proof of Theorem

2 concerning the behavior of R^{int} on D^{obs}_{int} still holds. But the fact that $\Sigma_1 R \Sigma_2$ holds may fail.

The characterization of observational equivalence in the general case of higher rank signatures is achievable via S-relations.

Main Theorem In order to have $M_1 \cong_{obs} M_2$, it suffices, and it is necessary that an S-relation R exists such that:

1. R_{nat} is the identity on D^{obs}_{nas}.

2. $\Sigma_1 R \Sigma_2$.

Proof Sufficience follows from the fundamental Theorem 3 (see 1.2).

Necessity. Define a relation $R \subseteq M_1^* \times M_1^*$ as follows:

$$\xi R \eta =_{def} \exists \text{term with indeterminates } (\xi = [\![\text{term}]\!]_{M_1^*} \wedge \eta = [\![\text{term}]\!]_{M_2^*}).$$

Note, that unlike in Theorem 2 above, the relation is defined directly for all types and not by induction from ground types, hence it is not clear automatically that it is logical. To terminate the proof, we need two lemmas.

Lemma 1 R is logical on $M_1^* \times M_2^*$.

Hence, its restriction on $M_1 \times M_2$ is an S-relation. Preserve for it the notation R.

Lemma 2 R is the identity on elements from D^{obs}_{int}.

Lemma 2 is proven as in Theorem 2 (above). Exercise.

Proof of Lemma 1 Check that

$$\phi_1 R \phi_2 \Leftrightarrow \forall u_1 u_2 (u_1 R u_2 \Rightarrow (\phi u_1) R (\phi_2 u_2)).$$

1. $\Rightarrow$ is immediate. Assume ϕ_1 $[\![\tau]\!]_{M_1^*}, [\![\tau]\!]_{M_2^*} = \phi_2$ and $u_1 = [\![\sigma]\!]_{M_1^*}, [\![\sigma]\!]_{M_2^*} = u_2$. Then

$$\phi_1 u_1 = [\![\tau\sigma]\!]_{M_1^*}, [\![\tau\sigma]\!]_{M_2^*} = \phi_2 u_2.$$

2. $\Leftarrow$ for each indeterminate x, we obviously have xRx (why?). Then take $u_1 \equiv u_2 \equiv x$ where x is a fresh indeterminate not occurring in $\phi_1 \phi_2$. Since $(\phi_1 x) R (\phi_2 x)$, there exists τ such that

$$\phi_1 x = [\![\tau]\!]_{M_1^*}, [\![\tau]\!]_{M_2^*} = \phi_2 x.$$

Now, consider $\lambda x, \tau$, then it is easy to check that

$$\phi_1 = [\![\lambda x, \tau]\!]_{M_1^*}, [\![\lambda x, \tau]\!]_{M_2^*} = \phi_2.$$

Though the criterion of observable equivalence sounds liberal enough, it may happen that in fact it is too restrictive. Remember that the elementary equivalence of models amounts to the fact that some of their ultrapowers are isomorphic (Scott's conjecture, proven by Shelah). Hence the informal.

Question 6 Isn't observable equivalence in fact too close to isomorphism?

6. INVARIANCE WITH RESPECT TO LOCATIONS

We mention here only the following features of the Algol-like language under consideration [2]:

About Types

There is an explicit type distinction between locations (addresses) and storable values. For simplicity consider values of only one type. Hence, the ground types loc and val and the use of the token cont for explicit dereferencing. Both the respective ground domains LOCATIONS and VALUES are infinite and flat. Use the notations LOCATIONS' and VALUES' for their essential parts (i.e., the bottom is not included). Another ground type is prog; the respective domain consists of partial mappings from the set STORE (of stores) into STORE, whereas a store is a total function from LOCATIONS' to VALUES'. Well, it looks as if we could deal with the types of program meanings (i.e., of store transformations) and of stores as with types derived from the only ground types loc and val. Nevertheless there are reasons (we shall return to them later in this section) for having a ground type prog and not to include at all a type for stores. Hence we consider the three ground types above (though in [2] some additional types are considered as well).

About Procedures and Declarations

Higher order procedures of all finite types formed from the ground types may be declared, passed as parameters and returned as values. Identifiers of type loc may be bound by var-declarations as in (#) below; though the type information is omitted there, it is hopefully clear that the bound (local) variable is of type loc and the call in the body is of a global (non declared) procedure whose type is loc → prog.

In comparison with pure functional languages the semantics of Algol-like languages rise additional problems due to the role of storage allocation and side-effects. A crucial point is to explain when a location is accessed by a procedure P and when it is free with respect to P. In the operational style semantics such an explanation relies on the inspection of the declaration body of P: the (meanings of the) free location identifiers in the body are considered as the accessible ones. But if P itself is global (i.e., it is not declared). Clearly we need a direct semantical explanation, which does not refer to a body text.

It is easy to give a pure denotational characterization of what it means to say that a mapping p from STORE to STORE does not have access to locations outside a given subset L $\subseteq$ LOCATIONS. The technical term we use is L supports p or L is a support of p and it is defined as the conjunction of two requirements:

1. p does not read from outside L: for all s,t in STORE (s coincides with t on L implies that ps = pt).

2. p does not write outside L: for all s in STORE (ps = s outside L or ps is not defined).

Now consider the block:

begin var x; P(x) end (#)

where P is a global (undeclared) procedure. We expect the meaning of the block to rely on the binding of the identifier x to a 'fresh' location which is not accessed by P itself. But in a pure denotational approach P should be interpreted as a mapping from locations to store transformers and a direct definition of what locations are not accessed by this mapping is still needed. What we are looking for is a precise and reasonable generalization of the notion that L supports P for P of all relevant types so that the locations outside L would be the 'non-accessed' or 'free' ones for P.

Below we consider some families of unary logical relations which are intended to provide such a generalization. As usually it suffices to define them for the ground types and then to lift them up to higher types. Serious technical problems arise in order to incorporate consistently these relations in a good working semantical model [2], but these problems are completely missed here and we concentrate only on the description of the logical relations involved there. As mentioned earlier as ground types we have to consider prog, loc, val, ignoring store as a type.

Consider the following families of logical relations:

Family 1 is directly suggested by the idea of support as explained above for the meaning of store transformation. It contains for each location ξ a unary logical relation R_ξ that behaves as the following:

For type val: $R_\xi(\alpha)$ is always true;
For type loc: $R_\xi(\alpha)$ is equivalent to $\{\xi = \bot \text{ or } \xi \neq \alpha\}$;

For type prog: $R_\xi(\alpha) \Leftrightarrow$ there exists L such that L supports α and ξ is not in L.

Family 2 is intended to capture an aspect of invariance that is relevant for higher order procedures though it was not mentioned explicitly in the explanation of support for store transformations. It contains for each permutation μ of locations (which preserves $\bot$) a binary logical relation M_μ that in fact is also a permutation, as described below (where for readability the subscript μ is omitted):

For type val: $M(a) = a$;
For type loc: $M(a) = \mu a$.

Since we do not have the type store in the language we are not committed to define the permutations for stores; nevertheless we need them as an intermediate stage in order to define the permutations for store transformations. Hence we require for s in STORE and a in LOCATIONS

$$(Ms)(\mu a) = sa.$$

Finally for a of type prog and s in STORE we require

$$(Ma)(Ms) = M(as).$$

Family 3 [9] contains for each location ξ a unary logical relation $R_\xi S$ that differs from R_ξ above only for type prog. Namely in this case

$$R_\xi S(\alpha) \Leftrightarrow \alpha \text{ does not change the content of } \xi.$$

Now the technical challenge is to construct an 'appropriate' model H which captures the intuitively expected meanings of Algol-constructs under consideration and which also exhibits a high degree of invariance with respect to the logical relations in the families listed above [2]. The following holds in H:

1. All these relations are inclusive.
2. The elements of H which interpret the specific constructs of the

language (assignment, sequencing, var-declaration, etc...) are invariant with respect to all these relations.

3. Each other functional F in the model respects almost all the relations in the following sense:

There is a finite set L of locations depending on F such that: (1) for each ξ outside L both $R_\xi(F)$ and $R_\xi S(F)$ hold; (2) for each permutation μ which fixes L there holds $M(F) = F$. Shortly - each functional in the model has a finite support.

The following simple example suggested by G. Plotkin may explain to some extent the reason to choose prog as a ground type. Consider the following typed λ-term t:

$$\lambda s^{loc \to val} \lambda l^{loc} . c^{val}$$

which has the type (loc → val) → (loc → val). If we decide to consider prog as a derived type with this presentation through loc and val, then we have also to admit that t is a program; but it is easy to see that t, being considered as a store transformer (where stores have type loc → val) has no finite support. Yet we are looking for a semantics where every program (and even every high-order procedure) has a finite support.

The properties of the model H, including the finite support features, are the basis of several axioms and rules for program correctness [2]. Let us give a hint to them using an example borrowed from [9]. Consider the following piece of program:

begin var x; x: = 1; begin E; P(R); R end (*)

where E is the procedure declaration R ⇐ y: = cont(x).

Note that P is global in (*) and therefore its meaning depends on the choice of the environment. Nevertheless one can show [9] that in the model H whatever this choice may be the following partial correctness assertion holds:

{true}(*){cont(y) = 1}

Here is a sketchy explanation of this fact:

a) The meaning of the block (*) in the model H is defined via a binding of x to a location ξ which is outside some support of the body of this block. Arguments which rely on the permutations from the family 2 assure that the meaning of (*) is indeed invariant with respect to the choice of such an ξ.

b) In particular ξ may be chosen outside a support of P, since whatever the meaning assigned to P might be, it has finite support. Therefore the (meaning of) P is invariant with respect to $R_\xi S$. That is true also for the procedure R which obviously is invariant with respect to the whole family 3. As application preserves invariance the call P(R) also respects $R_\xi S$, that is it does not change the content of ξ. Because x is initially assigned 1, this value is finally assigned to y by the call of R. That is just what the partial correctness formula asserts.

Since the definition of the relation R_ξ includes both requirements of not changing the content of ξ and not reading its content, it may look apparently that the family 3 is superfluous. But in fact one can construct (somehow an exotic example of) a frame in which some objects of type prog → prog respect all the relations in the first two families but do not respect any relation in the third one. In any case in the example above it is not clear how to justify the partial correctness using only relations from the families 1 and 2. Hence:

Question 7

Are there reasons to consider even additional logical relations in order to capture more location invariance?

REFERENCES

1. S. Abramsky, Strictness analysis via Logical Relations, Manuscript (1986).
2. J. Y. Halpern, A. R. Meyer and B. Trakhtenbrot, The semantics of local storage, or what makes the free-list free? in: "11th ACM Symp. on Principles of Programming Languages", 245-257 (1984).
3. J. Mitchell, Representation independence and data abstraction, in: "13th ACM Symp. on the Principles of Programming Languages", 263-276 (1986).
4. J. C. Mitchell and A. R. Meyer, Second-order logical relations (extended abstract), in: "Logics of Programs, Lect. Notes in Comp. Sci., R. Parikh, ed., Springer-Verlag, 193:225-236 (1985).
5. G. D. Plotkin, LCF considered as a programming language, Theoretical Computer Science, 5:223-257 (1977).
6. G. D. Plotkin, Lambda-definability in the full type hierarchy, in: "To H. B. Curry: Essays on Combinatory Logic, Lambda Calculus and Formalism", J. P. Seldin and J. R. Hindley, eds., Academic Press, 363-373 (1980).
7. J. C. Reynolds, Types, Abstraction, and Parametric Polymorphism, in: "Information Processing 83", R. E. A. Mason, ed., North-Holland, 513-523 (1983).
8. V. Yu. Sazonov, Expressibility of functions in D. Scott's LCF language, Algebra i Logika, 15:308-330 (1976) (Russian).
9. K. Sieber, A partial correctness logic for procedures, in: "Logics of Programs, Lect. Notes in Comp. Sci.", R. Parikh, ed., Springer-Verlag, 193:320-342 (1985).
10. R. Statman, Logical relations in the typed λ-calculus, Information and Control, 65:86-97 (1985).
11. B. A. Trakhtenbrot, Recursive program schemes and computable functionals, in: "Mathematical Foundations Computer Science Proceedings 1976, Lect. Notes in Comp. Sci.", A. Mazurkiewicz, ed., Springer-Verlag, 45:137-151 (1976).
12. B. Trakhtenbrot, J. Y. Halpern and A. R. Meyer, From denotational to operational and axiomatic semantics for Algol-like languages: an overview, in: "Logic of Programs, Proceedings 1983, Lect. Notes in Comp. Sci.", E. Clarke and D. Kozen, eds., Springer-Verlag, 164: 474-500 (1984).

Cambridge, Massachusetts, USA
28 June 1986

C O N F E R E N C E

(contributed papers)

SEARCH COMPUTABILITY AND COMPUTABILITY WITH NUMBERINGS ARE EQUIVALENT IN THE CASE OF FINITE SET OF OBJECTS

Angel V. Ditchev

High Pedagogical Institute
9700 Shumen, Bulgaria

In the paper [1] Moschovakis has defined a concept about relative computability, which he called search computability. D. Skordev has defined another concept about relative computability, which he called admissibility. That concept is connected with partial recursivness. Similar definition of admissibility has been given by Lacombe [2], which he called ∀-recursiveness.

In [3] Moschovakis and in [4] the author have proved that search computability implies admissibility. D. Skordev has stated a hypothesis that the admissibility will imply the search computability under some natural restrictions. In [3] Moschovakis has proved the admissibility in the Lacombe's sense implies search computability. Let us remark that in this paper Moschovakis supposes the numberings 1-1 and the "equality" is among the basic functions and predicates. In [5,6] some attempts have been done to establish Skordev's hypothesis. In this paper it will be proved that in the case of finite set of objects the admissibility implies the search computability.

Let us remind some definitios from [4,7].
By N we denote the set of all natural numbers. If β is Goedel's function, for every natural numbers $p_1, \dots, p_k$, $\langle p_1, \dots, p_k \rangle$ denotes $\mu p[\ \beta(p,0) = k\ \&\ \beta(p,1) = p_1\ \&\ \dots\ \&\ \beta(p,k) = p_k\ ]$; $(p)_i = \beta(p,i+1)$ for any natural numbers p, i. By DF we denote the domain of the function F and by RF - the range of values of F. If $DF \subseteq M_1$ and $RF \subseteq M_2$ we write $F: M_1 \dashrightarrow M_2$, but if $DF = M_1$, we write $F: M_1 \longrightarrow M_2$. Let B is an arbitrary set, 0 does not belong to B and $B_0 = B \cup \{0\}$. We define a set B^* in the following way:

a) If $s \in B_0$, then $s \in B^*$;
b) If $s \in B^*$ and $t \in B^*$ then $\langle s,t \rangle \in B^*$.

In other words B^* is the smallest set which contains B_0 and is closed in respect of the operation $\langle .,. \rangle$. We assume B_0 does not contain order pairs. If $x \in B^*$ we denote by $x^{\vee}$ the following function: $Dx^{\vee} = B^*$ and $Rx^{\vee} = \{x\}$. In the set of all functions defined in B^* partially, we define three operations: a composition, a Cartesian combination and an iteration in the following way:
a) We denote the composition of functions φ_1 and φ_2 by $\varphi_1\varphi_2$ and $\varphi_1\varphi_2 = \lambda s \varphi_1(\varphi_2(s))$.
b) We denote the Cartesian combination by $\Pi(\varphi_1,\varphi_2)$ and $\Pi(\varphi_1,\varphi_2) = \lambda s \langle \varphi_1(s), \varphi_2(s) \rangle$.
c) We denote the iteration of function φ controlled from Ψ by $[\varphi, \Psi]$ and we

define it by the equivalence:

$$[\varphi,\psi](s) = t \ \langle\text{--}\rangle\ \exists m \exists r_o \ldots \exists r_m \ (r_o = s \ \& \ r_m = t \ \& \ \psi(r_m) \in B_o \ \& \ \forall i_{i<m} (r_{i+1} = \varphi (r_i) \ \& \ \psi(r_i) \in B^* \setminus B_o)\}.$$

The function π and σ have been defined in the following way:

$$\pi(\langle s,t\rangle) = s \ ; \ \sigma(\langle s,t\rangle) = t \ ; \ \pi(s) = \sigma(s) = \langle 0,0\rangle \ ; \ \pi(0) = \sigma(0) = 0.$$

If k, n $\in$ N, $0 < i < k + 1$ by I_i^k we denote $\sigma\pi^{k-i}$ in the case $1 < i < k + 1$, but π^{k-1} in the case i = 1. We define order k-tuple $\langle\langle b_1 ,\ldots, b_k \rangle\rangle$ in the following natural way:

$$\langle\langle b_1\rangle\rangle = b_1 \ ; \ \langle\langle b_1 ,\ldots, b_k , b_{k+1} \rangle\rangle = \langle \langle\langle b_1 ,\ldots, b_k \rangle\rangle, b_{k+1} \rangle.$$

For each $b_1 ,\ldots, b_k$ belonging to B^*, $I_i^k (\langle\langle b_1 ,\ldots, b_k \rangle\rangle) = b_i$ is true. If $B_1 \subseteq B^* ,\ldots, B_k \subseteq B^*$ then by $B_1 \times\ldots\times B_k$ we denote the set

$$\{ \langle\langle b_1 ,\ldots, b_k \rangle\rangle : b_i \in B_i \ , \ i = 1,\ldots, k \}$$

and if $B_i = A$, $i = 1,\ldots, k$, then A^k denotes $B_1 \times\ldots\times B_k$.
$\varphi (b_1,\ldots, b_k)$ denotes $\varphi (\langle\langle b_1 ,\ldots, b_k \rangle\rangle)$. If $\psi_j : B^* \text{-·->} B^*$, $j = 1,\ldots, l$ by ψ^- we denote $\psi_1 ,\ldots, \psi_l$.

Let $\varphi : B^* \text{-·->} B^*$. We say φ is prime computable relatively (p.c.r.) ψ^- , if φ can be obtained from ψ^- , π, σ and a finite set of x^v by finite number applications of operations a composition, a Cartesian combination and an iteration. We say φ is search computable relative (s.c.r.) ψ^- , if there exists p.c.r. ψ^- function θ such that for every s and t the equivalence

$$\varphi (s) = t \ \langle\text{--}\rangle \ \exists r (\theta (\langle s, r \rangle) = t)$$

is true. Let us denote the definitions of p.c.r. ψ^- function and s.c.r. ψ^- function are similar to the definitions in [1] but are different from them. In spite of that, these definitions are equivalent and it is proved in [8] by D. Skordev. If the set B is finite or countable we say α is numbering of the set B if α is a mapping of N_1 onto B, where the set N_1 is infinite recursively enumerable (r.e.).

Let $\varphi : B^k \text{-·->} B$. We say φ is effective in respect of (e.r.) given numbering α ($D\alpha = N_1$) iff there exists such partially recursive function f, $f : N_1{}^k \text{-·->} N_1$ that for every $p_1 ,\ldots, p_k$ belonging to N_1 the conditional equality

$$\varphi (\alpha (p_1),\ldots, \alpha (p_k)) \simeq \alpha (f (p_1 ,\ldots, p_k))$$

is true.

Let $C \subseteq B^k$. We say C is effectively enumerable in respect of (e.e.r.) the numbering α if the set

$$\{ (p_1,\ldots, p_k) : \langle\langle \alpha (p_1),\ldots, \alpha (p_k) \rangle\rangle \in C \}$$

is r.e..

If $C_i \subseteq B$, $i = 1,\ldots, m$ by C^- we denote $C_1,\ldots, C_m$.

Let $\psi_i : B^{k_i} \text{-·->} B$, $i = 1,\ldots, l$ and $C_j \subseteq B^{r_j}$, $j = 1,\ldots, m$. We say the numbering α is coordinated with ψ^- ; C^- if ψ_i is e.r. α, $i = 1,\ldots, l$ and C_j is e.e.r. α , $j = 1,\ldots, m$. Let $\varphi : B^k \text{-·->} B$. We say φ is admissible relatively (a.r.) Ψ^- ; C^- if φ is e.r. every numbering coordinated with Ψ^- ; C^-. In [4,5,6] the definitions of effective function and effectively enumerable set in respect of given numbering, coordinatedness of a numbering with given functions and sets and admissible function relatively given functions and sets are different. It is not difficult to show, however, that in both cases the concepts admissibility coincide. If $C \subseteq B$ by $C^\frown$ we denote the following function: $DC^\frown = C$ and $RC^\frown = \{0\}$. By $C^{-\frown}$ we denote $C_1^\frown, \ldots, C_m^\frown$. In [4] it has been proved that if the function φ ($\varphi : B_o^k \dashrightarrow B_o$) is s.c.r. Ψ^- ; $C^{-\frown}$ then φ is a.r. Ψ^- ; C^- . In [5] it is shown that if the set B is finite, $C_i \subseteq B^{r_i}$, $i = 1, \ldots, m$ and φ is a.r. C^- , then φ is s.c.r. $C^{-\frown}$. In [6] it is shown if $\Psi_i : B^{k_i} \to B$, $i = 1, \ldots, l$ and φ is a.r. Ψ^- , then φ is s.c.r. Ψ^- . In this paper we assume the set B is finite. Let $\Psi_i: B^{k_i} \dashrightarrow B$, $C_j \subseteq B^{r_j}$, $i = 1, \ldots, l$; $j = 1, \ldots, m$. We define a collection $[\ \Psi^- ; C^-\ ]$ of subsets of B in the following way:

a) $B \in [\ \Psi^- ; C^-\ ]$; b) If $0 < i < m+1$, $0 < j < r_i+1$ and $a_1, \ldots, a_{j-1}, a_{j+1}, \ldots, a_{r_i}$ belong to B then

$$\{ x : \langle\langle a_1, \ldots, a_{j-1}, x, a_{j+1}, \ldots, a_{r_i} \rangle\rangle \in C_i \} \in [\ \Psi^- ; C^-\ ] ;$$

c) If $0 < i < l + 1$, $0 < j < k_i+1$, $A' \in [\ \Psi^- ; C^-\ ]$ and $a_1, \ldots, a_{j-1}, a_{j+1}, \ldots, a_{k_i}$ belong to B then

$$\{ x : \Psi_i (a_1, \ldots, a_{j-1}, x, a_{j+1}, \ldots, a_{k_i}) \in A' \} \in [\ \Psi^- ; C^-\ ].$$

1-component of $[\ \Psi^- ; C^-\ ]$ containing the point a ($a \in B$) is called the set $\cap \{ A : a \in A \ \&\ A \in [\ \Psi^- ; C^-\ ] \}$. We often say only " 1-component". If A is 1-component, then A' denotes the set $A \setminus \{ X : X \in [\ \Psi^- ; C^-\ ] \ \&\ \neg A \subseteq X \}$. If $k \in N$, $k > 1$ then k-component of $[\ \Psi^- ; C^-\ ]$ containing the point $\langle\langle a_1, \ldots, a_k \rangle\rangle$ is called the set $A_1 \times \ldots \times A_k$, where A_i is 1-component of $[\ \Psi^- ; C^-\ ]$ containing a_i , $i = 1, \ldots, k$. We often say only "k-component".

We shall prove the following

THEOREM. Let B be finite set, $\Psi_i : B^{k_i} \dashrightarrow B$, $i = 1, \ldots, l$; $C_j \subseteq B^{r_j}$, $j = 1, \ldots, m$ and $\varphi : B^k \dashrightarrow B$. If φ is admissible relatively Ψ^- ; C^- , then φ is search computable relatively Ψ^- ; $C^{-\frown}$.

Proof. Let $B = \{ b_1, \ldots, b_r \}$, where $b_1, \ldots, b_r$ are different, φ be a.r. Ψ^- ; C^- and f_i be defined by the equality $f_i (p_1, \ldots, p_{k_i}) = \langle i - 1, p_1, \ldots, p_{k_i} \rangle$, $i = 1, \ldots, l$. Evidently Rf_i is recursive, $i = 1, \ldots, l$ and if i and j are different, $0 < i, j < l + 1$ then $Rf_i \cap Rf_j = \emptyset$. The set $N_o = N \setminus (Rf_1 \cup \ldots \cup Rf_l)$ is also recursive.

Let $P_1, \ldots, P_r$ be subsets of N_o . We define the sequences $\{ P_1^n \}_{n=0}^{\infty}, \ldots, \{ P_r^n \}_{n=0}^{\infty}$ of sets in the following way:

a) $P_i^0 = P_i$, $i = 1, \ldots, r$; b) If $p \in P_i^n$ then $p \in P_i^{n+1}$, $i = 1, \ldots, r$; c) If $0 < i < l + 1$, $p \in P_{j_1}^n, \ldots, p_{jk} \in P_{jk_i}^n$ and $\Psi_i (b_{j_1}, \ldots, b_{jk_i}) = b_q$, then $f_i(p_1, \ldots, p_{k_i}) \in P_q^{n+1}$.

Let $[\ P_i\ ] = \cup_{n=0}^{\infty} P_i$, $i = 1, \ldots, r$.

Lemma 1. If $P_1, \ldots, P_r$ are disjoint sets such that $P_1 \cup \ldots \cup P_r \subseteq N_o$, then $[\ P_1\ ], \ldots, [\ P_r\ ]$ are disjoint sets. If $P_1, \ldots, P_r$ are recursive, then $[\ P_1\ ], \ldots, [\ P_r\ ]$ are recursive.

Proof See lemma 1 in [6]. To prove that [P_1] ,..., [P_r] are recursive one has to use that B is finite set.

Let B_1 ,..., B_r be 1-components of [Ψ^- ; C^-] containing respectively the points b_1 ,..., b_r .

Lemma 2. If $0 < i < l + 1$, Ψ_i (b_{j1} ,..., b_{jki}) = b_q and $a_1 \in B_{j1}$, ..., $a_{ki} \in B_{jki}$, then Ψ_i (a_1 ,..., a_{ki}) $\in B_q$.

Proof Let Ψ_i (b_{j1} ,..., b_{jki}) = b_q , $a_1 \in B_{j1}$,..., $a_{ki} \in B_{jki}$ and let us assume that either $\neg$ << a_1 ,..., a_{ki} >> $\in D\Psi_i$ or $\neg$ Ψ_i (a_1 ,..., a_{ki}) $\in B_q$. Then there exists such s that Ψ_i (a_1 ,..., a_{s-1} , b_{js} ,..., b_{jki}) $\in B_q$ and either $\neg$ << a_1 ,..., a_s , b_{js+1} ,..., b_{jki} >> $\in D\Psi_i$ or $\neg$ Ψ_i (a_1 ,..., a_s , b_{js+1} ,..., b_{jki}) $\in B_q$. We suppose $s = k_i - 1$. Let A_1 ,..., A_t be all A $\in$ [Ψ^- ; C^-] about which $b_q \in A$, i.e. $B_q = A_1 \cap \ldots \cap A_t$ and $A_j^- = \{ x : \Psi_i (a_1 ,\ldots, a_{ki-1} , x) \in A_j \}$, j = 1,..., t . Obviously $A_j \in$ [Ψ^- ; C^-] and $B_{jki} \subseteq A_1^- \cap \ldots \cap A_t^- = \{ x : \Psi_i (a_1 ,\ldots, a_{ki-1}, x) \in B \}$ i.e. $\neg$ $a_{ki} \in B_{jki}$. The received contradiction proves lemma 2.

Lemma 3 Let P_1 ,..., P_r be non empty disjoint subsets of N_0 such that the set $Q_i = \cup \{ P_j : b_j \in B_i \}$ is r.e., i = 1,..., r. If α is a mapping defined by the equality α^{-1} (b_i) = [P_i], i = 1,..., r, then α is a numbering coordinated with Ψ^- ; C^- .

Proof Let us note [Q_1],..., [Q_r] are r.e. sets like a minimal fixed point of some enumeration operator [9]. By lemma 2 one can easily prove inductively that $Q_i^n = \cup \{ P_j^n : b_j \in B_i \}$, i = 1,..., r for each number n, whence [Q_i] = $\cup$ { [P_j] : $b_j \in B_i$ }, i = 1,..., r. Since P_1 ,..., P_r are non empty and disjoint α is well defined. Besides α^{-1} (B) = α^{-1} ($B_1 \cup \ldots \cup B_r$) = α^{-1} (B_1) $\cup \ldots \cup$ α^{-1} (B_r) = [Q_1] $\cup \ldots \cup$ [Q_r] is r.e., i.e. α is a numbering. Since every set belonging to [Ψ^- ; C^-] is e.e.r. α , α is coordinated with C^- (see lemma 1 in [3]). Let f_i' is the restriction of f_i on f_i^{-1} ($D\alpha$) , i = 1,..., l. Evidently f_i' is p.r. function and one can easily show that α (f_i'(p_1 ,..., p_{ki})) = Ψ_i($\alpha(p_1)$,..., $\alpha(p_{ki})$) for every p_1 ,..., p_{ki} belonging to $D\alpha$, i = 1,...,l, i.e. α is coordinated with Ψ^- . Lemma 3 is proved.

Let now << a_1 ,..., a_k >> $\in D\varphi$, $A_1 \times \ldots \times A_k$ be k-component of [Ψ^- ; C^-] containing << a_1 ,..., a_k >> and $a_i' \in A_i$ for some i, $0 < i < k + 1$. We shall show << a_1 ,..., a_{i-1}, a_i', a_{i+1},..., a_k >> $\in D\varphi$. Let us suppose i = 1 and $\neg$ << a_1', a_2,..., a_k >> $\in D\varphi$. Let F be r.e. subset of N_0 about which $N_0 \setminus F$ is not r.e., { c_1,..., c_q, a_1 } = { x : << x, a_2,..., a_k >> $\in D\varphi$ }, { c_1',..., c_{r-q-2}', a_1'} = { x : $\neg$ << x, a_2,..., a_k >> $\in D\varphi$ }, p_1,..., p_r be members of $N_0 \setminus F$ but p_1',..., p_{r-q-2}' be members of F. We define the numbering α in the following way:

α^{-1} (x) = [$N_0 \setminus$ (F $\cup$ { p_1,..., p_q })] , if x = a_1,
α^{-1} (x) = [{ p_j }], if x = c_j, j = 1,...,q,
α^{-1} (x) = [F $\setminus$ { p_1',..., p_{r-q-2}' }] , if x = a_1',
α^{-1} (x) = [{ p_j' }] , if x = c_j', j=1,...,r-q-2.

According to lemma 3 α is coordinated with Ψ^- ; C^- . Since φ is a.r. Ψ^- ; C^- there exists such p.r.f. f, f : N^k -·-> N that for every q_1 ,..., q_k belonging to $D\alpha$ the equality φ (α (q_1),..., α (q_k)) $\simeq$ α (f (q_1 ,..., q_k)) is valid. Let p_2^0 ,..., p_k^0 be such natural numbers that α (p_2^0) = a_2 ,..., α (p_k^0) = a_k and h = λp f (p, p_2^0 ,..., p_k^0) is p.r.f.. It is corectly α (h (p)) = φ (α (p), a_2,..., a_k) for every p and Dh = α^{-1} (a_1) $\cup$ α^{-1} (c_1) $\cup \ldots \cup$ α^{-1} (c_q) is r.e. whence Dh $\cap$ N_0 = $N_0 \setminus F$ is r.e.. The received contradiction shows us that << a_1', a_2,..., a_k >> $\in D\varphi$. It is easy to prove $D\varphi$ is union of finite number k-components (see the theorem in [4]). We define the set F of functions in the following way [6]:

a) $f_i \in F$, $i = 1,\dots, l$; b) $I_i^n = \lambda p_1 \dots \lambda p_n . p_i \in F$, $i, n \in N$, $< i < n + 1$; c) If $f : N^n \dashrightarrow N$, $f \in F$, $g_i : N^m \dashrightarrow N$, $g_i \in F$, then $\lambda p_1 \dots \lambda p_m\ f\ (\ g_1\ (\ p_1 ,\dots, p_m\ ),\dots, g_n\ (\ p_1 ,\dots, p_m\ )\ ) \in F$.

Lemma 4 For each natural number p and for each different natural numbers $p_1 ,\dots, p_k$ of N_0 such $q_1 ,\dots, q_m$ of N_0 , $q_1 < \dots < q_m$ and $p_i \neq q_j$, $= 1,\dots, k$; $j = 1,\dots, m$ and such function $f \in F$, $f : N^{k+m} \dashrightarrow N$ exists, hat $p = f\ (\ p_1 ,\dots, p_k , q_1 ,\dots, q_m\ )$.

Proof See lemma 5 in [6]. In addition to the proof of lemma 5 [6] it s clear that for each k-tuple $(\ p_1 ,\dots, p_k\ )$ of $N_0{}^k$ we can associate idenically such function f that there does not exist any own subset $\{s_1,\dots,s_q\}$ f $\{\ q_1 ,\dots, q_m\ \}$ about which $p = g\ (\ p_1 ,\dots, p_k , s_1 ,\dots, s_q\ )$ for some $\in F$. Let this function which we identically associate to p, $p_1,\dots, p_k$ e denoted by $H\ (\ p, p_1,\dots, p_k\ )$.

Analogously we define a set F_1 of functios partially defined on some Cartesian degree of B :

a) $\psi_i \in F_1$, $i = 1,\dots, l$; b) $I_i^n /B^n \in F_1$, $i, n \in N$, $0 < i < n + 1$; c) If $\psi : B^n \dashrightarrow B$, $\psi \in F_1$, $\varphi_i : B^m \dashrightarrow B$, $\varphi_i \in F_1$, $i = 1,\dots, n$, then $\psi \sqcap (\ \dots \sqcap (\ \varphi_1 , \varphi_2\ ),\dots, \varphi_n\ ) \in F_1$.

Let F_1' be the following set of functions. If $\psi \in F_1$, $\psi : B^{k+n} \dashrightarrow B$ and $a_1 ,\dots, a_n$ belong to B then $\lambda x_1 \dots \lambda x_k\ \psi(x_1,\dots,x_k,a_1,\dots,a_n) \in F_1'$. Let us note that every function of F is general recursive, and every function of F_1' is p.c.r. ψ^-. We define mapping $\mathcal{A}$, $\mathcal{A} : F \dashrightarrow F_1$ in the following way:

a) $\mathcal{A}\ (\ f_i\ ) = \psi_i$, $i = 1,\dots, l$; b) $\mathcal{A}\ (\ I_i^n\ ) = I_i^n /B^n$, $i, n \in N$, $0 < i < n+1$;
c) $\mathcal{A}\ (\ \lambda p_1 \dots \lambda p_m\ f(g_1(p_1,\dots,p_m),\dots,g_n(p_1,\dots,p_m))) =$
$\mathcal{A}\ (\ f\ ) \sqcap (\ \dots \sqcap (\ \mathcal{A}\ (\ g_1\ ), \mathcal{A}\ (\ g_2\ )\ ),\dots, \mathcal{A}\ (\ g_n\ )\ )$.

Let us note if we want the mapping $\mathcal{A}$ to be well defined we must think that the members of F are values of some therms.

Lemma 5 Let $P_1 ,\dots, P_r$ be non empty disjoint subsets of N_0 such that $Q_i = \cup \{\ P_j : b_j \in B_i\ \}$ is r.e. $i = 1,\dots,r$, α be a numbering of B defined by the equalities $\alpha^{-1}\ (\ b_i\ ) = [\ P_i\ ]$, $i = 1,\dots,r$ and to every $f \in F$ we associate a function f' in the following way:
a) $f_i' = f_i/f_i^{-1}\ (D\alpha)$, $i = 1,\dots, l$; b) $(I_i^n)' = I_i^n/(D\alpha)^n$, $i,n \in N$, $0 < i < n+1$; c) $(\lambda p_1 \dots \lambda p_m\ f(g_1(p_1,\dots,p_m),\dots,g_n(p_1,\dots,p_m)))' =$
$\lambda p_1 \dots \lambda p_m\ f'(g_1'(p_1,\dots,p_m),\dots,g_n'(p_1,\dots,p_m))$.

Then if $f \in F$, $f : N^n \dashrightarrow N$ for any natural numbers $p_1,\dots, p_n$ the equality

$$\alpha\ (\ f'\ (\ p_1 ,\dots, p_n\ )\ ) \simeq \mathcal{A}\ (\ f\ )\ (\ \alpha\ (\ p_1\ ),\dots, \alpha\ (\ p_n\ )\ )$$

is correct.

Proof Analogously of lemma 6 in [6].

Lemma 6 Let $A_1 \times \dots \times A_k$ be k-component and φ_1 be such function that $A_1 \times \dots \times A_k \subseteq D\varphi_1$ and $\varphi_1/A_1\times\dots\times A_k$ is not a.r. ψ^- ; C^- then φ is not a.r. ψ^- ; C^- .

Proof It follows directly from the definitions.

Let $s_1 ,\dots, s_r$ be fixed members of N_0 and $N_1 ,\dots, N_k$ be disjoint infinite recursive sets such that $N_1 \cup \dots \cup N_k = N_0 \setminus \{\ s_1 ,\dots, s_r\ \}$.

Lemma 7 Let $A_1 \times \dots \times A_k$ be a k-component and φ_1 be such function

$D\varphi_1 = A_1 \times \ldots \times A_k$ and $\varphi_1 / A_1' \times \ldots \times A_k' \neq \Psi / A_1' \times \ldots \times A_k'$ for Ψ of F_1' and $\varphi_1 / A_1' \times \ldots \times A_k'$ not to be a constant. Then φ_1 is not a.r. Ψ^- ; C^- .

Proof We shall construct a numbering α which is coordinated with Ψ^- ; C^- and φ_1 is not e.r. α. We shall construct the numbering α constructing disjoint subsets $P_1, \ldots, P_r$ of N_0 such that $P_1 \cup \ldots \cup P_r = N_0$ and after that we fix $\alpha^{-1}(b_i) = [P_i]$, $i = 1, \ldots, r$. We shall construct the sets $P_1, \ldots, P_r$ on steps. By $P_{i,s}$ we shall denote the set of members of N_0 which are placed in P_i on step s. Then we fix $P_i = \cup_{s=0}^{\infty} P_{i,s}$, $i = 1, \ldots, r$. Let g_1 , g_2 , ... are the sequence of all p.r.f. g such that $Dg \subseteq N^k$ and if

$$(Dg)_i = \{ p_i : \exists p_1 \ldots \exists p_{i-1} \exists p_{i+1} \ldots \exists p_k ((p_1, \ldots, p_k) \in Dg) \}$$

then $(Dg)_i$ is infinite $i = 1, \ldots, k$. On step s we shall fulfil the condition

$$\alpha (g_s (p_1 , \ldots, p_k)) \neq \varphi_1 (\alpha (p_1), \ldots, \alpha (p_k)) .$$

Step 0. $P_{i,s} = \{ s_i \}$, $i = 1, \ldots, r$. In this way we provide the set to be non empty, $i = 1, \ldots, r$.

Step $s > 0$. Let the remainder of division of s to k be j-1 and p_j be the least number of N_j that for some $p_1, \ldots, p_{j-1}, p_{j+1}, \ldots, p_k$, $(p_1, \ldots, p_k) \in Dg_s$ $p_1 \in N_1$, ..., $p_k \in N_k$, $\{ p_1, \ldots, p_k \} \cap (P_1 \cup \ldots \cup P_k) = \emptyset$. For the fixed $p_1, \ldots, p_k$ let $g_s(p_1, \ldots, p_k) = f(p_1, \ldots, p_k, q_1, \ldots, q_m)$, where $f \in F$, $q_1, \ldots, q_m \in N_0 \setminus \{ p_1, \ldots, p_k \}$ and in addition $b_{ti} \in A_i'$, $i = 1, \ldots, k$. We fix $P_{ti,s} = P_{ti,s-1} \cup \{ q_j : \neg q_j \in (P_{1,s-1} \cup \ldots \cup P_{r,s-1}) \ \& \ q_j \in N_i \}$, $i = 1, \ldots, k$ and $P_{i,s} = P_{i,s-1}$, if $\neg i \in \{ t1, \ldots, tk \}$ and $0 < i < r + 1$. The following cases are possible:

a) $g_s (p_1, \ldots, p_k) \in [P_{io,s}']$, $0 < io < r + 1$;
b) $g_s (p_1, \ldots, p_k) = p_j$, $0 < j < k + 1$;
c) $g_s (p_1, \ldots, p_k) \in N \setminus (N_0 \cup [P_{1,s}'] \cup \ldots \cup [P_{r,s}'])$.

If $g_s (p_1, \ldots, p_k) \in [P_{io,s}']$, then $\varphi_1 / A_1' \times \ldots \times A_k'$ is not a constant, i.e. such $b_{j1}, \ldots, b_{jk}$ exist that $\varphi_1 (b_{j1}, \ldots, b_{jk}) \neq b_{io}$ and $\langle\langle b_{j1}, \ldots, b_{jk} \rangle\rangle \in A_1' \times \ldots \times A_k'$. For such j1, ..., jk we fix $P_{ji,s} = P_{ji,s}' \cup \{p_i\}$, $i = 1, \ldots, k$, and $P_{i,s} = P_{i,s}'$, if $\neg i \in \{ j1, \ldots, jk \}$.

If $g_s (p_1, \ldots, p_k) = p_j$ $(0 < j < k+1)$, then $\varphi_1 / A_1' \times \ldots \times A_k' \neq I_j^k / A_1' \times \ldots \times A_k'$ and such $b_{i1}, \ldots, b_{ik}$ exist that $\varphi_1 (b_{i1}, \ldots, b_{ik}) \neq b_{ij}$. In this case we fix $P_{i1,s} = P_{i1,s}' \cup \{p_1\}, \ldots, P_{ik,s} = P_{ik,s}' \cup \{p_k\}$ and $P_{i,s} = P_{i,s}'$, if $\neg i \in \{ i1, \ldots, ik \}$.

In the end if $g_s (p_1, \ldots, p_k) \in N \setminus (N_0 \cup [P_{1,s}'] \cup \ldots \cup [P_{r,s}'])$, let $\Psi = \lambda x_1 \ldots \lambda x_k \, \mathcal{A} (f) \, (x_1, \ldots, x_k, b_{j1}, \ldots, b_{jm})$ where $b_{j1}, \ldots, b_{jm}$ are such that $q_1 \in [P_{j1,s}'], \ldots, q_m \in [P_{jm,s}']$. Again there exists $\langle\langle b_{i1}, \ldots, b_{ik} \rangle\rangle \in A_1' \times \ldots \times A_k'$ such that $\varphi_1 (b_{i1}, \ldots, b_{ik}) \neq \Psi (b_{i1}, \ldots, b_{ik})$. We fix $P_{i1,s} = P_{i1,s}' \cup \{p_1\}, \ldots, P_{ik,s} = P_{ik,s}' \cup \{p_k\}$ and $P_{i,s} = P_{i,s}'$, if $\neg i \in \{ i1, \ldots, ik \}$. Let us note $P_{i,s} \neq \emptyset$ and $P_{i,s} \subseteq P_{i,s+1}$ for every $s \in N$ and every i ($0 < i < r+1$). In addition every member of N_0 is a member of some $P_{i,s}$ and $P_{i,s} \cap P_{j,s} = \emptyset$ if $i \neq j$, $0 < i,j < r+1$, $s \in N$. We fix $P_i = \cup_{s=0}^{\infty} P_{i,s}$, $i = 1, \ldots, r$. The set $Q_i = \cup \{ P_j : b_j \in B_i \}$ is r.e. $i = 1, \ldots, r$, since on every step we add some member p of such P_j about which $b_j \in A_i'$. Besides the sets $P_1, \ldots, P_r$ are disjoint. We define the numbering α in the following way: $\alpha^{-1} (b_i) = [P_i]$, $i = 1, \ldots, r$. According to lemma 3 α is coordinated with Ψ^- ; C^- . Let us suppose φ_1 is a.r. Ψ^- ; C^- . Then there exists such p.r.f. g, $g : N^k \dashrightarrow N$, that $\alpha (g (p_1, \ldots, p_k)) \simeq \varphi_1 (\alpha (p_1), \ldots, (p_k))$ for each natural numbers $p_1, \ldots, p_k$. Obviously $g = g_s$ for some s and let on step s we choose the members $p_1, \ldots, p_k$. Let us assume $g_s (p_1, \ldots, p_k) \in [P_{io,s}']$, i.e. $\alpha (g_s(p_1, \ldots, p_k)) = b_{io}$. On the other hand

$p_1 \in P_{j1,s} \subseteq [P_{j1}], \ldots, p_k \in P_{jk,s} \subseteq [P_{jk}]$, where j1,...,jk are such numbers that $\varphi_1(b_{j1},\ldots,b_{jk}) \risingdotseq b_{10}$, i.e. $\varphi_1(\alpha(p_1),\ldots,\alpha(p_k)) \risingdotseq \alpha(g_s(p_1,\ldots,p_k))$. So we obtain that this case is not possible. Analogously $g_s(p_1,\ldots,p_k) = p_j (0 < j < k+1)$ is not possible. Therefore, the case $g_s(p_1,\ldots,p_k) \in N \setminus (N_0 \cup [P_{1,s}'] \cup \ldots \cup [P_{r,s}'])$ and $g_s(p_1,\ldots,p_k) = g_s(p_1,\ldots,p_k) \in N \setminus (N_0 \cup [P_{1,s}'] \cup \ldots \cup [P_{r,s}'])$ and $g_s(p_1,\ldots,p_k) = f(p_1,\ldots,p_k,q_1,\ldots,q_m)$ where $f \in F$ and $q_1,\ldots,q_m \in N_0$. According to lemma 5 for fixed $p_1,\ldots,p_k,q_1,\ldots,q_m$ the equality

$$\mathcal{A}(f)(\alpha(p_1),\ldots,\alpha(p_k),\alpha(q_1),\ldots,\alpha(q_m)) = \alpha(f'(p_1,\ldots,p_k,q_1,\ldots,q_m))$$

is true and $g_s(p_1,\ldots,p_k) \in D\alpha$, i.e. $(p_1,\ldots,p_k,q_1,\ldots,q_m) \in Df'$. Then

$$\alpha(g_s(p_1,\ldots,p_k)) = \alpha(f'(p_1,\ldots,p_k,q_1,\ldots,q_m)) = \mathcal{A}(f)(\alpha(p_1),\ldots,\alpha(p_k),\alpha(q_1),\ldots,\alpha(q_m)) \risingdotseq \varphi_1(\alpha(p_1),\ldots,\alpha(p_k)).$$

The received contradiction proves that φ_1 is not a.r. Ψ^- ; C^- .

Lemma 8 Let $A_1, \ldots, A_k$, A_i^- $(0 < i < k+1)$ be 1-component $A_i^- \cap A_i' = \emptyset$, $A_i^- \cup A_i' = A_i$ and φ_1 is such function that $D\varphi_1 = A_1 \times \ldots \times A_k$, φ_1 not to be a constant and $\varphi_1 \risingdotseq \psi$ on $A_1' \times \ldots \times A_k' \cup A_1' \times \ldots \times A_{i-1}' \times A_i^{-\prime} \times A_{i+1}' \times \ldots \times A_k'$ for each $\psi \in F_1'$. Then φ_1 is not admissible relatively Ψ^- ; C^- .

Proof We can consider that i = 1. If φ_1 is not a constant and $\varphi_1 \risingdotseq \psi$ on $A_1' \times \ldots \times A_k'$ for each $\psi \in F_1'$, then φ_1 is not a.r. Ψ^- ; C^- , according to lemma 7. Analogously, if φ_1 is not a constant and $\varphi_1 \risingdotseq \psi$ on $A_1^{-\prime} \times A_2' \times \ldots \times A_k'$ for each $\psi \in F_1'$ then φ_1 is not a.r. Ψ^- ; C^- , according to lemmas 7 and 6. We shall consider the case when $\varphi_1 = \psi'$ on $A_1' \times \ldots \times A_k'$ where $\psi' \in F_1'$ or ψ' is a constant, and $\varphi_1 = \psi''$ on $A_1^{-\prime} \times A_2' \times \ldots \times A_k'$ where $\psi'' \in F_1'$ or ψ'' is a constant. Let $\langle\langle b_{t1},\ldots,b_{tk}\rangle\rangle \in A_1' \times \ldots \times A_k'$ and $\psi'(b_{t1},\ldots,b_{tk}) \risingdotseq \psi''(b_{t1},\ldots,b_{tk})$, but $\langle\langle b_{z1},\ldots,b_{zk}\rangle\rangle \in A_1' \times A_2' \times \ldots \times A_k'$ and $\psi'(b_{z1},\ldots,b_{zk}) \risingdotseq \psi''(b_{z1},\ldots,b_{zk})$.

We shall construct disjoint sets $P_1, \ldots, P_r$ such that they satisfy the suppositions of lemma 3 and after that we shall fix $\alpha^{-1}(b_i) = [P_i]$, $i = 1,\ldots, r$. The sets $P_1,\ldots, P_r$ will be constructed by steps as on step s we shall construct $P_{1,s}, \ldots, P_{r,s}$ and in the end

$$P_i = \{ p : \exists s_0 \forall s > s_0 \ (p \in P_{i,s}) \}, \quad i = 1,\ldots, r.$$

Let ξ_k be universal about p.r.f. of k variables, U =

$$\{ (n,p_1,\ldots,p_k,p) : \xi_k(n,p_1,\ldots,p_k) = p \ \& \ p_1 \in N_1 \ \& \ldots \& \ p_k \in N_k \},$$

X, $Y_1,\ldots, Y_k$, Z be such general recursive functions that $U = \{ (X(t),Y_1(t),\ldots,Y_k(t),Z(t)) : t \in N \}$, v_1 be such strictly monotonically increasing function that $Rv_1 = N_1$ and $v = \lambda p.\ v_1((p+1)\cdot(p+2)/2 + p)$ Our idea of constructing numbering α consists in the following: The numbering α is to be such that every function which is e.r. α, to be different from ψ' on $A_1' \times \ldots \times A_k'$ or to differ from ψ'' on $A_1^{-\prime} \times A_2' \times \ldots \times A_k'$. On step s we are going to satisfy one of the conditions:

$$(1)\quad \psi'(\alpha(p_1),\ldots,\alpha(p_k)) \risingdotseq \alpha(\xi_k((s)_0,p_1,\ldots,p_k)) \ \& \ \alpha(p_i)=b_{ti},\ i=1,\ldots,k$$

$$(2)\quad \psi''(\alpha(p_1),\ldots,\alpha(p_k)) \risingdotseq \alpha(\xi_k((s)_0,p_1,\ldots,p_k)) \ \& \ \alpha(p_i)=b_{zi},\ i=1,\ldots,k$$

For this aim if we find such $p_1,\ldots,p_k$ we shall place them in some $P_{j1,s},\ldots,P_{jk,s}$ and we remember that $(p_1,\ldots,p_k)$ satisfy (1) or (2) and we create $(s)_0$-condition. In addition some $q_1,\ldots,q_m$ will be connected with these $p_1,\ldots,p_k$. We place these $q_1,\ldots,q_m$ in some $P_{j1,s},\ldots,P_{jk,s}$ and we want $q_1,\ldots,q_m$ not to be placed in other sets. Because of that we name the set $\{ q_1,\ldots, q_m \}$ $(s)_o$-requirement created on step s. Since we want to obtain such numbering α that for every $(s)_o$ such that

$\lambda p_1 \ldots \lambda p_k \Phi_k((s)_o, p_1,\ldots,p_k)$ is defined for infinite set of different $(p_1,\ldots,p_k)$ to be satisfied (1) or (2) for some $(p_1,\ldots,p_k)$, we use priority method [10] as a priority will have less $(s)_o$. Let $(p_1,\ldots,p_k)$ is n-condition and $\{ q_1,\ldots,q_m \}$ is n-requirement created on step s. If $t > s$ and by step t p_i $(0 < i < k+1)$ which has been placed in some P_j, has not been removed in some other $P_{j'}$, and q_i $(0 < i < m+1)$, belonging to the n-requirement created on step s, which has been placed in some P_j has not been removed in some other $P_{j'}$, then we say that n-condition and n-requirement are active on step t. If a n-condition and a n-requirement created on step s are active on step t for each $t > s$, we say that n-condition and n-requirement are constant.

Let us describe now the step s: $s = 0$. $P_{i,o} = \{ s_i \}$, $i = 1,\ldots, r$.

$s > 0$. Let $(s)_o = n$. We verify whether an active n-condition exists. If such active n-condition exists we do nothing, i.e. $P_{i,s} = P_{i,s-1}, i=1,\ldots,r$ and we do not create any n-conditions and n-requirements. If such active n-condition does not exist, let for arbitrary t, $f = H(Z(t), Y_1(t),\ldots,Y_k(t))$ and $Z(t) = f\,(Y_1(t),\ldots,Y_k(t), q_1,\ldots,q_m)$. We fix $P_{ti,s}{}' = P_{ti,s-1} \cup ((N_i \setminus (P_{1,s-1} \cup \ldots \cup P_{r,s-1})) \cap \{ q_1,\ldots,q_m \})$, $i = 1,\ldots, k$, $P_{i,s}{}' = P_{i,s}$ for $\neg\ i \in \{ t1,\ldots,tk \}$ and let $q_i \in P_{ji}{}'$, $i = 1,\ldots, m$ and $\chi_t = \lambda x_1 \ldots \lambda x_k\, \mathcal{A}(f)\ (x_1,\ldots,x_k, b_{j1},\ldots,b_{jm})$. In this case we verify whether such $t < s+1$ exists that $X(t) = n$, $Y_i(t)$ does not belong to any active m-requirement, $m < n$; $i = 1,\ldots, k$ and $Y_i(t)$ is not j-th coordinate $(0<j<k+1)$ of any active m-condition, $m < n$; $i = 1,\ldots,k$ and at least one of the following two conditions is satisfied:

(i) $\neg\, Y_1(t) \in \bigcup_{j=0}^{s-1} (\cup \{ P_{i,j} : b_i \in A_1^{-}{}' \}) \ \&\ \langle\langle b_{t1},\ldots,b_{tk}\rangle\rangle \in D\chi_t$ $\&\ \chi_t\ (b_{t1},\ldots,b_{tk}) \not\simeq \psi'\ (b_{t1},\ldots,b_{tk})$.

(ii) $Y_1(t) > v(X(t)) \ \&\ \langle\langle b_{z1},\ldots,b_{zk}\rangle\rangle \in D\chi_t\ \&$ $\chi_t\ (b_{z1},\ldots,b_{zk}) \not\simeq \psi''(b_{z1},\ldots,b_{zk})$.

If such t does not exist we do nothing. In the oposite case let t_o be the least such t . We fix $P_{i,s}{}'' = P_{i,s}{}' \setminus \{ Y_1(t_o),\ldots,Y_k(t_o) \}$, $i=1,\ldots,r$. If the case (i) is valid we fix $P_{ti,s} = P_{ti,s}{}'' \cup \{ Y_i(t_o) \}$, $i = 1,\ldots,k$, $P_{i,s} = P_{i,s}{}''$ for $\neg\ i \in \{ t1,\ldots,tk \}$. If the case (ii) is valid we fix $P_{zi,s} = P_{zi,s}{}'' \cup \{ Y_i(t_o) \}$, $i = 1,\ldots,k$ and $P_{i,s} = P_{i,s}{}''$ for $\neg\ i \in \{ z1,\ldots,zk \}$. We create n-condition $(Y_1(t_o),\ldots,Y_k(t_o))$ and n-requirement $\{ q_1,\ldots,q_m \}$ where $q_1,\ldots,q_m$ are respective to t_o. Evidently the construction is recursive. $P_i = \{ p : \exists\, s_o\ \forall\, s > s_o\ (p \in P_{i,s})\}$, $\alpha^{-1}(b_i) = [P_i]$, $i = 1,\ldots,r$. Let us note, if $Q_{i,s} = \cup \{ P_{j,s} : b_j \in B_i \}$ then $Q_{i,s} \subseteq Q_{i,s+1}$, for each natural s and each i, $0 < i < r+1$. Therefore Q_i is r.e. $i = 1,\ldots,r$ and α is coordinated with ψ^{-} ; C^{-} . Let $M_1 = (\cup \{ P_i : b_i \in A_1{}' \}) \cap N_1$, $(N_1)_n = \{ v_1(0), v_1(1),\ldots, v_1((n+1)\cdot(n+2)/2 + n) \}$. We shall prove $N_1 \setminus M_1$ is infinite. For this aim we shall establish that $|\ (N_1 \setminus M_1) \cap (M_1)| > n$ for every natiral number n. Evidently, it is enough to establish only that $|M_1 \cap (N_1)_n| \leq (n+1)\cdot(n+2)/2$. Indeed, every member of M_1 is the first coordinate of some n-condition. If p_1 is the first coordinate of some (n+1)-condition and $p_1 \in M_1$, then $p_1 > v(n+1) > v(n) = v_1((n+1)\cdot(n+2)/2+n)$, i.e. $\neg\ p_1 \in M_1 \cap (N_1)_n$. For every m, however, the most (m+1) m-conditions exist. Therefore, $|\ M_1 \cap (N_1)_n| \leq (n+1)\cdot(n+2)/2$ for every n.

Lemma 9 If $M_1{}',\ldots, M_k{}'$ are infinite sets, $M_1{}' \subseteq N_1 \setminus M_1$, $M_i{}' \subseteq N_i$, $i = 2,\ldots,k$ and n is such natural number that $M_1{}' \times \ldots \times M_k{}' \subseteq D\lambda p_1 \ldots p_k\ \Phi_k(n, p_1,\ldots,p_k)$, then constant n-condition exists.

Proof Let us assume that such n-condition does not exist and let $(p_1^1,\ldots,p_k^1),\ldots,(p_1^z,\ldots,p_k^z)$ be all constant m-conditions such that $m < n$ and these conditions are created on steps $s_1,\ldots,s_z$ respectively. Let in addition $p_1,\ldots,p_k$ be such that $p_i \in M_i \setminus \{p_i^1,\ldots,p_i^z\}$, p_i does not belong to any m-requirement created on step s_j, $j = 1,\ldots,z$; $i = 1,\ldots,k$ and $p_1 > v(n)$. Then such t_o exists that $X(t_o) = n$ & $Y_1(t_o) = p_1$ & ... & $Y_k(t_o) = p_k$. Let s_o is the least s such that $(s)_o = n$, $s > s_i$, $i = 1,\ldots,z$ and $s > t_o$. It is easy to show that on step s_o appears an active n-condition whence lemma 9 is proved.

Lemma 10 The set $P_{t1} \cap N_1$ is infinite.

Proof. Let χ be such function, belonging to F_1 , that $\psi'' = \lambda x_1\ldots\lambda x_k\ \chi\ (x_1,\ldots,x_k,b_{j1},\ldots,b_{jk})$ and f is such function of F that $\mathcal{A}(f)=\chi$ We consider the function $g = \lambda p_1\ldots\lambda p_k\ f(p_1,\ldots,p_k,s_{j1},\ldots,s_{jm})$. Infinite set of such n exist that $g = \lambda p_1\ldots\lambda p_k\ \xi_k\ (n,p_1,\ldots,p_k)$ and g is defined on $N_1 \times\ldots\times N_k$. Therefore, infinite set of constant n-conditions $(p_1,\ldots,p_k)$, created on step s , exist. Evidently, on these steps the condition (i) is satisfied. Lemma 10 is proved.

Let us assume now that φ_1 is a.r. Ψ^- ; C^- . Since α is coordinated with Ψ^- ; C^- then such n exists that $\varphi_1\ (\alpha(p_1),\ldots,\alpha(p_k)) \simeq \alpha(\xi_k(n,p_1,\ldots,p_k))$ for any $p_1,\ldots,p_k$ of $D\alpha$. According to lemmas 9 and 10 constant n-condition $(p_1,\ldots,p_k)$, created on step s exists. In addition $n = X(t_o)$ & $p_1 = Y_1(t_o)$ & ... & $p_k = Y_k(t_o)$ and $Y_1(t_o)$ satisfies (i) or (ii). Let us suppose that $Y_1(t_o)$ satisfies (i) and $f = H(Z(t_o),Y_1(t_o),\ldots,Y_k(t_o))$ & $Z(t_o) = f(p_1,\ldots,p_k,q_1,\ldots,q_m)$. Obviously, $p_i \in P_{ti}$, i.e. $\alpha(p_i) = b_i$, $i = 1,\ldots,k$ and the following equalities are true:

$$\varphi_1\ (\ \alpha(p_1),\ldots,\alpha(p_k)\ ) \simeq \varphi_1\ (\ b_{t1},\ldots,b_{tk}\ ) \simeq \psi'\ (\ b_{t1},\ldots,b_{tk}\ ) \fallingdotseq$$
$$\chi_{to}\ (\ b_{t1},\ldots,b_{tk}\ ) \simeq \mathcal{A}\ (f)\ (\ b_{t1},\ldots,b_{tk},b_{j1},\ldots,b_{jm}\ ) \simeq$$
$$\alpha\ (\ f'(p_1,\ldots,p_k,q_1,\ldots,q_m)\ ) = \alpha\ (\ f(p_1,\ldots,p_k,q_1,\ldots,q_m)\ ) \simeq$$
$$\alpha\ (\ Z(t_o)\ ) = \alpha\ (\ \xi_k(n,p_1,\ldots,p_k)\ ).$$

The received contradiction proved lemma 8.

Corollary. Let $A_1\times\ldots\times A_k$ be k-component and φ_1 be such function that $D\varphi_1 = A_1\times\ldots\times A_k$, $R\varphi_1 \subseteq B$, φ_1 is not a constant and $\varphi_1 \fallingdotseq \psi$ for any $\psi \in F_1'$. Then φ_1 is not a.r. Ψ^- ; C^- .

Proof is obvious.

Now we can prove the theorem. Let $A_1^-,\ldots,A_n^-$ be k-components such that $D\varphi = A_1^- \cup\ldots\cup A_n^-$. As in [5] it can be proved such p.c.r. Ψ^- ; $C^{-\wedge}$ function χ_i exists, that $D\chi_i = A_i^-$ and if $s \in A_i^-$, then $\chi_i\ (\ \langle s,t\rangle\ ) = s$, $i = 1,\ldots,n$. Let $\varphi_i \in F_1'$ and $\varphi_i = \varphi/A_i^-$, $i = 1,\ldots,n$ and θ be the function

$$(\ \sigma \supset \varphi_1\psi_1,\ (\ \sigma^2 \supset \varphi_2\psi_2,\ldots,\ (\ \sigma^n \supset \varphi_n\psi_n,\ \alpha_\varnothing\ )\ldots)),$$

Where $\alpha_\varnothing$ is a function defined nowhere. Evidently, θ is p.c.r. Ψ^- ; $C^{-\wedge}$ and

$$\varphi\ (\ s\ ) = t \longleftrightarrow \exists\, r\ (\ \theta\ (\ \langle s,r\rangle\ ) = t\ )$$

(see in [4]), i.e. φ is s.c.r. Ψ^- ; $C^{-\wedge}$. The theorem is proved.

R E F E R E N C E S

[1] Y. N. Moschovakis, Abstract first order computability, I, Trans. Amer. Math. Soc., 138, 1969, 427 - 464

[2] D. Lacombe, Deux généralisations de la notion de récursivité relative, Computs Rendus de l' Academie des Scienses de Paris, 258, 1964, 3410 - 3413
[3] Y. N. Moschovakis, Abstract computability and invariant definability, J. Symb. Log., 34 :4 , 1969
[4] A. V. Ditchev, Computability in the sense of Moschovakis and its connection with partial recursiveness by the numberings (in Russian), Serdica, 7, 1981, 117 - 130
[5] A. V. Ditchev, On the computability in the sense of Moschovakis and its connection with partial recursivness by the numberings (in Russian), Mathematical logic: Proceedings of the Conf. on Math. Logic, dedicated to the memory of A.A. Markov, 1903-1979, Sofia, Sept. 22-23, 1980
[6] A. V. Ditchev, On Skordev's hypothesis (in Russian), Algebra and Logic, 24, 1985, 379 - 391
[7] J. R. Shoenfield, Mathematical logic, Addisson-Wesley publishing company, 1967
[8] D. G. Skordev, Combinatory spaces and recursiveness in them (in Russian), Sofia, 1981
[9] H. Rogers, Jr, Theory of recursive functions and effective computability, Mc Graw-Hill Book Company, 1967
[10] J. R. Shoenfield, Degrees of unsolvability, North-Holland Publishing Company, American Elsevier Publishing Company, 1971

CUT-ELIMINATION THEOREM FOR HIGHER-ORDER CLASSICAL LOGIC: AN INTUITIONISTIC PROOF

A.G. Dragalin

Debrecen University
Debrecen, 4010/ Hungary

I. It is not difficult to see that usual inductive cut-elimination proof fails for higher-order logics. The cause is that the induction goes to the ruin in the case of quantifier rules in logics with the impredicative comprehension shema. In fact, it follows from one Takeuti's result, that finite proof of cut-elimination is impossible in this case (see, for example,[1], chapter 5, point 4). At the end of sixties some nonelementary set-theoretical proofs was worked out for higher-order logics by Tait, Prawitz, Takahasi, Girard (see[2] and[3] for the further information). Especially remarkable success was reached in the case of higher-order intuitionistic logic, where owing to Girard's invention developed by Prawitz, Martin-Löf et al. there is an intuitionistic proof of the cut elimination result.

In contrast to the intuitionistic case, all these proofs for classical logic are founded on considerations of maximal consistent sets of formulas and, hence, are nonconstructive. Below we give a proof of cut-elimination theorem for classical higher-order logic completely in frame of intuitionistic metamathematics (but, of course, it is nonelementary). We use Girard's trick in the proof of the theorem 5.2. below but the main technical novelty of our paper is in using the complete Boolean algebra $\mathcal{N}$ in I.9. reaching an intuitionistic proof in the case of classical logic.

For the sake of brevity we limit ourself by second-order logic but a generalization of our proof for the simple type theory (with or without extensionality) is straightforward along the line directed in [3]or in [1], chapter 5. We put aside for the future also results about strong normalization of deductions. Such results can be worked out by our technique as well.

The sign $\leftrightharpoons$ below means "is by definition". By $\triangleright$ we mark the beginning of a proof and by $\square$ we mark its end.

I.I. Let T be the set $\{0,1\}^*$ of all finite sequences of 0 and I, including the empty sequence Λ. By $p*q$ we denote the concatenation p and q. The number of members of p we denote by ∂p. Instead of $p*\langle i\rangle$ we shall write sometimes $p*i$ simply.

For $p,q\in T$, q is said to be an extension of p

(in symbols $p \leq q$) iff p is an initial segment of q, i.e. iff $(\exists z \in T)(p * z = q)$. The relation $\leq$ gives an evident tree-like order on T with the smallest element (root) Λ.

For $x, y \subseteq T$ let us define an <u>open implication</u>

$$x \supset_\circ y \leftrightharpoons \{p \in T \mid (\forall q \geq p)(q \in x \Rightarrow q \in y)\}.$$

A set $x \subseteq T$ is (order) <u>open</u> iff for all $p, q \in T$:

$$p \in x,\ p \leq q \Rightarrow q \in x.$$

Let $\mathcal{O}$ be the family of all open subsets of T . Note that for every $x, y \subseteq T$ we have $(x \supset_\circ y) \in \mathcal{O}$

Below we use some simple properties of complete Heyting algebras. All necessary information about Heyting algebras can be found, for example, in [4], where these algebras are named as pseudo-Boolean algebras.

I.2. <u>Fact</u> The structure $\langle \mathcal{O}, \subseteq \rangle$ is a complete Heyting algebra. In this algebra

$$\mathbb{1} = T, \quad \mathbb{0} = \emptyset,$$
$$a \mathbin{\mathbb{\wedge}} b = a \cap b, \quad a \mathbin{\mathbb{\vee}} b = a \cup b,$$
$$a \mapsto b = (a \supset_\circ b),$$
$$\neg a = (a \supset_\circ \mathbb{0}) = \{p \in T \mid (\forall q \geq p)(q \notin a)\}.$$

Further, if $Q \subseteq \mathcal{O}$, then $\mathbb{\Lambda} Q = \cap Q$, $\mathbb{V} Q = \cup Q$.

Here and below in analogous cases

$$\cap Q = \{p \in T \mid (\forall a \in Q)(p \in a)\},$$

so $\cap Q = T$, if $Q = \emptyset$.

I.3. A set $x \subseteq T$ is said to be <u>complete</u> iff

$$(\forall p \in T)(p * 0 \in x \wedge p * 1 \in x \Rightarrow p \in x).$$

Let $\mathcal{C}$ denotes the family of all complete subsets of T.

<u>Fact</u> $Q \subseteq \mathcal{C} \Rightarrow \cap Q \in \mathcal{C}$.

Let us introduce the <u>completion operator</u>. Namely, if $x \subseteq T$,

$$\mathfrak{D} x = \cap \{b \in \mathcal{C} \mid x \subseteq b\}.$$

The following list contains some simple properties of $\mathfrak{D}$.

I.4. <u>Fact</u> For all $a, b \subseteq T$ we have

(i) $a \subseteq \mathfrak{D} a$; $\mathfrak{D} a \in \mathcal{C}$;

(ii) $a \subseteq b,\ b \in \mathcal{C} \Rightarrow \mathfrak{D} a \subseteq b$;

(iii) $a \subseteq b \Rightarrow \mathfrak{D} a \subseteq \mathfrak{D} b$;

(iv) $\mathfrak{D}\mathfrak{D} a = \mathfrak{D} a$;

(v) $a \in \mathcal{C} \Leftrightarrow \mathfrak{D} a = a$;

(vi) $\mathfrak{D}(a \cup b) = \mathfrak{D}(a \cup \mathfrak{D} b)$;

(vii) $a \in \mathcal{O} \Rightarrow a \cap \mathfrak{D} b \subseteq \mathfrak{D}(a \cap b)$.

I.5. <u>Lemma</u> (i) $x \in \mathcal{O} \Rightarrow \mathfrak{D} x \in \mathcal{C} \cap \mathcal{O}$;

(ii) $x \in \mathcal{O}, y \in \mathcal{C} \Rightarrow (x \supset_\circ y) \in \mathcal{C} \cap \mathcal{O}$;

(iii) $x, y \in \mathcal{O} \Rightarrow \mathcal{D}(x \cap y) = \mathcal{D}x \cap \mathcal{D}y$.

▷ (i) Let $x \in \mathcal{O}$, we show $\mathcal{D}x \in \mathcal{O}$. Let us consider the set

$$c = \{p \in T \mid (\forall q \geqslant p)(q \in \mathcal{D}x)\}.$$

Now it should be checked that $x \subseteq c$, $c \in \mathcal{b}$. Hence, $\mathcal{D}x \subseteq c$ (I.4., (ii)) and therefore $\mathcal{D}x \in \mathcal{O}$.

(ii) Let $p*0, p*1 \in (x \supset_0 y)$ and prove $p \in (x \supset_0 y)$. Let us consider $q \geqslant p$, $q \in x$ and conclude $q \in y$. If $q \geqslant p*0$ or $q \geqslant p*1$, then $q \in (x \supset_0 y)$ and $q \in y$ follows from $q \in x$. It remains only the case $q = p$. We have $q*0, q*1 \in x$ (as $x \in \mathcal{O}$), and $q*0, q*1 \in (x \supset_0 y)$ (in view $q = p$), so $q*0, q*1 \in y$. Now $q \in y$ immediately in view of $y \in \mathcal{b}$.

(iii) Nontrivial is only the inclusion

$$\mathcal{D}x \cap \mathcal{D}y \subseteq \mathcal{D}(x \cap y).$$

It follows from (i) with help of I.4. and pure algebraic inclusions in algebra $\mathcal{O}$:

$x \cap y \subseteq \mathcal{D}(x \cap y)$; $x \subseteq (y \supset_0 \mathcal{D}(x \cap y))$;

$\mathcal{D}x \subseteq (y \supset_0 \mathcal{D}(x \cap y))$; $y \cap \mathcal{D}x \subseteq \mathcal{D}(x \cap y)$;

$y \subseteq (\mathcal{D}x \supset_0 \mathcal{D}(x \cap y))$; $\mathcal{D}y \subseteq (\mathcal{D}x \supset_0 \mathcal{D}(x \cap y))$;

$\mathcal{D}x \cap \mathcal{D}y \subseteq \mathcal{D}(x \cap y)$. □

I.6. <u>Theorem</u>. The structure $\langle \mathcal{b} \cap \mathcal{O}, \subseteq \rangle$ is a complete Heyting algebra. In this algebra

$\mathbb{1} = T$; $\mathbb{0} = \emptyset$; $a \wedge b = a \cap b$; $a \vee b = \mathcal{D}(a \cup b)$;

$a \mapsto b = (a \supset_0 b)$; $\neg a = (a \supset_0 \mathbb{0})$.

Further, if $Q \subseteq \mathcal{b} \cap \mathcal{O}$, then $\bigwedge Q = \bigcap Q$; $\bigvee Q = \mathcal{D}(\bigcup Q)$.

▷ It is corollary of I.2., I.4., I.5. □

I.7. Let us fix now some set $v \in \mathcal{b} \cap \mathcal{O}$. Let $\mathcal{N}$ be the family of all sets $a \in \mathcal{b} \cap \mathcal{O}$, such that

$$a = ((a \supset_0 v) \supset_0 v).$$

Note, that $v \in \mathcal{N}$.

I.8. <u>Fact</u>. (i) $a \in \mathcal{O} \Rightarrow a \subseteq ((a \supset_0 v) \supset_0 v)$;

(ii) $a \in \mathcal{O}, b \in \mathcal{N} \Rightarrow (a \supset_0 b) \in \mathcal{N}$;

(iii) $a \in \mathcal{O} \Rightarrow (a \supset_0 v) \in \mathcal{N}$;

(iv) $a \in \mathcal{N} \Rightarrow v \subseteq a$;

(v) $Q \subseteq \mathcal{N} \Rightarrow \bigcap Q \in \mathcal{N}$.

I.9. <u>Theorem</u>. The structure $\langle \mathcal{N}, \subseteq \rangle$ is a complete <u>Boolean</u> algebra. In this algebra

$\mathbb{1} = T$; $\mathbb{0} = v$;

$a \wedge b = a \cap b$; $a \vee b = ((a \supset_0 v) \cap (b \supset_0 v)) \supset_0 v$;

$a \mapsto b = (a \supset_0 b)$; $\neg a = (a \supset_0 v)$.

Further, if $Q \subseteq \mathcal{N}$, then

$$\bigwedge Q = \bigcap Q; \quad \bigvee Q = (\bigcap \{a \supset_0 v \mid a \in Q\}) \supset_0 v.$$

▷ It is corollary of I.2. and I.8. □

2. Let us describe a language of our logic. Three natural

numbers 0, 1, 2 is said to be types. Let us fix for every type τ an infinite set $Var(\tau)$ of variables of the type τ. Variables of type 0 are considered as variables for elements of an individual domain, variables of type 2 are considered as variables for some subsets of a given individual domain, at last, variables of type 1 are considered as variables for sentences.

It is unessential for our proof, but for simplicity we suppose that our language does not contain any constants and functional symbols. Our language maybe contains some predicate symbols, such that all argument places of these symbols are of type 0.

2.1. Now we give the inductive definition of an expression of type τ. The set of all expressions of type τ we denote as $Exp(\tau)$.

1) If $x \in Var(\tau)$, then $x \in Exp(\tau)$.

2) If $t_1, \ldots, t_n \in Exp(0)$ and P is a n-place predicate symbol, then $P(t_1, \ldots, t_n) \in Exp(1)$; such expressions are said to be atoms.

3) If $A, B \in Exp(1)$, $x \in Var(\tau)$, then $(A \wedge B)$, $(A \vee B)$, $(A \supset B)$, $\neg A$, $\forall x A$, $\exists x A$ are elements of $Exp(1)$.

4) If $A \in Exp(1)$ and $x \in Var(0)$, then $\{x | A\} \in Exp(2)$.

5) If $t_1 \in Exp(0)$ and $t \in Exp(2)$, then $(t_1 \varepsilon t) \in Exp(1)$.

Elements of $Exp(1)$ are said to be formulas.

A sequent is a formal string of the form $(\Gamma \to \Delta)$ where Γ and Δ are finite (maybe empty) collections of formulas. The order of members in Γ as well as in Δ is unessential. If S is $(\Gamma \to \Delta)$, then let us denote

$$(S)^0 = \Gamma, \quad (S)^1 = \Delta .$$

3. Now we develop an apparatus essentially equivalent to deducibility in classical sequent calculi without cuts.

If S_1, S_2, S_3 are sequents, then let us define a three place relation: $S_1 \prec S_2, S_3$ (in words: S_1 branches into S_2 and S_3).

To begin with, for an arbitrary sequent S we put:

$$S \prec S, S.$$

Further, we list the rest cases for this relation. Every case has a special symbolic name, depending on the construction of S_1. The notation $S_1 \prec S_2$ is an abbreviation for $S_1 \prec S_2, S_2$.

1) $(\wedge \to)$ $(A \wedge B)\Gamma \to \Delta \prec AB(A \wedge B)\Gamma \to \Delta$;

2) $(\to \wedge)$ $\Gamma \to \Delta(A \wedge B) \prec \Gamma \to \Delta(A \wedge B)A$, $\Gamma \to \Delta(A \wedge B)B$;

3) $(\vee \to)$ $(A \vee B)\Gamma \to \Delta \prec A(A \vee B)\Gamma \to \Delta$, $B(A \vee B)\Gamma \to \Delta$;

4) $(\to \vee)$ $\Gamma \to \Delta(A \vee B) \prec \Gamma \to \Delta(A \vee B)AB$;

5) $(\supset \to)$ $(A \supset B)\Gamma \to \Delta \prec (A \supset B)\Gamma \to \Delta A$, $B(A \supset B)\Gamma \to \Delta$;

6) $(\to \supset)$ $\Gamma \to \Delta(A \supset B) \prec A\Gamma \to \Delta(A \supset B)B$;

7) $(\neg \to)$ $\neg A\Gamma \to \Delta \prec \neg A\Gamma \to \Delta A$;

8) $(\to \neg)$ $\Gamma \to \Delta\neg A \prec A\Gamma \to \Delta\neg A$;

9) $(\forall\rightarrow)$ $\forall x A(x)\Gamma\rightarrow\Delta \prec A(t)\forall x A(x)\Gamma\rightarrow\Delta$;

10) $(\rightarrow\forall)$ $\Gamma\rightarrow\Delta\forall x A(x) \prec \Gamma\rightarrow\Delta\forall x A(x)\ A(y)$, where y is not free in $\Gamma\rightarrow\Delta\forall x A(x)$;

11) $(\exists\rightarrow)$ $\exists x A(x)\Gamma\rightarrow\Delta \prec A(y)\exists x A(x)\Gamma\rightarrow\Delta$, where y is not free in $\exists x A(x)\Gamma\rightarrow\Delta$;

12) $(\rightarrow\exists)$ $\Gamma\rightarrow\Delta\exists x A(x) \prec \Gamma\rightarrow\Delta\exists x A(x) A(t)$;

13) $(\varepsilon\rightarrow)$ $(t\varepsilon\{x|A(x)\})\Gamma\rightarrow\Delta \prec A(t)(t\varepsilon\{x|A(x)\})\Gamma\rightarrow\Delta$;

14) $(\rightarrow\varepsilon)$ $\Gamma\rightarrow\Delta(t\varepsilon\{x|A(x)\}) \prec \Gamma\rightarrow\Delta(t\varepsilon\{x|A(x)\})A(t)$.

Here in 9), 12), 13), 14) $x\in Var(\tau)$, $t\in Exp(\tau)$ for the same τ, so in 13) and 14) $\tau=0$. $A(t)$ is a result of substitution t instead of free occurrences x in $A(x)$ with necessary renaming of bound variables of $A(x)$ avoiding collisions of variables.

Let us denote by $\vdash^+ S$ the fact of deducibility S in some higher-order classical sequent calculus <u>without cuts</u> (about such calculus cf., for example [2]).

3.1. <u>Fact</u> If $S_1 \prec S_2, S_3$, then

(i) $(S_1)^i \subseteq (S_2)^i$, $(S_1)^i \subseteq (S_3)^i$, for $i=0,1$;

(ii) if $\vdash^+ S_2$ and $\vdash^+ S_3$, then $\vdash^+ S_1$.

4. A <u>sequent tree</u> is a function h defined on T, such that for every $p\in T$, $h(p)$ is a sequent and, moreover, $h(p) \prec h(p*0), h(p*1)$.

4.1. A <u>zero</u> of a sequent tree h is a set

$$v = \{p\in T \mid (\exists q\leq p)(\vdash^+ h(q))\}.$$

4.2. <u>Lemma</u> $v \in \mathcal{C}\cap\mathcal{O}$.

$\triangleright$ Use 3.1. (ii). $\square$

For every formula A we define two sets $L(A)$ and $R(A)$:

$$L(A) = \{p\in T \mid A\in (h(p))^0\},$$
$$R(A) = \{p\in T \mid A\in (h(p))^1\}.$$

4.3. <u>Lemma</u> $L(A), R(A) \in \mathcal{O}$.

$\triangleright$ Use 3.1. (i). $\square$

4.4. A sequent tree h is said to be <u>systematic</u>, if the following conditions are fulfilled:

1) $L(A)\cap R(A) \subseteq v$;

2) $L(A\wedge B) \subseteq \mathfrak{D}(L(A)\cap L(B))$;

3) $R(A\wedge B) \subseteq \mathfrak{D}(R(A)\cup R(B))$;

4) $L(A\vee B) \subseteq \mathfrak{D}(L(A)\cup L(B))$;

5) $R(A\vee B) \subseteq \mathfrak{D}(R(A)\cap R(B))$;

6) $L(A\supset B) \subseteq \mathfrak{D}(R(A)\cup L(B))$;

7) $R(A \supset B) \subseteq \mathcal{D}(L(A) \cap R(B))$;

8) $L(\neg A) \subseteq \mathcal{D}(R(A))$;

9) $R(\neg A) \subseteq \mathcal{D}(L(A))$;

10) $L(\forall x A(x)) \subseteq \mathcal{D}(L(A(t)))$, $x \in Var(\tau)$, $t \in Exp(\tau)$;

11) $R(\forall x A(x)) \subseteq \mathcal{D}(\bigcup_{y \in Var(\tau)} R(A(y)))$, $x \in Var(\tau)$;

12) $L(\exists x A(x)) \subseteq \mathcal{D}(\bigcup_{y \in Var(\tau)} L(A(y)))$, $x \in Var(\tau)$;

13) $R(\exists x A(x)) \subseteq \mathcal{D}(R(A(t)))$, $x \in Var(\tau)$, $t \in Exp(\tau)$;

14) $L(t \varepsilon \{x \mid A(x)\}) \subseteq \mathcal{D}(L(A(t)))$;

15) $R(t \varepsilon \{x \mid A(x)\}) \subseteq \mathcal{D}(R(A(t)))$.

4.5. Theorem For every sequent S a systematic sequent tree can be constructed, such that $h(\Lambda) = S$.

▷ A systematic sequent tree h is constructed by standard methods of constructing of refutation trees; $h(p)$ is constructed by induction on the length ∂p of p. Details can be found, for example, in[3] or in[5], point 4.2., this volume. Note the main difference between our sequent tree and refutation tree in[3]. In[3] the refutation tree is constructed by consistent way, therefore the corresponding function is essentially nonconstructive. In our case there is no necessity for consistency providing systematic function h, so we can build up h in constructive (in fact, primitive-recursive) manner.□

5. Let h be a systematic sequent tree. Now we construct some algebraic model for our logic. To begin with, for every type τ we define a domain $I(\tau)$ and some relation $a \approx t$, where $a \in I(\tau)$ and $t \in Exp(\tau)$. Namely:

For $\tau = 0$. $a \approx t \leftrightharpoons (a \text{ is } t)$, $I(0) = Var(0)$.

For $\tau = 1$. If $a \in \mathcal{N}$ (cf. I.7.) and $t \in Exp(1)$, let us define $a \approx t \leftrightharpoons (L(t) \subseteq a) \wedge (R(t) \cap a \subseteq v)$.
Further, $a \in I(1) \leftrightharpoons (a \in \mathcal{N}) \wedge (\exists t \in Exp(1))(a \approx t)$, (see 4.1., 1.7. and 1.9.).

For $\tau = 2$. Let a be a function $a : I(0) \to \mathcal{N}$ and $t \in Exp(2)$. We define $a \approx t$ iff for every $a_1 \in I(0)$, $t_1 \in Exp(0)$, we have

$$a_1 \approx t_1 \Rightarrow a(a_1) \approx (t_1 \varepsilon t).$$

Further, $a \in I(2) \leftrightharpoons (a : I(0) \to \mathcal{N}) \wedge (\exists t \in Exp(2))(a \approx t)$.

5.1. If $t(x_1, \ldots, x_n) \in Exp(\tau)$ and $x_1, \ldots, x_n$ are all of parameters t of types $\tau_1, \ldots, \tau_n$ respectively and $a_1, \ldots, a_n$ are elements of $I(\tau_1), \ldots, I(\tau_n)$ respectively, then the result $t(a_1, \ldots, a_n)$ of substitution $a_1, \ldots, a_n$ instead of free occurrences of $x_1, \ldots, x_n$ in t respectively is said to be a valued expression (of type τ), corresponding to t

For a given valued expression e of type τ sometimes can be defined a natural value $\|e\| \in I(\tau)$ by induction on the building of e according 2.1.

Namely, for a valued atom P we put

$$\|P\| \leftrightharpoons ((\mathfrak{D}(L(P) \cup v) \supset_0 v) \supset_0 v).$$

Note, that argument places in P are 0-sorted and $I(0) = Exp(0)$, so a valued atom is a simple atom in our language and $L(P)$ is defined. Further, it is not difficult to see that $\|P\| \in \mathcal{N}$ and, moreover, $\|P\| \in I(1)$.

In general situation we define $\|e\|$ by inductive way and $\|e\|$ is defined only if the all "previous" parts of e are defined already. Formulas' values are defined accordingly operations in algebra $\mathcal{N}$.

For example, if A, B are valued formulas, $\|A\|$ and $\|B\|$ are defined and

$\|A\| \cap \|B\| \in I(1)$, $((\|A\| \supset_0 v) \cap (\|B\| \supset_0 v)) \supset_0 v \in I(1)$,

then we put

$$\|A \wedge B\| \leftrightharpoons (\|A\| \cap \|B\|),$$
$$\|A \vee B\| \leftrightharpoons ((\|A\| \supset_0 v) \cap (\|B\| \supset_0 v)) \supset_0 v.$$

If $\forall x A(x)$ is a valued formula, $x \in Var(\tau)$ and for every $a \in I(\tau)$ the value $\|A(a)\|$ is defined and, moreover,

$$\bigcap_{a \in I(\tau)} \|A(a)\| \in I(1),$$

then we put

$$\|\forall x A(x)\| \leftrightharpoons \bigcap_{a \in I(\tau)} \|A(a)\|.$$

If $(t_1 \varepsilon t)$ is a valued formula $\|t_1\|$, $\|t\|$ are defined (so $\|t\| \in I(2)$ and $\|t\|$ is a function $\|t\| : I(0) \to \mathcal{N}$), and $\|t\|(\|t_1\|) \in I(1)$, then we put $\|t_1 \varepsilon t\| \leftrightharpoons \|t\|(\|t_1\|)$.

If $\{x \mid t(x)\}$ is a valued expression, $\|t(a)\|$ is defined for every $a \in I(0)$ and the function $f : I(0) \to \mathcal{N}$, such that $f(a) = \|t(a)\|$ for $a \in I(0)$, belongs to $I(2)$, then we put $\|\{x \mid t(x)\}\| = f$.

5.2. <u>Theorem</u> Let $t(x_1, \ldots, x_n) \in Exp(\tau)$, where $x_1, \ldots, x_n$ is the list of all parameters of t, $x_i \in Var(\tau_i)$. Let $t_1, \ldots, t_n$ is a list of expressions $t_i \in Exp(\tau_i)$ and $a_1, \ldots, a_n$ is a list of elements, $a_i \in I(\tau_i)$. Let us denote $t' = t(t_1, \ldots, t_n)$ and $t'' = t(a_1, \ldots, a_n)$. Now let us suppose, that $a_i \approx t_i$ for all $i = 1, \ldots, n$.

Then $\|t''\|$ is defined and $\|t''\| \approx t'$.

$\triangleright$ By straightforward induction on the building of t (2.I.) Note, that in case of higher-order logic (i.e. in our case) the expression t' can be more complicate in sense of building 2.I., than t and it is very essential that we carry out our induction on t, but not on t'. This way of proving origins from one Girard's construction. Here we consider only two representative cases of this induction.

Let, for example, t be $A \supset B$. On inductive supposition $\|A''\|$ and $\|B''\|$ are defined and, moreover, $\|A''\| \approx A'$ and $\|B''\| \approx B'$. So $\|A''\|$ and $\|B''\|$ are elements of $I(1)$. We claim that $(\|A''\| \supset_0 \|B''\|) \approx (A' \supset B')$, therefore $\|A'' \supset B''\| = (\|A''\| \supset_0 \|B''\|)$ is defined and is an element of $I(1)$.

Indeed, in view of $L(B') \subseteq \|B''\|$ we have $L(B') \cap \|A''\| \subseteq \|B''\|$. Acting in $\mathcal{O}$ (4.3., I.2.), we get $L(B') \subseteq (\|A''\| \supset_0 \|B''\|)$. Further, $R(A') \cap \|A''\| \subseteq v$, so $R(A') \cap \|A''\| \subseteq \|B''\|$ and again in $\mathcal{O}$, $R(A') \subseteq (\|A''\| \supset_0 \|B''\|)$. From this we have $L(B') \cup R(A') \subseteq (\|A''\| \supset_0 \|B''\|)$. Using I.4., $\mathfrak{D}(R(A') \cup L(B')) \subseteq \|A''\| \supset_0 \|B''\|$. Now by 4.4.6) we get $L(A' \supset B') \subseteq (\|A''\| \supset_0 \|B''\|)$.

On the other hand, in $\mathcal{O}$, $\|A''\| \cap (\|A''\| \supset_0 \|B''\|) \subseteq \|B''\|$. From $R(B') \cap \|B''\| \subseteq v$ we have $\|A''\| \cap R(B') \cap (\|A''\| \supset_0 \|B''\|) \subseteq v$.

Further, $L(A') \subseteq \|A''\|$, hence, $L(A') \cap R(B') \cap (\|A''\| \supset_0 \|B''\|) \subseteq v$.
Using I.4., 4.2., I.5., I.6. we get

$$\mathfrak{D}(L(A') \cap R(B')) \cap (\|A''\| \supset_0 \|B''\|) \subseteq v,$$

and in view of 4.4.7 at last

$$L(A' \supset B') \cap (\|A''\| \supset_0 \|B''\|) \subseteq v.$$

Let now t be $\exists x A(x)$, where $x \in Var(\tau)$. We show that $\|\exists x A''(x)\|$ is defined and $\|\exists x A''(x)\| \approx \exists x A'(x)$. On inductive supposition (using $A(x)$ as t) $\|A''(b)\|$ is defined and $\|A''(b)\| \approx A'(z)$ for all $b \in I(\tau)$, $z \in Exp(\tau)$, $b \approx z$. Note, that $A(x)$ is simpler, than $\exists x A(x)$, but $A'(z)$ maybe more complcate, than $A'(x)$. It is the very place, where we use the Girard's trick.

Let us consider in the complete Boolean algebra $\mathcal{N}$ (I.9.) the following elements:

$$a = \bigcap_{b \in I(\tau)} (\|A''(b)\| \supset_0 v),$$

$$c = (a \supset_0 v) = \bigvee_{b \in I(\tau)} \|A''(b)\|.$$

We show $c \approx \exists x A'(x)$. Hence, $c = \|\exists x A''(x)\|$, $c \in I(1)$.
It is necessary to show $L(\exists x A'(x)) \subseteq c$ and $R(\exists x A'(x)) \cap c \subseteq v$.

Let $y \in Var(\tau)$. We find $b \in I(\tau)$ such that $b \approx y$. In $\mathcal{O}$ we have $\|A''(b)\| \cap (\|A''(b)\| \supset_0 v) \subseteq v$.
On inductive supposition $\|A''(b)\| \approx A'(y)$ and hence $L(A'(y)) \subseteq \|A''(b)\|$.
Further, $L(A'(y)) \cap (\|A''(b)\| \supset_0 v) \subseteq v$.
So much the more $L(A'(y)) \cap a \subseteq v$, and acting in $\mathcal{O}$ we conclude $L(A'(y)) \subseteq (a \supset_0 v) = c$. Hence,

$$\bigcup_{y \in Var(\tau)} L(A'(y)) \subseteq c.$$

Using $c \in \mathfrak{b}$ we conclude

$$\mathfrak{D}\Big(\bigcup_{y \in Var(\tau)} L(A'(y))\Big) \subseteq c,$$

and using 4.4.12), we get $L(\exists x A'(x)) \subseteq c$.

Let us consider now $b \in I(\tau)$, $z \in Exp(\tau)$, $b \approx z$. On inductive supposition $\|A''(b)\| \approx A'(z)$, hence, $R(A'(z)) \cap \|A''(b)\| \subseteq v$.
Acting in $\mathcal{O}$, $R(A'(z)) \subseteq (\|A''(b)\| \supset_0 v)$. But on the right of this inclusion we have an element of $\mathfrak{b}$, so

$$\mathfrak{D}(R(A'(z))) \subseteq (\|A''(b)\| \supset_0 v).$$

Using 4.4.13) we get $R(\exists x A'(x)) \subseteq \|A''(b)\| \supset_0 v$. Intersecting by all $b \in I(\tau)$ we get $R(\exists x A'(x)) \subseteq a$. In $\mathcal{O}$ we have $a \cap (a \supset_0 v) \subseteq v$, i.e. $a \cap c \subseteq v$. Hence,

$$R(\exists x A'(x)) \cap c \subseteq v. \ \square$$

5.3. <u>Corollary</u> Let a sequent

$$A_1, \ldots A_n \to B_1, \ldots B_m$$

is deducible in classical sequent calculus <u>with cuts</u>. Let $x_1, \ldots, x_n$ is the list of all parameters of this sequent and $a_1 \approx x_1$, ..., $a_n \approx x_n$. Let
$A'_i = A_i = A_i(x_1, \ldots, x_n)$ and $A''_i = A_i(a_1, \ldots, a_n)$.
Then $L(A_1) \cap \ldots \cap L(A_n) \cap R(B_1) \cap \ldots \cap R(B_m) \subseteq v$.

$\triangleright$ The sequent $A_1 \ldots A_n \to B_1 \ldots B_m$ is deducible in classical logic, hence, in $\mathcal{N}$ we have

$$\|A''_1\| \cap \ldots \cap \|A''_n\| \cap (\|B''_1\| \supset_0 v) \cap \ldots \cap (\|B''_m\| \supset_0 v) \subseteq v.$$

Now in accordance with 5.2. it is true that $L(A_i) \subseteq \|A''_i\|$ and $R(B_j) \cap \|B''_j\| \subseteq v$ (i.e. $R(B_j) \subseteq (\|B''_j\| \supset_0 v)$).
Hence we get the desired inclusion. $\square$

6. Theorem Let a sequent $A_1 \ldots A_n \to B_1 \ldots B_m$ is deducible in classical sequent calculus with cuts, then it is deducible and without cuts.

▷ For a given sequent we construct a systematic sequent tree h with a given sequent in the root (4.5.). Then (5.3.)

$$\Lambda \in L(A_1) \cap \ldots \cap L(A_n) \cap R(B_1) \cap \ldots \cap R(B_m) \subseteq v.$$

According to the definition 4.I. the sequent considered is deducible without cuts □.

REFERENCES

[I] A.G.Dragalin, Mathematical intuitionism. Introduction to proof theory in Russian, Nauka pbl., Moscow, I979.

[2] Takahashi Moto-o, A system of simple type theory of Gentzen style with inference of extensionality and cut-elimination in it, Comment. math. Univ. St. Pauli, I970, I8, p. I29-I47.

[3] Takahashi Moto-o, Cut-elimination theorem and Brouwerian-valued models for intuitionistic type theory, Comment. math. Univ. St. Pauli, I970, I9, p. 55-72.

[4] H.Rasiowa, R.Sikorski, The Mathematics of Metamathematics, second ed., PWN pbl., Warszawa, I968.

[5] A.G.Dragalin, A completeness theorem for higher-order intuitionistic logic. An intuitionistic proof, this volume.

March I986, revised version July I986.

MODAL ENVIRONMENT FOR BOOLEAN SPECULATIONS

(preliminary report)

George Gargov,
Solomon Passy, and
Tinko Tinchev

Faculty of Mathematics at Sofia University
boul. Anton Ivanov 5, Sofia 1126, Bulgaria

ABSTRACT The common form of a mathematical theorem consists in that "the truth of some properties for some objects is necessary and/or sufficient condition for other properties to hold for other objects". To formalize this, one happens to resort to Kripke modal logic K which, having in the syntax the notions of 'property' and 'necessity', appears to provide a reliable metamathematical fundament. In this paper we challenge this reliability. We propose two different approaches each claiming better formal treatment of the state of affairs. The first approach is in formalizing the notion of 'sufficiency' (which remains beyond the capacities of K), and consequently of 'sufficiency' and 'necessity' in a joint context. The second is our older idea to formalize the notion of 'object' in the same modal spirit. Having 'property, object, sufficiency, necessity', we establish some basic results and profess to properly formalize the everyday metamathematical reason.

1. INTRODUCTION: sufficiency

Modal logic extends syntactically the ordinary propositional language with new, as a rule unary, operators known as modalities, a typical one being the necessity modal operator $\Box$. On the semantical side one has the Kripke "possible worlds" interpretation of the extended language: frames $F = (W, R)$ with $W \neq \emptyset$ and $R \subseteq W^2$; and models based on them, i.e. (W, R, V) where V, denoted as $\models$, assignes to each formula A a (truth) set $V(A)$, or $\{s/\ s \models A\}$, of possible worlds. The truth set of a Boolean junction is the respective set-theoretic junction of the truth set(s). The truth set for the modality, i.e. $V(\Box A)$, is usually given as a particular first-order condition on the relationship between $V(A)$ and R.

Consider the simplest case when just $\Box$ has been added, and denote this modal language by $\mathcal{ML}(\Box)$. To each $\mathcal{ML}(\Box)$-formula A, a formula $St(A)$ corresponds in the language $\mathcal{L}_1$, cf. van Benthem (1977): a first-order language with one binary predicate Rxy, and infinitely many unary predicates P_ix.

<u>Definition</u> $St(A)$ is defined inductively as follows:

1. $St(p_i) = P_ix$, for propositional variables p_i
2. $St(0) = 0$ (0 is the falsity)
3. $St(A \longrightarrow B) = St(A) \longrightarrow St(B)$
4. $St(\Box A) = \forall y(Rxy \longrightarrow [y/x]St(A))$, where y does not occur in $St(A)$. #

Since $St(A)$ reflects the semantics of the modal formula A, a model $\mathcal{M}$ is a first-order structure for the language $\mathcal{L}_1$, and

$\mathfrak{M}, s \models A$ iff $\mathfrak{M} \models St(A)[s]$.

Focus now on clause 4, which is the point of Kripke's approach to necessity. The truth of $\Box A$ depends on the truth set of A and the relation R, in a way given by a formula of one bound variable. In this paper we study modalities for which the corresponding truth condition on x is given by an arbitrary formula of one bound variable y, containing Rxx, Rxy, Ryx, Ryy and $[y/x]St(A)$ and find a basis for these modalities, i.e. a "small" subset of them sufficient for defining each of the remaining.

Kripke's mathematical interpretation of "p is necessary (true) in x", $R(x) \subseteq V(p)$, only sharpens but does not satisfy one's desire to formally handle the "sufficiency" phenomena as well. The first and trivial attempt is to grammatically reduce the "sufficiency" to "necessity" saying that "x is sufficient for p" iff "p is necessary in x", and this surely will not enrich our knowledge. The first non-trivial suggestion is to interpret the sufficiency more Kripkely: "p is sufficient for (accessibility from) x" iff $V(p) \subseteq R(x)$. This leads to an "alternative" modal logic K^*, Tehlikeli (1985) which formally at least, is equal in rights with Kripke's K.

Language of K^* is $\mathcal{ML}(\boxplus)$, i.e., $\Box$ and $\Diamond$ are replaced by $\boxplus$ and Φ (named by Slavjan Radev as "window" and "kite").

Semantics of K^*. Kripke models with: $x \models \boxplus A$ iff $\forall y(y \models A \rightarrow Rxy)$.

Axiomatics of K^*: Besides the Boolean tautologies, we have also the scheme

$\vdash \boxplus A \wedge \boxplus(\neg A \wedge B) \rightarrow \boxplus B$, and the inference rule

If $\vdash A$, then $\vdash \boxplus\neg A$.

Common validity notions (possibly in a frame or model) will be freely used. Mathematically, the equivalence between K and K^* is justified by the

Correspondence Theorem (Tehlikeli, 1985) Take the bijective translation * from $\mathcal{ML}(\Box)$ onto $\mathcal{ML}(\boxplus)$, which uniformly replaces $\Box$ by $\boxplus\neg$. For a K-model $\mathfrak{M} = (W, R, V)$, let $\mathfrak{M}^*$ denote the K^*-model $(W, W^2\backslash R, V)$.

Then: (a) $K \vdash A$ iff $K^* \vdash A^*$.

(b) $\mathfrak{M}, s \models A$ iff $\mathfrak{M}^*, s \models A^*$, whence

(c) $K \models A$ iff $K^* \models A^*$.

Proof (a): straightforward induction on $\vdash$.

(b): straightforward induction on $\models$. #

Informally, the window $\boxplus$ may be pretty well interpreted as 'sufficiency', to the same extent at least to which 'necessity' is $\Box$ and 'possibility' is $\Diamond$.

Question (L. Ivanov) Is there in this line a natural (or, at least philosophical) interpretation of kite Φ: $x \models \Phi A$ iff $\exists y(\neg Rxy \ \& \ y \models \neg A)$? #

Metaphysically, R ($\neg$R) is the (in)accessibility, whence $\{A/ \ x \models \Box A\}$ captures the eternities for x, while $\{A/ \ x \models \boxplus A\}$ subsums the falsities of the eternit beyond x. Computationally, window-α-A, $[\![\alpha]\!]A$, states A's sufficiency as post-condition in a state, for program α's termination this last in.

By the Correspondence and the respective theorems for K, one gets:

Theorem K^* is sound, (finitely) complete, decidable, compact, etc. #

Short historical notes on K^* are left to a large discussion in the Epilogue.

So K^* shares all traditional virtues of K, and foreseeably, cf. the Correspondence, all its deficiencies as well. For illustration, take the poly-modal case, where questions about modal-axiomatizability of relations between binary relations are of traditional interest. The class of three-relational K-frames determined by the property $R = S \cup T$ is modal-axiomatic (over K), by the scheme $[R]p \leftrightarrow [S]p \wedge [T]p$, whereas the properties $R = S \cap T$, or $R = \neg S$ do not determine such classes, after Theorem 8 of Goldblatt & Thomason (1975). Such a discrimination between intersection $\cap$ an union $\cup$ in K is a bit shocking for the democratic spirit of a Boolean consciousness. As expected, K^* supports similar partiality, just reversing colours: $R = S \cap T$ is modally-axiomatizable over K^*, via the scheme $[\![R]\!]p \leftrightarrow [\![S]\!]p \wedge [\![T]\!]p$, whereas union and complement turn out not to be.

This observation can be generalized. For a class $\mathfrak{F}$ of K-frames, let $\mathfrak{F}^* = \{F^*/ F \in \mathfrak{F}\}$, where $(W, R)^* = (W, W^2\backslash R)$ is regarded as a K*-frame. Again by the Correspondence, we have:
<u>Theorem</u> $\mathfrak{F}^*$ is modal axiomatic over K* iff $\mathfrak{F}$ is modal-axiomatic over K. #

This and Goldblatt & Thomason's Theorem 8 give necessary & sufficient conditions for a class of K*-frames to be modal-axiomatic. Due to lack of space we do not explicitly mention these conditions here; just say that, not surprizingly, these last when compared to Goldblatt & Thomason's only transpose the relation R and its complement ¬R.

Thus necessity and sufficiency split the modal theory into two dual branches each of which spreads over less than a half of the Boolean realm. The complement ¬R, remaining outside the scope of both branches cannot be framed before uniting them: $[\neg R]A \leftrightarrow [\![R]\!]\neg A$, and $[\![\neg R]\!]A \leftrightarrow [R]\neg A$. Consequently, in this environment the "iff" modality is definable:
"p is necessary & sufficient in x", i.e. $V(p) = R(x)$, iff $x \models \Box p \wedge \boxdot p$.

So the union of K and K* appears to suggest a reliable base for governing the Boolean kingdom. This will be established in the next section.

2. THE BOOLEAN MODAL LOGIC K~

The language of K~ is $\mathcal{ML}(\Box, \boxdot)$, i.e. with $\{\Box, \boxdot\}$ as a modal fragment.
Semantics. Models for K~ are Kripke models $\mathcal{M} = (W, R, V)$ with
$x \models \Box A$ iff $\forall y(Rxy \longrightarrow y \models A)$, and
$x \models \boxdot A$ iff $\forall y(y \models A \longrightarrow Rxy)$.
Abbreviations: $N(A, B) =_{DF} \Box A \wedge \boxdot\neg B$, $[\![U]\!]A =_{DF} N(A, A)$, $\langle\!\langle U\rangle\!\rangle A =_{DF} \Diamond A \vee \Diamond\!\!\!\cdot\,\neg A$.
Alternatively, in the language $\mathcal{ML}(N)$, one has $\Box A = N(A, 1)$ and $\boxdot B = N(1, \neg B)$.
Axiomatics (the non-Boolean part):

1. $N(A, B) \wedge N(A \rightarrow A', B \rightarrow B') \longrightarrow N(A', B')$
2. $N(1, 1)$
3. $[\![U]\!]A \longrightarrow A$
4. $[\![U]\!]A \longrightarrow [\![U]\!][\![U]\!]A$
5. $A \longrightarrow [\![U]\!]\langle\!\langle U\rangle\!\rangle A$

i.e. $[\![U]\!]$ is an S5-modality, and the rule
(RN) If $\vdash A \longrightarrow A'$ & $\vdash B \longrightarrow B'$, then $\vdash N(A, B) \longrightarrow N(A', B')$.

Omitting the trivial proof, we note that the axiomatics of K~ is sound, i.e. all theorems of K~ are valid. Such a system has been discussed earlier by van Benthem (1979): see Epilogue.

<u>Definition</u> A <u>generalized model</u> for K~ is a quadruple (W, R, S, V), where
$R \cup S = W^2$, and
$x \models \Box A$ iff $\forall y(Rxy \longrightarrow y \models A)$, and
$x \models \boxdot A$ iff $\forall y(y \models A \longrightarrow \neg Sxy)$. #
In generalized models we still have: $x \models [\![U]\!]A$ iff $\forall y(y \models A)$; though in general $R \cap S \neq \emptyset$, what makes them "generalized". The axiomatics of K~ is sound with respect to generalized models as well.

<u>"Generalized" Completeness Theorem</u> If $K^\sim \not\vdash A$, then A is refuted in a generalized model.
<u>Proof</u> By the familiar, at least since Segerberg, canonical model techniques, consider the set W of maximal K~ theories, and for $x \in W$ define:
$\Box x = \{B/ \Box B \in x\}$, $\boxdot\neg x = \{B/ \boxdot\neg B \in x\}$, and $[\![U]\!]x = \{B/ [\![U]\!]B \in x\}$.
Note that $[\![U]\!]x = \Box x \cap \boxdot\neg x$. Define also three relations R, S, T on W by:
Rxy iff $\Box x \subseteq y$; Sxy iff $\boxdot\neg x \subseteq y$; Txy iff $[\![U]\!]x \subseteq y$.
By the axioms, one immediately obtains:
1. $T = R \cup S$, and 2. T is reflexive, transitive, symmetric.

Take now a maximal theory x such that $A \neg\!\in x$ and consider the <u>generated</u>

model $\mathfrak{M}_x = (W_x, R_x, S_x, V_x)$, where: $W_x = \{y \in W \,/\, Txy\}$, $R_x = R \cap W_x^2$, $S_x = S \cap W_x^2$, $V_x(B) = \{y / \ y \in W_x \ \&\ B \in y\}$. Clearly, $R_x \cup S_x = W_x^2$, so $\mathfrak{M}_x$ is a generalized model, refuting A at x. #

<u>Important Lemma</u> Each generalized model $\mathfrak{M} = (W, R, S, V)$ is modally equivalent to some (generalized) model $\underline{\mathfrak{M}} = (\underline{W}, \underline{R}, \underline{S}, \underline{V})$ with $\underline{R} \cap \underline{S} = \emptyset$.
<u>Proof</u> Take a disjoint copy $\mathfrak{M}' = (W', R', S', V')$ of the initial model $\mathfrak{M}$, where $x' \in W'$ is the image of $x \in W$, and construct $\underline{\mathfrak{M}}$ as follows:

$\underline{W} =_{DF} W \cup W'$, $\underline{V}(p) =_{DF} V(p) \cup V'(p)$, for propositional variable p.

And $\underline{R}$, $\underline{S}$ are defined by cases, according to the following "important" construction (if no $\underline{R}st$ or $\underline{S}st$ is specified, then $\neg\underline{R}st$ or $\neg\underline{S}st$ is assumed):

If $Rxy \ \&\ Sxy$, then: $\underline{R}xy'$, $\underline{R}x'y$, $\underline{S}xy$, $\underline{S}x'y'$.
If $Rxy \ \&\ \neg Sxy$, then: $\underline{R}xy$, $\underline{R}xy'$, $\underline{R}x'y$, $\underline{R}x'y'$.
If $\neg Rxy \ \&\ Sxy$, then: $\underline{S}xy$, $\underline{S}xy'$, $\underline{S}x'y$, $\underline{S}x'y'$.
If $\neg Rxy \ \&\ \neg Sxy$, ..., but this cannot be the case, since $R \cup S = W^2$.

So we obtain a generalized model $\underline{\mathfrak{M}} = (\underline{W}, \underline{R}, \underline{S}, \underline{V})$. The construction also gives: $\underline{R} \cup \underline{S} = \underline{W}^2$, $\underline{R} \cap \underline{S} = \emptyset$, i.e. $\underline{R} = \underline{W}^2 \backslash \underline{S}$. On the other hand, inducting on the complexity of B, using the "construction" on the modal step, we obtain: $V(B) \cup V'(B) = \underline{V}(B)$. Since $\mathfrak{M}'$ copies $\mathfrak{M}$, $\mathfrak{M}$ and $\underline{\mathfrak{M}}$ are modally equivalent. #
<u>Note</u> We owe this Important construction to Dimiter Vakarelov (see Epilogue)

<u>Completeness Theorem</u> $K^\sim$ is complete.
<u>Proof</u> By the "Generalized" Completeness, and the Important lemma. #

<u>Theorem</u> $K^\sim$ has the finite model property, and consequently is decidable.
<u>Proof</u> It is a routine task, cf. Segerberg (1971). Take the generated generalized model $\mathfrak{M}_x$ (where x is the filter which does not contain the disprovable formula A) from the "Generalized" Completeness theorem. By the minimal filtration on $\mathfrak{M}_x$ one obtains a finite generalized model refuting A. Then the Important construction leads to the finite countermodel desired. #
<u>Consequence</u> $K^\sim$ is conservative over K, and over K^*. #

Now we come to the point. Let π_o be some index set, with $\nu \in \pi_o$, and le π be π_o's inductive closure under $\cup$, $\cap$, $-$. Let, for each $\alpha \in \pi$, N_α be a modality respecting the $K^\sim$ axioms. Let also the uniform substitution rule be assumed: $\vdash A(p)$ only if $\vdash A(B)$, for each formulae A, B, and proposition p.
<u>Theorem</u> The 5 axiom schemes $[\alpha\cup\beta]A \leftrightarrow [\alpha]A \wedge [\beta]A$, $N_{\bar{\alpha}}(A, B) \leftrightarrow N_\alpha(B, A)$, $[\![\alpha\cap\beta]\!]A \leftrightarrow [\![\alpha]\!]A \wedge [\![\beta]\!]A$, $[\nu]A \leftrightarrow [U_\alpha]\!]A$, $[\![\nu]\!]1$, in addition to the rule:

$\vdash [\beta]p \rightarrow ([\alpha]p \rightarrow [\tau]p)$ only if $\vdash [\beta]p \rightarrow ([\![\tau]\!]\neg p \rightarrow [\![\alpha]\!]\neg p)$

yield a complete axiomatization for set-theoretic union, complement, intersection, and universe ($R(\nu) = W^2$), respectively.
<u>Proof</u> Via important constructions. (We thank to D. Vakarelov for pointining an error in a previous version of this theorem.) #

While the systems K and K^* share mirror-image advantages and drawbacks, the above theorem shows that $K^\sim$ enjoys the advantages of both avoiding the typical shortcomings of either, thus presenting a necessary and sufficient basis for Boolean speculations. So from the Boolean point of view, a bi-moda language, e.g. $\mathfrak{ML}(\Box, \boxminus)$ or $\mathfrak{ML}(N)$, seems more natural to deal with, at least as natural as, say, to develop arithmetics of all natural numbers, and not the odd one of solely the odd ones. We do not specify here the expressivenes capacities of the language of $K^\sim$, leaving this job to the next section.

3. THE PREDICATE <u>loop</u> IN $K^\sim$

We extend the language of $K^\sim$ to $\mathfrak{ML}(\Box, \boxminus, \underline{loop})$ adding a propositional constant (or, possibly, a null-ary modal operator) <u>loop</u> with the semantics:

$x \models \underline{loop}$ iff Rxx .

Axioms for <u>loop</u> over $K^\sim$: $\vdash \underline{loop} \longrightarrow (\Box A \longrightarrow A)$
$\vdash \neg\underline{loop} \longrightarrow (\boxminus\neg A \longrightarrow A)$.

Theorem The axiomatics for $K^{\sim}_{LOOP}$ is sound and complete.

Proof Soundness is obvious. For the completeness we repeat the generalized canonical model construction from the "generalized" completeness theorem in the notation of which one has:

$\underline{loop} \in x$ implies Rxx, and $\neg\underline{loop} \in x$ implies Sxx.

We repeat now the important construction, modifying it only in the case when:

$x = y$ & Rxx & Sxx & $x \models \underline{loop}$.

In this case we exchange the places of $\underline{R}$ and $\underline{S}$ obtaining:

$\underline{R}xx$, $\underline{R}x'x'$, $\neg\underline{R}xx'$, $\neg\underline{R}x'x$, and

$\underline{S}xx'$, $\underline{S}x'x$, $\neg\underline{S}xx$, $\neg\underline{S}x'x'$. #

Theorem $K^{\sim}_{LOOP}$ has the fmp, and is decidable.

Proof Via the minimal filtration. #

Having $\Box$, $\boxdot$ and $\underline{loop}$ around, we reach a reliable base to express arbitrary "Boolean" modalities, which last are defined through a suitable sublanguage of $\mathfrak{L}_1$. We first take four examples, representative enough, for such modalities $St(\underline{M})$ and their expressions $\underline{M}$ in $\mathcal{ML}(\Box, \boxdot, \underline{loop})$.

$St(\underline{M}A) =$	$\underline{M}A =$
Rxx	$\underline{loop}$
$\forall y(Rxy \vee Ryy \vee \neg[y/x]St(A))$	$\boxdot\neg(\underline{loop} \vee \neg A)$
$\forall y(\neg Rxy \vee Ryy \vee \neg[y/x]St(A))$	$\Box(\underline{loop} \vee \neg A)$
$\forall y(\neg Ryy \vee [y/x]St(A))$	$[U](\neg\underline{loop} \vee A)$

$\mathfrak{L}_1$**-Sublanguage Definition** $\forall y\text{-}\mathfrak{L}_1(Rxx, Rxy, Ryy) =_{DF} \{\varphi(x) \in \mathfrak{L}_1 / \varphi$ is in prenex form with one free variable, x, at the most and one bound variable, y, at the most, the quantifier being $\forall$, and R occurs in the matrix only in Rxx, Rxy, Ryy (and not in Ryx)$\}$. #

Variations of this definition will reasonably reflect on the denotation.

Expressiveness theorem for $\mathcal{ML}(\Box, \boxdot, \underline{loop})$ Let $\varphi(x) \in Qy\text{-}\mathfrak{L}_1(Rxx, Rxy, Ryy)$, where $Q \in \{\forall, \exists\}$. Then there exists a modal formula $\varphi^{\sim} \in \mathcal{ML}(\Box, \boxdot, \underline{loop})$ with

$St(\varphi^{\sim}) = \varphi$, hence with

$\mathcal{M}, s \models \varphi^{\sim}$ iff $\mathcal{M} \models \varphi[s]$, for each model $\mathcal{M}$ and state $s \in \mathcal{M}$.

Proof We shall only construct $\varphi^{\sim}$, the remaining being left to the reader. Let $\varphi(x) \in \forall y\text{-}\mathfrak{L}_1(Rxx, Rxy, Ryy)$ and φ's matrix be in conjunctive normal form. (For an existential formula Ψ we shall have $\Psi^{\sim} = \neg(\neg\Psi)^{\sim}$.) Now, distributing the quantifier $\forall y$ over the conjuncts, we obtain φ as a conjunction of $\forall$-quantified elementary disjunctions: $\varphi = \varphi_1 \wedge \ldots \wedge \varphi_k$.

Then we define $\varphi_i^{\sim}$ taking the sample of the four examples above and, finally, set $\varphi^{\sim}$ to be $\varphi_1^{\sim} \wedge \ldots \wedge \varphi_k^{\sim}$. #

This proof leads to the above-promissed expressiveness of $K^{\sim}$.

Expressiveness theorem for $\mathcal{ML}(\Box, \boxdot)$ Drop Rxx, Ryy, $\underline{loop}$ from last theorem. #

Another small demonstration of the capacities of our language, provided the identity, or dummy, relation δ with $x\delta y$ iff $x=y$ is present, is in expressing Bull's (1968) operator Q:

$\mathcal{M} \models QA$ iff $\forall y\forall z(y \models A$ & $z \models A \longrightarrow y = z)$, and consequently

$\mathcal{M} \models Q'A$ iff $\exists! y(y \models A)$, i.e. iff $\exists y\forall z(y \models A$ & $(z \models A \rightarrow z = y))$.

We have: $QA \leftrightarrow \langle U\rangle[\![\delta]\!]A$, and

$Q'A \leftrightarrow \langle U\rangle(A \wedge [\![\delta]\!]A)$.

Questions What is the expressiveness of $\mathcal{ML}(\langle U\rangle, \delta)$, and of $\mathcal{ML}(\Box, \boxdot, \delta)$? #

Concluding Remarks. We reached in these sections a kind of universal language, in which all Boolean operations are axiomatizable, and almost all modalities are definable. Out of "almost all" remain the cases in which Ryx (y the bound variable) occurs in φ. Here we touch the converse relation $R^{\cup}$, $R^{\cup}xy$ iff Ryx, which leads out of the Boolean realm into purely relational considerations, and consequently to the familiar tense logic. Therefore, take a new modality $\Box^{\cup}$ (over K or $K^{\sim}$) with the tense semantics

$x \models \Box^{\cup}A$ iff $\forall y(Ryx \longrightarrow y \models A)$, and the usual converse axioms

$A \to \Box\Diamond^{\smile}A$,
$A \to \Box^{\smile}\Diamond A$, and
$[U]A \to \Box^{\smile}A$, if over $K^{\sim}$.

This, although not so necessary, is sufficient for axiomatization of the non-Boolean relational operations composition and converse:

$\langle \alpha \circ \beta \rangle A \leftrightarrow \langle \alpha \rangle \langle \beta \rangle A$,
$\langle \alpha^{\smile} \rangle A \leftrightarrow \langle \alpha \rangle^{\smile} A$, and $\langle \alpha^{\smile} \rangle^{\smile} A \leftrightarrow \langle \alpha \rangle A$.

For expressibility, however, although necessary, this means does not suffice: $\Box^{\smile}$ covers only the negative occurrences of Ryx in φ, the positive ones being manageable by the remaining, the fourth possible, modality $\boxplus^{\smile}$: $x \vDash \boxplus^{\smile}A$ iff $\forall y (y \vDash A \to Ryx)$. This exhausts modal Q-$\mathcal{L}_1$-definability:

<u>Q-$\mathcal{L}_1$ Expressiveness Theorem</u> For every $\varphi(x) \in$ Qy-$\mathcal{L}_1$(Rxx, Ryy, Rxy, Ryx) there is a modal formula $\varphi^{\sim} \in \mathcal{ML}(\Box, \boxplus, \underline{\text{loop}}, \Box^{\smile}, \boxplus^{\smile})$ with $St(\varphi^{\sim}) = \varphi$, hence with $\mathcal{M}, x \vDash \varphi^{\sim}$ iff $\mathcal{M} \vDash \varphi[x]$. #

In the long run, all this is aimed at explicit description of the sets: $\{St(\varphi) \in \mathcal{L}_1 / \varphi \in$ some $\mathcal{ML}\}$ and $\{St^{-1}(\Psi) \in$ some $\mathcal{ML} / \Psi \in$ some portion of $\mathcal{L}_1\}$, i.e., at first-order-definability and modal-expressibility results. And this can be embedded in, call it, "general modal program" which asks: what model condition under what truth definition responds to what modal axiom, where "responds to" means "guarantees" or "is guaranteed by", or both. This general modal program is in the spirit of van Benthem's (1984) "perhaps most basic question" concerning the interplay of the two 'degrees of freedom' in semantic explanation: truth definition and model condition, leaving a third parameter free - the modal axioms to be satisfied.

4. MODELS WITH NAMES FOR POSSIBLE WORLDS

The above expressiveness results state a relationship between one unmovable predicate language $\mathcal{L}_1$ and several flexile modal languages. On the modal side we examine also <u>loop</u> which at first glance appears to be a modal counterpart of the usual first-order equality =, and so it appeals to the "modal program" for the source language $\mathcal{L}_1$ to be replaced by its "equalized" version $\mathcal{L}_1^{=}$. Revised like this, the modal program however immediately fails: <u>loop</u> contains only an equality relevant to R. Indeed, take the simplest $\mathcal{L}_1^{=}$-formula $\Psi = \forall y(x=y)$, whose truth in a (state of) model or frame is equivalent with the universe's cardinality to be 1. There is no modal formula $\varphi \in \mathcal{ML}(\Box, \boxplus, \underline{\text{loop}}, \Box^{\smile}, \boxplus^{\smile})$ with $\forall \mathcal{M}(\mathcal{M} \vDash \varphi$ iff $card(\mathcal{M}) = 1)$. (For, e.g., the models $(\{x\}, \emptyset, V)$ and $(\{y,z\}, \emptyset, U)$, where $V^{-1}(x) = U^{-1}(y) = U^{-1}(z)$, are modally indiscernible in that language.)

In fact, it can hardly be expected to express the equality of states while no special means are available identifying the states themselves. All we have at our disposal in $\mathcal{ML}$'s are propositional letters interpreted as subsets of the universe, $V(p) \subseteq W$, and not even a syntactical hint is there for particular individuals the equality of which is the target. In this section we enhance the expressive power of $\mathcal{ML}$'s by adding to the syntax names for the states (or constants) with the natural for "name" semantics, and appropriate axiomatics. Such move is, modulo traditional virtues, quite a natural one: the traditional modal theory of anonymous worlds looks as unnatural as, say, an arithmetics in the language of which predicates are only available (e.g. 'even', 'odd', 'prime', '=0', etc) instead of individual variables. Thus we continue our works initiated in Passy & Tinchev (1985a,b) and settle in a modal background an idea (called "combinatory") which proved curious, if not even useful, in the ambience of the dynamic logic.

<u>Definition</u> The language $\mathcal{ML}_N(\Box)$ of named models contains two sorts of propositional variables: ordinary ones $p_1, p_2, \ldots$ and names (or, constants) $c_1, c_2, \ldots$ Formulae are built starting from variables and names applying the Boolean connectives and the modalities $\Box$ and $\Diamond$. #

Semantics. Models for $\mathcal{ML}_N$ are triples (W, R, V) where (W, R) is a frame, and the valuation V, besides the ordinary truth conditions, satisfies also: $V(c) \subseteq W$ is either empty or a singleton, for each name c. #

In modal-axiomatic capacities $\mathcal{ML}_N(\Box)$ is closer to $\mathcal{ML}(\Box, \boxminus)$:

1. R is irreflexive iff $(W, R) \models (c \to \Box\neg c)$.
2. In three-relational case,
   $R = S \cap T$ iff $(W, R) \models \langle R\rangle c \leftrightarrow \langle S\rangle c \wedge \langle T\rangle c$.

However, we leave open the following

Questions Describe, in the spirit of Goldblatt & Thomason's theorems, the classes of frames modal-axiomatic over the respective $\mathcal{ML}$'s and $\mathcal{ML}_N$'s. #

Definition A model is total, if $\forall c(V(c) \neq \emptyset)$, i.e. when each name names some world. A model is surjective, if $\forall x_{x\in W}\exists c(V(c) = \{x\})$, i.e. if all worlds have names. A model is standard, if it is both total and surjective. These notions yield respectively total, surjective and standard validity, denoted by $\models_{TOT}$, $\models_{SUR}$, $\models_{STAND}$. #

Notes 1. The original "combinatory" models from our previous papers are standard, even very standard having an extra S5 modality [U] interpreted as the Cartesian square of the universe. The very standard language suffices for modal-axiomatization of one-world universes - fixing a name c, one has: card($\mathcal{M}$) = 1 iff $\mathcal{M} \models [U]c$.
2. Surjective (hence also standard) models are based on frames which are at most countable. On finite and countable frames validity and surjective validity coincide. On uncountable frames surjective validity is trivially fulfiled: all formulae are of course valid. #

We extend the translation St: $\mathcal{ML}(\Box) \longrightarrow \mathcal{L}_1$ to ST: $\mathcal{ML}_N(\Box) \longrightarrow \mathcal{L}_1^{=}$ defining $ST(c_i) = (x = y_i)$, for each name c_i, where $y_1, y_2, \ldots$ are, say, "half" of the individual variables of $\mathcal{L}_1^{=}$, and other than x.

Fact $(W, R) \models_{TOT} A$ iff $(W, R) \models \forall P_1 \ldots \forall y_1 \ldots \forall x ST(A)$. #

In order to axiomatize the set K_N of valid formulae we introduce, following Goldblatt (1982), the notions of necessity form ($\Box$-form) and possibility form ($\Diamond$-form).

Definition 1. \$ is a $\Box$-form. If L is a $\Box$-form, A - a formula, then 2. $A \longrightarrow L$ is a $\Box$-form, and 3. $\Box L$ is a $\Box$-form.

1'. \$ is a $\Diamond$-form. If M is a $\Diamond$-form, A - a formula, then 2'. $A \wedge M$ is a $\Diamond$-form, and 3'. $\Diamond M$ is a $\Diamond$-form. #

Each form L or M has a unique occurrence of the symbol \$; if it is replaced by a formula A of $\mathcal{ML}_N(\Box)$ a formula results, denoted by L(A) or M(A).

Axiomatics of K_N. We add to the deductive system of K over $\mathcal{ML}_N(\Box)$, also:
Ax_N. $M(c \wedge A) \longrightarrow L(c \longrightarrow A)$, for each name c, $\Diamond$-form M, and $\Box$-form L, which reflects the behaviour of V(c).

Fact If $\vdash A \longrightarrow B$, them $\vdash L(A) \longrightarrow L(B)$ and $\vdash M(A) \longrightarrow M(B)$. #

Theorem The axiomatics for K_N is sound and complete.

Proof Let, for the completeness, A be disprovable. Consider the standard canonical model construction and let x be a maximal theory such that $\neg A \in x$. Take the submodel (W_x, R_x, V_x) generated by x. We have:

Lemma For a name c, and states $y, z \in W_x$, $c \in y$ & $c \in z$ imply $y = z$.

Proof of the lemma Assume the contrary, i.e. $y \neq z$. Then there is B such that: $B \in y$ & $\neg B \in z$. So $c \wedge B \in y$ and since $y \in W_x$, for some $\Diamond$-form M, $M(c \wedge B) \in x$. By Ax_N $L(c \longrightarrow B) \in x$, for any $\Box$-form L. But then clearly $c \longrightarrow B \in z$. Thus $B \in z$ - a contradiction.

By the above, (W_x, R_x, V_x) is a model where A is refuted. #

The notion of surjective validity can be captured by the following axiomatic system K_N^{SUR}: add to K_N the (infinitary) inference rule

COV If $\vdash L(\neg c)$, for all names c, then $\vdash L(0)$ (here L is any $\Box$-form).

$\underline{\text{Soundness theorem for } K_N^{SUR}}$. If $K_N^{SUR} \vdash A$, then $\models_{SUR} A$.
Proof The rule COV preserves truth in surjective models. #

Notes 1. In fact, the rule COV is not infinitary; it is interchangeable with COV^*: If $\vdash L(\neg c)$, for some c not occurring syntactically in L, then $\vdash L(0)$.
2. Both Ax_N and COV look more neat in the context of dynamic logic, where no explicit reference to □- and ◇-forms is needed (α, β are PDL programs):
(Ax_N') $\langle\alpha\rangle(c\wedge A) \longrightarrow [\beta](c \rightarrow A)$
COV' If $\vdash [\alpha]\neg c$, for all c, then $\vdash [\alpha]0$. #

Next we give completeness proof for K_N^{SUR} which proceeds in two steps. On the first step, as usual, we place the disprovable formula in a maximal theory, cf. e.g. Goldblatt (1982), or Rasiowa-Sikorski's (1963) lemma on Q-filters. Secondly, instead of taking other maximal theories, we use the names in a typical Henkin way to build the counter-model.
Definition A theory is any set of formulae containing all K_N^{SUR} theorems which is closed under MP and COV. Unless otherwise specified, theories will be consistent. For a set of formulae X let Th(X) be the smallest theory containing X, and let $Th(X, A) =_{DF} Th(X \cup \{A\})$. #
Deduction lemma $B \in Th(T, A)$ iff $A \longrightarrow B \in T$, for each theory T. #
Lindenbaum lemma Any theory T can be extended to a maximal one.
Proof Enumerate all formulae $A_0, A_1, \ldots$ Let T_0 be T. Assume T_n defined and consistent. If $Th(T_n, A_n)$ is consistent, then $T_{n+1} =_{DF} Th(T_n, A_n)$.
If no, then study the graphical form of A_n: a) if $A_n = L(0)$ for some □-form L, then for at least one name c, $L(\neg c) \neg\in T_n$ – otherwise, by the COV-closeness of T_n we would get $A_n \in T_n$. (By the Deduction lemma this leads to a contradiction.) In this case let $T_{n+1} = Th(T_n, \neg L(\neg c))$.
b) if A has any other graphical form, then let $T_{n+1} = T_n$.
This construction produces an infinite chain of growing theories. Their union T' is a theory, too, and moreover T' is maximal theory containing T. #

For a maximal theory T, let $N_T =_{DF} \{c/ M(c) \in T$ for some ◇-form M$\}$.
For $c,d \in N_T$, let $c \sim d$ iff for some M, $M(c\wedge d) \in T$.
Lemma (maximal theory forms) Where M is a ◇-form, L – a □-form, and $c \in N_T$.
(*) $M(c\wedge A) \in T$, for some M iff $L(c \longrightarrow A) \in T$, for all L.
(*) $M(c\wedge A) \neg\in T$, for all M iff $M'(c\wedge\neg A) \in T$, for some M'.
(*) $M(c\wedge\Diamond A) \in T$ iff $M(c\wedge\Diamond d) \in T$ & $M'(d\wedge A) \in T$, for some M' and $d \in N_T$.
Proof By Ax_N and COV-closeness of T. #
Lemma $\sim$ is an equivalence relation on N_T .
Proof Use the above lemma. #

Definition For a maximal theory T, let $\mathcal{M}_T = (W_T, R_T, V_T)$, where $W_T = N_T/\sim$, $R_T = \{(|c|, |d|)/ \exists M(M(c\wedge\Diamond d) \in T)\}$, and $V_T = \{|c| / \exists M(M(c\wedge A) \in T)\}$. #
Henkin model lemma $\mathcal{M}_T$ is a surjective model.
Proof The truth conditions have to be checked, i.e. $|c| \in V_T(\Box A)$ iff $\forall|d|(R_T|c||d|$ implies $|d| \in V_T(A))$ etc, and they follow from the lemma about forms. Surjectivity is clear: $V_T(c) = \{|c|\}$, for $|c| \in W_T$. #
$\underline{\text{Completeness theorem for } K_N^{SUR}}$ If $\models_{SUR} A$, then $K^{SUR} \vdash A$.
Proof For a disprovable formula A, there is a maximal theory T with $\neg A \in T$, and it can be easily verified that A is refuted in the model $\mathcal{M}_T$. #

The next proves the rule COV redundant for this basic system (but not for the extensions).
Lemma K_N has the finite model property.
Proof Standard filtration. #

Note now that the finite model refuting A (obtained by the above filtration) can be transformed into a surjective model (refuting A) by redefining the valuations of names not occurring in A. Moreover, the model can be "totalized" (hence "standardized"), by adding, if necessary, one new world in order to ensure totality. Thus we have

Theorem K_N and K_N^{SUR} coincide as sets of theorems. #

Corollaries 1. Both K_N and K_N^{SUR} are decidable.

2. For a fixed formula, all four kinds of validity coincide, and the problem whether it is valid is decidable. #

Adding names with standard interpretation to the language of $K^\sim$, one obtains a very standard language $\mathfrak{ML}_N(\Box,\boxdot)^{STAND}$ in which the equality can be spoken of. We have, for a model $\mathfrak{M}$ with $V(c) = s$ and $V(d) = t$, that $s = t$ iff $\mathfrak{M} \models \langle U\rangle(c \wedge d)$. Indeed, identifying individuals, the language $\mathfrak{ML}_N(\Box,\boxdot)^{STAND}$ serves as the modal analogue sought, of the first-order language with equality $\mathfrak{L}_1^-$, and one has:

Expressiveness theorems for named modal languages Let us replace, in the expressiveness theorems, section 3, $\mathfrak{L}_1$ by $\mathfrak{L}_1^-$, and the $\mathfrak{ML}$'s - by the respective very standard $\mathfrak{ML}_N$'s. These are still the cases. #

Such a nice language deserves to be axiomatized. We propose the following axioms and rules for $K_N^{\sim\ STAND}$.

1. Axioms and rules of $K^\sim$ (over the new language)
2. $\langle U\rangle(c \wedge A) \longrightarrow [U](c \longrightarrow A)$ (instead of Ax_N)
3. $\langle U\rangle c$ (guaranteeing totality)
4. $\Diamond c \leftrightarrow \boxdot c$ (the implication $\leftarrow$ is in fact a theorem)
5. The rule COV for the suitably extended notion of a modal form.

Standard-completeness theorem $\vdash_{STAND} A$ iff $\models_{STAND} A$.

Proof Soundness is clear. If A is not a "standard" theorem, then the Henkin model construction from above will extract a model (W_T, R_T, V_T) out of a maximal theory not containing A. Here W_T is the set of all names factorized by the relation $\langle U\rangle(c \wedge d) \in T$. Further $R_T =_{DF} \{(|c|, |d|) / \langle U\rangle(c \wedge \Diamond d) \in T\}$, and $V_T(B) =_{DF} \{|c| / \langle U\rangle(c \wedge B) \in T\}$. Axiom 4 guarantees the correct relationship between $\Box$ and $\boxdot$. Axiom 3 yields totality of the model. So we have a standard model where A is refuted. #

Concluding remarks. The names on the modal soil provide an effective tool for a first-order quantification, cf. Passy & Tinchev (1985b). Let the Quantified $\mathfrak{ML}_N$ extend the respective $\mathfrak{ML}_N$, allowing, on the inductive step, formulae of the type $\forall cA$, where c is a name, and A is a formula with the semantics: $\mathfrak{M}, s \models \forall cA$ iff for each d, $\mathfrak{M}, s \models [d/c]A$. Thus at long last we reach on a modal level the expressibility of the entire $\mathfrak{L}_1^-$ language.

The $\mathfrak{L}_1^-$-Expressiveness Theorem For every formula φ in $\mathfrak{L}_1^-$ of one free variable there is a closed formula $\varphi^\sim$ in Quantified $\mathfrak{ML}_N(\Box,\boxdot)^{STAND}$ with: $\mathfrak{M} \models \varphi[s]$ iff $\mathfrak{M}, s \models \varphi^\sim$.

Proof Let φ be in prenex form with bound variables $y_1, y_2, \ldots$, and free variable x. Each atomic subformula of φ's matrix has one of the forms $z=y$, Rzy, $P(z)$ or the negations of these, where $z, y \in \{x, y_1, y_2, \ldots\}$. Let Ψ be the quantified modal formula obtained by uniform replacement of all variables z in φ's prefix by a name c_z, and all occurrences of $z=y$, Rzy, $P(z)$ in φ's matrix by $\langle U\rangle(c_z \wedge c_y)$, $\langle U\rangle(c_z \wedge \langle R\rangle c_y)$, $\langle U\rangle(c_z \wedge p)$, respectively. Finally, define $\varphi^\sim$ to be $\exists c_x(c_x \wedge \Psi)$. #

Another impact of the names is in first-order definability. Call an $\mathfrak{ML}_N$-formula <u>pure</u> if it does not contain propositional variables, i.e. consists only of names, 0 and connectives. Clearly, in terms of van Benthem (1977), pure formulae are first-order definable. In particular, the "pure" instant $\Box\Diamond c \to \Diamond\Box c$ of the famous first-order undefinable formula $\Box\Diamond p \to \Diamond\Box p$, is already $\mathfrak{L}_1^-$-definable, via the sentence: $\forall z(\forall y(Rxy \to Ryz) \to \exists y(Rxy \ \& \ \forall t(Ryt \to t=z)))$.

Theorem If A is a pure formula, then A is complete.

Sketch of the proof A defines a first-order condition φ true in every frame where A is standardly valid (frames are at most countable). Now translating φ into $\varphi^\sim \in$ Quantified $\mathfrak{ML}_N(\Box, \boxdot)^{STAND}$ we can obtain a quantified theory containing $\varphi^\sim$ which is conservative over standard $K_N^\sim + \{A\}$. Now the Henkin model of the theory is based on a frame for which we can check φ. #

Conjecture If A is first-order definable, then A is complete. #

In the long run, all this aims at a "general named modal program", which is a matter of another, probably longer, discussion.

EPILOGUE: The Ghost of the Modality vs. the Spirit of Kripke

The present paper gives another flavour to the series of "combinatory" investigations of the three of us, initiated in Passy (1984), Passy & Tinchev (1985a,b), Tinchev (1986). The names in the modal logic give a satisfactory solution to some problems, and supplying the deficiences of Kripke nature, they as if put a (first-) order in the modal atonality. The combinatory solutions given, however, do not explain the reason and the entity of this last. Such an explanation in the person of K^*, together with some other solutions among which $K^\sim$, $K^\sim_{LOOP}$, $K^\sim_N$ is probably what is gained here.

In particular, we claim that K^* presents itself as an equipollent counterpart of K, and being such it gives (a partial, at least) answer to van Benthem's (1984, p. 385) query about truth definitions alternative to Kripke's and working equally well (hence, equally bad). So it is not to be expected the idea of K^* to be a novelty in the modal field, and the referees agree on this: the semantics of $\boxplus$ repeats the semantics for negation in quantum logic given in Goldblatt (1974); moreover, our translation * and the Correspondence theorem have their similitudes in Goldblatt's definition 4.2 and lemma 4.3. Reportedly, Humberstone has also axiomatized similar modalities. D. Vakarelov (1974) has a semantics for the negation which is the semantics of $\Box\neg$ from K. The essay however closest to our own also belongs to van Benthem (1979), where he studies some modal operators in a deontic context. Modulo philosophical background, our K^* and $K^\sim$ turn to be van Benthem's K_d - logic of permissions and K_D - permissions & obligations, his mixing principle being a theorem of $K^\sim$. So the completeness theorems for K^* and $K^\sim$, up to the proof strategies, can be attributed to van Benthem.

Concerning the completeness proof strategy presented for $K^\sim$ and $K^\sim_{LOOP}$, we use the construction called Important, which was invented by Dimiter Vakarelov (for some other completeness results). This construction replaces one of our own, which - although of smaller size - is less transparent. And all these constructions originate probably from Sahlqvist. Concluding the first part of the paper, we frame the largely discussed in modal logic "reflexivity" by adding the <u>loop</u> predicate. Such a step can be thought of as inspired by the familiar in dynamic logic constructs δ (identity program), or by cycle-predicate or by iteration, and when so it also appears not to be something very original. Moreover, the natural extensions of $K^\sim_{LOOP}$ from section 3 lead to the ordinary tense logic, as notes one of the referees. Thus the first part of the paper suggests just a new arrangement of common notions and is only a step towards fulfilling the "general modal program".

In the second part, we fuse together these and the idea of the names, which, cf. the Postscriptum of Passy (1984), on its part can also be thought of as rearrangement of folklore speculations. Hence we join here some roving notions proving some theorems for them, some of which one may even find inadmissibly simple. However, our aim is not at all notions' introduction or theorems' proving: such are quite abundant in modal logic. Our goal is to propose another viewpoint towards the latter.

The viewpoint sought is, as opposed to Kripke's, not discriminating first-order phenomena when regarded thereof. Consequently introducing several $\mathfrak{ML}$'s and their named versions, we are approaching, and hopefully encircling, Hermann Weyl's 1940 "ghost of the modality". The question unfortunately still remains open: which of all these logics is better, and is there a best one, with "better" and "best" referring to some aesthetical order predicted by the pure mathematical reason.

ACKNOWLEDGEMENTS

Van Benthem, Goldblatt, Segerberg, Vakarelov, the two referees: these six, or at least four, persons, incidentally members of the same Program Committee, contributed - probably unknown to them - in some way to our understanding of modality. Rumi Draganova, Christo Kazasov, Ivan Soskov and, of course, Kuzyo Kuzev were very helpful, and made our work on the computer more than a pleasure. Special thanks are due to Iliana Sherkova, for polishing the language and encouraging the style of this paper.

REFERENCES

van Benthem, J.F.A.K., 1977, Modal Logic as Second-Order Logic, Report 77-04, Dept. of Mathematics, Univ. of Amsterdam, March.

van Benthem, J.F.A.K., 1979, Minimal Deontic Logics (abstract), Bull. of Sec. of Logic, 8, No.1 (March), 36-42.

van Benthem, J.F.A.K., 1984, Possible Worlds Semantics: A Research Prorgam that Cannot Fail?, Studia Logica, 43, No.4, 379-393.

Bull, R.A., 1968, On Possible Worlds in Propositional Calculi, Theoria, 34.

Goldblatt, R.I., 1974, Semantic Analysis of Orthologic, J. Philos. Logic, 3, 19-35.

Goldblatt, R.I., 1982, Axiomatizing the Logic of Computer Programming, Springer LNCS 130, Berlin.

Goldblatt, R.I. & S.K. Thomason, 1975, Axiomatic Classes in Propositional Modal Logic, in: Springer LNM 450, 163-173.

Humberstone, I.L., 1983, Inaccessible Worlds, Notre Dame J. of Formal Logic, 24, No.3 (July), 346-352.

Humberstone, I.L., 1985, The Formalities of Collective Omniscience, Philos. Studies 48, 401-423.

Passy, S.I., 1984, Combinatory Dynamic Logic, Ph.D. Thesis, Mathematics Faculty, Sofia Univ., October.

Passy, S. & T. Tinchev, 1985a, PDL with Data Constants, Inf. Proc. Lett., 20, No.1, 35-41.

Passy, S. & T. Tinchev, 1985b, Quantifiers in Combinatory PDL: Completeness, Definability, Incompleteness, in: Springer LNCS 199, 512-519.

Rasiowa, H. & R. Sikorski, 1963, Mathematics of Metamathematics, PWN, Warsaw.

Segerberg, K., 1971, An Essay in Classical Modal Logic, Uppsala Univ.

Tehlikeli, S. (S. Passy), 1985, An Alternative Modal Logic, Internal Semantics and External Syntax (A Philosophical Abstract of a Mathematical Essay), manuscript, December.

Tinchev, T.V., 1986, Extensions of Propositional Dynamic Logic (in Bulgarian), Ph.D. Thesis, Mathematics Faculty, Sofia Univ., June.

Vakarelov, D., 1974, Consistency, Completeness and Negation, in: Essays on Paraconsistent Logics, Philosophia Verlag, to appear.

Weyl, H., 1940, The Ghost of the Modality, in: Philosophical Essays in Memory of Edmund Husserl, Cambridge (Mass.), 278-303.

ADDED IN PROOF

On November 14, 1986, after having prepared the above - then hoped to be final - version of this exhaustible typescript, we received offprints of the papers of Lloyd Humberstone mentioned, kindly sent to us by the author. In the former paper, essentially, $K^{\sim}$ is considered: by curious coincidence, on one hand, the axioms suggested there are exactly as van Benthem's mixing principles, and, on the other, the completeness proof goes through Beth's semantic tableaus (as this also is claimed in van Benthem's (1979) abstract); some other valuable observations on "complementary" modalities are stated. In the latter paper, the "intersection" of modalities is mentioned, in an epistemic context. We do regret for not having these worthy papers earlier.

DISTRIBUTIVE SPACES

Ljubomir L. Ivanov

Faculty of Mathematics
Sofia University
1126 Sofia, Bulgaria

INTRODUCTION

There are two different classical concepts of relative effective computability in Ordinary Recursion Theory (cf. [5,7]), namely μ-recursiveness and partial recursiveness. The latter can be described either (i) as μ-recursiveness in a certain multiple-valued function, say $U = \lambda s.\{2s, 2s+1\}$, or (ii) as existentially quantified primitive recursiveness.

The abstract notion of recursiveness in operative spaces [1-4] generalizes, among others, the μ-recursiveness and the prime computability of Moschovakis. Here we introduce and study the so called distributive operative spaces in which it is possible to generalize also the ordinary partial recursiveness and Friedman's computability by effectively definitional schemes. Our abstract definition of partial recursiveness is modeled on (i), while an abstract analog to (ii) is the central result of this work.

1. DISTRIBUTIVE SPACES: DEFINITIONS

Operative spaces are algebraic systems $\mathcal{S} = (\mathcal{F}, I, \Pi, L, R)$, where $\mathcal{F} = (\mathcal{F}, \cdot, \leq)$ is a partially ordered semigroup with a unit I, L and R are fixed distinct members of $\mathcal{F}$ and Π is a monotone binary operation over $\mathcal{F}$ called pairing, s.t. $(\varphi, \psi)\cdot\chi = (\varphi\cdot\chi, \psi\cdot\chi)$, $L\cdot(\varphi, \psi) = \varphi$ and $R\cdot(\varphi, \psi) = \psi$ for all $\varphi, \psi, \chi \in \mathcal{F}$, writing (φ, ψ) for $\Pi(\varphi, \psi)$.

In this paper we consider spaces with a zero O which is also the bottom element, and a distributor U s.t., writing $\varphi + \psi$ for $U\cdot(\varphi, \psi)$ and L_1, R_1 respectively for (I, O), (O, I), the following additional axioms are satisfied:

$$L + (L\cdot R + R^2) = (L + L\cdot R) + R^2,$$
$$L + R = R + L,$$
$$(L, R) = L_1\cdot L = R_1\cdot R.$$

Multiplying to the right by (φ, ψ, χ) (right grouping of brackets) or (φ, ψ), one gets the more general equalities

$$\varphi + (\psi + \chi) = (\varphi + \psi) + \chi,$$
$$\varphi + \psi = \psi + \varphi,$$
$$(\varphi, \psi) = L_1 \cdot \varphi + R_1 \cdot \psi.$$

We assume without loss of generality that $U = U \cdot (L, R)$, i.e. $U = L + R$.

An element φ is <u>distributive</u> iff $\varphi \cdot (L + R) = \varphi \cdot L + \varphi \cdot R$; equivalently, iff $\varphi \cdot (\psi + \chi) = \varphi \cdot \psi + \varphi \cdot \chi$ for all ψ, χ. The space $\mathcal{G}$ itself is <u>distributive</u> iff so are all φ in $\mathcal{F}$.

We further assume that $\mathcal{G}$ is <u>iterative</u>, which in the present work will mean that the following μ-induction principle holds:

(@) Every mapping $\Gamma: \mathcal{F} \to \mathcal{F}$ has a fixed point $\mu\theta.\Gamma(\theta)$ which belongs to all subsets $\mathcal{E}$ of $\mathcal{F}$ closed under Γ, provided $0 \in \mathcal{E}$, $\mathcal{E}$ is of the form $\{\theta / \forall n(\Gamma_n(\theta) \leq \tau_n)\}$, Γ and all Γ_n are of the form $\lambda\theta.(\varphi + \psi \cdot \theta \cdot \chi)$ (equivalently, of the form $\lambda\theta.\varphi \cdot (\psi, \theta \cdot \chi)$).

It follows easily that the element $\mu\theta.\Gamma(\theta)$ in question is the least fixed point of Γ and also the least solution to the inequality $\Gamma(\theta) \leq \theta$. (Supposing $\Gamma(\tau) \leq \tau$, show that $\mathcal{E} = \{\theta / \theta \leq \tau\}$ is closed under Γ; this set is of the required form since $0 + \theta = \theta$ by the proof of 2.2.)

The operations $\langle\rangle = \lambda\varphi.\mu\theta.(\varphi \cdot L, \theta \cdot R)$ and $[\,] = \lambda\varphi.\mu\theta.(I, \varphi \cdot \theta)$ are called respectively <u>translation</u> and <u>iteration</u>. Together with $\cdot$, $\sqcap$ they form a complete collection, because all least fixed points of mappings constructed by means of $\cdot$, $\sqcap$, $\langle\rangle$, $[\,]$ are expressible by the same operations. (The axiom (@) actually guarantees this by proposition 3 [2] a detailed proof to which is given in [1].) The operation $\Delta = \lambda\varphi\psi.\mu\theta.(\varphi, \theta \cdot \psi) = \lambda\varphi\psi.\langle\varphi\rangle \cdot [\langle\psi\rangle]$ is called <u>primitive</u> <u>recursion</u>. Translation, primitive recursion and iteration will be described more explicitly in section 3.

2. DISTRIBUTIVE SPACES: EXAMPLES

In this section we give several examples of distributive spaces and suggest some methods for constructing such spaces.

Distributive spaces are oftenly introduced via their companion semirings. By <u>semiring</u> we mean a triple $(\mathcal{F}, +, \cdot)$, where $\mathcal{F}$ is a set, $+, \cdot$ are associative binary operations over $\mathcal{F}$, $+$ is commutative and the distributive laws $(\varphi + \psi) \cdot \chi = \varphi \cdot \chi + \psi \cdot \chi$, $\chi \cdot (\varphi + \psi) = \chi \cdot \varphi + \chi \cdot \psi$ hold; assumed are also zero 0, unit I and fixed elements L, R, L_1, R_1 s.t. $L \cdot L_1 = R \cdot R_1 = I$ and $L \cdot R_1 = R \cdot L_1 = 0$. Notice that whenever $(\mathcal{F}, +, \cdot)$ is a semiring, then so is $(\mathcal{F}, +, \lambda\varphi\psi.\psi \cdot \varphi)$.

<u>Proposition 2.1</u>. Let $(\mathcal{F}, +, \cdot)$ be a semiring. Set $\varphi \leq \psi$ iff $\exists\chi \cdot (\varphi + \chi = \psi)$, and $\sqcap = \lambda\varphi\psi.(L_1 \cdot \varphi + R_1 \cdot \psi)$. Then $\mathcal{G} = ((\mathcal{F}, \cdot, \leq), I, \sqcap, L, R)$ is a distributive space with distributor $U = L + R$.

Follows by the corresponding definitions.

Comments. The semigroup $(\mathcal{F}, \cdot, \leq)$ should in general be factorized, because $\leq$ is a quasi-order rather than a partial order.

Conversely, all distributive spaces have companion semirings.

Proposition 2.2. Whenever $\mathcal{G} = (\mathcal{F}, I, \Pi, L, R)$ is a distributive space, then $(\mathcal{F}, +, \cdot)$ is a semiring.

We adduce only the proof to $\varphi + 0 = \varphi$:

$$\varphi + 0 = L\cdot L_1\cdot\varphi + L\cdot R_1 = L\cdot(L_1\cdot\varphi + R_1) = L\cdot(\varphi, I) = \varphi.$$

Comments It follows that in distributive spaces $\varphi = \varphi + 0 \leq \varphi + \chi$ for all χ. If we call dense those spaces in which $\varphi \leq \psi$ iff $\exists\chi(\varphi + \chi = \psi)$, then not even all iterative distributive spaces are dense, e.g. the dense spaces of examples 1-3 below have minimal iterative distributive subspaces (consisting of all elements partial recursive in the sense of section 3) which are not dense.

Let us return to the construction of 2.1. In order to obtain an iterative $\mathcal{G}$ it suffices to assume that the given semiring is complete, i.e. $\Sigma_n\varphi_n = \sup_n\Sigma_{i\leq n}\varphi_i$ always exists and $\Sigma_n\varphi_n + \Sigma_n\psi_n = \Sigma_n(\varphi_n + \psi_n)$, $(\Sigma_n\varphi_n)\cdot\psi = \Sigma_n(\varphi_n\cdot\psi)$ and $\psi\cdot\Sigma_n\varphi_n = \Sigma_n\psi\cdot\varphi_n$ take place.

Proposition 2.3. Let $(\mathcal{F}, +, \cdot)$ be a complete semiring. Then the space $\mathcal{G}$ constructed by 2.1 is iterative.

Proof. (Outline) Suppose that $\Gamma = \lambda\theta.(\varphi + \psi\cdot\theta\cdot\chi)$, $\mathcal{E}$ is of the form specified in (@), $0 \in \mathcal{E}$ and $\Gamma(\mathcal{E}) \subseteq \mathcal{E}$. Take by definition $\mu\theta.\Gamma(\theta) = \Sigma_n\psi^n\cdot\varphi\cdot\chi^n$, then it follows by making use of the semiring completeness that $\Gamma(\mu\theta.\Gamma(\theta)) = \mu\theta.\Gamma(\theta)$. An easy induction on n gives that $\Sigma_{i\leq n}\psi^i\cdot\varphi\cdot\chi^i \in \mathcal{E}$ for all n, which implies $\mu\theta.\Gamma(\theta) \in \mathcal{E}$ by the semiring completeness. Thereby the proof is completed. (Compare with Remark 1 of [8], chapter 3, section 1.1.)

The following three examples correspond to examples 2, 17 and a modification of example 13 [8], chapter 3.

Example 1 (relational). Take an arbitrary infinite set M with a splitting scheme f_1, f_2 (injective functions f_1, f_2: $M \to M$ with disjoint ranges), then take $\mathcal{F} = \{\varphi/\ \varphi \subseteq M^2\}$, $\varphi + \psi = \varphi \cup \psi$, $\varphi\cdot\psi = \lambda s.\psi(\varphi(s))$ (relations = multiple-valued functions), $0 = \emptyset = \lambda s.\uparrow$, $I = \lambda s.s$, $L = f_1$, $R = f_2$, $L_1 = f_1^{-1}$ and $R_1 = f_2^{-1}$. The space $\mathcal{G}$ is obtained from the complete semiring $(\mathcal{F}, +, \cdot)$ by 2.3.

Example 2 (fuzzy). Take M, f_1, f_2 as above and a complete lattice E s.t. $a \wedge \vee E_1 = \vee\{a \wedge b/\ b \in E_1\}$ for all $a \in E$, $E_1 \subseteq E$, then take $\mathcal{F} = \{\varphi/\ \varphi: M^2 \to E\}$, $\varphi + \psi = \lambda st.\varphi(s, t) \vee \psi(s, t)$, $\varphi\cdot\psi = \lambda st.\vee_r\varphi(s, r) \wedge \psi(r, t)$, $I(s, s) = \top = \vee E$ and $I(s, t) = \bot = \wedge E$ otherwise, $0 = \lambda st.\bot$, $L = \lambda st.I(f_1(s), t)$, $R = \lambda st.I(f_2(s), t)$, $L_1 = L^{-1} = \lambda st.L(t, s)$ and $R_1 = R^{-1}$. The space $\mathcal{G}$ is obtained from the complete semiring $(\mathcal{F}, +, \cdot)$ by 2.3.

Example 3 (probabilistic). Take M, f_1, f_2 as in example 1, then take $\mathcal{F} = \{\varphi/\ \varphi: M^2 \to [0,\infty]\}$, $\varphi + \psi = \lambda st.\varphi(s, t) + \psi(s, t)$, $\varphi\cdot\psi = \lambda st.\Sigma_r\varphi(s, r)\psi(r, t)$, $0 = \lambda st.0$, $I(s, s) = 1$ and $I(s, t) = 0$ otherwise, $L = \lambda st.I(f_1(s), t)$, $R = \lambda st.I(f_2(s), t)$, $L_1 = L^{-1} = \lambda st.L(t, s)$ and $R_1 = R^{-1}$. The space $\mathcal{G}$ is obtained from the complete semiring $(\mathcal{F}, +, \cdot)$ by 2.3 again.

Comments We do not adduce the necessary verifications that the structures $(\mathcal{F}, +, \cdot)$ in hand are complete semirings; arguments to this effect can be found in [4] or [8]. Example 1 is isomorphic with the subspace of example 2 consisting of all $\bot,\top$-valued elements. Besides that, example 1 is example 2 with $E = \{\bot, \top\}$. An aspect in which examples 1, 2 differ from example 3 is that $I + I = I$ holds in the former but fails in the latter. Of course, this equality is equivalent to $\forall\varphi(\varphi + \varphi = \varphi)$.

Properly probabilistic is the iterative subspace of example 3 consisting of all $[0, 1]$-valued elements, but it is not distributive. An interesting iterative distributive subspace is that of all $\omega \cup \{\infty\}$-valued elements. Intuitively, $\varphi(s, t) \in \omega \cup \{\infty\}$ is the number of computational paths from s to t in the processing of the nondeterministic program φ, while in general $\varphi(s, t) \in [0, \infty]$ can be interpreted [8] as an average number of such paths.

There are non-distributive spaces with distributors, e.g. the space $\mathcal{G}'$ obtained from a given distributive space $\mathcal{G}$ by the construction of proposition 8 [2] (19.10 [4]). However, spaces with distributors can be shown under certain assumptions to have maximal distributive subspaces.

<u>Proposition 2.4</u>. Let $\mathcal{G} = (\mathcal{F}, I, \sqcap, L, R)$ be an iterative distributive space. Take $\mathcal{F}^\sim = (\mathcal{F}, \lambda\varphi\psi.\psi\cdot\varphi, \leq)$, $\sqcap^\sim = \lambda\varphi\psi.(\varphi\cdot L + \psi\cdot R)$, $L^\sim = (I, O)$ and $R^\sim = (O, I)$. Then $\mathcal{G}^\sim = (\mathcal{F}^\sim, I, \sqcap^\sim, L^\sim, R^\sim)$ is an iterative distributive space with distributor $U^\sim = (I, I)$.

Proof. Straightforward.

We complete this section with another general construction yielding an iterative distributive space from a given space which has a bottom-zero but need be neither iterative nor distributive; such is for instance the subspace of example 1 consisting of all elements constructed from O, I, L, R by $\cdot, \sqcap$.

<u>Proposition 2.5</u>. Let $\mathcal{G} = (\mathcal{F}, I, \sqcap, L, R)$ be an operative space with a bottom element O s.t. $\psi\cdot O = O\cdot\psi = O$ for all ψ. For all $\mathcal{A} \subseteq \mathcal{F}$ write $\mathcal{A}^\wedge$ for $\{\theta / \forall\varphi \in \mathcal{A}(\theta \leq \varphi)\}$. Take $\mathcal{F}' = \{\varphi' = \mathcal{A}^\wedge / \emptyset \neq \mathcal{A} \subseteq \mathcal{F}\}$, $\varphi' + \psi' = \varphi' \cup \psi'$, $\varphi'\cdot\psi' = \{\varphi\cdot\psi / \varphi \in \varphi', \psi \in \psi'\}^\wedge$, $O' = \{O\}^\wedge$, $I' = \{I\}^\wedge$, $L' = \{L\}^\wedge$, $R' = \{R\}^\wedge$, $L_1' = \{(I, O)\}^\wedge$ and $R_1' = \{(O, I)\}^\wedge$. Then obtain the space $\mathcal{G}' = (\mathcal{F}', I', \sqcap', L', R')$ from the complete semiring $(\mathcal{F}', +, \cdot)$ by 2.3.

Proof. Straightforward.

3. ABSTRACT PARTIAL RECURSIVENESS

This section studies partial recursiveness in the abstract and as related to classical concepts of effective computability in particular spaces.

Assume that an iterative distributive space $\mathcal{G} = (\mathcal{F}, I, \sqcap, L, R)$ is given. An element φ is said to be <u>partial recursive in</u> a subset $\mathcal{B}$ of $\mathcal{F}$ iff φ is recursive in $\{U\} \cup \mathcal{B}$. We recall that φ is <u>recursive</u> (<u>primitive recursive</u>, <u>primitive</u>) in $\mathcal{B}$ iff it can be constructed from L, R and members of $\mathcal{B}$ by means of the operations $\cdot, \sqcap, \langle\rangle, [\,]$ (respectively, $\cdot, \sqcap, \Delta$ and $\cdot, \sqcap, \langle\rangle$).

<u>Lemma 3.1</u> (Basic Lemma). Let $\Gamma = \lambda\theta.(\varphi + \psi\cdot\theta\cdot\chi)$ and $\theta_o = \mu\theta.\Gamma(\theta)$. Then $\delta\cdot\theta_o\cdot\sigma = \Sigma_n(\delta\cdot\psi^n\cdot\varphi\cdot\chi^n\cdot\sigma)$ for all δ, σ, where $\Sigma_n\varphi_n = \sup_n\Sigma_{i \leq n}\varphi_i$ by definition.

Proof. The equality $\Gamma(\theta_o) = \theta_o$ implies for all n

$$\delta\cdot\Gamma^n(0)\cdot\sigma \leq \delta\cdot\Gamma^n(\theta_o)\cdot\sigma = \delta\cdot\theta_o\cdot\sigma.$$

Suppose that $\delta\cdot\Gamma^n(0)\cdot\sigma \leq \tau$ for all n. The set $\mathcal{E} = \{\theta/\ \forall n(\delta\cdot\Gamma^n(\theta)\cdot\sigma \leq \tau)\}$ is of the form required by (@) since $\Gamma^n(\theta) = \Sigma_{i<n}\psi^i\cdot\varphi\cdot\chi^i + \psi^n\cdot\theta\cdot\chi^n$. It follows that $0 \in \mathcal{E}$ and whenever $\theta \in \mathcal{E}$, then

$$\delta\cdot\Gamma^n(\Gamma(\theta))\cdot\sigma = \delta\cdot\Gamma^{n+1}(\theta)\cdot\sigma \leq \tau$$

for all n, hence $\Gamma(\theta) \in \mathcal{E}$. Therefore, $\theta_o \in \mathcal{E}$ by (@), in particular $\delta\cdot\theta_o\cdot\sigma \leq \tau$. One gets

$$\delta\cdot\theta_o\cdot\sigma = \sup_n\delta\cdot\Gamma^n(0)\cdot\sigma = \sup_n\Sigma_{i\leq n}\delta\cdot\psi^i\cdot\varphi\cdot\chi^i\cdot\sigma,$$

which completes the proof.

Comments. The element θ_o is partial recursive in φ, ψ, χ, i.e. recursive in U, φ, ψ, χ, by proposition 3 [2] or the weaker statement 1.22 [3] (6.39 [4]). The latter, to which we shall refer as Second Recursion Lemma, asserts that $\mu\theta.\varphi\cdot(\psi, \theta\cdot\chi)$ is recursive in φ, ψ, χ.

The Basic Lemma gives explicit characterizations of the operations $\langle\rangle$, Δ, $[\,]$ in distributive spaces.

<u>Corollary 3.2</u>. $\langle\varphi\rangle = \Sigma_nR_1{}^n\cdot L_1\cdot\varphi\cdot L\cdot R^n$, $\Delta(\varphi, \psi) = \Sigma_nR_1{}^n\cdot L_1\cdot\varphi\cdot\psi^n$, $[\varphi] = \Sigma_n(R_1\cdot\varphi)^n\cdot L_1$.

The Basic Lemma can be generalized as follows.

<u>Lemma 3.3</u>. $\Sigma_{i\leq m}\delta_i\cdot(\mu\theta.(\varphi + \psi\cdot\theta\cdot\chi))\cdot\sigma_i = \Sigma_n\Sigma_{i\leq m}\delta_i\cdot\psi^n\cdot\varphi\cdot\chi^n\cdot\sigma_i$.

Proof. (Outline) Writing D, $\underline{n}$, $\underline{\omega}$ respectively for $\Delta(L, R^2)$, $L\cdot R^n$, $\mu\theta.(L + \theta\cdot R)$, and making use of 3.1 and the general equalities $\underline{n}\cdot\Delta(\varphi, \psi) = \varphi\cdot\psi^n$, $\underline{n}\cdot\langle\varphi\rangle = \varphi\cdot\underline{n}$, one shows that

$$\begin{aligned}\Sigma_{i\leq m}\Sigma_n\delta_i\cdot\psi^n\cdot\varphi\cdot\chi^n\cdot\sigma_i &= \Sigma_{i\leq m}\Sigma_n\underline{n}\cdot D\cdot\langle\Delta(\delta_i, \psi)\rangle\cdot\Delta(\varphi, \chi)\cdot\sigma_i\\ &= \Sigma_{i\leq m}\underline{\omega}\cdot D\cdot\langle\Delta(\delta_i, \psi)\rangle\cdot\Delta(\varphi, \chi)\cdot\sigma_i = \underline{\omega}\cdot\Sigma_{i\leq m}D\cdot\langle\Delta(\delta_i, \psi)\rangle\cdot\Delta(\varphi, \chi)\cdot\sigma_i\\ &= \Sigma_n\underline{n}\cdot\Sigma_{i\leq m}D\cdot\langle\Delta(\delta_i, \psi)\rangle\cdot\Delta(\varphi, \chi)\cdot\sigma_i = \Sigma_n\Sigma_{i\leq m}\delta_i\cdot\psi^n\cdot\varphi\cdot\chi^n\cdot\sigma_i.\end{aligned}$$

<u>Corollary 3.4</u>. Let $\theta_o = \mu\theta.(\varphi + \psi\cdot\theta\cdot\chi)$. Then

$$(\delta_o\cdot\theta_o\cdot\sigma_o, \ldots, \delta_m\cdot\theta_o\cdot\sigma_m) = \Sigma_n(\delta_o\cdot\psi^n\cdot\varphi\cdot\chi^n\cdot\sigma_o, \ldots, \delta_m\cdot\psi^n\cdot\varphi\cdot\chi^n\cdot\sigma_m).$$

Follows by 3.3 and the equality $(\varphi, \psi) = L_1\cdot\varphi + R_1\cdot\psi$.

We proceed on with several lemmas on the element $\underline{\omega} = \mu\theta.(L + \theta\cdot R)$ which is to play a key role in our considerations. This element is partial recursive, i.e. recursive in $U = L + R$ by the Second Recursion Lemma. Explicitly, $\underline{\omega} = \Sigma_n\underline{n}$ by 3.1.

<u>Lemma 3.5</u>. $\underline{\omega}\cdot R_1{}^n\cdot L_1 = I$, $\underline{\omega}\cdot(L, R, 0) = U$.

Follows by 3.1.

<u>Lemma 3.6</u>. $\varphi\cdot\underline{\omega} = \underline{\omega}\cdot\langle\varphi\rangle$.

Proof. One gets by 3.1 and $\varphi\cdot\underline{n} = \underline{n}\cdot\langle\varphi\rangle$ that $\varphi\cdot\underline{\omega} = \Sigma_n\varphi\cdot\underline{n} = \Sigma_n\underline{n}\cdot\langle\varphi\rangle = \underline{\omega}\cdot\langle\varphi\rangle$.

Lemma 3.7. $\underline{\omega}\cdot C\cdot L_1 = L_1\cdot\underline{\omega}$ and $\underline{\omega}\cdot C\cdot R_1 = R_1\cdot\underline{\omega}$, where $C = \Delta((L^2, L\cdot R), (R\cdot L, R^2))$.

Proof. By the equality $\underline{n}\cdot C = (\underline{n}\cdot L, \underline{n}\cdot R)$ and 3.1

$$\underline{\omega}\cdot C\cdot L_1 = \Sigma_n\underline{n}\cdot C\cdot L_1 = \Sigma_n(\underline{n}\cdot L, \underline{n}\cdot R)\cdot L_1 = \Sigma_n(n, 0) = \Sigma_n L_1\cdot\underline{n} = L_1\cdot\underline{\omega},$$

and similarly $\underline{\omega}\cdot C\cdot R_1 = R_1\cdot\underline{\omega}$.

Lemma 3.8. $(\underline{\omega}\cdot\varphi, \underline{\omega}\cdot\psi) = \underline{\omega}\cdot C\cdot(\varphi, \psi)$.

Proof. It follows by 3.7 that

$$(\underline{\omega}\cdot\varphi, \underline{\omega}\cdot\psi) = L_1\cdot\underline{\omega}\cdot\varphi + R_1\cdot\underline{\omega}\cdot\psi = \underline{\omega}\cdot C\cdot L_1\cdot\varphi + \underline{\omega}\cdot C\cdot R_1\cdot\psi$$
$$= \underline{\omega}\cdot C\cdot(L_1\cdot\varphi + R_1\cdot\psi) = \underline{\omega}\cdot C\cdot(\varphi, \psi).$$

Lemma 3.9. $\langle\underline{\omega}\rangle = \underline{\omega}\cdot G$, where $G = \Delta(\langle L\rangle, \langle R\rangle)$. Therefore, $\langle\underline{\omega}\cdot\varphi\rangle = \langle\underline{\omega}\rangle\cdot\langle\varphi\rangle = \underline{\omega}\cdot G\cdot\langle\varphi\rangle$.

Proof. $\underline{\omega}\cdot\underline{m} = \Sigma_n\underline{n}\cdot\underline{m} = \Sigma_n\underline{m}\cdot\underline{n}\cdot G = \underline{m}\cdot\underline{\omega}\cdot G$ for all m, hence $\langle\underline{\omega}\rangle = \langle I\rangle\cdot\underline{\omega}\cdot G = \underline{\omega}\cdot\langle\langle I\rangle\rangle\cdot G = \underline{\omega}\cdot G$. The proof is completed.

Consider the bijective pair coding $\langle m, n\rangle = 2m(2n + 1) - 1$ with decoding functions $g = \lambda\langle m, n\rangle.m$, $h = \lambda\langle m, n\rangle.n$ which are primitive recursive. Take by proposirion 1 [2] (8.1 [4]) primitive recursive elements Y_1, Y_2 s.t. $\underline{n}\cdot Y_1 = \underline{g(n)}$, $\underline{n}\cdot Y_2 = \underline{h(n)}$ for all n. Set $Y = D\cdot\langle Y_1\rangle\cdot Y_2$ with $D = \Delta(L, R^2)$, then Y is primitive recursive and $\underline{n}\cdot Y = \underline{g(n)}\cdot\underline{h(n)}$ for all n.

Lemma 3.10. $\underline{\omega}\cdot\underline{\omega} = \underline{\omega}\cdot Y$.

Proof. It follows that $\Sigma_{k\leq n}\underline{g(k)}\cdot\underline{h(k)} = \Sigma_{i\leq g(n)}\underline{i}\Sigma_{j\leq h(n)}\underline{j} \leq \underline{\omega}\cdot\underline{\omega}$ for all n, hence $\underline{\omega}\cdot Y = \Sigma_n\underline{n}\cdot Y = \Sigma_n\underline{g(n)}\cdot\underline{h(n)} \leq \underline{\omega}\cdot\underline{\omega}$.

On the other hand, $\Sigma_{i\leq m}\underline{i}\Sigma_{j\leq n}\underline{j} = \Sigma_{k\leq\langle m,n\rangle}\underline{g(k)}\cdot\underline{h(k)} \leq \Sigma_n\underline{g(n)}\cdot\underline{h(n)} = \underline{\omega}\cdot Y$ for all m, n, hence $\underline{\omega}\cdot\Sigma_{j\leq n}\underline{j} \leq \underline{\omega}\cdot Y$ for all n, which implies $\underline{\omega}\cdot\underline{\omega} \leq \underline{\omega}\cdot Y$. Thereby the proof is completed.

Lemma 3.11. $[\underline{\omega}\cdot\varphi] = \underline{\omega}\cdot C\cdot[Y\cdot\langle\varphi\rangle\cdot C]\cdot L_1$.

Proof. Making use of 3.7,

$$\underline{\omega}\cdot C\cdot[Y\cdot\langle\varphi\rangle\cdot C]\cdot L_1 = \Sigma_n\underline{\omega}\cdot C\cdot(R_1\cdot Y\cdot\langle\varphi\rangle\cdot C)^n\cdot L_1^2$$
$$= \Sigma_n(R_1\cdot\underline{\omega}\cdot\varphi)^n\cdot\underline{\omega}\cdot C\cdot L_1^2 = \Sigma_n(R_1\cdot\underline{\omega}\cdot\varphi)^n\cdot L_1 = [\underline{\omega}\cdot\varphi].$$

Lemma 3.12. Whenever φ is partial recursive in $\mathcal{B}$, then $\varphi = \underline{\omega}\cdot\psi$ with a certain ψ recursive in $\mathcal{B}$.

Proof. By induction on the construction of φ, making use of 3.5, 3.6, 3.8–3.11.

Comments. Lemma 3.12 is a "lightface" version of proposition 3.3.9 [8], chapter 5.

Lemma 3.13 (Iteration Elimination Lemma). $[\varphi] = \underline{\omega}\cdot\Delta(I, R_1\cdot\varphi)\cdot L_1$.

Proof. By 3.5

$$[\varphi] = \Sigma_n(R_1\cdot\varphi)^n\cdot L_1 = \Sigma_n\underline{\omega}\cdot R_1{}^n\cdot L_1\cdot(R_1\cdot\varphi)^n\cdot L_1 = \underline{\omega}\cdot\Delta(I, R_1\cdot\varphi)\cdot L_1.$$

Proposition 3.14 (Normal Form Theorem). Whenever φ is partial recursive in $\mathcal{B}$, then $\varphi = \underline{\omega}\cdot Y\cdot\Delta(\underline{1}, R_1\cdot\psi)\cdot L_1$ with a certain ψ primitive in $\mathcal{B}$.

Proof. It follows by 3.12 that $\varphi = \underline{\omega}\cdot\varphi_1$ with φ_1 recursive in $\mathfrak{B}$, while $\varphi_1 = \underline{1}\cdot[\psi]$ with ψ primitive in $\mathfrak{B}$ by proposition 9.3 [4]. Making use of the identity $\langle\delta\rangle\cdot\Delta(I, \sigma) = \Delta(\delta, \sigma)$, one finally gets

$$\varphi = \underline{\omega}\cdot\underline{1}\cdot[\psi] = \underline{\omega}\cdot\underline{1}\cdot\underline{\omega}\cdot\Delta(I, R_1\cdot\psi)\cdot L_1$$
$$= \underline{\omega}\cdot\underline{\omega}\cdot\langle\underline{1}\rangle\cdot\Delta(I, R_1\cdot\psi)\cdot L_1 = \underline{\omega}\cdot Y\cdot\Delta(\underline{1}, R_1\cdot\psi)\cdot L_1.$$

The proof is completed.

Comments. A similar normal form result takes place for mappings, where $\Gamma: \mathfrak{F} \to \mathfrak{F}$ is <u>partial recursive in</u> $\mathfrak{B}$ iff for all θ the element $\Gamma(\theta)$ is partial recursive in $\{\theta\} \cup \mathfrak{B}$ uniformly in θ. Namely, it can be shown that whenever Γ is partial recursive in $\mathfrak{B}$, then

$$\Gamma = \lambda\theta.\underline{\omega}\cdot Y\cdot\Delta(\underline{1}, R_1\cdot\psi\cdot(I, \langle\theta\rangle))\cdot L_1$$

with ψ primitive in $\mathfrak{B}$.

Proposition 3.14 corresponds to the following alternative definition of the ordinary partial recurrsiveness: Given a set $\mathfrak{F}$ of partial number-theoretic functions, a n-ary function f is partial recursive in $\mathfrak{F}$ iff there is a n+1-ary function g primitive recursive in $\{\lambda s.\uparrow\} \cup \mathfrak{F}$ s.t. $f(s_1, \ldots, s_n) = t$ iff $\exists r(g(s_1, \ldots, s_n, r) = t)$.

If α is a positive integer or ω, then we write $\alpha\cdot\varphi$ for $\Sigma_{i<\alpha}\underline{i}\cdot[I]\cdot\varphi$. An element φ is said to <u>quasi-represent</u> a function $f: \omega^n \to \omega$ iff for all $s_1, \ldots, s_n$ there is an α s.t. $\underline{s_1}\cdot\ldots\cdot\underline{s_n}\cdot\varphi = \alpha\cdot\underline{f(s_1, \ldots, s_n)}$, if $f(s_1, \ldots, s_n)\downarrow$, and $\underline{s_1}\cdot\ldots\cdot\underline{s_n}\cdot\varphi = 0$ otherwise. In spaces with $I + I = I$ it follows that $\alpha\cdot\varphi = \varphi$ for all α, φ, and quasi-representability does not differ from the representability of [2,4]. In general however representability is inadequate, as far as partial recursivenes is concerned.

<u>Proposition 3.15</u> (Quasi-Representation Theorem). Let $\mathfrak{B} \subseteq \mathfrak{F}$ and $\mathfrak{F}$ be a set of number-theoretic functions quasi-representable by members of $\mathfrak{B}$. Then every function f partial recursive in $\mathfrak{F}$ is quasi-representable by an element φ partial recursive in $\mathfrak{B}$.

Proof. (Outline) Let g be primitive recursive in $\{\lambda s.\uparrow\} \cup \mathfrak{F}$ s.t. $f(s_1, \ldots, s_n) = t$ iff $\exists r(g(s_1, \ldots, s_n, r) = t)$. Then by the proof of 8.1 [4] g is quasi-represented by an element ψ primitive recursive in $\{0\} \cup \mathfrak{B}$. Take $\varphi = \underline{\omega}\cdot\psi$, then $\underline{s_1}\cdot\ldots\cdot\underline{s_n}\cdot\varphi = \Sigma_r\underline{s_1}\cdot\ldots\cdot\underline{s_n}\cdot\underline{r}\cdot\psi$ for all $s_1, \ldots, s_n$, which implies quite easily that φ quasi-represents f. (It is this last point of the proof where representability does not fit.)

In the space of example 1 recursiveness in U, i.e. partial recursiveness in the present terms, has been studied in some detail in [4]. The notions of recursiveness and partial recursiveness account respectively for the prime computability of Moschovakis and Friedman's computability by effectively definitional schemes; covered is also the modification of the latter suggested by [6, 9]. Taking in particular $M = \omega$ with a splitting scheme $f_1 = \lambda s.2s$, $f_2 = \lambda s.2s + 1$ and adding $Z = \lambda s.(2s + 1)sgs$ to the initial elements, one gets the ordinary μ-recursiveness and partial recursivenes--both for single-valued and multiple-valued functions.

ACKNOWLEDGEMENTS

The author is indebted to Ivan Soskov for a good deal of discussion on computability.

REFERENCES

[1] L. L. Ivanov, Iterative operative spaces, Ph. D. thesis, Sofia University (1980), in Bulgarian.

[2] L. L. Ivanov, Iterative operative spaces, C. R. Acad. Bulg. Sci. 33 (1980) 735-738, in Russian.

[3] L. L. Ivanov, Iterative operative spaces and the system of Scott and de Bakker, Serdica Bulg. Math. Publ. 9 (1983) 275-288.

[4] L. L. Ivanov, Algebraic Recursion Theory (Ellis Horwood, Chichester, 1986), to appear.

[5] J. Myhill, Note on degrees of partial functions, Proc. Amer. Math. Soc. 12 (1961) 519-521.

[6] J. C. Shepherdson, Computation over abstract structures, in: Logic Colloquium '73, eds. H. E. Rose, J. C. Shepherdson (North-Holland, Amsterdam, 1975) 445-513.

[7] D. G. Skordev, Computable and μ-recursive operators, Izv. Mat. Inst. BAN 7 (1963) 5-43, in Bulgarian, Russian summary.

[8] D. G. Skordev, Combinatory spaces and recursiveness in them (BAS Publishers, Sofia, 1986), in Russian, English summary.

[9] I. N. Soskov, Computability in algebraic systems, C. R. Acad. Bulg. Sci 36 (1983) 301-304, in Russian.

APPROXIMATING THE PROJECTIVE MODEL

Evangelos Kranakis

Centrum voor Wiskunde en Informatica
P.O. Box 4079, 1009 AB Amsterdam, The Netherlands

Abstract

One of the fundamental questions in the calculus of communicating processes is determining if a given system of fixed point equations has a solution in the projective model. The present paper provides an approximation principle for the projective model, which makes it posssible to prove assertions in this model by proving them in an infinite sequence of certain finite process algebras. Motivated from this principle a new model for process algebras is defined and its relationship to the projective model is studied.

Mathematics Subject Classification: 68B05.
Key Words and Phrases: process algebra, process, projective model, polynomial operator, metric space, approximation principle, positive formulas, ultrafilter, ultraproduct.

Acknowledgements: This research was carried out while the author was visiting the Computer Science Department of the University of Amsterdam and it was partially supported by Esprit under contract no. 432, Meteor. I would like to thank all the participants of the P.A.M. seminar for their numerous comments. However, I am particularly indebted to J. Baeten, R. van Glabbeek, J. W. Klop and F. Vaandrager, whose valuable criticisms helped me correct my numerous errors.

1. Introduction

In the formal analog of Milner's work on **Calculus of Communicating Systems** (see [M]), as described by Bergstra and Klop in [BK], one builts large systems of processes by assembling together **atomic processes** (or **actions**) chosen from a **finite** set A of such atomic processes (see [H]). These systems of processes satisfy a set of equational laws, called the axioms of the **theory of the algebras of communicating processes** (or **theory of process algebras**). The models of this theory are called **process algebras**. Its axioms are described in a signature that includes: $+$ (**alternative composition** or **sum**), $\cdot$ (**sequential composition** or **product**), $\|$ (**parallel composition** or **merge**), $\lfloor\!\lfloor$ (**left merge**), $|$ (**communication merge**), ∂_H (**encapsulation**, where H is a subset of the set A of atoms), the atomic process δ (**deadlock** or **failure**) and the atom a, for each $a \in A$.

In the table below the equational laws for process algebras are given. The communication function $| : A_\delta \times A_\delta \rightarrow A_\delta$ (where A_δ consists of the atoms in A including δ) is initially defined only on atomic processes. Then it is extended to all finite terms using the merge and communication axioms. In the absence of communication, the axiom $x \parallel y = y \mathbin{\lfloor\!\lfloor} x + x \mathbin{\lfloor\!\lfloor} y + x \mid y$ should be replaced with the new axiom $x \parallel y = y \mathbin{\lfloor\!\lfloor} x + x \mathbin{\lfloor\!\lfloor} y$. The theory consisting of the first five basic axioms together with the first four communication axioms is known as **(basic) process algebra** and is abbreviated by *PA*. *ACP*, the **algebra of communicating processes** consists of the basic, merge, communication and encapsulation axioms. As usual, the universal quantifiers, which quantify the variables x, y, z in the axioms below are omitted. In addition, the letters a, b range over A. The axioms of process algebras are the following:

Basic Axioms	**Communication-Merge Axioms**
$x + y = y + x$	$x \parallel y = y \mathbin{\lfloor\!\lfloor} x + x \mathbin{\lfloor\!\lfloor} y + x \mid y$
$x + (y + z) = (x + y) + z$	$a \mathbin{\lfloor\!\lfloor} x = a{\cdot}x$
$x + x = x$	$(a{\cdot}x) \mathbin{\lfloor\!\lfloor} y = a{\cdot}(x \parallel y)$
$(x + y){\cdot}z = x{\cdot}z + y{\cdot}z$	$(x + y) \mathbin{\lfloor\!\lfloor} z = x \mathbin{\lfloor\!\lfloor} z + y + \mathbin{\lfloor\!\lfloor} z$
$(x \cdot y) \cdot z = x \cdot (y \cdot z)$	$(a{\cdot}x) \mid b = a \mid (b{\cdot}x) = (a \mid b){\cdot}x$
$x + \delta = x$	$(a{\cdot}x) \mid (b{\cdot}y) = (a \mid b){\cdot}(x \parallel y)$
$\delta{\cdot}x = \delta$	$(x + y) \mid z = x \mid z + y \mid z$
Encapsulation Axioms	$z \mid (x + y) = z \mid x + z \mid y$
$\partial_H(a) = a$ if $a \in H$	**Communication Axioms**
$\partial_H(a) = \delta$ if $a \in A - H$	$a \mid b = b \mid a$
$\partial_H(x + y) = \partial_H(x) + \partial_H(y)$	$(a \mid b) \mid c = a \mid (b \mid c)$
$\partial_H(x{\cdot}y) = \partial_H(x){\cdot}\partial_H(y)$	$\delta \mid a = \delta$

In this axiomatic framerwork one can define the so-called **term** (or **initial**) **model** A_ω, i.e. the least set S of finite strings such that S contains all the constants of the given signature, and S is closed under the operations of the given signature. (The reader should be aware of all the possible signatures arising in the present study; practically every subset of $+, \cdot, \parallel, \lfloor\!\lfloor, \mid, \partial_H, \delta, a$ $(a \in A)$ is a possible signature and hence it can give rise to a different term model A_ω. It would be very cumbersome however to keep a different notation for A_ω for each possible signature. Instead, it will be left to the reader to derive from the context what the proper signature in each case is.)

Given any term t in A_ω and any positive integer n let $(t)_n$ be the subtree of t of height at most n obtained from t by deleting all those nodes which are located at height bigger than n. More formally define:

$$
\begin{aligned}
(a)_n &= a,\\
(at)_1 &= a,\\
(at)_n &= a(t)_{n-1}, \text{ for } n > 1,\\
(t + t')_n &= (t)_n + (t')_n, \text{ for } n > 0.
\end{aligned}
$$

Now A_n is defined to be the set $\{(t)_n : t \in A_\omega\}$. In a sense, $(\cdot)_n$ can be considered as the **projection** of the term model A_ω onto the model A_n. For each binary operation $\Box$ on A_ω define an operation $\Box_n$ on A_n by $t \Box_n s = (t \Box s)_n$ (the case of the operation ∂_H is treated similarly). This makes each A_n into a process algebra. The **projective** (or **standard**) **model**, denoted by A^∞ consists of all infinite sequences $(p_1, p_2, \ldots, p_n, \ldots)$ such that $p_n \in A_n$ and $(p_{n+1})_n = p_n$, for all $n > 0$. The operations are defined on A^∞ in a natural way; thus, following [BK], if $\Box$ is any binary operation on A_ω one defines a new binary operation $\square'$, which for convenience will also be denoted by $\square$, as follows:

$$(p_1,...,p_n,...) \square (q_1,...,q_n,...) = ((p_1 \square q_1)_1,...,(p_n \square q_n)_n,...)$$

Remark on Notation: Throughout the present paper $T(x_1,...,x_n)$, $S(x_1,...,x_n)$, etc. with or without subscripts and superscripts, will always denote (polynomial) operators, i.e. terms built up from the atomic processes, the variables $x_1,...,x_n$, the atoms in A and the operations of the given signature.

The class P of **positive formulas** is the smallest class of well founded formulas in the signature $+, \cdot, \|, \lfloor\!\lfloor, |, \partial_H, \delta, a$ $(a \in A)$, which satisfies the following properties:

(a) For all polynomial operators T,S, $T(v_1,...,v_n) = S(v_1,...,v_n)$ is in P.

(b) If $\Phi,\Psi \in P$ then $\Phi \vee \Psi \in P$.

(c) For any countable $\Delta \subseteq P$, the formula $\bigwedge \Delta$ is in P.

(d) If $\Phi(v_1,...,v_n,...) \in P$ and the variables $u_1,...,u_k,...$ are from the set $\{v_1,...,v_n,...\}$ then both formulas $(\exists u_1 \cdots \exists u_k \cdots)\Phi$, $(\forall u_1 \cdots \forall u_k \cdots)\Phi$ are in P.

The class P_0 of **finite positive formulas** is the smallest class of well founded formulas in the signature $+, \cdot, \|, \lfloor\!\lfloor, |, \partial_H, \delta, a$ $(a \in A)$, which satisfies the following properties:

(a) For all polynomial operators T,S, $T(v_1,...,v_n) = S(v_1,...,v_n)$ is in P.

(b) If $\Phi,\Psi \in P$ then $\Phi \vee \Psi, \Phi \wedge \Psi \in P$.

(c) If $\Phi(v_1,...,v_n) \in P$ and $u_1,...,u_k \in \{v_1,...,v_n\}$ then both formulas $(\exists u_1 \cdots \exists u_k)\Phi$, $(\forall u_1 \cdots \forall u_k)\Phi$ are in P.

Most of section 3 will be dedicated to a proof of the following theorem.

Theorem 1.1. [Approximation Theorem] *Any formula $\Phi(v_1,...,v_n,...) \in P$ satisfies the following approximation principle: for any convergent sequences $\{x_{1,n}\},...,\{x_{k,n}\},...$ such that $x_{k,n} \in A_n$ for all k,n, if the set $\{n > 0 : A_n \models \Phi(x_{1,n},...,x_{k,n},...)\}$ is infinite then it is true that $A^\infty \models \Phi(\lim_{n\to\infty} x_{1,n},..., \lim_{n\to\infty} x_{k,n},...)$.*

Such formulas Φ occur when one wants to prove that a system of fixed point equations has a solution. As an example, consider the infinite system

$$\Sigma = \{x_k = T_k(x_1,...,x_{n(k)}) : k > 0\},$$

where each T_k is a polynomial operator in the indicated variables. The assertion Σ *has a solution in* A^∞ can be expressed by the formula: $(\exists x_1 \cdots \exists x_k \cdots)\Psi$, where Ψ is the countable conjunction of all the formulas $x_k = T_k(x_1,...,x_{n(k)})$, for $k > 0$. Theorem 1.1 states that in order to prove that Σ has a solution in A^∞, it is enough to show that Σ has a solution in infinitely many A_n's. For more specific examples of systems the reader is referred to [BK], [H] and [K].

It is also possible to prove a partial converse of the approximation principle.

Theorem 1.2. [Converse of the Approximation Principle] *For any formula* $\Phi(v_1,\ldots,v_k,\ldots) \in P$ *and any* $p_1,\ldots,p_k,\ldots \in A^\infty$ *the following statements are equivalent:*

(i) $A^\infty \models \Phi(p_1,\ldots,p_k,\ldots)$.
(ii) $\{n > 0 : A_n \models \Phi((p_1)_n,\ldots,(p_k)_n,\ldots)\}$ is infinite.
(iii) $\forall n > 0[A_n \models \Phi((p_1)_n,\ldots,(p_k)_n,\ldots)]$.

Motivated from the approximation theorem one can define a new process algebra, which is an extension of the projective algebra A^∞. To state the next theorem the notion of ultrafilter on the set N of positive integers will be required. Call D a **(nonprincipal) ultrafilter** on N if D is a nonempty set of subsets of N satisfying the following properties for all $X,Y \subseteq N$: (i) $\varnothing$ is not a member of D, (ii) if $X \subseteq Y$ and $X \in D$ then $Y \in D$, (iii) $X \in D$ or $N-X \in D$, (iv) if $X,Y \in D$ then $X \cap Y \in D$ and (v) if $X \in D$ then X is infinite. Notice that the existence of such ultrafilters requires the axiom of choice (see [E] or [CK]).

The main theorem of section 4 is the following.

Theorem 1.3. *For any ultrafilter D on the set N of positive integers there exists a process algebra* A^D, *which is a proper extension of the projective algebra* A^∞. *Moreover, for any finite, positive formula* $\Phi(v_1,\ldots,v_k) \in P_0$ *and any* $p_1,\ldots,p_k \in A^\infty$ *the following statements are equivalent:*

(i) $A^D \models \Phi(p_1,\ldots,p_k)$.
(ii) $A^\infty \models \Phi(p_1,\ldots,p_k)$.
(iii) $\{n > 0 : A_n \models \Phi((p_1)_n,\ldots,(p_k)_n)\} \in D$.
(iv) $\{n > 0 : A_n \models \Phi((p_1)_n,\ldots,(p_k)_n)\}$ *is infinite.*
(v) $\forall n > 0[A_n \models \Phi((p_1)_n,\ldots,(p_k)_n)]$.

2. Topology of the Projective Model

As explained above the projective model A^∞ consists of all sequences $(p_1,\ldots,p_n,\ldots)$ such that each $p_n \in A_n$ and $(p_{n+1})_n = p_n$, for all $n > 0$. The term model A_ω can be embedded in a natural way in the projective model A^∞; for any finite term t associate the infinite sequence $p(t) = ((t)_1,\ldots,(t)_n,\ldots)$. Because of this it is identified with a subset of the projective model (this also explains why the same symbol is used for the corresponding operations in A_ω, A^∞). Extend the projection functions to A^∞ by defining $(p)_n = p_n$, for all $p = (p_1,\ldots,p_n,\ldots) \in A^\infty$ and all $n > 0$. For any two distinct elements p, q of A^∞ let $k(p,q) =$ the least $n > 0$ such that $(p)_n$ is not equal to $(q)_n$. This definition makes it possible to endow A^∞ with a **metric space structure.** Indeed, define the **distance** $d(p,q)$ between p, q by

$$d(p,q) = \begin{cases} 2^{-k(p,q)} & \text{if } p \neq q \\ 0 & \text{if } p = q. \end{cases}$$

This metric was used by Arnold and Nivat (see [AN]) in the context of Denotational Semantics of Concurrency. An essentially equivalent metric was also defined by de Bakker and Zucker (see [dBZ]). For additional information and further properties of this metric the reader is advised to consult [L] and [Ro].

The following results summarize all the basic properties of the metric space (A^{∞},d) and will be used frequently in the sequel. Their proof is omitted, but the interested reader can easily derive the essential details from [Du], [L], [AN] and [K].

Theorem 2.1. [In the signature $+, \cdot, \|, \|\!\_, |, \partial_H, \delta, a\ (a \in A)$**]**

(i) (A^{∞},d) *is an ultrametric space, i.e. it satisfies the following three properties for all elements* $p,q,r \in A^{\infty}$,

(a) $d(p,q) = 0$ *if and only if* $p = q$,

(b) $d(p,q) = d(q,p)$,

(c) $d(p,q) \leqslant \max\{d(p,r),d(r,q)\}$.

(ii) $p^{(r)} \to p$ *if and only if* $\forall n \exists m \forall k \geqslant m\ [(p^{(k)})_n = (p)_n]$.

(iii) (A^{∞},d) *is the metric completion of the metric space* (A_{ω},d'), *where* d' *is the restriction of* d *on* A_{ω}.

(iv) *For all* $p \in A^{\infty}$ *and each* $n > 0$, $d(p,(p)_n) \leqslant 2^{-n}$. Hence, $\lim_{n \to \infty} (p)_n = p$.

(v) *The operations* $(\cdot)_n : A^{\infty} \to A_n$ *are continuous.*

(vi) *Any operator* $T(x_1,...,x_n)$ *is continuous in the variables* $x_1,...,x_n$. *In fact, for all* $p_1,...,p_n,q_1,...,q_n \in A^{\infty}$,

$$d(T(p_1,...,p_n),T(q_1,...,q_n)) \leqslant \max\{d(p_1,q_1),...,d(p_n,q_n)\}.$$

(vii) *A is finite if and only if* (A^{∞},d) *is compact.* ●

In view of this last theorem from now on and for the rest of the paper it will be always assumed that A **is finite. This will guarantee that** A^{∞} **is compact.**

3. The Approximation Principle

Intuitively, the approximation principle enables one to verify assertions in the projective model by proving that the same assertion is valid in infinitely many A_n's. To be more specific, a formula Φ is said to satisfy the approximation principle, and this will be abbreviated by $A(\Phi)$, if the following property holds: for any convergent sequences $\{x_{1,n}\},...,\{x_{k,n}\},...$ such that $x_{k,n} \in A_n$ for all k,n, if the set

$$\{n > 0 : A_n \models \Phi(x_{1,n},...,x_{k,n},...)\}$$

is infinite then it is true that

$$A^{\infty} \models \bar{\Phi}(\lim_{n \to \infty} x_{1,n},..., \lim_{n \to \infty} x_{k,n},...).$$

Now it is possible to prove theorem 1.1.

Proof of theorem 1.1: It is enough to show that each formula $\Phi \in P$ satisfies $A(\Phi)$. The proof is by induction on the construction of Φ.

Case 1: $\Phi \equiv T(v_1,...,v_m)=S(v_1,...,v_m)$, where T,S are polynomial operators.

For any operator $T(v_1,...,v_m)$ let $T^n(v_1,...,v_m)$ denote the interpretation of T in the model A_n. Using induction on the construction of T and the definitions of the operations in the process algebras A_n (see section 1) it is easy to show that

Lemma 3.1. *For all* $x_1,...,x_m \in A_n$, $T^n(x_1,...,x_m) = (T(x_1,...,x_m))_n$. ●

Now it is required to show that the formula $T(v_1,...,v_m) = S(v_1,...,v_m)$ satisfies the above approximation principle. Indeed, let $\{x_{1,n}\},...,\{x_{k,n}\},...$ be any convergent sequence such that $x_{k,n} \in A_n$ for all k,n and the set

$$J = \{n > 0 : A_n \models T(x_{1,n},...,x_{m,n}) = S(x_{1,n},...,x_{m,n})\}$$

is infinite. It is enough to show that

$$A^\infty \models T(\lim_{n\to\infty} x_{1,n},..., \lim_{n\to\infty} x_{m,n}) = S(\lim_{n\to\infty} x_{1,n},..., \lim_{n\to\infty} x_{m,n}).$$

Put $(x_1,...,x_m) = (\lim_{n\to\infty} x_{1,n},..., \lim_{n\to\infty} x_{m,n})$. It is clear that for all $n \in J$,

$$T^n(x_{1,n},...,x_{m,n}) = S^n(x_{1,n},...,x_{m,n}).$$

Using this last equation and lemma 3.1 it is clear that for all $n \in J$,

$$(T(x_{1,n},...,x_{m,n}))_n = (S(x_{1,n},...,x_{m,n}))_n. \tag{1}$$

The following result is an easy consequence of the definition of convergence in A^∞:

Lemma 3.2. *For any sequence* $\{u_n\}$ *of terms in* A_ω, *if* $u_n \to u$ *then* $(u_n)_n \to u$. ●

Using the continuity of the operators T,S (see theorem 2.1) it follows that

$$T^n(x_{1,n},...,x_{m,n}) \to T(x_1,...,x_m),\ S^n(x_{1,n},...,x_{m,n}) \to S(x_1,...,x_m). \tag{2}$$

Hence, using lemma 3.2 as well as (1) and (2) it follows that

$$A^\infty \models T(x_1,...,x_m) = S(x_1,...,x_m),$$

which completes the proof in case 1.

Case 2: $\Phi \equiv \Theta \vee \Psi$.

Let $\{x_{1,n}\},...,\{x_{k,n}\},...$ be a convergent sequence such that $x_{k,n} \in A_n$ for all k,n, and the set $J = \{n > 0 : A_n \models \Phi(x_{1,n},...,x_{k,n},...)\}$ is infinite. To show that

$$A^\infty \models \Phi(\lim_{n\to\infty} x_{1,n},..., \lim_{n\to\infty} x_{k,n},...).$$

Put $K = \{n > 0 : A_n \models \Theta(x_{1,n},...,x_{k,n},...)\}$, $L = \{n > 0 : A_n \models \Psi(x_{1,n},...,x_{k,n},...)\}$. Since

$J = K \cup L$ it is clear that at least one of the sets K,L (say K) is infinite. It follows from the induction hypothesis that

$$A^{\infty} \models \Theta(\lim_{n\to\infty} x_{1,n},\ldots, \lim_{n\to\infty} x_{k,n},\ldots)$$

and hence also

$$A^{\infty} \models \Phi(\lim_{n\to\infty} x_{1,n},\ldots, \lim_{n\to\infty} x_{k,n},\ldots)$$

which completes the proof of case 2.

Case 3: $\Phi \equiv \bigwedge\{\Phi_i : i > 0\}$.

Let $\{x_{1,n}\},\ldots,\{x_{k,n}\},\ldots$ be a convergent sequence such that $x_{k,n} \in A_n$ for all k,n, and the set $J = \{n > 0 : A_n \models \Phi(x_{1,n},\ldots,x_{k,n},\ldots)\}$ is infinite. To show that

$$A^{\infty} \models \Phi(\lim_{n\to\infty} x_{1,n},\ldots, \lim_{n\to\infty} x_{k,n},\ldots).$$

For each $i > 0$ put $J_i = \{n > 0 : A_n \models \Phi_i(x_{1,n},\ldots,x_{k,n},\ldots)\}$. Clearly, each J_i is infinite and hence the induction hypothesis implies that for all $i > 0$,

$$A^{\infty} \models \Phi_i(\lim_{n\to\infty} x_{1,n},\ldots, \lim_{n\to\infty} x_{k,n},\ldots)$$

which completes the proof of case 3.

Case 4: $\Phi \equiv (\exists u_1 \cdots \exists u_k \ldots)\Psi(u_1,\ldots,u_k,\ldots,v_1,\ldots,v_n,\ldots)$.

Actually this is the only part of the proof which requires the compactness of A^{∞}. Let $\{x_{1,n}\},\ldots,\{x_{k,n}\},\ldots$ be a convergent sequence such that $x_{k,n} \in A_n$ for all k,n, and the set $J = \{n > 0 : A_n \models \Phi(x_{1,n},\ldots,x_{k,n},\ldots)\}$ is infinite. To show that

$$A^{\infty} \models \Phi(\lim_{n\to\infty} x_{1,n},\ldots, \lim_{n\to\infty} x_{k,n},\ldots).$$

By assumption, for each $n \in J$ there exist elements $y_{k,n} \in A_n$ such that

$$A_n \models \Psi(y_{1,n},\ldots,y_{k,n},\ldots,x_{1,n},\ldots,x_{k,n},\ldots).$$

The sequences $(\{y_{k,n}\} : k > 0)$ need not be convergent. However, using the compactness of the metric space A^{∞} (in fact one needs the compactness of the cartesian product space of countable many copies of A^{∞}) there exists an infinite subset L of J such that each of the sequences $\{y_{k,n}\}_{n \in L}$ is convergent. For each k let y_k be the limit of this last sequence. Now apply the induction hypothesis to the formula Ψ and the convergent sequences $\{x_{k,n}\}_{n \in L}$ $\{y_{k,n}\}_{n \in L}$, for $k > 0$, to obtain that

$$A^{\infty} \models \Psi(y_1,\ldots,y_k,\ldots,x_1,\ldots,x_k,\ldots).$$

This completes the proof of case 4.

Case 5: $\Phi \equiv (\forall u_1 \cdots \forall u_k \cdots)\Psi(u_1,\ldots,u_k,\ldots,v_1,\ldots,v_n,\ldots)$.

This is similar to the cases above and its proof is left to the reader. Now the proof of theorem 1.1 is complete. ●

The examples given below indicate that theorem 1.1 is best possible, in the sense that the approximation principle is not invariant neither under infinitary disjunctions nor under negations.

Example 3.3. [Noninvariance under infinitary disjunctions] Consider the formula $\Phi \equiv \bigvee\{x = a^n : n > 0\}$. For each $n > 0$ $A_n \models \Phi(a^n)$; however, $A^\infty \models \Phi(a^\omega)$ (a^ω is the limit of the sequence a^n in A^∞). An even better example involving a sentence (due to H. Mulder) is $\Phi \equiv \bigvee\{\exists x(xa = a^n) : n > 0\}$.

Example 3.4. [Noninvariance under negations] Consider the formula $\Psi(x,y) \equiv x \neq y$. Clearly, for all $n > 0$ $A_n \models \Psi(a^n, a^{n-1})$; however, $A^\infty \models \neg\Psi(a^\omega, a^\omega)$.

Proof of theorem 1.2: In view of theorem 1.1 it is enough to prove that (i) implies (iii). In fact, it is enough to show by induction on positive formulas Φ that for all $p_1,\ldots,p_k,\ldots \in A^\infty$,

$$A^\infty \models \Phi(p_1,\ldots,p_k,\ldots) \Rightarrow \forall n > 0[A_n \models \Phi((p_1)_n,\ldots,(p_k)_n,\ldots)]$$

The initial step of the proof is for formulas of the form $T(v_1,\ldots,v_k) = S(v_1,\ldots,v_k)$. Suppose that $p_1,\ldots,p_k \in A^\infty$ such that $A^\infty \models T(p_1,\ldots,p_k) = S(p_1,\ldots,p_k)$. Then for all $n > 0$,

$$(T^n((p_1)_n,\ldots,(p_k)_n)_n = (T(p_1,\ldots,p_k))_n = (S(p_1,\ldots,p_k))_n = (S^n((p_1)_n,\ldots,(p_k)_n)_n,$$

as desired. The rest of the proof is much like the proof of theorem 1.1 and is left to the reader. ●

4. The Ultraproduct Model

The most natural way to interpret the approximation principle is via the ultraproduct model. Details of its definition and fundamental properties can be found in [CK] and [E]. Given an ultrafilter D on N, define the equivalence relation $\equiv_D$ on the product set $\Pi\{A_n : n > 0\}$, as follows: $f \equiv_D g$ if and only if $\{n > 0 : f(n) = g(n)\} \in D$. Call $[f]_D$ the equivalence class of f modulo $\equiv_D$ and let A^D be the set of these equivalence classes. A^D can be turned into a process algebra by defining an operation $\Box_D$, for any binary operation $\Box$, as follows:

$$[f]_D \Box_D [g]_D = [f(n) \Box_n g(n) : n > 0]_D.$$

(The unary operation ∂_H is treated similarly.) It turns out that the mapping $p \to [(p)_n : n > 0]_D$ is a homomorphic embedding of A^∞ into A^D. This makes it possible to identify the elements of A^∞ with their corresponding images in A^D via the above embedding, and hence consider A^∞ as a subset of A^D.

Proof of theorem 1.3: Let $\Phi(v_1,\ldots,v_k)$ be a finite positive formula in the given signature. The equivalence of (i), (iii) is a consequence of the fundamental theorem for ultraproducts (see [E]). Since the ultrafilter D is nonprincipal (i.e. all its elements are infinite sets) the implications (v) $\Rightarrow$ (iii) $\Rightarrow$ (iv) are also immediate. The implication (iv) $\Rightarrow$ (ii) is a consequence of the approximation theorem and the fact that $\lim_{n \to \infty} (p)_n = p$. Finally the implication (ii) $\Rightarrow$ (v) is a special case of theorem 1.2. ●

Example 4.1. Theorem 1.3 cannot be extended to a set of formulas which is closed under negation. To see this consider the formula $\phi \equiv (\exists x,y)[x = ax \wedge y^2 = ay^2 \wedge x \neq y]$. The fundamental theorem for ultraproducts implies that $A^D \models \phi$ (this is because in A_n the equation $x = ax$ has exactly one solution, namely a^n, while the equation $y^2 = ay^2$ is satisfied by any of $a^k, a^{k+1}, \ldots, a^n$, where $k = \lfloor (n-1)/2 \rfloor$). However, since the only solution of the system $x = ax$, $y^2 = ay^2$ (in A^∞) is (a^ω, a^ω) the sentence ϕ cannot be valid in A^∞.

It is also possible to define nonstandard processes in the ultraproduct model. Inded, for any function $\sigma : N \rightarrow N$ such that for all $n > 0$ $\sigma(n) \leqslant n$ define the element

$$a^\sigma = [a^{\sigma(n)} : n > 0]_D.$$

For such functions σ it is possible to show that

Proposition 4.2. $a^\sigma \in A^\infty$ if and only if $(\exists X \in D)$ [σ is either constant on X or the identity on X]. ●

Example 4.3. [In the signature $+, \cdot, \|, \lfloor\!\lfloor, |, \partial_H, a$ $(a \in A)$**]:** For different A's the models A^∞ are not necessarily elementarily equivalent. To see this consider two distinct atoms a,b and define the formula $\phi \equiv (\exists x,y)[x = x^2 \wedge y = y^2 \wedge x \neq y]$. Then it is easy to show that $\{a\}^\infty \models \neg\phi$, while $\{a,b\}^\infty \models \phi$.

5. Discussion and Open Problems

The proof of the approximation principle requires the compactness of the topological space A^∞. This not only forces the set A of atoms to be finite, but it also excludes the possibility of using τ (silent or internal action). It is not known, however, if the approximation principle could be proved for the same class of positive formulas without these restrictions. The ultraproduct construction is quite general and it seems it would be interesting to study the ultraproduct obtained when one takes countably many copies of the finite term model A_ω (which, by the way, is no longer an extension of the projective model), as well as its relation to the so called graph models (see [BK]). It might also be possible to use the ultraproduct construction in order to prove that certain concepts in process algebra are undefinable in a given signature (see [E], corollary 3.4).

REFERENCES

[AN] Arnold, A. and Nivat, M., *The Metric Space of Infinite Trees: Algebraic and Topological Properties,* Fundamenta Informatica, 3, 4(1980), pp. 445-476.

[dBZ] de Bakker, J. W. and Zucker, J. I., *Denotational Semantics of Concurrency,* Proceedings 14th STOC, pp. 153-158, 1982.

[BK] Bergstra, J. A. and Klop, J. W., *Algebra of Communicating Processes,* in: Proceedings of the CWI Symposium on Mathematics and Computer Science, J. W. de Bakker, M. Hazenwinkel and J. K. Lenstra, eds., 1986.

[CK] Chang, C. C. and Keisler, H. J., *Model Theory,* North-Holland, 1973.

[Du] Dugundji, J., *Topology,* Allyn and Bacon, 1966.

[Di] Dieudonne, J., *Elements d' Analyse, Tom 1,* Gauthier-Villars, Paris, 1968.

[E] Eklof. P. C., *Ultraproducts for Algebraists,* in: Handbook of Mathematical Logic, Barwise, J., ed., pp. 105-137, North-Holland, 1977.

[H] Hoare, C. A. R., *Communicating Sequential Processes,* Prentice/Hall, 1985.

[K] Kranakis, E., *Fixed Point Equations with Parameters in the Projective Model,* CWI Technical Report, Computer Science/Department of Algorithms and Architectures, Report CS-R8606, January 1986, to appear.

[L] Lloyd, J., *Foundations of Logic Programming,* Springer Verlag, 1984.

[M] Milner, R., *A Calculus of Communicating Systems,* Springer Verlag Lecture Notes in Computer Science, Vol. 92, 1980.

[Ro] Rounds, W. C., *Applications of Topology to Semantics of Communicating Processes, in: Seminar in Concurrency,* Springer Verlag Lecture Notes in Computer Science, Vol. 197, 1985, pp. 360-372.

PROJECTION COMPLETE GRAPH PROBLEMS CORRESPONDING TO A BRANCHING-PROGRAM-BASED CHARACTERIZATION OF THE COMPLEXITY CLASSES $\mathcal{NC}^1$, $\mathcal{L}$ AND $\mathcal{NL}$

Christoph Meinel

Sektion Mathematik
Humboldt-Universität
DDR-1086 Berlin, PSF 1297

ABSTRACT

The p-projection completeness of some restricted graph accessibility problems for the (nonuniform) complexity classes $\mathcal{NC}^1$, $\mathcal{L}$ and $\mathcal{NL}$ will be proved by means of branching-program-based characterizations of these classes. A simulation result concerning polynomial-size, bounded-width disjunctive branching programs and polynomial-size, bounded-width usual ones yields that $\mathcal{NC}^1 = \mathcal{L}$ implies $\mathcal{L} = \mathcal{NL}$. Some consequences of these results for separating these classes are discussed.

INTRODUCTION

One of the major goals in complexity theory is to separate complexity classes such as L , NL , P or NP (or to prove their coincidence) or, equivalently, to show that nondeterministic Turing machines with certain ressource restrictions are more powerful than deterministic ones (or not). Since combinatorial techniques and counting arguments which are expected to be of fundamental importance in doing this can be applied more directly to circuitry-based computation devices such as Boolean circuits and branching programs than to the very complex types of Turing machines circuitry-based characterizations of the mentioned complexity classes gain more and more importance. While the nonuniform counterparts $\mathcal{P}$ and $\mathcal{NP}$ of the classes P and NP can be represented

in terms of polynomial-size Boolean circuits and polynomial-size nondeterministic Boolean circuits the class $\mathcal{L}$ of functions computable by nonuniform logarithmic space-bounded deterministic Turing machines can be represented in terms of polynomial-size branching programs. Moreover, very recently the nondeterministic counterpart $\mathcal{NL}$ of $\mathcal{L}$ could be characterized by polynomial-size disjunctive branching programs /Me86/ and the class $\mathcal{NC}^1$ of all functions computable by families of Boolean circuits with fan-in 2 and depth O(log n) could be described by polynomial-size, bounded-width branching programs /Ba86/. By means of these branching-program-based characterizations given in section 1 questions such as $\mathcal{NC}^1 \overset{?}{=} \mathcal{NL}$, $\mathcal{NC}^1 \overset{?}{=} \mathcal{L}$ or $\mathcal{L} \overset{?}{=} \mathcal{NL}$ can be formulated as the question whether polynomial-size disjunctive branching programs or polynomial-size usual ones are more powerful than polynomial-size branching programs of bounded width or whether polynomial-size disjunctive branching programs are more powerful than polynomial-size usual ones. However, polynomial-size disjunctive branching programs of bounded width turn out to be not more powerful than usual polynomial-size, bounded-width branching programs (Prop. 1). One of the consequences of this result is that $\mathcal{NC}^1 = \mathcal{L}$ would imply $\mathcal{L} = \mathcal{NL}$.
In section 2 the p-projection completeness of some restricted graph accessibility problems for the complexity classes $\mathcal{NC}^1$, $\mathcal{L}$ and $\mathcal{NL}$ will be proved by means of the given branching-program-based characterizations of these classes. Finally, some consequences of these results for separating these classes are discussed.

1. BRANCHING-PROGRAM-BASED CHARACTERIZATIONS OF THE COMPLEXITY CLASSES $\mathcal{NC}^1$, $\mathcal{L}$ AND $\mathcal{NL}$

As usual we denote by L and NL the complexity classes of languages $A \subseteq \{0,1\}^*$ which are accepted by deterministic and nondeterministic log n space-bounded Turing machines, respectively. The nonuniform counterparts $\mathcal{L}$ and $\mathcal{NL}$ are the languages $A \subseteq \{0,1\}^*$ for which there is a polynomial p(n) and an advice $\alpha_n \in \{0,1\}^*$ for all n of length $|\alpha_n| \leq p(n)$ polynomial in n such that a deterministic or nondeterministic log n space-bounded Turingmachine accepts $w\S\alpha_n$ iff $w \in A$, $|w| = n$ (§ is the blank tape symbol) /KL80/. Further, $\mathcal{NC}^1$

denotes the (nonuniform) complexity class of languages $A \subseteq \{0,1\}^*$ whose instances $A^n = A \cap \{0,1\}^n$ can be computed by Boolean circuits with fan-in 2 and depth $O(\log n)$ (and hence polynomial size). A problem is an infinite sequence of Boolean functions F_n, $n \in I \subseteq \mathbb{N}$, $|I| = \infty$

$$F = \{F_n\}$$

such that F_n has n variables and no two members of the sequence have the same number of variables. Via the usual correspondence of binary languages $A \subseteq \{0,1\}^*$ to YES-NO problems $F(A)$, namly $w \in A$ iff $F_{|w|} \in F(A)$ and $F_{|w|}(w) = 1$, the mentioned complexity classes can be regarded as classes of problems. A branching program is a directed acyclic graph where each node has outdegree 2 or 0. Nodes with outdegree 0 are called sinks and are labelled by Boolean constants. The remaining nodes are labelled with Boolean variables taken from a set $X = \{x_1, \ldots, x_n\}$. There is a distinguished node, called the starting node, which has indegree 0. A branching program computes an n-argument Boolean function as follows: starting in the starting node, the value of the variable labelling the current node is tested. If it is 0(1) the next node tested is the left(rigth) descendant of the current node. The branching program accepts $A \subseteq \{0,1\}^n$ if and only if for all $w \in \{0,1\}^n$ the path traced from the starting node under w halts at a sink labelled $\chi_A(w)$ where χ_A denotes the characteristic function of A. The complexity measure for a branching program is the number of non-sink nodes. Let us denote by $\mathcal{P}_{BP}$ the class of all languages $A \subseteq \{0,1\}^*$ whose instances $A^n = A \cap \{0,1\}^n$ will be accepted by polynomial-size branching programs. Applying a famous idea of Savitch /Sa70/ one obtains

THEOREM 1: $\mathcal{L} = \mathcal{P}_{BP}$. ☒

In order to describe $\mathcal{NL}$ by means of polynomial-size branching-program-like devices disjunctive (or 1-time-only nondeterministic) branching programs where introduced in /Me86/. A branching program is called a disjunctive branching program if some of its non-sink nodes are labelled by $\vee$ (OR) instead of Boolean variables. A disjunctive branching program accepts its input w iff there is one accepting path in the set of all paths from the starting node traced under w. If $\mathcal{P}_{dBP}$ again denotes the class of all languages $A \subseteq \{0,1\}^*$ whose instances A^n will be accepted by polynomial-size disjunctive branching programs, then we can prove

THEOREM 2 /Me86/: $\mathcal{NL} = \mathcal{P}_{dBP}$. ∅

Very recently Barrington succeeded in describing $\mathcal{NC}^1$ by means of polynomial-size, bounded-width branching programs. Thereby, the width of a branching program can be defined as the maximal number, over all d, of vertices at distance d from the root taking the length of a longest path from the root to a vertex v as the distance d(v) of v . If $\mathcal{P}_{bw-BP}$ denotes the class of all languages $A \subseteq \{0,1\}^*$ whose instances A^n will be accepted by polynomial-size branching programs of bounded width, then it holds

THEOREM 3 /Ba86/: $\mathcal{NC}^1 = \mathcal{P}_{bw-BP}$. ∅

Moreover, Barrington proved

$$\mathcal{NC}^1 = \mathcal{P}_{k-BP} \quad \text{for all } k \geq 5 ,$$

where $\mathcal{P}_{k-BP}$ denotes the class of all languages whose instances will be accepted by polynomial-size branching programs of width k .

One of the major goals of complexity theory is to separate complexity classes or to prove their coincidence. In terms of the branching-program-based characterization of the classes $\mathcal{L}$ and $\mathcal{NL}$ given above the outstanding $\mathcal{L} \stackrel{?}{=} \mathcal{NL}$ question can be formulated as the question whether polynomial-size disjunctive branching programs are more powerful than usual polynomial-size ones. However, in the restricted case of polynomial, bounded-width branching programs the occurence of disjunctive OR-nodes does not increase the power of these devices.

PROPOSITION 1: $\mathcal{P}_{bw-dBP} = \mathcal{P}_{bw-BP}$.

PROOF: Let $A \in \mathcal{P}_{bw-dBP}$. Then there exists a $k \in \mathbb{N}$ such that all instances A^n of A will be accepted by a polynomial size disjunctive branching program P' of width k'. Handling the disjunctive OR-nodes as nodes labelled with fictive Boolean variables, one can transform P' into a levelled disjunctive branching program P (i.e. all paths leading from the root to a point v have length d(v)) of polynomial size and width $k \leq 2k'$ with the additional property that all nodes of a level are either labelled with the same variable or are OR-nodes /Ba86/.

Obviously, a level j of P labelled with a Boolean variable x_i is completely described by two functions

$$f_j , g_j : [k] \longrightarrow [k]$$

($[k] := \{1,\ldots,k\}$) giving the end points $f_j(v)$, $g_j(v)$ in the level $j+1$ of the two edges leaving a node $v \in [k]$ of the level j in the cases $x_i = 0$ and $x_i = 1$. If the level j consists of OR-nodes, then it is completely described by a relation

$$R_j \subseteq [k]^2$$

giving the end points $\{s,t\} = R_j(v)$ in the level $j+1$ of the edges leaving v of the level j .

Now we will simulate the disjunctive branching program P level by level by a (usual) branching program $\tilde{P}$ of width 2^k whose length equals that of P , which implies $A \in \mathcal{P}_{\text{bw-BP}}$, too. In order to do this take the elements M of $2^{[k]}$ as the nodes of the levels of $\tilde{P}$. If the level j of P is labelled with the variable x_i , then label the nodes of the level j of $\tilde{P}$ by x_i , too, and define

$$\tilde{f}_j , \tilde{g}_j : 2^{[k]} \longrightarrow 2^{[k]}$$

by

$$\tilde{f}_j(M) = \{f_j(m) \mid m \in M\} \quad \text{and} \quad \tilde{g}_j(M) = \{g_j(m) \mid m \in M\} .$$

But if the level j of P consists of disjunctive OR-nodes, then label the nodes of the level j of $\tilde{P}$ by any **one of** the variables $x_1,\ldots, x_n$ and define $\tilde{f}_j$, $\tilde{g}_j$ by

$$\tilde{f}_j(M) = \tilde{g}_j(M) = \bigcup_{m \in M} R_j(m) .$$

Obviously, $\tilde{P}$ is of polynomial size and accepts exactly A^n . Conversely, since every bounded-width branching program is a disjunctive one of bounded width, we are finished. $\varnothing$

Obviously, in the case of unbounded-width disjunctive branching programs the construction above leads to exponential-size branching programs.

As a consequence of Proposition 1 and Theorem 2 one obtains

COROLLARY: $\mathcal{NC}^1 = \mathcal{L}$ implies $\mathcal{L} = \mathcal{NL}$ ($= \mathcal{NC}^1$) .

PROOF: From $\mathcal{NC}^1 = \mathcal{L}$ it follows that for all polynomial-size branching programs P there is a polynomial-size branching program P' of bounded width which accepts the same set as P does. Assigning to each of the disjunctive OR-nodes of a polynomial-size disjunctive branching program P (which is polynomial in the number n of arguments of P) a new Boolean variable one obtains a branching program $\tilde{P}$ which is of polynomial size in n , too. Now, this branching program can be transformed into a polynomial-size, bounded-width branching program $\tilde{P}'$ which accepts the same set as P does when all new Boolean variables are substituted back into disjunctive OR-nodes. From this observation one obtains $\mathcal{NL} = \mathcal{P}_{dBP} = \mathcal{P}_{bw-dBP} = \mathcal{P}_{bw-BP} = \mathcal{NC}^1 = \mathcal{L}$ by means of Proposition 1 and Theorem 2. ☒

2. SOME p-PROJECTION COMPLETE GRAPH ACCESSIBILITY PROBLEMS FOR $\mathcal{NC}^1$, $\mathcal{L}$ AND $\mathcal{NL}$

It is standard in complexity theory to relate the complexity of a particular problem to that of a larger class as a whole by showing that it is complete with respect to certain reducibilities. In the following section we prove the p-projection completeness of same restricted graph accessibility problems for the classes $\mathcal{NC}^1$ and $\mathcal{L}$ by means of the branching-program-based characterization of these complexity classes given in section 1. We use these characterizations also to make well known facts as the completeness of the (general) graph accessibility problem for $\mathcal{NL}$ /Sa70/,/CSV84/ more transparent.

Recall, a problem $G = \{G_n\}$ is <u>p-projection reducible</u> /SV81/ to a problem $F = \{F_n\}$, and this will be abbreviated $G \leq F$, if there is a function $p(n)$ bounded above by a polynomial in n , and if for every $G_n \in G$ there is a mapping

$$\sigma_n : \{y_1,\ldots,y_{p(n)}\} \longrightarrow \{x_1,\bar{x}_1,\ldots,x_n,\bar{x}_n,0,1\}$$

such that

$$G_n(x_1,\ldots,x_n) = F_{p(n)}(\sigma_n(y_1),\ldots,\sigma_n(y_{p(n)})) .$$

F and G are p-projection equivalent $G \equiv F$ if $G \leq F$ and $F \leq G$. A problem $F = F_n$ is <u>p-projection complete</u> ($\leq$-complete) in a complexity class $\mathcal{K}$ if all problems $G = \{G_n\}$ of $\mathcal{K}$ are p-projections of F , and if F itself belongs to $\mathcal{K}$.

At first we consider the graph accessibility problem $GAP^{(1,k)} = \{GAP_n^{(1,k)}\}_{n \in \mathbb{N}}$ for monotone graphs of outdegree 1 and bandwidth k . Thereby, a directed graph $G = (V,E)$ with vertex set $V = \{v_1,\ldots,v_t\}$ is said to be of <u>bandwidth</u> k if $|j-i| \leq k/2$ holds for all its edges (v_i,v_j) . G is said to be <u>monotone</u> if its edges (v_i,v_j) always lead from nodes v_i with a smaller index to nodes v_j with a higher one, $i \leq j$.

ne outdegree of G is the maximal outdegree of one of its odes. It is clear that monotone graphs $G = (V,E)$ can be niquely represented by the rigth upper part of an adjacency atrix $A_G = (a_{ij})_{1 \le i < j \le |G|}$ with

$$a_{ij} = \begin{cases} 1 & (v_i, v_j) \in E \\ 0 & \text{otherwise} \end{cases}$$

aking into account the vertex enumeration of V.

$$GAP_n^{(1,k)} : \{0,1\}^{(n-1)(n-2)/2} \longrightarrow \{0,1\}$$

$$A_G = (a_{ij})_{i<j} \mapsto \begin{cases} 1 & \text{there exists a path from vertex } v_1 \text{ to vertex } v_n \text{ in the monotone graph of outdegree } 1 \text{ and bandwidth } k \text{ defined by the first } 1 \text{ within the first } k \text{ bits of every row of } (a_{ij})_{i<j} \\ 0 & \text{otherwise .} \end{cases}$$

COROLLARY TO THEOREM 3: $GAP^{(1,k)}$ is $\le$-complete for $\mathcal{NC}^1$ for all $k \geq 10$.

PROOF: Let $F = \{F_n\}_{n \in \mathbb{N}}$ be a problem in $\mathcal{NC}^1$. From Theorem 3 we know that every $F_n \in F$ can be computed by a width-$\lceil k/2 \rceil$ branching program P_n of polynomial size $p(n)$. Without loss of generality we can again assume that P_n is levelled. Since P_n is based on an acyclic graph with $p(n)$ vertices we can enumerate its vertices by $0,1\ldots, p(n)-1$ in such a way that

- the starting node is numbered by 0,
- the accepting node is numbered by $p(n)-1$, and
- all nodes of a lower level get smaller numbers than those of a higher one. (Consequently, all edges lead always from nodes with a lower number to nodes with a higher one.

Now, to every input $(x_1,\ldots,x_n) \in 2^n$ of F_n a graph $G_{(x_1,\ldots,x_n)} = (x_{ij})_{0 \le i,j \le p(n)-1}$ can be assigned

$$x_{ij} = \begin{cases} x_k & \text{if vertex } i \text{ is labelled } x_k \text{ and if there is an edge } (i,j) \text{ in } P_n \text{ labelled } 1 \text{ ;} \\ \bar{x}_k & \text{if vertex } i \text{ is labelled } x_k \text{ and if there is an edge } (i,j) \text{ in } P_n \text{ labelled } 0 \text{ ;} \\ 0 & \text{otherwise .} \end{cases}$$

By definition, $G_{(x_1,\ldots,x_n)}$ is of outdegree 1. Further, it is monotone and of bandwidth k by means of the special nature of the enumeration of the vertices of P_n. Finally, since $F_n(x_1,\ldots,x_n) = 1$ if and only if $GAP_{p(n)}^{(1,k)}((x_{ij})_{i<j}) = 1$ we have

$$F \leq GAP^{(1,k)} .$$

Moreover, $GAP^{(1,k)}$ is $\leq$-complete for $\mathcal{NC}^1$, too, because its instances $GAP_n^{(1,k)}$ can be computed by the following po nomial-size branching program P_n of width **k** which is bui from the following stages $S_{i,r}$, $r = \lceil i/k \rceil k+1$ presented to gether with their joining conditions.

$S_{i,r}$, $i+k < n$:

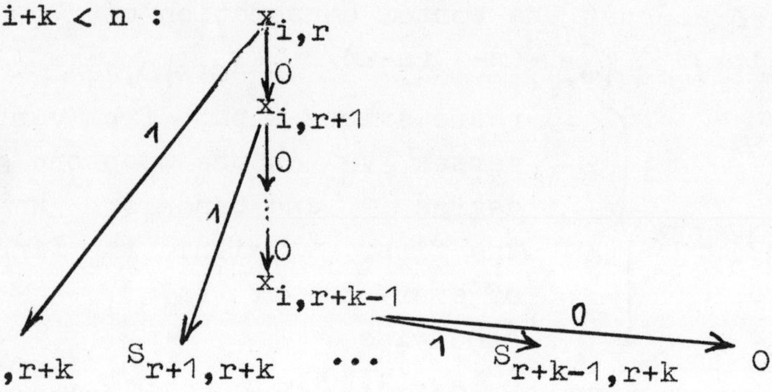

If $r = \lceil i/k \rceil k+1 = n$ then $S_{i,r}$ degenerates to an accepting sink, and if $r > n$ then $S_{i,r}$ degenerates to a rejecting sink. Assigning these stages in the following manner

P_n :

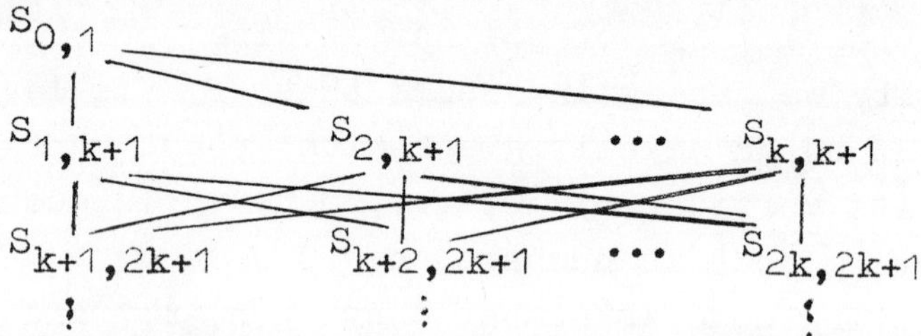

one obtains a polynomial-size, width $-$ k branching program which computes $GAP_n^{(1,k)}$. ☒

REMARK: Since the parity functions belong to $\mathcal{NC}^1$ one obtair from /FSS81/ and the last Corollary that every constant-depth Boolean circuit computing a $GAP_n^{(1,k)}$ has to be of exponentia size in n. ☒

Similar arguments show that the graph accessibility problem $GAP^{(1)} = \{GAP_n^{(1)}\}_{n \in \mathbb{N}}$ of monotone graphs of outdegree 1

$$GAP_n^{(1)} : \{0,1\}^{(n-1)(n-2)/2} \longrightarrow \{0,1\}$$

$$A_G = (a_{ij})_{i<j} \longmapsto \begin{cases} 1 & \text{there exists a path from vertex } v_1 \text{ to ver-} \\ & \text{tex } v_n \text{ in the monotone graph of outdegree} \\ & \text{1 defined by the first 1 of every row of} \\ & (a_{i,j})_{i<j} \\ 0 & \text{otherwise} \end{cases}$$

is p-projection complete for $\mathcal{L}$ /Me86/.

COROLLARY TO THEOREM 1: $GAP^{(1)}$ is $\leq$-complete for $\mathcal{L}$. ☒

The classical result that the graph accessibility problem GAP and the more restricted graph accessibility problem $GAP^{(2)} = \{GAP^{(2)}_n\}_{n \in \mathbb{N}}$ concerning monotone graphs G of outdegree 2 is $\leq$-complete /Sa70/, /CSV84/ we easily obtain from the characterization of $\mathcal{NL}$ by polynomial-size disjunctive branching programs by Theorem 2 /Me86/.

$$GAP^{(2)}_n : \{0,1\}^{(n-1)(n-2)/2} \longrightarrow \{0,1\}$$

$$(a_{ij})_{i<j} \mapsto \begin{cases} 1 & \text{if there is a path from vertex } v_1 \text{ to vertex } v_n \\ & \text{in the monotone graph of outdegree 2 defined by} \\ & \text{taking the first two 1's of every row of } (a_{ij}) \\ 0 & \text{otherwise .} \end{cases}$$

COROLLARY TO THEOREM 2: $GAP^{(2)}$ is $\leq$-complete for $\mathcal{NL}$. ☒

After having seen that disjunctive branching programs correspond to the graph accessibility problem for monotone graphs of outdegree 2 we immediately obtain the $\leq$-completeness of the graph accessibility problem $GAP^{(2,k)}$ of monotone graphs of outdegree 2 and bandwidth k from Proposition 1 and the Corollary of Theorem 3

COROLLARY 2 TO THEOREM 3: $GAP^{(2,k)}$ is $\leq$-complete for $\mathcal{NC}^1$ for all $k \geq 3$. ☒

Reversily, the construction of a polynomial-size, bounded-width branching program for the instances $GAP^{(2,k)}_n$ of $GAP^{(2,k)}$ which must be omitted here gives another proof for Proposition 1.

3. $\mathcal{NC}^1$, $\mathcal{L}$ AND $\mathcal{NL}$ AS p-PROJECTION CLOSURES OF THE CLASSES NC^1, L AND NL

By means of the complete problems for the classes $\mathcal{L}$ and $\mathcal{NL}$ given in section 2 and their branching-program-based characterizations of sction 1 in /Me86/ it was proved that $\mathcal{L}$ and $\mathcal{NL}$ are closed under p-projection reducibility

$$\mathcal{L}^* = \mathcal{L} \qquad \text{and} \qquad \mathcal{NL}^* = \mathcal{NL} \quad ,$$

i.e. every problem $\leq$-reducible to a problem of one of these classes is itself a member of this class. Similar arguments applied to a graph accessibility problem $GAP^{(1,k)}$, $k \geq 10$

show that $\mathcal{NC}^1$ is $\leq$-closed, too

$$\mathcal{NC}^{1*} = \mathcal{NC}^1 .$$

Moreover, $\mathcal{NC}^1$, $\mathcal{L}$ and $\mathcal{NL}$ are the $\leq$-closures of the uniform complexity classes NC^1, L and NL, respectively, since their $\leq$-complete problems $GAP^{(1,k)}$, $GAP^{(1)}$ and $GAP^{(2)}$ belong to NC^1, L and NL, respectively.

PROPOSITION 2: $NC^{1*} = \mathcal{NC}^1$,

$L^* = \mathcal{L}$

and $NL^* = \mathcal{NL}$.

However, in order to separate $\mathcal{NC}^1$, $\mathcal{L}$ and $\mathcal{NL}$ it would be enough to prove that $GAP^{(1)} \not\leq GAP^{(1,k)}$ $(k \geq 10)$ or $GAP^{(2)} \not\leq GAP^{(1)}$, respectively.

REFERENCES

/Ba86/ D.A.Barrington, 'Bounded width polynomial-size branching programs recognizing exactly those languages in NC^1', Proc. 18-th STOC, 1986, 1-5

/CSV84/ A.K.Chandra, L.Stockmeyer, U.Vishkin, 'Constant depth reducibility', SIAM J.Comput. 13, No. 2,1984, 423-439

/FSS81/ M.Furst, J.B.Saxe, M.Sipser, 'Parity, circuits and the polynomial-time hierarchy', Proc. 22-th FOCS, 1981, 260-270

/KL80/ R.M.Karp, R.J.Lipton, 'Some connections between nonuniform and uniform complexity classes', Proc. 12-th STOC, 1980, 302-309

/Me86/ Ch.Meinel, '$\mathcal{L}$(nonuniform) and $\mathcal{NL}$(nonuniform)', Proc. MFCS-86 (Bratislava), 1986

/Sa70/ W.Savitch, 'Relations between nondeterministic and deterministic tape complexities', J. Comp. and Sys.Sc. 4, 1970, 177-192

/SV81/ S.Skyum, L.G.Valiant, 'A complexity theory based on Boolean algebra', Proc. 22-th FOCS, 1981, 244-253

CONSTRUCTIVE THEORIES WITH ABSTRACT DATA TYPES FOR PROGRAM SYNTHESIS

Pierangelo Miglioli, Ugo Moscato, and Mario Ornaghi

Dept. of Information Science
University of Milan
Milan, Italy

1. GENERAL DISCUSSION

The research explained in this paper originates from program synthesis in the frame of intuitionistic logic [6] and has been furtherly developed as a study involving, on the one hand, constructive proofs as programs [12], on the other hand the possibility of providing axiomatizations of mathematical structures (abstract data types) compatible with constructive logical principles [3].

As for the first aspect, in the recent years various ways of looking at proofs as programs have been taken into account with an increasing interest in Computer Science. In the constructive attitude, Martin-Löf's approach [10] proposes a strictly intuitionistic point of view; other approaches, including also implementative efforts, are described, e.g., in Goad [7] and in [1]. In this area, the emphasis is on the computational interpretation of the logical rules and on the various information structures making this interpretation possible, even if the problem of the axiomatization of specific data structures is not disregarded (induction principles are taken into account [1,7] and the use of Harrop axioms is proposed, e.g., in [7]).

As for the second aspect, the problem of the axiomatization of mathematical structures adequate to program construction and analysis has been systematically investigated in the area of abstract data types.In this frame the underlying logic is classical. The main approach, originated from works such as [8], is the algebraic one and uses tools of Universal Algebra and Category Theory; it has given rise to a number of interesting developments. Another approach, corresponding to the author's point of view, is the "isoinitial models as abstract data types" one: it uses (classical) Model Theory and has been compared with the algebraic approach in various papers [2,3].

Whether or not the algebraic approach is more appropriate to other fields of Computer Science, the authors believe that for the above quoted purpose the one based on isoinitiality is more fruitful. Indeed, in [3] some results are given in orde to show that the classical characterization of the isoinitia abstract data types fits well in a constructive attitude; thi is intended as a starting point to handle in a unified wa "abstract data types" and "abstract programs", the latte being the proofs of a constructive system.

In the present paper we will describe some classificatio schemes and some general techniques to study constructive pro perties of large and effective systems S=T+L consistent wit classical logic, where:

1) T is a theory with Abstract Data Types (with ADT), i.e. theory completely formalizing an isoinitial model [2,3] a specified below;

2) L is a "superintuitionistic" logic with identity;

3) effectiveness is taken as synonymous of axiomatizability.

We will be mainly interested in the two constructive properties explained below, where "$T \mid\!\overset{L}{-}\, A$" indicates the provability of A in the system S=T+L and "CL" indicates classical logic:

4) A system S = T+L is said to be "constructive" iff it satisfies the disjunction property DP, i.e.
$T \mid\!\overset{L}{-}\, A \vee B \Rightarrow T \mid\!\overset{L}{-}\, A$ or $T \mid\!\overset{L}{-}\, B$ for A and B closed
and the explicit definability property EDP, i.e.
$T \mid\!\overset{L}{-}\, \exists x A(x) \Rightarrow T \mid\!\overset{L}{-}\, A(t)$ for $\exists x A(x)$ and $A(t)$ closed.

5) A system S = T+L is said to be "semi-constructive" iff it satisfies the weak disjunction property WDP, i.e.
$T \mid\!\overset{L}{-}\, A \vee B \Rightarrow T \mid\!\overset{CL}{=}\, A$ or $T \mid\!\overset{CL}{=}\, B$ for A and B closed
and the weak explicit definability property WEDP, i.e.
$T \mid\!\overset{L}{-}\, \exists x A(x) \Rightarrow T \mid\!\overset{CL}{=}\, A(t)$ for $\exists x A(x)$ and $A(t)$ closed.

The semi-constructiveness of various effective systems S can be obtained by showing that they are contained in non effective sets of classically provable formulas closed under modus ponens MP, DP and EDP. As for the constructive systems, a syntactical technique to extract information from proofs and called by the authors the "collection technique" [3,12] can be applied; it turns out to be useful in order to obtain (fully) constructive sub-systems from semi-constructive systems.
We will treat with some detail two of the non effective sets of formulas quoted above: the sets WG(T) and EVF(T).
These sets can be viewed as upper bounds corresponding to two different kinds of constructivism: WG(T) directly comes as an upper bound from the collection technique itself;EVF(T), based on the notion of "valuation form", comes from researches on constructive semantics (we quote in this field the various notions of realizability, more or less connected with the intuitionistic school [14]) and has been heavily inspired by the works on Medvedev's propositional logic of finite problems [11].

Aspects related to our research concern the simultaneous "constructive compatibility" of logical and mathematical principles; according to these aspects, our results are oriented to a classification.

The above general setting gives also a theoretical frame which is being used in the implementation of the system PAP; in particular it is used in the part of PAP related to the definition and to the extension of abstract data types (along the lines explained in [4]). A short outline of PAP will conclude this paper.

2. THEORIES WITH ISOINITIAL MODEL AND THE LOGIC IKA

A theory T will be intended in the usual (classical) model theoretic sense [5].

According to our purposes, a theory T <u>completely formalizing</u> an isoinitial model will be so characterized:

1) for any closed quantifier-free formula F, $T \vdash^{CL} F$ or $T \vdash^{CL} \neg F$

2) there is a model M of T, called isoinitial, such that every element of its carrier is denoted by a closed term.

The theories with an isoinitial model, according to the semantical characterization given, e.g., in [2,3], can be extended, without affecting their axiomatizability and their class of models, into theories completely axiomatizing an isoinitial model: the addition of a recursive diagram is sufficient [3].

The logic IKA consists of all the rules of the intuitionistic predicate calculus with identity, of the (superintuitionistic) Kuroda Principle

(K) $\forall x \neg\neg A(x) \rightarrow \neg\neg \forall x\, A(x)$

and of the rule of the double negation restricted to the atomic formulas

(AT) $\neg\neg A \rightarrow A$ for every atomic formula A

Both (K) and (AT) are essential in proving the following fact (see [3]):

I) If T completely formalizes an isoinitial model, then, for every closed quantifier-free formula F,
$T \vdash^{IKA} F$ or $T \vdash^{IKA} \neg F$.

3. INDUCTION PRINCIPLES, DESCENDING CHAIN PRINCIPLE AND OTHER MATHEMATICAL AXIOMS

A theory completely formalizing an isoinitial model may well satisfy an induction principle and a descending chain principle. These principles will be distinguished from the other mathematical axioms: unless otherwise specified, the term "axiom" will denote the latter.
Induction principles (which can also be taken in generalized forms such as e.g. Structural Induction [9]) are well known. A descending chain principle, for a T with a binary relation < in its signature, can be given as follows:

(DC) $\exists x A(x) \wedge \forall x(A(x) \rightarrow (\exists y(A(y) \wedge y < x) \vee B)) \rightarrow B$
for A and B any formulas, B not containig x free.
We say that (DC) is in the appropriate context if T completely

formalizes an isoinitial model, $<$ is axiomatized in T as an irreflexive partial ordering which is well founded in any isoinitial model of T.

As concerns the (properly said) axioms, restrictions in their forms are needed in order to use them in constructive systems; we will be interested in the following classes of formulas:

A $\forall$-formula is of the kind $\forall \underline{x}\ H$, an $\exists$-formula is of the kind $\exists \underline{x}\ H$, a $\forall\exists$-formula is of the kind $\forall \underline{x}\ \exists \underline{y}\ H$, with H quantifier-free and $\underline{x}$, $\underline{y}$ possibly empty.

A $\forall\exists\neg$-formula is inductively so defined:
every $\forall\exists$-formula is a $\forall\exists\neg$-formula; every formula such as $\neg A$ is a $\forall\exists\neg$-formula;
if A and B are $\forall\exists\neg$-formulas and C is <u>any</u> formula, then $A \wedge B$, $C \rightarrow A$, $\forall x A$ are $\forall\exists\neg$-formulas.
A $\forall\neg$-formula is defined likewise, taking in the basic clause the $\forall$-formulas instead of the $\forall\exists$-formulas.

4. THE CONSTRUCTIVE FRAMES WG(T) AND EVF(T) FOR T COMPLETELY FORMALIZING AN ISOINITIAL MODEL

In order to define the non effective set WG(T), we introduce the notion of a "formula A well given in a collection $\mathcal{C}$ of closed formulas" (abbreviated by "A is w.g. in $\mathcal{C}$").

We say that A is w.g. in $\mathcal{C}$ iff $A \in \mathcal{C}$ and one of the following inductive clauses holds:
1) A is atomic, or $A = \neg B$ for some B;
2) $A = B \wedge C$, and B and C are w.g. in $\mathcal{C}$;
3) $A = B \vee C$, and B is w.g. in $\mathcal{C}$ or C is w.g. in $\mathcal{C}$;
4) $A = B \rightarrow C$, and if B is w.g. in $\mathcal{C}$ then C is w.g. in $\mathcal{C}$;
5) $A = \exists x\ B(x)$, and B(t) is w.g. in $\mathcal{C}$ for some closed term t;
6) $A = \forall x\ B(x)$, and B(t) is w.g. in $\mathcal{C}$ for any closed term t.

Now, let $\mathcal{C}(T) = \{ A \ / \ A \text{ closed and } T \mid\!\frac{CL}{}\ A \}$ (we recall that CL is classical logic); then WG(T) is so defined:

$$WG(T) = \{ A \ / \ A \text{ is w.g. in } \mathcal{C}(T) \}$$

One can prove the following maximality result:

<u>Theorem 1</u> Let $\mathcal{C}$ be any set of classically T-provable closed <u>formulas</u> satisfying MP, DP and EDP and such that $WG(T) \subseteq \mathcal{C}$; then $WG(T) = \mathcal{C}$. □

In order to define the frame EVF(T) we will associate, with every model M of a theory T completely formalizing an isoinitial model, a language L(M) extending the language L of T: this extension is made by introducing new constants in order to represent all elements of the carrier of M by closed terms; so, for any isoinitial model I we have L(I) = L. In the following, Term(L(M)) will be the set of all closed terms of L(M) and Term(L) will be the set of all closed terms of L = L(I).

Now, for every closed $A \in L(M)$, we associate with A a set F(M,A) of formal objects (called <u>M-valuation forms</u> of A) inductively defined as follows:

1) $F(M,A) = \{A\}$ for A an atomic or a negated formula (here there is a single valuation form);
2) $F(M,B \wedge C) = \{\langle b,c\rangle \ / \ b \in F(M,B)$ and $c \in F(M,C)\}$
3) $F(M,B \vee C) = \{\langle 1,b\rangle \ / \ b \in F(M,B)\} \cup \{\langle 2,c\rangle \ / \ c \in F(M,C)\}$
4) $F(M,B \rightarrow C) = \{f \ / \ f : F(M,B) \rightarrow F(M,C)\}$ (i.e., $F(M,B \rightarrow C)$ is the set of all the functions from $F(M,B)$ to $F(M,C)$);
5) $F(M, \exists x\, B(x)) = \{\langle t, b\rangle \ / \ t \in Term(L(M))$ and $b \in F(M,B(t))\}$;
6) $F(M, \forall x\, B(x)) = \{f \ / \ f$ is a function associating, with every $t \in Term(L(M))$, an element of $F(M,B(t))\}$.

If I is an isoinitial model of T, $F(I,A)$ will be simply represented by "$F(A)$" and the elements of $F(A)$ will be called basic valuation forms (or, simply, valuation forms).

The *satisfaction in a model* M of $a \in F(M,A)$ (denoted by $M \models a$) *is inductively* so *defined*:
1) $M \models A$ iff A is true in M, for A atomic or negated (in this case, A is the unique M-valuation form of A);
2) $M \models \langle b, c\rangle$ (with $\langle b, c\rangle$ $F(M,B \wedge C)$) iff $M \models b$ and $M \models c$;
3) $M \models \langle 1,b\rangle$ (with $\langle 1,b\rangle \in F(M,B \vee C)$) iff $M \models b$ and $M \models \langle 2,c\rangle$ (with $\langle 2,c\rangle \in F(M,B \vee C)$) iff $M \models c$;
4) $M \models f$, with f $F(M,B \rightarrow C)$, iff, for every $b \in F(M,B)$, if $M \models b$ then $M \models f(b)$ (where $f(b) \in F(M,C)$);
5) $M \models \langle t,b\rangle$ (with $\langle t,b\rangle \in F(M, \exists x B(x))$) iff $M \models b$;
6) $M \models f$, with $f \in F(M, \forall x B(x))$, iff, for every $t \in Term(L(M))$, $M \models f(t)$ (where $f(t) \in F(M,B(t))$).

The following fact relates the satisfaction in M of a formula $A \in L(M)$ and of its M-valuation forms (justifying the sense of "M-valuation forms"):

(II) $M \models A$ iff there is an $a \in F(M,A)$ such that $M \models a$.

To obtain something different from classical truth, new notions are to be attached to the valuation forms.

We say that $a' \in F(M,A)$ is an extension of $a \in F(A)$ iff A is atomic or negated (in which case $a' = a = A$) or one of the following inductive clauses is satisfied:
1) $a' = \langle b',c'\rangle$ and $a = \langle b,c\rangle$ and b' is an extension of b and c' is an extension of c;
2) $a' = \langle 1,b'\rangle$ and $a = \langle 1,b\rangle$ and b' is an extension of b, or $a' = \langle 2,c'\rangle$ and $a = \langle 2,c\rangle$ and c' is an extension of c;
3) $a' \in F(M,B \rightarrow C)$ and $a \in F(B \rightarrow C)$ and, for every $b' \in F(M,B)$ and every $b \in F(B)$, if b' is an extension of b, then $c' = a'(b')$ is an extension of $c = a(b)$;
4) $a' = \langle t,b'\rangle$ and $a = \langle t,b\rangle$ (with $t \in Term(L)$) and b' is an extension of b;
5) $a' \in F(M, \forall x B(x))$ and a $F(\forall x B(x))$ and, for every $t \in Term(L)$, $b' = a'(t)$ is an extension of $b = a(t)$.

Now we say that any isoinitial model I of T *extensively satisfies* $a \in F(A)$ (and we denote this by "$I \models^{e} a$") iff, for every model M of T, there is $a' \in F(M,A)$ such that a' is an extension of a and $M \models a'$. With this definition we can introduce $EVF(T)$ as follows:

$$EVF(T) = \{A \ / \ A \in L \text{ and } A \text{ is closed, and there is } a \in F(A) \text{ such that } I \models^{e} a \text{ for any isoinitial } I \text{ of } T\}$$

The following fact is easily proved:

(III) EVF(T) is a set of classically T-provable formulas closed under MP, DP and EDP.

We do not know whether or not EVF(T) satisfies a maximality result such as the one quoted for WG(T); we believe that the answer is negative.

5. CONSTRUCTIVE SYSTEMS IN THE FRAME OF WG(T)

The logic EMIKA is obtained by adding to IKA the well known Markov Principle (Ma) and the two principles (E1) and (E2) given below:

(Ma) $\forall x(A(x) \vee \neg A(x)) \wedge \neg\neg\exists x A(x) \rightarrow \exists x A(x)$ for any A;

(E1) $(\forall x(\neg\neg A(x) \rightarrow A(x)) \rightarrow \exists x(\neg\neg A(x) \vee B(x)))$
$\rightarrow \exists x(\neg\neg A(x) \vee B(x))$ for any A and B;

(E2) $(\forall x(\neg\neg A(x) \rightarrow A(x)) \rightarrow \forall x(\neg\neg A(x) \vee B(x))$
$\rightarrow \exists x \neg\neg A(x) \vee B(t)$ for any A, B and any term t.

Now, let T be a (mathematical) theory possibly _including induction and_ (_DC_); then one has:

Theorem 2 If T completely formalizes an isoinitial model and all the (properly said) axioms of T are $\forall\exists\neg$-formulas and A is closed, then T $\vdash_{EMIKA}$ A implies $A \in WG(T)$. □

The proof of Theorem 2 requires the choice of a formal system for EMIKA and is a lengthy induction on the structure of a proof in the system.

Now, let (WE1) and (WE2) be obtained, respectively, from (E1) and (E2) by allowing B(x) to be only a negated formula and let WEMIKA (weak EMIKA) be the logic obtained by adding to IKA (WE1) and (WE2); then WEMIKA satisfies the following (full) constructiveness result:

Theorem 3 With the same hypotheses on T as in Theorem 2, the formal system S=T+WEMIKA is constructive. □

The proof of 3 is rather involved and can be made by proving that all the formulas of $\mathcal{C}$(WEMIKA,T) = {A/T $\vdash_{EMIKA}$ A with A closed} are w.g. in $\mathcal{C}$(WEMIKA,T). To obtain this result, one has to follow the schema outlined in the proof of Theor. 5.2 of [3]; the above Theorem 2 is to be used to treat the critical rules (WE1) and (WE2).

6. EFFECTIVE SYSTEMS IN THE FRAME OF EVF(T)

We say that a formula $A \in L$ is _T,F(A)-evaluable_ iff, for every model M of T:
if M $\models$ A then there are $a \in F(A)$ and $a' \in F(M,A)$ such that a' extends a and M $\models$ a'.

With this definition we can state the following theorem:

Theorem 4 Let A,B,C,D be closed formulas such that:

1) for every $a \in F(A)$ and every model M of T, there is a unique extension a' of a such that $a' \in F(M,A)$;

2) for every model M of T such that $M \models A$ one of the following conditions holds:

2.1) $M \models b'$ for every $b' \in F(M,B)$;

2.2) $M \not\models b'$ for every $b' \in F(M,B)$;

2.3) if there are a model M* of T, $c^* \in F(C)$ and $c^{**} \in F(M^*,C)$ such that c^{**} extends c^* and $M^* \models c^{**}$, then there is some $c^{*\prime} \in F(M,C)$ such that $c^{*\prime}$ extends c^* and $M \models c^{*\prime}$;

3) B is T,F(B)-evaluable.

Then the following formula belongs to EVF(T):

$$\text{(COSTR} \vee\text{)}: (A \to (B \to C \vee D)) \to (A \to (B \to C) \vee (B \to D))$$

Moreover (with the above hypotheses on A,B,C), if $C = \exists wC'(w)$, then the following formula belongs to EVF(T):

$$\text{(COSTR} \exists\text{)}: (A \to (B \to \exists wC'(w))) \to (A \to \exists w(B \to C'(w))).\square$$

The set of the T-stable formulas is inductively so defined:
1) if A is atomic or negated then A is T-stable; if A is a $\exists$-formula and $T \vdash^{CL} A$, then A is T-stable;
2) if B and C are T-stable then $B \wedge C$, $B \vee C$, $B \to C$ and $\forall xB$ are T-stable. One can prove:

(IV) If A is T-stable, then A is T,F(A)-evaluable.

Let $U(\underline{y})$ and $V(\underline{y},\underline{z})$ be formulas containing free at most the indicated variables; we say that C is a formula in U, U -> V, $\underline{y}$ iff C is inductively so defined:

1) $C=U(\underline{t})$, where the (possible) variables of the terms $\underline{t}$ are in $\underline{x}$;

2) $C=U(\underline{t}) \to V(\underline{t},\underline{t}')$ where the (possible) variables of $\underline{t}$ and $\underline{t}'$ are in $\underline{x}$;

3) $C= \neg C'$, or $C=C1 \wedge C2$, or $C=C1 \vee C2$, or $C=C1 \to C2$, with C', C1 and C2 formulas in U, U -> V, $\underline{y}$;

4) $C= \exists vC'(v)$, or $C= \forall vC'(v)$, with $v \in \underline{x}$ and C' a formula in U, U -> V, $\underline{y}$.

With the above definitions we give the two following families of principles:

$$\text{(FP} \vee\text{)} \quad (\forall x(\neg\neg H(x) \wedge \neg\neg K(x) \to \forall \underline{y} \neg U(\underline{y})) \to (\forall x(H(x) \to K(x)) \to C \vee D)) \to (\forall x(\neg\neg H(x) \wedge \neg\neg K(x) \to \forall \underline{y} \neg U(\underline{y})) \to (\forall x(H(x) \to K(x)) \to C) \vee (\forall x(H(x) \to K(x)) \to D)),$$

where K(x) is T-stable and C is any formula in U, U -> V, $\underline{y}$ (D is arbitrary);

$$\text{(FP} \exists\text{)} \quad (\forall x(\neg\neg H(x) \wedge \neg\neg K(x) \to \forall \underline{y} \neg U(\underline{y})) \to (\forall x(H(x) \to K(x)) \to \exists wC(w))) \to (\forall x(\neg\neg H(x) \wedge \neg\neg K(x) \to \forall \underline{y} \neg U(\underline{y})) \to \exists w(\forall x(H(x) \to K(x)) \to C(w))),$$

where K(x) is T-stable and C(w) is any formula in U, U -> V, $\underline{y}$.

The closed instances of the principles (FP $\vee$) and (FP $\exists$) can be proved to belong to EVF(T) according to the above Theorem 4. We call FPIKA the logic obtained by adding to IKA the principles (FP $\vee$) and (FP $\exists$); in line with the above, we have the following semi-constructiveness result for a theory T possibly including induction and (DC):

Theorem 5. If T completely formalizes an isoinitial model and all the (properly said) axioms of T are $\forall\exists$-formulas or $\forall\neg$-formulas and A is closed, then $T \vdash^{FPIKA} A$ implies $A \in EVF(T)$.$\square$

The Kreisel and Putnam Principles

(KP$\vee$): $(\neg A \to B \vee C) \to (\neg A \to B) \vee (\neg A \to C)$ for any A,B,C,

(KP$\exists$): $(\neg A \to \exists x B(x)) \to \exists x(\neg A \to B(x))$ for any A and B and for A not containing x free ,

can be deduced from the logic FPIKA; other deducible principles are e.g.:

(P1) $((A \to B) \to C \vee D) \to ((A \to B) \to C) \vee (\neg A \to D)$, with B T-stable;

(P2) $(\forall x(\neg\neg A(x) \to A(x)) \to \exists x(\neg\neg A(x) \vee B(x)))$
$\to \neg\neg\exists x A(x) \vee \neg\neg\exists x B(x)$, with A(x) T-stable;

(P3) $(\forall x(\neg\neg A(x) \to A(x)) \to \forall x(\neg\neg A(x) \vee B(x)))$
$\to \neg\neg\exists x A(x) \vee \neg\neg B(t)$, with A(x) T-stable and t any term.

Remark that (P2) and (P3) are weak versions of the principles (E1) and (E2) of EMIKA.

To obtain from Theorem 5 a (full) constructiveness result, the following definitions are in order. We say that a formula A is <u>negatively saturated</u> iff every occurrence in A of a quantifier is in the scope of a negation. We say that B is <u>almost negatively saturated</u> iff B= $\exists \underline{z}\, A(\underline{z})$ and $A(\underline{z})$ is negatively saturated.
Let WFPIKA (weak FPIKA) be the logic obtained from FPIKA by allowing the application of (FP1) and (FP2) only when C is almost negatively saturated; let KKPWFPIKA be the logic obtained by adding to WFPIKA the Kreisel and Putnam Principles above quoted; then we can prove the following theorem:

<u>Theorem 6</u> With the same hypotheses on T as in Theorem 5, the system S=T+KKPWFPIKA is constructive. □

While (KP$\vee$) and (KP$\exists$) can be proved without restrictions from FPIKA, the same doesn't hold for WFPIKA: thus, in order to obtain a stronger result, in the definition of the logic of Theorem 6 we have added the Kreisel and Putnam Principles. With this addition, the proof of the theorem becomes very involved (see [13],where an appropriate variant of the collection technique is introduced).

7. THE SYSTEM PAP

Here we describe the main features of the system PAP (Proofs as Programs), we are developing. The project PAP began in 1984 as an "experimental" counterpart of the theoretical work on program synthesis.

In PAP, proofs are interpreted as "abstract programs" handling "data" whose structure is inspired by the notion of "valuation form". More precisely, "data" are seen as "structured pieces of information" on the truth of the related logical formulas; a datum may represent (in a given classical interpretation) a "true" or a "false" information; if such an information is true, it "constructively" explains why the related formula is true; if the information is false, it may happen that the related formula is true, since there are, generally, many structured pieces of information associated with a formula, depending on its structure; the set info(F) of the pieces of information associated with a formula F is inductively so defined:

$$\begin{aligned}
&\text{info}(E) &&= \{\text{true}\} \quad \text{if E is elementary (*)}\\
&\text{info}(A1 \wedge \ldots \wedge An) &&= \{<a1,.,an> \ / \ ai \in \text{info}(Ai),\ i \text{ in } 1..n\}\\
&\text{info}(A1 \vee \ldots \vee An) &&= \{<j,aj> \ / \ j \text{ in } 1..n \text{ and } aj \in \text{info}(Aj)\}\\
&\text{info}(A \to B) &&= \text{a class F1 of functions f: info(A)} \to \text{info(B)} \quad (**)\\
&\text{info}(\exists x\ A) &&= \{<c,a> \ / \ c \in \text{CONSTR and } a \in \text{info}(A)\} \quad (***)\\
&\text{info}(\forall x\ A) &&= \text{a class F2 of functions f: CONSTR} \to \text{info(A)} \quad (**)
\end{aligned}$$

(*) in PAP, one may choose different notions of an "elementary formula"; e.g., one may choose the Harrop formulas as elementary;
(**) we do not specify here the classes F1 and F2; up to now, in PAP, any f of F1 or of F2 may be an "input function" (in this case f is realized by an interactive colloquium with the user) or a function associated with a proof (in this case it can be computed, when needed, by a "procedure call" of the proof itself with the appropriate input parameters);
(***) we do not specify here the set CONSTR of the "constructors" of the individuals of the domains of an interpretation; we are implementing two general cases:

1) abstract data types: in this case one has an axiomatization T admitting an isoinitial model and the constructors may be closed terms in "normal form" (any individual is represented by a unique normal constructor) or in "free form" (any closed term is a free form);
2) "data bases": in this case one has any kind of "axioms" and the involved domains are finite; the constructors are names given by the user (eg., the names of the books of a library).

As we said, the pieces of information may be "true" or "false"; we omit the inductive clauses since they are similar to the ones considered for the satisfaction of the valuation forms.

Finally, we present some features of PAP:

= if one does not use special logical rules (corresponding to some constructive principles previously discussed), the logical part of the system PAP is equivalent to IKA

= the synthesis of a program is given by the following steps:

1) choice or definition of an "environment" (e.g., a data type with the corresponding axioms and possible previously proved theorems);

2) specification of the problem to be solved by a logical formula of the language of the environment (e.g., the "problem-formula" PF = $\forall x(A(x) \vee \neg A(x))$ specifies the problem of deciding A);

3) proof of the problem-formula, whose correctness is automatically checked by PAP;

4) translation of the proof into a program P written in a compilable language (at the moment, PAP uses MODULA-2);

5) compilation and execution of the program P, which correctly solves all the instances of the problem specified by the problem formula (e.g., with the previously seen PF, the program P accepts as an input any constructor x and answers "A(x) holds" if the result of the computation is <1,true> and " A(x) holds" if the result is <2,true>).

= the rules considered in PAP have also a "computational meaning" as "control structures" and "actions", to be applied when a proof is translated into a program; e.g., the rule of the or-elimination is translated into a "case" control structure and iteration and recursion are related to the induction rules.

A detailed report on PAP is in progress.

REFERENCES

[1] Bates J., Constable R. - Proofs as programs - ACM Transactions on Programming Languages and Systems, Vol. 7, n.1, 1985.

[2] Bertoni A., Mauri G., Miglioli P. - On the power of model theory to specify abstract data types and to capture their recursiveness - Fundamenta Informaticae IV.2, 1983.

[3] Bertoni A., Mauri G., Miglioli P., Ornaghi M. - Abstract data types and their extension within a constructive logic - Lect. Notes in Comp. Sci., n.173, Springer, 1984.

[4] Bresciani P., Miglioli P., Moscato U., Ornaghi M. - PaP: Proofs as Programs - abstract presented at the ASL meeting,Stanford University,July 15-19,1985.To appear in JSL.

[5] Chang C.,Keisler H. - Model theory - North Holland, 1973.

[6] Degli Antoni G., Miglioli P., Ornaghi M. - Top-down approach to the synthesis of programs - Proc. Collo. sur la Programmation, Paris 1974, Lect. Notes in Comp. Sci. n. 19, Springer, 1974.

[7] Goad C. - Computational uses of the manipulation of formal proofs - Rep. STAN-CS-80-819,Stanford University,1980.

[8] Goguen J.A., Thatcher J.W., Wagner E.G. - An initial algebra approach to the specification, correctness and implementation of abstract data types - IBM Res. Rep. RC6487, Yorktown Heights, 1976.

[9] Manna Z. - Mathematical theory of computation - Mc Grow Hill, 1973.

[10] Martin-Lof P. - Constructive Mathematics and Computer Programming - Presented at the 6-th Congress for Logic, Methodology and Philosophy of Sciences, Hannover, 1979.

[11] Medvedev T. - Finite problems - Sov.Math.Dok. vol.3,1962.

[12] Miglioli P., Ornaghi M. - A logically justified model of computation I,II - Fundamenta Informaticae IV.1,2, 1981.

[13] Miglioli P., Moscato U., Ornaghi M. - Constructive theories with superintuitionistic deductive systems: some results and techniques - Internal report, 1986.

[14] Troelstra A.S. - Metamathematical investigation of Intuitionistic Arithmetic and Analysis - Lect. Notes in Math., n.344, Springer, 1973.

This research has been partially supported by the Italian Ministero della Pubblica Istruzione.

A FIRST ORDER LOGIC FOR LOGIC PROGRAMMING

(Extended Abstract)

J. J. Moreno Navarro and
M. Rodriguez Artalejo

Departamento de Algorítmica
Facultad de Informática
Universidad Politécnica de Madrid
28031 Madrid, Spain

1. INTRODUCTION

Clark and Tärnlud[4] have proposed a methodology for the specification,desing and verification of logic programs in the framework of first order logic. Their main idea is to derive the correctness of a logic program from its own clauses and suitable induction axioma on the data. This approach has been also advocated by Cartwright[3] for the case of recursive, functional programs. He has contributed some theoretical results about the semantics of programs in nonstandard structures, using the notion of least definable fixpoint of the operator naturally associated to a given recursive program. This idea is also implicit in the independent work of Andréka, Németi and Sain[1] and Hajék[7] about nonstandard dynamic logic for flowchart-like programs and regular programs; see also Sain[10].

In this paper we propose a nonstandard programming logic, called NPL, for logic programs defined as finite sets of definite clauses with equality and primitive atoms. Following an approach parallel to Andréka, Németi and Sain[1] , we obtain a sound and complete finitary calculus for NPL. Our semantics uses 3-sorted structures to model time, data and finite sequences of data.

Keeping with van Emden and Kowalski[6], we define model-theoretic and least fixpoint semantics for our logic programs and prove them equivalent to NPL-semantics in standard structures. Then we introduce so called admissible structures (similar to the weakly arithmetic domains supporting elementary syntax in Cartwright[3]) and prove a least definable fixpoint characterization of NPL-semantics in admissible structures.

Moreover, already known results of Apt an van Emdem[6] and Fribourg[5] guarantee the existence of practical computation procedures for NPL programs without primitive atoms.

Regular programs and recursive functional programs can be translated to logic programs while preserving the meaning in all admissible structures. In this sense, NPL embodies the already quoted results by Andréka, Németi and Sain[1], Hajék[7] and Cartwright[3].

Admissible structures are axiomatizable in the language of NPL. Formal deduction in NPL, based on admissibility and induction axioms, can be seen as a formalization of Clark and Tarnlud's methodology in [4].

2. THE LOGIC NPL. COMPLETENESS THEOREM

In this section we introduce a logic NPL to reason about logic programs and prove a completeness theorem.

Similarity types and structures

2.1. Definition.

Many-sorted similarity types are tuples $\langle \mathbf{S},\mathbf{C},\mathbf{F},\mathbf{P}, \text{rank}\rangle$ where

- **S** is the set of *sorts*,
- **C**,**F**, **P** are the sets of *constants*, *function symbols* and *predicate symbo*
- rank: $\mathbf{C} \cup \mathbf{F} \cup \mathbf{P} \to \mathbf{S}^+$ assigns *ranks* to the symbols.

In this paper we use only 1- and 3-sorted similarity types of the following special shapes:

t: 1 sorted *similarity type of time*

Sort: t

Symbols and ranks: (see fig. 1)

d: 1-sorted *similarity type of the data*

Sort: d

Symbols and ranks: Arbitrary, but restricted to share no symbols with **t**.

td: 3 sorted *similarity type of NPL's structures*

Sorts: t,d,s (for time, data and sequences)

Symbols and ranks: (see fig. 2) (**comp** is intended to compute a data item as the i-th member of a given sequence) □

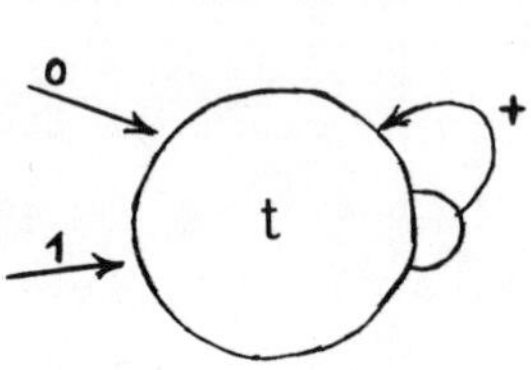

Fig. 1

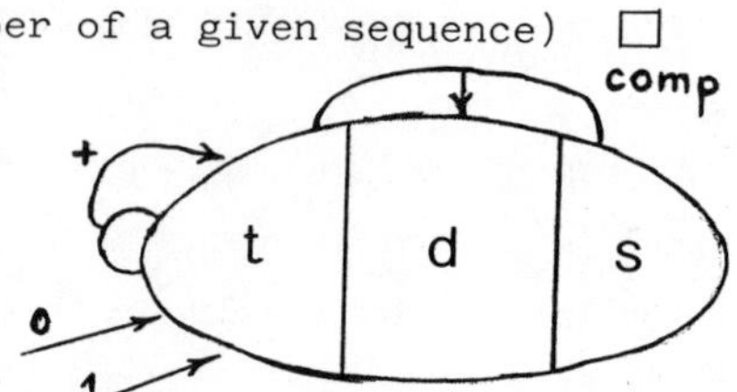

Fig. 2

We forced **d** to be 1-sorted only to simplify the notation ; all our results could be easily generalized to many-sorted data types.

2.2. Definition.

t-*structures*, **d** -*structures* and **td**-*structures* consist of nonempty domains for the sorts and suitable interpretations for the symbols:

t-structures: $\mathcal{T} = \langle T, 0^{\mathcal{T}}, 1^{\mathcal{T}}, +^{\mathcal{T}} \rangle$

d-structures: $\mathcal{D} = \langle D, (c^{\mathcal{D}})_{c \in \mathbf{C}}, (f^{\mathcal{D}})_{f \in \mathbf{F}}, (p^{\mathcal{D}})_{p \in \mathbf{P}} \rangle$

td-structures: $\mathcal{M} = \langle \mathcal{T}, \mathcal{D}, S, \mathbf{comp}^{\mathcal{M}} \rangle$

where $\mathcal{T}$ is a **t**-structure, $\mathcal{D}$ is a **d**-structure, S is the domain of $\mathcal{M}$-sequences, and $\mathbf{comp}^{\mathcal{M}}: S \times T \to D$

$\mathcal{N} = \langle \mathbf{N}, 0, 1, + \rangle$ is the *standard* **t**-*structure*.

For a given **d**-structure $\mathcal{D}$, the *standard* **td**-*structure over* $\mathcal{D}$ is

$\mathcal{M}_{\mathcal{D}} = \langle \mathcal{N}, \mathcal{D}, D^+, \text{comp} \rangle$ where

- $D^+ = \{\langle a_0, \ldots, a_n \rangle : n \in \mathbf{N};\ a_0, \ldots, a_n \in D\}$

- $\text{comp}(\langle a_0, \ldots, a_n \rangle, m) \begin{cases} a_m & \text{if } m \leq n \\ a_n & \text{otherwise} \end{cases}$

Notice that we use letter such as a, x, f, p, etc. for the interpretation of symbols **a, x, f, p**,etc. , denoted by boldface letters. □

<u>Syntax of NPL</u>

2.3. Definition.

Using $V_t = \{\mathbf{i}, \mathbf{j}, \mathbf{k}, \ldots\}$, $V_d = \{\mathbf{x}, \mathbf{y}, \mathbf{z}, \ldots\}$, $V_s = \{\mathbf{u}, \mathbf{v}, \mathbf{w}, \ldots\}$ and the usual logical and auxiliary symbols, we build:

$\mathbf{T_t}$ terms of type **t** $\qquad$ $\mathbf{T_d}$ terms of type **d**

$\mathbf{T_{td,d}}$ terms of types **td** and sort d

$\mathbf{F_t}$ first order formulas of type **t**

$\mathbf{F_d}$ first order formulas of type **d**

$\mathbf{F_{td}}$ first order formulas of type **td**

The equality sign $\doteq$ is allowed in formulas. We put τ , σ , etc. for terms and φ , ψ , etc. for formulas. We also feel free to use such abbreviations as **u**(i) for **comp** (**u**,**i**), $\mathbf{i} \leq \mathbf{j}$ for $\exists \mathbf{k}\ (\mathbf{i}+\mathbf{k} \doteq \mathbf{j})$ etc. □

2.4. Definition

Let **PV** be a denumerable collection of *predicative variables* X, Y, Z, etc. where each X ∈ **PV** is uniquely determined by two parameters rank (X), ord(X) ∈ **N**. Fix a similarity type **d** for the data.

- *Primitive atoms* are atomic or negated atomic formulas from $\mathbf{F_d}$:

$\tau_1 \doteq \tau_2 \qquad \neg\, \tau_1 \doteq \tau_2 \qquad \mathbf{p}(\tau_1, \ldots, \tau_n) \qquad \neg\, \mathbf{p}(\tau_1, \ldots, \tau_n)$

- *Nonprimitive atoms* are predicative expressions:

$X(\tau_1,\ldots,\tau_n)$ where $X \in \mathbf{PV}$, rank(X) = n, $\tau_j \in \mathbf{T_d}$

- *Definite clauses* are implications

$\beta_1 \wedge \ldots \wedge \beta_m \rightarrow \alpha$ usually written as $\alpha \leftarrow \beta_1,\ldots,\beta_m$ (read: α if β_1 and ... and β_m), where α is a nonprimitive atom and $\beta_1,\ldots,\beta_m$ are atoms.

- A *logic program* is any finite set Π of definite clauses. Let $\mathbf{LP_d}$ stand for the set of all logic programs of type **d**. □

2.5. Definition

The set $\mathbf{LF_d}$ of formulas (of type **d**) of the *nonstandard programing logic* NPL is recursively defined by:

- $\varphi \in \mathbf{LF_d}$ if $\varphi \in \mathbf{F_{td}}$
- $\Pi X(\tau_1,\ldots,\tau_n) \in \mathbf{LF_d}$ if X occurs in $\Pi \in \mathbf{LP_d}$, rank (X) = n and $\tau_1,\ldots,\tau_n \in \mathbf{T_d}$
- $\mathbf{LF_d}$ is closed under boolean operations and first order cuantifications over the three sorts t,d,s. □

2.6. Convention

For the rest of the paper , we assume the existence of countably many different ground terms $\ulcorner n \urcorner$ ($n \in \mathbb{N}$) of type **d**, called *labels*.

Semantics of NPL

2.7. Definition.

Let $\Pi \in \mathbf{LP_d}$ and a **td**-structure $\mathcal{M} = \langle \mathcal{T}, \mathcal{D}, S, comp \rangle$ be given. Take

$$N = \mathrm{rank}(\Pi) := \max\{\mathrm{rank}(X) : X \text{ occurs in } \Pi\}$$

A tuple $\langle \overset{N}{u}, v, w, i\rangle = \langle u_1,\ldots,u_n, v, w, i\rangle \in S^{N+2} \times T$ is an *intern proof* of $_N\Pi X(x_1,\ldots,x_n)$ in $\mathcal{M}$ (where X occurs in Π and $x_1,\ldots,x_n \in D$) iff

- $\langle \overset{N}{u}, v, w\rangle(i)$ codes $X(x_1,\ldots,x_n)$ in the sense that

$\langle u_1(i),\ldots,u_{\mathrm{rank}(X)}(i), v(i), w(i)\rangle = \langle x_1,\ldots,x_n, \ulcorner \mathrm{rank}(X)\urcorner^{\mathcal{D}}, \ulcorner \mathrm{ord}(X)\urcorner^{\mathcal{D}}\rangle$,

and

- for every $j \leq^{\mathcal{T}} i$ there is some D-instance $\alpha \leftarrow \beta_1,\ldots,\beta_m$ of a clause in Π , such that $\langle \overset{N}{u}, v, w\rangle(j)$ codes α and for $1 \leq r \leq m$, β_r is either primitive and true in $\mathcal{D}$ or nonprimitive and coded by $\langle u, v, w\rangle(k)$ for some $k <^{\mathcal{T}} j$, where "D-instance" has the obvious meaning of substituting members of D for variables of sort d. □

2.8. Definition

- *States* over $\mathcal{M}$ are mappings μ assigning values of the appropiate sort to all individual variables.
- $\mathcal{M} \models \Pi X(\tau_1,\ldots,\tau_n)(\mu)$ (read: $\Pi X(\tau_1,\ldots,\tau_n)$ is true in state μ of $\mathcal{M}$) iff $\Pi X(x_1,\ldots,x_n)$ has an intern proof in $\mathcal{M}$, where x_i is the value of τ_i in state μ of $\mathcal{M}$ ($1 \leq i \leq n$).
- $\mathcal{M} \models \varphi(\mu)$ gets now an obvious recursive definition, for all $\varphi \in \mathbf{LF_d}$.
- $\mathcal{M} \models \varphi$ (read: φ is *valid* in $\mathcal{M}$) iff $\mathcal{M} \models \varphi(\mu)$ for all states μ over $\mathcal{M}$.

- $\mathcal{M} \models \Phi$ iff $\mathcal{M} \models \varphi$ for all $\varphi \in \Phi$ (where $\Phi \subseteq \mathbf{LF_d}$)

- $\Phi \models \varphi$ (read : φ is a logical consequence of Φ) iff $\mathcal{M} \models \varphi$ for all $\mathcal{M}$ such that $\mathcal{M} \models \Phi$. □

A sound and complete calculus for NPL

2.9. Definition

Let $\vdash_{NPL}$ be the formal derivability relation of the following calculus for NPL:

Axioms

(A_o) All $\mathbf{LF_d}$ -formulas which are instances of propositional tautologies.

(A_1) Suitable axiom schemes for a complete calculus for $\mathbf{F_{td}}$.

(A_2) The axiom scheme:

$$\Pi X(\mathbf{x}_1,\ldots,\mathbf{x}_n) \leftrightarrow \exists \mathbf{u}_1 \ldots \exists \mathbf{u}_N \exists \mathbf{v} \exists \mathbf{w} \exists \mathbf{i} \; (\mathbf{proof}_\Pi (\mathbf{u}_1,\ldots, \mathbf{u}_N, \mathbf{v}, \mathbf{w}, \mathbf{i}) \wedge \mathbf{v}(\mathbf{i}) \doteq \ulcorner \mathrm{rank}(X) \urcorner \wedge \mathbf{w}(\mathbf{i}) \doteq \ulcorner \mathrm{ord}\ (X) \urcorner \wedge \mathbf{u}_1 (\mathbf{i}) \doteq \mathbf{x}_1 \wedge \ldots \wedge \mathbf{u}_n(\mathbf{i}) \doteq \mathbf{x}_n)$$

where N = rank (Π), n = rank (X), X occurs in $\Pi \in \mathbf{LP_d}$, and $\mathbf{proof}_\Pi \in \mathbf{F_{td}}$ asserts "$\langle \mathbf{u}_N, \mathbf{v}, \mathbf{w}, \mathbf{i} \rangle$ is an intern proof of $\Pi X(\mathbf{x}_1,\ldots,\mathbf{x}_n)$".

Rules

(MP) Modus ponens: $\dfrac{\varphi \quad \varphi \rightarrow \psi}{\psi}$ (φ , $\psi \in \mathbf{LF_d}$)

(G) Generalization: $\dfrac{\varphi \rightarrow \psi}{\forall \mathbf{i} \varphi \rightarrow \forall \mathbf{i} \psi}$, $\dfrac{\varphi \rightarrow \psi}{\exists \mathbf{i} \varphi \rightarrow \exists \mathbf{i} \psi}$ (φ , $\psi \in \mathbf{LF_d}$)

and analogously for the sorts d,s. □

2.10. Theorem

For arbitrary $\Phi \cup \{ \psi \} \subseteq \mathbf{LF_d}$ we have

- $\Phi \models \psi$ if $\Phi \vdash_{NPL} \psi$ (*soundness*), and
- $\Phi \vdash_{NPL} \psi$ if $\Phi \models \psi$ (*completeness*).

Sketch of proof

Soundness is easy to check. For the completeness, use formulas $\mathbf{proof}_\Pi$ to translate each $\varphi \in \mathbf{LF_d}$ to an equivalent $\nu(\varphi) \in \mathbf{F_{td}}$, in such a way that $\vdash_{NPL} \varphi \leftrightarrow \nu(\varphi)$; this uses essentially (A_2). As (A_o), (A_1), (MP) and (G) embody a complete calculus for $\mathbf{F_{td}}$, completeness of $\vdash_{NPL}$ follows easily.

This results follows the work of Andréka, Németi and Sain[1], who have applied the same technique to flowchart-like programs. The idea of using **td**-structures is also their's. □

3. NPL -SEMANTICS VERSUS CLASSICAL SEMANTICS FOR LOGIC PROGRAMS

From the computational point of view, NPL-semantics for logic programs behaves unproperly in many **td**-structures. In this section we prove model-theoretic and least fixpoint characterizations valid in standard **td**-structures, and even extend the second one to some well behaved nonstandard **td**-structures.

3.1. Definition

Let $\Pi \in \mathbf{LP}_{\mathbf{d}}$ and a **d**-structure $\mathcal{D}$ be given. Form an extended type $\mathbf{d}(\mathcal{D})$ by adding to **d** a new constant **a** for each element $a \in D$.

- The *Herbrand base* $B_{\Pi}^{\mathcal{D}}$ of Π over $\mathcal{D}$ consists of all atoms $X(\mathbf{a}_1, \ldots \mathbf{a}_n)$ where X occurs in Π, rank (X) = n, and $a_1, \ldots, a_n \in D$.

- *Herbrand interpretations* for Π over $\mathcal{D}$ are subsets $I \subseteq B_{\Pi}^{\mathcal{D}}$.

- The *diagram* $\Delta(\mathcal{D})$ of $\mathcal{D}$ is the colletion of all primitive ground atoms of type $\mathbf{d}(\mathcal{D})$ true in $\mathcal{D}$.

- $I \subseteq B_{\Pi}^{\mathcal{D}}$ is a *Herbrand model* of Π (in symbols, $I \models \Pi$) iff $C^+ \in I$ for every D-instance C of a clause in Π such that $C^- \subseteq \Delta(\mathcal{D}) \cup I$, where $C^+ = \alpha$ and $C^- = \{\beta_1, \ldots, \beta_m\}$ if $C = \alpha \leftarrow \beta_1, \ldots, \beta_m$.

- The *operator* $T_{\Pi}^{\mathcal{D}}: \mathcal{P}(B_{\Pi}^{\mathcal{D}}) \rightarrow \mathcal{P}(B_{\Pi}^{\mathcal{D}})$ associated to Π over $\mathcal{D}$ is defined by:

$$T_{\Pi}^{\mathcal{D}}(I) = \{ C^+ : C \text{ is D-instance of a clause in } \Pi \text{ such that } C^- \subseteq \Delta(\mathcal{D}) \cup I \}$$

Notice that $\mathcal{P}(B_{\Pi}^{\mathcal{D}})$ is a *complete lattice* relative to the partial ordering of set inclusion. □

3.2. Theorem

The operator $T_{\Pi}^{\mathcal{D}}$ is monotonic and continuous is the sense of Scott[11]. It has a *least fixpoint* $\mu T_{\Pi}^{\mathcal{D}} = I_{\Pi}^{\mathcal{D}}$ which admits the following characterizations

(a) $I_{\Pi}^{\mathcal{D}} = \bigcap \{ I \subseteq B_{\Pi}^{\mathcal{D}} : T_{\Pi}^{\mathcal{D}}(I) \subseteq I \}$

(b) $I_{\Pi}^{\mathcal{D}} = \bigcup_{n<\omega} (T_{\Pi}^{\mathcal{D}})^{n}(\emptyset)$

(c) $I_{\Pi}^{\mathcal{D}}$ is the *least Herbrand model* of Π over $\mathcal{D}$, in the sense of set inclusion.

(d) $I_{\Pi}^{\mathcal{D}} = \{ X(\mathbf{a}_1, \ldots, \mathbf{a}_n) : X \text{ occurs in } \Pi;\ \text{rank}(X) = n;\ a_1, \ldots, a_n \in D;\ \Pi \cup \Delta(\mathcal{D}) \models X(\mathbf{a}_1, \ldots, \mathbf{a}_n) \}$

Proof

The monotonicity and continuity of $T_{\Pi}^{\mathcal{D}}$ are easily checked. Then (a) and (b) follow from well known result, see for instance Lloyd[8].

For $I \subseteq B_{\Pi}^{\mathcal{D}}$, it is clear that $I \models \Pi$ iff $T_{\Pi}^{\mathcal{D}}(I) \subseteq I$. Then (a) says that $I_{\Pi}^{\mathcal{D}}$ is the intersection of the nonempty family of all Herbrand models of Π over $\mathcal{D}$ (notice that $B_{\Pi}^{\mathcal{D}} \models \Pi$). As Π is a set of Horn clauses, this gives a Herbrand model of Π (see again Lloyd[8]) and we get (c).

$\Pi \cup \Delta(\mathcal{D}) \models X(\mathbf{a}_1, \ldots, \mathbf{a}_n)$ iff $\Pi \cup \Delta(\mathcal{D}) \cup \{ \leftarrow X(\mathbf{a}_1, \ldots, \mathbf{a}_n) \}$ is unsatisfactible iff $\Pi \cup \{ \leftarrow X(\mathbf{a}_1, \ldots, \mathbf{a}_n) \}$ has no Herbrand model over $\mathcal{D}$ (by an adapted version of Herbrand's Theorem).

iff $X(\mathbf{a}_1, \ldots, \mathbf{a}_n) \in \bigcap \{ I \subseteq B_{\Pi}^{\mathcal{D}} : I \models \Pi \} = I_{\Pi}^{\mathcal{D}}$ (by (c))

This prove part (d), lastly. □

3.3. Corollary

The least fixpoint (and least Herbrand model) I_Π^D corresponds to NPL-semantics in the standard structure M_D.

Proof

$I_\Pi^D \models X(a_1,\ldots,a_n)$ iff $X(a_1,\ldots,a_n) \in I_\Pi^D$

iff there is some $n < \omega$ such that $X(a_1,\ldots,a_n) \in (T_\Pi^D)^n\ (\emptyset)$ (by Theorem 3.2 (b))

iff there is an intern proof of $\Pi X(a_1,\ldots,a_n)$ in M_D (because M_D is standard)

iff $\Pi X(a_1,\ldots,a_n)$ is true in M_D (by Definition 2.8) □

Procedural semantics

Computationally feasible deduction procedures, giving rise to interpreters and procedural semantics, are already known for logic programs without primitive atoms.

3.4. Theorem

(a) Assume that Π is a *Horn program* using neither equality nor primitive atoms. Let D be the Herbrand structure corresponding to the Herbrand universe U_Π of Π. Then, for every X occurring in Π with rank(X) = n and arbitrary ground terms $\tau_1,\ldots,\tau_n \in U_\Pi$ we have:

$X(\tau_1,\ldots,\tau_n) \in I_\Pi^D$ iff

$\Pi \cup \{\leftarrow X(\tau_1,\ldots,\tau_n)\}$ has a SLD-refutation.

(b) Assume that Π is an *equational logic program* using no primitive atoms (but possibly using equality). Let D be the quotient of the Herbrand structure corresponding to U_Π modulo the least model congruence E_Π of Π the sense of Fribourg[5]. Then for X ocurring in Π with rank (X) = n and $\tau_1,\ldots,\tau_n \in U_\Pi$, with E_Π-classes $[\tau_1],\ldots,[\tau_n]$, we have

$X([\tau_1],\ldots,[\tau_n]) \in I_\Pi^D$

iff $\Pi \cup \Pi^f \cup \{\leftarrow X(\tau_1,\ldots,\tau_n)\}$ has an RS-refutation, where RS is Fribourg's procedure based on *reflection* and *clausal superposition* and

$\Pi^f = \{\mathbf{f}(\mathbf{x}_1,\ldots,\mathbf{x}_n) \doteq \mathbf{f}(\mathbf{x}_1,\ldots,\mathbf{x}_n)\leftarrow : \mathbf{f}$ occurs in Π, rank(f) = n$\}$
is the set of *functional reflexive* axioms.

Proof.

See Apt and van Emden[2] and Fribourg[5]. Notice a minor syntactical difference : Fribourg writes $X(\tau_1,\ldots,\tau_n) \doteq$ **true** instead of $X(\tau_1,\ldots,\tau_n)$. He also defines the operator T_Π^D in a different way.

The occurrences of primitive predicates in actual computations could perhaps be satisfactorily handled in a LISP environment, in the style of Robinson and Sibert[9]. We hope to pursue this topic in a future work. □

Fixpoint semantics in nonstandard structures

If $\mathcal{M} = \langle \mathcal{T}, \mathcal{D}, \mathcal{S}, \text{comp} \rangle$ is nonstandard, the least Herbrand model $I_{\Pi}^{\mathcal{D}}$ is no longer guaranteed to correspond to NPL-semantics.

3.5. Example

Take $\mathcal{M} = \mathcal{M}_{\mathcal{D}}$ with $\mathcal{D} = \langle \mathbf{N}, 0, \text{suc} \rangle$. Let $\mathcal{M}^* = \langle \mathcal{T}^*, \mathcal{D}^*, S^*, \text{comp}^* \rangle$ be a nonstandard model of $\text{Th}(\mathcal{M}_{\mathcal{D}})$, such that $\mathcal{D}^*$ is also a nonstandard model of $\text{Th}(\mathcal{D})$.

For $\Pi = \{ X(\mathbf{0}) \leftarrow , X(\mathbf{suc}(\mathbf{x})) \leftarrow X(\mathbf{x}) \}$ we have $\mathcal{M} \models \forall \mathbf{x}\, \Pi X(\mathbf{x})$. As $\mathcal{M}^* \equiv \mathcal{M}$, we get also $\mathcal{M}^* \models \forall \mathbf{x}\, \Pi X(\mathbf{x})$. On the other side:

$$I_{\Pi}^{\mathcal{D}^*} = \bigcup_{n<\omega} (T_{\Pi}^{\mathcal{D}^*})^n (\emptyset) = \{ X(\mathbf{suc}^n(0)) : n \in \mathbf{N} \}$$

This kind of difficulty was encountered by Cartwright[3] while studying a first order programming logic for recursive functional programs. He proposed to use the notion of *least definable fixpoint*, also implicit in the work of Hajék[7] about nonstandard dynamic logic. Cartwright's development of this idea is technically quite involved. In NPL, things go on more easily.

3.6. Definition

Fix a similarity type **d**. The *admissibility axioms* consist of

- *Basic axioms on time*

$\forall i\, \neg(i + 1 \doteq 0)$ $\qquad$ $\forall i\, \forall j (i + 1 \doteq j + 1 \rightarrow i \doteq j)$

$\forall i\, (i + 0 \doteq i)$ $\qquad$ $\forall i\, \forall j (i + (j + 1) \doteq (i + j) + 1)$

- *Induction axioms on time*

For each $\varphi \in \mathbf{F}_{\mathbf{td}}$:

$\mathbf{ind}_t(\varphi): \varphi[0/i] \wedge \forall i (\varphi \rightarrow \varphi[i + 1/i]) \rightarrow \forall i\, \varphi$

- *Basic axioms on sequences*

unit: $\forall \mathbf{x}\, \exists \mathbf{u}\, \forall i\, \mathbf{u}(i) \doteq \mathbf{x}$

conc: $\forall \mathbf{u}\, \forall i\, \forall \mathbf{v}\, \forall j\, \exists \mathbf{w}\, (\forall k (k \leq i \rightarrow \mathbf{w}(k) \doteq \mathbf{u}(k)) \wedge \forall l (l \leq j \rightarrow \mathbf{w}(l+i+1) \doteq \mathbf{v}(l)))$

- *Basic axioms on labels*

$\neg \ulcorner m \urcorner \doteq \ulcorner n \urcorner$ for all $0 \leq m < n$

Additionally, we can define

- *Induction axioms on the data*

For each $\varphi \in \mathbf{F}_{\mathbf{td}}$:

$$\mathbf{ind}_d(\varphi) : \bigwedge_{c \in \mathbf{C}_0} \varphi[c|\mathbf{x}] \wedge \bigwedge_{\substack{f \in \mathbf{F}_0 \\ n = \text{rank}(f)}} \forall \mathbf{y}_1 \ldots \forall \mathbf{y}_n (\varphi[\mathbf{y}_1/\mathbf{x}] \wedge \ldots \wedge \varphi[\mathbf{y}_n/\mathbf{x}] \rightarrow \varphi[\mathbf{f}(\mathbf{y}_1, \ldots, \mathbf{y}_n)/\mathbf{x}]) \rightarrow \forall \mathbf{x}\, \varphi$$

where variables $\mathbf{y}_j$ don't occur free in φ and $\mathbf{C}_0 \cup \mathbf{F}_0$ is the set of *generating symbols* ($\mathbf{C}_0 \subseteq \mathbf{C}$, $\mathbf{F}_0 \subseteq \mathbf{F}$).

Given a **td**-structure $\mathcal{M}$, we say

- $\mathcal{M}$ is *admissible* iff the admissibility axioms are valid in $\mathcal{M}$.
- $\mathcal{M}$ is *inductive* iff the induction axioms on the data are valid in $\mathcal{M}$. □

Inductive admissible structures are similar to the "weakly arithmetic domains supporting elementary syntax" in Cartwright[3] but presented in the style of Andréka, Németi and Sain[1].

3.7. Examples

Given any **d**-structure $\mathcal{D}$, the standard **td**-structure $\mathcal{M}_{\mathcal{D}}$ is admissible. $\mathcal{M}_{\mathcal{D}}$ is also inductive, provided that $\mathcal{D}$ is finitely generated from some generating subset of its similarity type. Of course, there are many nonstandard inductive admissible structures.

3.8. Definition

Let $X_1, \dots, X_r$ be the predicate variables occurring in $\Pi \in \mathbf{LP_d}$, where rank $(X_i) = n_i$ for $1 \leq i \leq r$.

A Herbrand interpretation $I \subseteq B_{\Pi}^{\mathcal{D}}$ is *definable* in the **td**-structure. $\mathcal{M} = \langle \mathcal{T}, \mathcal{D}, \mathcal{S}, \text{comp} \rangle$ iff there are formulas $\varphi_{X_i}(\mathbf{x}_1, \dots, \mathbf{x}_n)$, $1 \leq i \leq r$, such that for any $1 \leq i \leq r$, $a_1, \dots, a_{n_i} \in D$:

$X_i(\mathbf{a}_1, \dots, \mathbf{a}_{n_i}) \in I$ iff $\varphi_{X_i}(a_1, \dots, a_{n_i})$ is true in $\mathcal{M}$. □

3.9. Theorem

Let $\mathcal{M}$ be an admissible **td**-structure. For any $\Pi \in \mathbf{LP_d}$, the operator $T_{\Pi}^{\mathcal{D}}$ has a least fixpoint definable in $\mathcal{M}$ which corresponds to the NPL-semantics of Π in $\mathcal{M}$.

<u>Sketch of Proof</u>

Let $N = \text{rank}(\Pi)$. Take as $\varphi_{X_j}(\mathbf{x}_1, \dots, \mathbf{x}_{n_j})$ the formula
$\exists \mathbf{u}_1 \dots \exists \mathbf{u}_N \exists \mathbf{v} \exists \mathbf{w} \exists \mathbf{i}\ (\mathbf{proof}_{\Pi}(\mathbf{u}_1, \dots, \mathbf{u}_N, \mathbf{v}, \mathbf{w}, \mathbf{i}) \wedge \mathbf{v}(\mathbf{i}) \doteq \ulcorner n_j \urcorner \wedge \mathbf{w}(\mathbf{i}) \doteq \ulcorner \text{ord}(X_j) \urcorner \wedge \mathbf{u}_1(\mathbf{i}) \doteq \mathbf{x}_1 \wedge \dots \wedge \mathbf{u}_{nj}(\mathbf{i}) \doteq \mathbf{x}_{n_j})$

Let $I_{\Pi}^{\mathcal{M}}$ be the Herbrand interpretation over $\mathcal{D}$ defined in $\mathcal{M}$ by these formulas, Then:

(i) $I_{\Pi}^{\mathcal{M}} \subseteq T_{\Pi}^{\mathcal{D}}(I_{\Pi}^{\mathcal{M}})$

(ii) $T_{\Pi}^{\mathcal{D}}(I_{\Pi}^{\mathcal{M}}) \subseteq I_{\Pi}^{\mathcal{M}}$

(iii) $I_{\Pi}^{\mathcal{M}} \subseteq I$ for any $I \subseteq B_{\Pi}^{\mathcal{D}}$ definable in $\mathcal{M}$ and such that $T_{\Pi}^{\mathcal{D}}(I) \subseteq I$.

in fact, we can informally argue:

(i) Let $X(\mathbf{a}_1, \dots, \mathbf{a}_n) \in I_{\Pi}^{\mathcal{M}}$. By definition $I_{\Pi}^{\mathcal{M}}$, there is an intern proof of $\Pi X(a_1, \dots, a_n)$ in $\mathcal{M}$. Using the fact any intern prefix of an intern proof is also an intern proof, we conclude that
$X(\mathbf{a}_1, \dots, \mathbf{a}_n) \in T_{\Pi}^{\mathcal{D}}(I_{\Pi}^{\mathcal{M}})$.

(ii) Let $X(\mathbf{a}_1, \dots, \mathbf{a}_n) \in T_{\Pi}^{\mathcal{D}}(I_{\Pi}^{\mathcal{M}})$. There must be some D-instance $\mathbf{C}$ of a clause in Π with $\mathbf{C}^+ = X(\mathbf{a}_1, \dots, \mathbf{a}_n)$ and $\mathbf{C}^- \subseteq \Delta(\mathcal{D}) \cup I_{\Pi}^{\mathcal{M}}$. Composing several internal proofs through internal concatenation, we can obtain an internal proof of $\Pi X(a_1,\ldots,a_n)$, and $X(\mathbf{a}_1,\ldots,\mathbf{a}_n) \in I_\Pi^{\mathcal{M}}$.

(iii) If $X(\mathbf{a}_1,\ldots,\mathbf{a}_n) \in I_\Pi^{\mathcal{M}}$ there is an internal proof of $\Pi X(a_1,\ldots,a_n)$ in $\mathcal{M}$, and by induction on its internal length we can prove that $X(\mathbf{a}_1,\ldots,\mathbf{a}_n) \in I$. This needs induction on time and both hypothesis about I.

This informal reasoning can be formalized on the sole basis of the admissibility axioms. □

Notice how this result generalizes Corollary 3.3: $I_\Pi^{\mathcal{D}} = I_\Pi^{\mathcal{M}\mathcal{D}}$.

4. ITERATIVE AND RECURSIVE PROGRAMS VIEWED AS LOGIC PROGRAMS

Sain[10] studies a nonstandard dynamic logic for structured, nondeterministic regular programs. Cartwright[3] investigates a first order logic for recursive funcional programs. Both approaches can be embodied into our's via the following result. Due to lack of space, we must omit details.

4.1. Theorem

Regular programs and recursive functional programs can be effectively translated to logic programs, while "preserving the meaning" (in a appropiate sense) in all admissible structures. □

5. NPL AS A VERIFICATION TOOL.

Clark and Tärnlud[4] propose a methodology which can be summarized as follows:

(i) Isolate some characteristic axioms Ax_d valid in the data type, including appropiate induction axioms.

(ii) Use case analysis, guided by the structure of data, to build a first order specification Ax_p of the predicates to be computed (eventually including auxiliary predicates).

(iii) Try to use the if-parts of double implications in Ax_p to construct a logic (Horn) program Π for the predicates.

(iv) Derive any desired properties of the predicates by first order reasoning based on $Ax_d \cup Ax_p$.

Although they don't give any formal justification, the soundness of this method depends from the fact axioms $Ax_d \cup Ax_p$ are valid in the least Herbrand model of Π.

We get a rigorous foundation of essentially the same methodology by reasoning in NPL based on admissibility axioms and induction axioms on the data. Let us illustrate this at one of the examples given by Clark and Tärnlud: *Insertion in an ordered tree.* The data type of ordered trees is assumed to include *atoms* (provided with an ordering $<$), an *empty tree* denoted by **empty** and nonempty trees builded by the *constructor* **tree** which takes two trees (sons) and an atom (labelled root) as arguments. The program is :

$$\text{Insert}\ (\mathbf{empty}, \mathbf{z}, \mathbf{tree}\ (\mathbf{empty}, \mathbf{z}, \mathbf{empty})) \leftarrow$$

$$\text{Insert}\ (\mathbf{tree}(\mathbf{x}, \mathbf{z}, \mathbf{y}), \mathbf{z}, \mathbf{tree}\ (\mathbf{x}, \mathbf{z}, \mathbf{y})) \leftarrow$$

$$\text{Insert}\ (\mathbf{tree}\ (\mathbf{x}, \mathbf{z}, \mathbf{y}), \mathbf{z'}, \mathbf{tree}\ (\mathbf{x'}, \mathbf{z}, \mathbf{y})) \leftarrow \mathbf{z'} < \mathbf{z}, \text{Insert}\ (\mathbf{x}, \mathbf{z'}, \mathbf{x'})$$

$$\text{Insert}\ (\mathbf{tree}\ (\mathbf{x}, \mathbf{z}, \mathbf{y}), \mathbf{z'}, \mathbf{tree}(\mathbf{x}, \mathbf{z}, \mathbf{y'})) \leftarrow \mathbf{z} < \mathbf{z'}, \text{Insert}\ (\mathbf{y}, \mathbf{z}, \mathbf{y'})$$

If the properties to establish are now

(1) $\forall \mathbf{x}\ \forall \mathbf{z}\ (\text{Ordered-tree}(\mathbf{x}) \wedge \text{Atom}(\mathbf{z}) \rightarrow \exists \mathbf{x'}\ \Pi\, \text{Insert}(\mathbf{x},\mathbf{z},\mathbf{x'}))$

(2) $\forall \mathbf{x}\ \forall \mathbf{z}\ \forall \mathbf{x}\ \forall \mathbf{x'}\ (\text{Ordered-tree}(\mathbf{x}) \wedge \text{Atom}(\mathbf{z}) \wedge \Pi\, \text{Insert}\ (\mathbf{x},\mathbf{z},\mathbf{x'}) \rightarrow (\text{Ordered-tree}(\mathbf{x'}) \wedge \forall \mathbf{z'}\ (\text{Occurs}(\mathbf{z'},\mathbf{x'}) \leftrightarrow \mathbf{z'} \doteq \mathbf{z} \vee \text{Occurs}\ (\mathbf{z'}, \mathbf{x}))))$

we can reason in NPL, using **unit, conc,** induction on **x** and case analysis for (1) , and induction on the lenght of a derivation and case analysis for (2). Of course, some axioms about the auxiliary predicates Ordered-tree, Atom and Occurs are also needed.

Reasoning in NPL can also be compared with the work of Cartwright[3]. He uses first order deduction based on axioms about the data, the equations of the recursive functional program, and the induction axiom on the data for first order formulas of the extended similarity type, including the new symbols $\mathbf{f}_1, \ldots, \mathbf{f}_r$. This method is not strong enough is some cases. Cartwright notices the following example (adapted to our terminology):

- **d**-structure: $\mathcal{D} = \langle \mathbf{N}, \mathbf{0}, \mathbf{suc} \rangle$
- Program: $\mathbf{f}(\mathbf{x}) \doteq \mathbf{f}(\mathbf{x})$

The assertion that $\mathbf{f}(\mathbf{x})$ is undefined for all **x** holds in the least fixpoint model over $\mathcal{D}$. But in cannot be derived from the axioms quoted above, which also hold for the identity function and infinitely many other fixpoint. Carwright surmounts the difficulty by giving an effective procedure to translate programs to equivalent *complete* programs with a unique fixpoint.

In NPL the program above can be written as

$$\Pi = \{ F(\mathbf{x},\mathbf{y}) \leftarrow F(\mathbf{x},\mathbf{y}) \}$$

and the assertion $\forall \mathbf{x}\ \forall \mathbf{y} \neg \Pi F(\mathbf{x},\mathbf{y})$ can be easily proved: By applying the axioms on time, it turns out hat no internal proof exists. In fact, Cartwright's complete programs do also incorporate internal sequences, but in a more cumbersome way.

6. CONCLUSION

We have presented a formal system NPL to reason about logic program with equality and primitive atoms.

This logic admits a sound and complete finitary calculus. Although program semantics is nonstandard, it has model-theoretic and least fixpoint characterizations in standard structures. Moreover , previous results of other authors provide a computation procedure for programs without primitive atoms, which works in the least standard model.

We have also established a least definable fixpoint characterization of NPL-semantics in admissible structures. Finally, we have shown that NPL-deduction on the basis of admissibility and induction axioms is an adequate verification tool.

All this applies also to iterative and recursive programs, because they can be effectively translated to logic programs.

In a future work, we would like to investigate computation procedures for logic programs with equality and primitive atoms. We also plan to show how NPL can be applied to the reasoning about PROLOG programs, taking into account the search strategy of PROLOG interpreters.

REFERENCES

1. Andréka, H., I. Németi and I. Sain, A Complete Logic for Reasoning about Programs via Nonstandard Model Theory, Parts I and II, TCS 17 (1982), 193-212 and 259-278.
2. Apt, K.R. and M.H. van Emden, Contributions to the Theory of Logic Programming, J.ACM 29(1982), 841-862.
3. Cartwright, R., Recursive Programs as Definitions in First Order Logic, SIAM J. on Computing 13(1984), 374-408.
4. Clark, K.L. and S.A. Tärnlud, A First Order Theory of Data and Programs Proceedings IFIP'77, North-Holland (1977), 939-944.
5. Fribourg, L., Oriented Equational Clauses as a Programming Language, Procedings ICALP'84, Lecture Notes in Comp. Sci. 172(1984), 162-173.
6. van Emden, M.A. and R.A. Kowalski, The Semantics of Predicate Logic as a Programming Language, J.ACM 23(1976), 733-742.
7. Hajék, P., Making Dynamic Logic Firts Order, in Mathematical Foundations of Comp. Sci., Lecture Notes in Comp. Sci. 118(1981), 287-295.
8. Lloyd, J.W., Foundations Of Logic Programming, Springer Verlag (1984).
9. Robinson, J.A. and E.E. Sibert, LOGLISP: and alternative to PROLOG, in Machine Intelligence 10, J. Wiley & Sons (1982), 399-419.
10. Sain,I., Structured Nonstandard Dynamic Logic, ZMLGM 30(1984), 481-497.
11. Scott, D.S., Outline of a Mathematical Theory of Computation ,Tchn. Monograph PRG-2, Oxford University Computing Laboratory (1970).

THE RELATIONAL SEMANTICS FOR BRANCHED QUANTIFIERS

Marcin Mostowski

Warsaw University
Bialystok's Branch

This paper concerns a new second order semantics for branched quantifiers (independent on AC). The proof theory for a logic of branched quantifiers (LB) is formulated, and it is proved that this logic is complete for some very natural semantics.

1. SECOND ORDER SEMANTICS FOR BRANCHED QUANTIFIERS

Henkin considered branched quantitifers in [1]. These quantifiers can be written in two dimensional way as follows:

$$\begin{matrix} \forall x & \exists y \\ \forall z & \exists v \end{matrix} \; \alpha(x, y, z, v)$$

which was explicated by him as $\exists f,g\ \forall x,y\ \alpha(x, f(x), y, g(y))$. Such quantifiers are rather strong. For instance it was shown by Ehrenfeut that the sentence

$$\exists u \begin{matrix} \forall x & \exists y \\ \forall z & \exists t \end{matrix} \; ((x = z \equiv y = t) \ \& \ u \neq y)$$

is true exactly in infinite models. It means that branched quantifiers are stronger than the usual linear ones because the class of infinite models is not elementarily definable.

Several papers were devoted to investigations of different properties of branched quantifiers, Enderton [2], Walkoe [3], Krynicki and Lachlan [4], but only Henkin's explication for them was considered and investigated (the sole exception is Harel [5]). It seems to me that this explication is a bit problematic. Let us observe that the equivalence of $\forall x\ \exists y\ \alpha(x, y)$ and $\exists f\ \forall x\ \alpha(x, f(x))$ for every α is true in every model if and only if the Axiom of Choice is true.

Using branched quantifiers we can write a statement true exactly when the universum has a linear ordering. It was proved in [2] and [3] that every Σ_1-sentence can be expressed in the logic of branched quantifiers with identity as an only predicate. But the sentence $\exists R$ (R is a total linear ordering) is Σ_1. Assuming AC, this statement is true but we know that it is independent statement of the set theory without AC.

It seems to me that it is not so obvious that the Axiom of Choice is wrong in the Foundation of Mathematics, but it is obviously wrong as a logical axiom. Therefore I propose another (independent on AC) explication for branched quantifiers. Let us observe that for any α we have an equivalnce (not assuming AC):

$$\forall x\, \exists y\, \alpha(x, y) \text{ if and only if } \exists R\, (\forall x\, \exists y\, xRy \text{ and } \forall x, y(xRy \Rightarrow \alpha(x, y)).$$

The implication from the right to the left is obvious. On the other hand, assuming $\forall x\, \exists y\, \alpha(x, y)$, we can take $R = \{(x, y) : \alpha(x, y)\}$.

Based on the above idea we can explicate a statement:

$$\begin{matrix} \forall x\ \exists y \\ \forall z\ \exists v \end{matrix}\ \alpha(x, y, z, v)$$

as follows:

$$\exists Q, R(\forall x\, \exists y\, xQy\ \&\ \forall x\, \exists y\, xRy\ \&$$

$$\forall x, y, z, t(xQy\ \&\ zRt \Rightarrow \alpha(x, y, z, t))).$$

Let us consider what it means from the point of view of the theory of relations. A concatenation $Q\hat{\ }R$ of two relations Q and R we define as follows: $Q\hat{\ }R = a\hat{\ }b : a \in Q$ and $b \in R$, where $a\hat{\ }b$ means usual concatenation of two finite sequences a and b. According to the explication given above we can say that

$$\begin{matrix} \forall x\ \exists y \\ \forall z\ \exists v \end{matrix}\ \alpha(x, y, z, v)$$

is equivalent to:

$$\exists Q, R(\forall x\, \exists y\, xQy \text{ and } \forall x\, \exists y\, xRy \text{ and } Q\hat{\ }R \subseteq \{(x, y, z, t) : \alpha(x, y, z, t)\}).$$

It means that some sub-relation of the relation defined by α can be divided into relations Q and R satisfying some conditions. Choosing different conditions we can obtain other quantifiers possibly having some interesting properties, but I do not know any good interpretation for such quantifiers.

It would be interesting to compare a power of two above discussed explications for branched quantifiers. We call a branched quantifier with Henkin's semantics a H-quantifier, and a quantifier with semantics here proposed we call a R-quantifier.

Of course, assuming AC, both explications are equivalent. Then the only interesting case is when we do not assume AC. For any formula α and branched prefix Q, if α has no branched quantifiers then $Q\alpha$ interpreted according to Henkin implies the same according to semantics for R-quantifiers

H-quantifiers can be defined in terms of R-quantifiers. The idea is simple. For instance a formula $\begin{matrix} \forall x\ \exists y \\ \forall z\ \exists v \end{matrix}\ \alpha(x, y, z, v)$ can be translated into:

can be translated into:

$$\begin{matrix} \forall x & \exists y \\ \forall x' & \exists y' \\ \forall z & \exists v \\ \forall z' & \exists v' \end{matrix}\ (x = x' \Rightarrow y = y') \text{ and } (z = z' \Rightarrow v = v') \text{ and } \alpha(x, y, z, v)).$$

The sets of tautologies for H- and R-quantifiers would be different. Particularly if AC does not hold, then the sentence:

$$\forall x(A(x) \Rightarrow \exists y\ E(y, x)) \Rightarrow \begin{matrix} \forall x\ \exists y \\ \\ \forall z\ \exists t \end{matrix}\ (A(x) \Rightarrow E(y, x))$$

is a tautology for relational semantics but not for Henkin's one.

2. MORE REALISTIC SEMANTICS FOR BRANCHED QUANTIFIERS

Using branched quantifiers we can define the standard model for arithmetic uniquely up to isomorphism (see [4]). It means that the logic of such quantifiers does not allow any complete finite proof procedure (it follows from the theorem of Goedel). It seems that the power of branching quantifiers follows from quantification on arbitrary functions or relations. However, we cannot use all relations (or functions), but only some of them. Because relations used for an interpretation of branched quantifiers correspond to some connections between variables, then it seems to be more reasonable to take connections expressible in a given language than all set-theoretical entities. Set-theoretical relations could be considered rather as possible connections, and only connections expressible in our language can be treated as actual connections.

In this section, based on these intuitions, I shall describe the Logic of Branched quantifiers (LB).

<u>Syntax for a Logic of Branched Quantifiers</u>

Let k be a signature for first order language of predicate calculus (k_i is an arity of a predicate P_i). The set of formulae F we define as usual, adding the following inductive condition: if Q is a quantifier prefix, and $\alpha \in F$, and no variable bound in α occurs in Q, then $Q\alpha \in F$.

The set of quantifier prefixes PREF is defined as the smallest set X such that:

(1) if x is a variable then $\forall x$, and $\exists x \in X$;

(2) if $Q, Q' \in X$, and Q and Q' have no common variables then $QQ' \in X$, and $\begin{matrix} Q \\ Q' \end{matrix} \in X$.

Actually, we shall treat quantifier prefixes as elements of an algebra free in the class of algebras with two associative compositions (vertical and horizontal) and a neutral element (empty prefix), the same for both operations. This algebra is generated by the set $\{\forall x, \exists x : x \text{ is a variable}\}$.

Reading the above definition, it is easy to see why we do not use a well-known term "partially ordered quantifiers". However, it follows from characterization theorems of Walkoe [3] that partially ordered H-quantifiers can be written using only prefixes belonging to the set PREF.

Let FE be a set of formulae defined inductively in the same way as F, but constructed, based on more atomic formulae obtained, and adding so-called free predicates (for any arity countably there are many new predicates).

<u>Interpretation</u>

In this section we shall describe the interpretation for all formulae from the set FE, translating them into some theory in many-sorted first order predicate calculus. At first let us define the <u>Semantical Theory</u> (ST).

The language of Semantical Theory of signature k has sorts U, U_1, U_2, ..., where U corresponds to the only sort of language of predicate calculus with signature k; U_i corresponds to the sort of i-ary relations on U. All relations from signature k have arguments of sort U. For i = 1, 2, ..., we have a predicate E_i, it is the predicate with first i arguments of sort U, and the last argument of sort U_i, its arity is i + 1, the identity is defined only for the sort U.

Let FST be a set of formulae of first order many-sorted language with the above described sorts and predicates. Here, and in what follows we omit the parameter k, because a correspondence for any concrete signature will not be used.

For any formula $\alpha \in$ FE we define $*\alpha \in$ FST by induction as follows:

1. $*(P(\underline{t})) = E_i(\underline{t}, P_i)$, where P is a free predicate, i is the arity of P and $\underline{t}$ is a string of terms (for different free predicates P, we choose different free variables P_i);
2. $*\alpha = \alpha$ for other atomic formulae;
3. $*\neg\alpha = \neg *\alpha$;
4. $*(\alpha \,\&\, \beta) = (*\alpha \,\&\, *\beta)$, and for other binary connectives in a similar way;
5. $*(\exists x\ \alpha) = \exists x\ *\alpha$;
6. $*(\forall x\ \alpha) = \forall x\ *\alpha$;
7. $*(^{Q}_{Q'}\ \alpha) = \exists P_i\ \exists P_j(*(Q\ E_i(\underline{x}, P_i)) \,\&\, *(Q'\ E_j(\underline{y}, P_j)) \,\&\, \forall x,y(E_i(\underline{x}, P_i) \,\&\, E_j(\underline{y}, P_j) \Rightarrow *\alpha))$,

where i is a number of variables in Q, j is a number of variables in Q', $\underline{x}$ is a list consisting of all variables from Q, and $\underline{y}$ is a list consisting of all variables from Q', and P_n is a variable of sort U_n.

Constructing $*\alpha$ we pass through a language containing all constructions from F and FST, but at the end we obtain $*\alpha \in$ FST.

Now we can define the theory ST. ST has as axioms all formulae:

$$\exists P_i\ \forall x_1 \ldots \forall x_i(E_i(x_1,\ldots,x_i, P_i) \equiv *\alpha),$$

where $\alpha \in$ FE and i = 1, 2,...

Semantical Validity for the Logic LB

For any $\alpha \in$ FE: α is a tautology in LB ($\models \alpha$) if and only if ST $\models *\alpha$. Of course we are interested only in validity of formulae from F, but for comparing semantical validity with provability, it would be better to have a definition for all formulae from FE.

3. PROOF SYSTEM FOR LB

We are interested only in proofs for formulae from F, but to construct such proofs we shall need all formulae from FE. Besides traditional deduction rules, LB has two schemes LB1 and LB2:

LB1

$$\frac{Q_1\,\alpha_1,\ldots,Q_n\,\alpha_n,\ \forall\underline{x}(\alpha_1\ \&\ \ldots\ \&\ \alpha_n \Rightarrow \beta)}{\begin{matrix} Q_1 \\ \cdot \\ \cdot \\ \cdot \\ Q_n \end{matrix}\ \beta}$$

for n = 1,2,..., where $\underline{x}$ is a list of all variables from $Q_1,\ldots,Q_n$, and Q_i and Q_j have no common variables for $i = j$.

LB2

$$\frac{\begin{matrix} Q \\ Q' \end{matrix}\ \alpha}{Q\ P(\underline{x})\ \&\ Q'\ \forall\underline{x}(P(\underline{x}) \Rightarrow \alpha)}$$

where x is a list of all variables occuring in Q, and P is a new free predicate symbol with suitable arity; Q and Q' can be empty prefixes.

We need also a proof system for the first order predicate calculus with natural extension of axiom-schemes and rules to formulae from the set FE. We can use the Natural Deduction System of Slupecki and Borkowski [6] or a version of Mizar-MSE as described in [7,8].

As an example I shall prove the implication $\begin{matrix} Q \\ Q' \end{matrix}\alpha \Rightarrow \begin{matrix} Q' \\ Q \end{matrix}\alpha$ (usual logical rules will be used without any special reference). Assume:

1. $\begin{matrix} Q \\ Q' \end{matrix}\alpha$;
2. $Q\ P(\underline{x})\ \&\ Q'\ \forall\underline{x}(P(\underline{x}) \Rightarrow \alpha)$ by 1 (LB2);
3. $Q\ P(\underline{x})$ by 2;
4. $Q'\ \forall\underline{x}(P(\underline{x}) \Rightarrow \alpha)$ by 2;
5. $\forall\underline{x},\underline{y}(\forall\underline{x}(P(\underline{x}) \Rightarrow \alpha)\ \&\ P(\underline{x}) \Rightarrow \alpha)$ (tautology);

thus $\begin{matrix} Q' \\ Q \end{matrix}\alpha$ by 4, 3, 5 (LB1).

In the above proof $\underline{x}$ represents a list of variables in Q, and $\underline{y}$ represents a list of variables in Q'.

Provability in LB

We say that $\alpha \in F$ is provable in LB ($\vdash \alpha$) if α has a proof in the above described sense. For instance we have proved that for any $\alpha \in F$ and for any disjoint prefixes Q and Q', the following $\begin{matrix} Q \\ Q' \end{matrix}\alpha \Rightarrow \begin{matrix} Q' \\ Q \end{matrix}\alpha$ holds.

The next natural question is about the relation between semantical validity and provability, that is between relations $\vdash$ and $\models$.

The Completeness Theorem for LB

$\models \alpha$ if and only if $\vdash \alpha$, for any $\alpha \in F$.

Proof. Let $\alpha \in F$. It is rather obvious that $\vdash \alpha$ implies $\models \alpha$. It suffices to check *-translation of LB1 and to remark that LB2 does not provide us beyong formulae provable in ST (this fact may be justified in a similar way to the well-known uncreativity of the choice rule in the predicate calculus).

Now let us assume that $\models \alpha$. It means by the definition and the completeness of many-sorted predicate calculus that ST $\vdash *\alpha$. From the known Gentzen's theorem (see Takeuti [9]) it follows that for some axioms of ST $\alpha_1,\ldots,\alpha_n$ a sequence $\{\alpha_1,\ldots,\alpha_n\} \to \{*\alpha\}$ has a proof G without cutting. Proofs without cutting have a very nice property that formulae occuring in premisses are subformulae or subformulae of instantations (and so on) of formulae occuring in a conclusion. Using this property we can eliminate from a proof all formulae which are not *-translations of formulae from FE.

Not every formula in G is *-translation of any formula, one of the causes of this is that not every subformula of $*\alpha$ needs to be a *-translation of any formula. All exceptions are of the form:

$$\exists P_1 \ldots \exists P_{n-k}(Q_1\beta_1 \;\&\; \ldots \;\&\; Q_k\beta_k \;\&\; Q_{k+1}E_{j(1)}(\underline{x}_1, P_1)$$

$$\&\; \ldots \;\&\; Q_n E_{j(n-k)}(\underline{x}_{n-k}, P_{n-k}) \;\&\; \forall \underline{x}_1 \ldots \underline{x}_{n-k}(\beta_1 \;\&\; \ldots$$

$$\&\; \beta_k \;\&\; E_{j(1)}(\underline{x}_1, P_1) \;\&\; \ldots \;\&\; E_{j(n-k)}(\underline{x}_{n-k}, P_{n-k}) \Rightarrow \gamma),$$

where $n > k > 0$. If μ is such a formula then only two kinds of rules can be applied to it in G: (1) absorbtion or (2) introduction of $\exists$. For the application of other rules μ is untouched. Then we can apply all $\exists$-introductions to μ, and then we can eliminate all mediate steps, and replace all occurences such as mediate formulae by μ with all necessary existential quantifiers. The only possible restriction is in the case when an unbound variable (which has to be quantified) occurs differently than in μ formulae. This trouble can arise only in the case when μ is in an antecedent, because in Gentzen's system there are some restrictions only for $\exists$-introduction on the left. In that situation we delete all necessary quantifiers in the beginning of μ and in all its premisses and formulae cooperating with them in absorbtion. Then we introduce all necessary quantifiers in a place where in the original proof the last relevant existential quantifier was introduced. It cannot cause any trouble because free variables introduced in such an obtained proof can occur only in formulae obtained from μ and in those from which μ was obtained. Let us remark that we can assume that no free variable used to $\exists$-introduction on the left is used to $\exists$-introduction on the left to obtain some other formula. We shall use this fact at the end of our proof.

In what follows, eliminating other possible irregularities, we shall obtain a tree built up only from formulae belonging to FE.

At the first step we eliminate each unbound variable P_j^i such that α_i is obtained binding P_j^i by $\exists$ in an antecendent, for $i = 1,2,\ldots,n$. That is if α_i was obtained from $\forall x_1 \ldots \forall x_j \; (E_j(x_1,\ldots,x_j, P_j^i) \equiv \beta_i)$ then we replace all occurences of $E_j(t_1,\ldots,t_j, P_j^i)$ by $\beta_i[x_1/t_1,\ldots,x_j/t_j]$ in G. In this way we obtain G_1 from G.

At the second step we delete all occurences of $\alpha_1,\ldots,\alpha_n$. Then we obtain G_2 from G_1.

At the third step we eliminate all free variables of sorts $U_1,U_2,\ldots$, replacing them by mutually different free predicates. That is for every P_j unbound variable occuring in G_2, we have introduced a new free predicate P with arity j, and we replace every atomic formula $E_j(\underline{t}, P_j)$ occuring in G_2 by $P(\underline{t})$. In this way we obtain G_3 from G_2.

Every formula occuring in G_3 is *-translation of some formula from FE. It is easy to see that every FE-formula is uniquely determined by its *-translation. Then we obtain G_4 from G_3 replacing all formulae β in G_3 by corresponding to the FE-formulae. In this way we obtain a tree with the sequences of formulae belonging to FE on the nodes.

Now we obtain G_5 from G_4 replacing any sequence:

$$\{\beta_1,\ldots,\beta_k\} \to \{\gamma_1,\ldots,\gamma_s\} \text{ by } (\beta_1 \,\&\, \ldots \,\&\, \beta_k) \Rightarrow (\gamma_1 \vee \ldots \vee \gamma_s),$$

$$\{\beta_1,\ldots,\beta_k\} \to \{\} \text{ by } \neg(\beta_1 \,\&\, \ldots \,\&\, \beta_k),$$

$$\{\} \to \{\gamma_1,\ldots,\gamma_s\} \text{ by } (\gamma_1 \vee \ldots \vee \gamma_s).$$

G_5 is a tree with α at the bottom and with formulae from FE on branchings and leaves.

Now we are going to prove by a tree-induction that all formulae on the tree G_5 are provable in LB. Formulas corresponding to Gentzen's axioms are of course logical tautologies. In the case of structural rules an induction step is obvious, similarly for rules of introduction of sentential connectives. Cases of introduction $\forall$, $\exists$ for variables of the basic sort are also trivial.

We have to investigate also an introduction of existential quantifiers binding variables of non basic sorts (these variables can be bound only by existential quantifiers!).

Let us assume that on some branching we have a formula $\beta \Rightarrow (\gamma \vee \delta)$ and $*\delta$ was obtained introducing a sequence of existential quantifiers. Then $\beta \Rightarrow (\gamma \vee \delta)$ can be justified by LB1 and the previous formula in the tree.

Now we wish to show that $(\delta \text{ and } \beta) \Rightarrow \gamma$ is provable in LB if $(\sigma \text{ and } \beta) \Rightarrow \gamma$ is provable in LB and $*\delta$ was obtained from $*\sigma$ by introduction of a sequence of existential quantifiers of non basic sorts. Let us note that the rule of $\exists$-introduction to antecedent is restricted to cases where existentially generalized variables do not occur in a conclusion. Using a proof of $(\sigma \text{ and } \beta) \Rightarrow \gamma$, we construe a proof for $(\delta \text{ and } \beta) \Rightarrow \gamma$ as follows. Assume:

1. $\alpha \,\&\, \beta$;
2. σ by 1 (applying suitable number of times LB2);
3. $\sigma \,\&\, \beta$ by 1, 2

   .
   . a proof for $(\sigma \,\&\, \beta) \Rightarrow \gamma$
   .

n: $(\sigma \,\&\, \beta) \Rightarrow \gamma$; thus γ by 3, n. QED.

REFERENCES

1. L. Henkin, "Some Remarks on Infinitely Long Formulae", Infinitistic Methods, Pergamon Press, New York and Polish Scientific Publishers, Warsaw, pp. 167-183 (1961).
2. H. B. Enderton, Finite partially-ordered quantifiers, Zeitschrift fur Mathematische Logik und Grundlagen der Mathematik, vol. 16:397- , (1970).
3. W. J. Wolkoe, Jr., Finite partially-ordered quantification, Journal of Symbolic Logic, vol. 35:535-555 (1970).

4. M. Krynicki and A. H. Lachlan, On the semantics of the Henkin quantifier, Journal of Symbolic Logic, vol. 44:184-200 (1979).
5. D. Harel, Characterfying second order logic with first order quantifier, Zeitschrift fur Mathematische Logik und Grundlagen der Mathematik, vol. 25:419-422 (1979).
6. J. Slupecki and L. Borkowski, "Elementy Logiki Matematycznej i Teorii Mnogosci" (Elements of Mathematical Logic and Set Theory), Polish Scientific Publishers, Warsaw (1963).
7. A. Trybulec and H. Blair, Computer assisted reasoning with Mizar, Proceedings of the Ninth International Joint Conference on Artificial Intelligence, Los Angeles (1985).
8. M. Mostowski, Textbook of Logic based on Mizar-MSE, in preparation.
9. G. Takeuti, "Proof Theory", North-Holland Pub. Co., Amsterdam-London and American Elsevir Co., New York (1975).

PROPOSITIONAL DYNAMIC LOGIC IN TWO- AND MORE DIMENSIONS

Assen Petkov

Inform. Comp. Centre
State Committee for Research and Technologies
ul. Chapaev 55A, Sofia 1574, Bulgaria

INTRODUCTION

The Propositional Dynamic Logic PDL extends the Kripke modal logic with the intension to frame the behavior and formalize the properties of computer programs. A program α (as a rule non-deterministic) and a property A are interpreted respectively as a binary relation $R(\alpha)$ and a subset $V(A)$ of some presupposed universe $U=\{u, v, \ldots\}$. Its elements are refered to as possible worlds or memory states and are traditionally regarded as some indivisible units. The aim of this paper is to suggest a reliable framework and to establish some basic results for that case, where the memory states have themselves been divided and granted the structure of ordered pairs (or generally, ordered N-tuples). In this case, formally $R(\alpha)$ and $V(A)$ will be quaternary and binary relations, respectively. Informally this corresponds to the case of computation with two-cell memory device.

The first succesful attempt towards splitting the states was probably Segerberg's [3]. This attempt, provoked by purely philosophical reasons (to express properties concerned simultaneousely to two different moments in the time) is in the pure modal logic; the dynamic logic by that time, had not been yet invented. Segerberg axiomatizes the two-dimentional modal logic **B**, the proof technique being intricate enough, and somehow coined ad hoc: it is aplicable neither to higher dimensions, nor easily to extensions of the basic system, and according to the autor, usatisfactory long.

The present study overcomes all these shortcommings; it is an answer to a problem, posed by Dimiter Vakarelov, appealing to a natural axiomatization of N-dimensional dynamic (hence modal) logic. Such a natural axiomatization is achieved here by setting the question in the environment of the combinatory PDL (CPDL), introduced by Passy and Tinchev [1,2], which extends the ordinary dynamic logic by adding in the syntax proper names for the states, appropriately axiomatized. So the present study, besides having its philosophical and computational applications, can be motivated as itself an application of the combinatory approach to modal logic. Technically, the paper proceeds as follows.

The language of CPDL is extended by adding a particular atomic formula **E** and two atomic programs **x** and **y**. Semantically the predicate E verifies the identity of states: (s, t) ⊨ E if s=t. The program letters x and y are interpreted as total functions X, Y with X((s, t))=(s, s) and Y((s, t))=(t, t). The semantic structures, called **nets**, being similar to Segerberg's **frames** are isomorphic to a Cartesian square of some set and when so, X and Y play the role of decoding functions: u=(Xu, Yu). Sintactically, however, Segerberg's logic B is much more wasteful: besid x and y (denoted by ⦶ and ⊖), it also includes the modal operators ⊗, ◫, ⊟, □ with the following meaning:

⊗: (s, t) --> (t, s) ; □: (s, t) --> (@, @) ,
◫: (s, t) --> (s, @) ; ⊟: (s, t) --> (@, t) ,

where "@" stands for any state. **B** is axiomatized and proven to have the finite model property.

Our logic resulting, is called Square CPDL (SqCPDL). We show that the deductive system for SqCPDL is complete for the class of all Kripke models with universes Cartesian squares. In a certain sense, what **B** is in the modal logic, SqCPDL is in the dynamic one. This is the basic system, framing the two-dimensional properties: it needs only x, y and E. Further we enlarge SqCPDL by atomic programs ⊗, ◫, ⊟ (with meaning as above and by ⧄, ⧅, with semantics:

⧄: (s, t) --> (@, s) ; ⧅: (s, t) --> (t, @) .

The system obtained is denoted as SqCPDL^{-}.

Evidently, properties expressed in PDL and CPDL by programs (binary relations) are represented in the Square logics by formulas. In particular, the linear ordering is proven to be axiomatizable over SqCPDL^{-}.

At the end of the paper the techniques is straightforward extended to spread over the n-dimensional dynamic logic, where n is an arbitrary fixed integer.

NETS

Definition A quadruple **N**=(U, D, X, Y) is a **net** if U is a set and D⊆U, and X, Y ⊆ UxD, and the following conditions are me

(m1) X and Y are total functions.
(m2) For each u, v ∈ D there is some w, such that Xw = u , Yw = v (where Xw, Yw denote X(w), Y(w)).
(m3) If Xu = Xv and Yu = Yv, then u = v .
(m4) If u∈D, then u = Xu = Yu .
(m5) If Xu = Yu, then u ∈ D . #

A special kind of nets are the **Cartesian** one, for which there is some set M with U = M^2, and for each s, t ∈ M:

X(s, t) = (s, s) , Y(s, t) = (t, t).

We say that the nets (U, D, X, Y) and (U', D', X', Y') are **isomorphic**, if an one to one mapping f: U-->U' exists, with:

(D) $u \in D$ iff $f(u) \in D'$
(X) $f(Xu) = X'(f(u))$
(Y) $f(Yu) = Y'(f(u))$.

<u>Net Representation Lemma</u> Each net is isomsrphic to some Cartesian one.

<u>Proof.</u> Let **N** = (U, D, X, Y) be a net. We set $U' = D^2$, $D' = D$, $f: U \dashrightarrow D^2$ with $f(u) = (Xu, Yu)$. Let X' and Y' be functions onto D^2 defined as in the above Cartesian net definition. Now (m1) justifies totality of f. Conditions (m2) and (m3) show that f is surjective and injective; (m4) and (m5) prove (D).

$$f(Xu) = (XXu, YXu) = (Xu, Xu) = X'(Xu, Yu) = X'(f(u)),$$

wherefrom (X) and analogically (Y). Then **N'**=(U', D', X', Y') is a Cartesian net isomorphic to **N**. #

So having this Represantation, we shall further consider only Cartesian nets, not loosing generality.

SYNTAX AND SEMANTICS FOR SQUARE CPDL

Let Φ_0, Σ and Π_0 be tree countable, infinite and pairwise disjoint sets (of **atomic formulas, state constants** and **atomic programs**, respectively). Inductively defined are the sets Φ of **formulas** and Π of **programs** of SqCPDL:

1. $\Phi_0 \cup \Sigma \cup \{E\} \subseteq \Phi$; $\Pi_0 \cup \{\nabla\} \cup \{x, y\} \subseteq \Pi$

2. If $A, B \in \Phi$; $\alpha, \beta \in \Pi$, then:
   $\neg A, A \wedge B, [\alpha]A \in \Phi$; $\alpha \circ \beta, \alpha \cup \beta, \alpha^*, A? \in \Pi$.

As typical representatives we shall use $A, B \in \Pi$; $c, d, e \in \Phi$; $\alpha, \beta \in \Pi$. The Boolean functions $\vee$, $\rightarrow$, $\leftrightarrow$, **0**, 1 and program operator $\langle\rangle$ are introduced as usually. We abbreviate $[\nabla]$, $\langle\nabla\rangle$, $\langle\nabla \circ c?\rangle$ as $\Box$, $\Diamond$, $\underline{c}$. $\exists$ and $\forall$ are used informally. P(M) is the power-set of M.

<u>Definition</u> A **CPDL-model** is any triple **M**=(U, V, R), where U is nonempty set with:

$$V : \Pi_0 \cup \Sigma \cup \{E\} \dashrightarrow P(U) \quad ; \quad R : \Pi_0 \cup \{\nabla\} \cup \{x, y\} \dashrightarrow P(U^2) \quad ,$$

where $R(\nabla) = U^2$, for each $c \in \Sigma$ $V(c)$ is singleton and $\bigcup_{c \in \Sigma} V(c) = U$.

M=(U, V, R) is a **Square CPDL-model**, if:

1. It is a CPDL-model.
2. (U, V(E), R(x), R(y)) is a net.

For a given model **M**=(U, V, R), V and R are extended to cover all formulas and programs, these extensions being also denoted by R and V. And X, Y stand for R(x), R(y). As usual $u \in V(A)$ and $(s, t) \in R(a)$ are denoted as $u \models A$ and s.Ra.t . $\mathbf{M} \models A$ means, that $u \models A$ for all $u \in U$. $\models A$ denotes, that $\mathbf{M} \models A$ for each SqCPDL-model **M** and we read this as "A is valid".

Of course, a fixed square model has always exactly k^2 points, for some $k \leq \omega$.

THE DEDUCTIVE SYSTEM FOR SQUARE CPDL

Axioms:

A1) All PDL axioms.

A2) All proper CPDL axioms:
$\Box$ is an S5 modality
$\Box A \rightarrow [\alpha]A$
$\underline{c}1$
$\underline{c}A \rightarrow [\vee c?]A$
And for $SCPDL^{\cap}$ additionally:
$\langle \alpha \cap \beta \rangle c \leftrightarrow \langle \alpha \rangle c \wedge \langle \beta \rangle c$

B) The proper "square" axioms, where z runs over {x, y}:

a1.1) $\langle z \rangle A \rightarrow [z]A$
a1.2) $\langle z \rangle E$
a2) $\underline{c}E \wedge \underline{d}E \rightarrow \Diamond(\langle x \rangle c \wedge \langle y \rangle d)$
a3) $\underline{c}\langle x \rangle e_1 \wedge \underline{d}\langle x \rangle e_1 \wedge \underline{c}\langle y \rangle e_2 \wedge \underline{d}\langle y \rangle e_2 \rightarrow \underline{c}d$
a4) $E \wedge A \rightarrow [z]A$
a5) $\langle x \rangle d \wedge \langle y \rangle d \rightarrow E$

Rules of SqCPDL are, besides Modus ponens (MP) :

(Nec) If $\vdash A$, then $\vdash [\alpha]A$
(Ind) If $\vdash [\beta][\alpha^i]A$ for all $i \leq \omega$, then $\vdash [\beta][\alpha*$
(Cov) If for all $c \in \Sigma$ ($\vdash [\alpha]\neg c$), then $\vdash [\alpha]0$

Soundnes Theorem If $\vdash A$, then $\models A$.
Proof Straightforward. #

COMPLETENESS THEOREM FOR SqCPDL

By a **theory** we understand in this paper any set T of formulas, containing all axioms and closed under the rules (MP), (Ind and (Cov). T is a **maximal** theory if $\forall A$ ($A \in T$ iff $\neg A \notin T$). T is **consistent** if $0 \notin T$. Obviously the maximal theories are consistent ones.

A **logic** is any theory, closed under (Nec). The least logic obviously consists of all SqCPDL theorems. Let, until end of thi section, L and T be fixed logic and theory with $L \subseteq T$.

The proof of the Completeness theorem follows as in [2]. We use three facts from this paper:

($\wedge$) $\vdash \underline{c}(A \wedge B) \leftrightarrow \underline{c}A \wedge \underline{c}B$
($\neg$) $\underline{c}A \in T$ iff $\underline{c}\neg A \notin T$
($\langle \alpha \rangle$) $\underline{c}\langle \alpha \rangle A \in T$ iff $\exists d$ ($\underline{c}\langle \alpha \rangle d \in T$ and $\underline{d}A \in T$)

Lemma Let T be a maximal theory. Then:

($\exists$) $\Diamond A \in T$ iff $\exists c$ ($\underline{c}A \in T$).
($\forall$) $\Box A \in T$ iff $\forall c$ ($\underline{c}A \in T$). #

Obviously if $L \vdash A$ then $\exists c$ ($\underline{c}A \in T$).

We define: $c \approx d$ iff $\Diamond(c \wedge d) \in T$, $c^{\sim} = \{d \in \Sigma \mid d \approx c\}$, $U = \Sigma_{/\approx}$, $V(A) = \{ c^{\sim} \mid \underline{c}A \in T \}$, $V(c) = \{c^{\sim}\}$, $R(\alpha) = \{ (c^{\sim}, d^{\sim}) \mid \underline{c}\langle \alpha \rangle d \in T \}$.
Then $\mathbf{M}_T = (U, R, V)$ is called **canonical model** generated by T.

Proposition **Mr** is a SqCPDL-model.

Proof In [2] it is proven, that **Mr** is a CPDL model. We have to prove, that (U, V(E), R(x), R(y)) is a net, but this follows easy from the corresponding axioms (a1)-(a5). For instance:

(m2) Let c˜∈V(E) and d˜∈V(E). We have to show that there exists such e, for which: e˜.Rx.c˜ and e˜.Ry.d˜. From definition c̲E∈T and d̲E∈T, hence c̲E∧d̲E∈T and because of (a2) ◇(<x>c∧<y>d)∈T. From Lemma(∃) it follows ∃e: e̲(<x>c∧<y>d)∈T, which combined with (∧) yields e̲<x>c∈T and e̲<y>d∈T. #

Completeness theorem If ⊨L A, then ¦-L A .

Two lemmas are used:

Deduction lemma L ¦- ◇A->B iff L, ◇A ¦- B . #

Lindenbaum lemma If T is consistent theory, then a maximal theory T' exists with T⊆T'. #

Proof(of completeness). Let L not ¦- A, i.e. A∉L. Then L ∪ {◇¬A} is consistent and there exists a maximal theory T: L∪{◇¬A}⊆T. From (∃) it follows, that ∃c (c̲◇¬A∈T), i.e. **M**, c˜ ⊨ ◇¬A, from where ∃d: d˜ ⊨ ¬A, i.e. A fails. #

Proposition SqCPDL is not a conservative extension of CPDL.

Proof As noted above, there is no Square model with exactly 2 points, i.e. Card(**M**)≤2 implies Card(**M**)=1, hence the formula

$A_{\square}$ = □(c∨d)->□c

is valid. By the Completeness for SqCPDL and Soundness of CPDL, $A_{\square}$ turns to be a theorem of SqCPDL, disprovable in CPDL. #

EXTENSIONS OF SqCPDL

By means of basic programs **x** and **y** we may express the relations ⊗, ◫, ⊟ (in [3] they are not used indenpendentely each other), and the new programs ⧄, ⧅ in such a way: The basic language of Square logic is extended by program symbols ⊗, ◫, ⊟, ⧄, ⧅. We shall call that **M**=(U, V, R) is a SqCPDL$^{\blacksquare}$ model if it is a SqCPDL model and new relations are defined:

(m6)	u.R⊗.v	iff	Xu=Yv and Yu=Xv
(m7)	u.R◫.v	iff	Xu=Xv
	u.R⊟.v	iff	Yu=Yv
(m8)	u.R⧄.v	iff	Xu=Yv
	u.R⧅.v	iff	Yu=Xv , for all u, v∈U.

The corresponding axioms added to (a1)-(a5) are:

(a6)	c̲<⊗>d <-> <∀∘c?∘x ∩ ∀∘c?∘y>1 ∧ <∀∘c?∘y ∩ ∀∘c?∘x>1
(a7)	c̲<◫>d <-> <∀∘c?∘x ∩ ∀∘c?∘x>1
	c̲<⊟>d <-> <∀∘c?∘y ∩ ∀∘c?∘y>1
(a8)	c̲<⧄>d <-> <∀∘c?∘x ∩ ∀∘c?∘y>1
	c̲<⧅>d <-> <∀∘c?∘y ∩ ∀∘c?∘x>1

Now the axiom (a3) may be written shorter as:

(a3) <◫>d ∧ <⊟>d -> d

If we denote the canonical model generated by some maximal teory T by $\mathbf{M}^{\blacksquare}_{T}$, it is true that

<u>Proposition</u> ($\mathbf{M}^{\blacksquare}_{T}$) $\mathbf{M}^{\blacksquare}_{T}$ is a SqCPDL$^{\blacksquare}$-model. #

Having this, the Completeness theorem for SqCPDL$^{\blacksquare}$ immediatelly follows.

Not surprisingly, notions dealt with in CPDL (or PDL) as programs, can be treated in the Square logics as predicates. For instance we shall take the linear ordering predicate $\leq\ \subseteq U^2$. And now the reflexiveness, asymetry, transitivity and linearity of $\leq$, expressed in SqCPDL$^{\blacksquare}$ models as:

(rf) If $Xu=Yu$, then $Xu\leq Yu$
(as) If $Xu\leq Yu$ and $Yu\leq Xu$, then $Xu=Yu$
(tr) If $Xu\leq Yu$, $Yu=Xv$, $Xv\leq Yv$, $Xw=Xu$, $Yw=Yv$, then $Xw\leq Yw$
(lin) $Xu\leq Yu$ or $Yu\leq Xu$

are axiomatized by adding the atomic formula H (corresponding to $\leq$) to the SqCPDL$^{\blacksquare}$ language by the axioms:

(a9) $E \to H$
(a10) $H \wedge \langle\otimes\rangle H \to E$
(a11) $\underline{c}H \wedge \underline{d}H \wedge \underline{c}\langle ⧅ \rangle d \wedge \underline{c}\langle ◫ \rangle e \wedge \underline{d}\langle ⊟ \rangle e \to \underline{e}H$
(a12) $H \vee \langle\otimes\rangle H$

<u>Theorem</u> The above axiomatic system is sound and complete for lineary ordered SqCPDL$^{\blacksquare}$-models.

N-DIMENSIONAL DYNAMIC LOGIC

It is not difficult to generalize all considerations of preceding sections. Until end of this section let n be arbitrary fixed integer. We say, that $\mathbf{M}=(U, D, X_1,\dots,X_n)$ is an **n-net** if the following conditions are satisfied:

(n1) X_k: $U \to D$ is a total function, where $D\subseteq U$ for every k ($1\leq k\leq n$).
(n2) For all $u_1, \dots, u_n \in D$ there is some v, such that $X_1v=u_1, \dots, X_nv=u_n$
(n3) If for each k ($1\leq k\leq n$) $X_ku=X_kv$, then $u=v$.
(n4) If $u\in D$, then for every k ($1\leq k\leq n$) $X_ku=u$.
(n5) If $X_1u=X_2u=, \dots, =X_nu$, then $u\in D$.

The **Cartesian n-net** is defined as: $U=M^n$, $D=\{(s,\dots,s) \mid s\in M\}$. For all $(s_1,\dots,s_n)\in M^n$ and for each k ($1\leq k\leq n$) $X_k(s_1,\dots,s_n)=(s_k,\dots,s_k)$. It is easy to verify, that each n-net is isomorphic to some Cartesian one. The language of CPDL is extended with n atomic program symbols: $x_1,\dots,x_n$ and atomic formula E (interpreted as n-dimensional diagonal). n-dimensional models are build on n-nets in the same way as in two-dimensional case. A deductive system, called n-CPDL is obtained adding to CPDL the following five "proper" n-dimensional axioms:

(a#1.1) $\langle x_k\rangle A \to [x_k]A$
(a#1.2) $\langle x_k\rangle E$
(a#2) $\underline{c}_1E\wedge \dots \underline{c}_nE \to \Diamond(\langle x_1\rangle c_1\wedge \dots \wedge\langle x_n\rangle c_n)$
(a#3) $\underline{c}\langle x_1\rangle e_1 \wedge \underline{d}\langle x_1\rangle e_1\wedge \dots \wedge\underline{c}\langle x_n\rangle e_n \wedge \underline{d}\langle x_n\rangle e_n \to \underline{c}d$

(a#4) $E \wedge A \rightarrow \langle x_k \rangle A$

(a#5) $\langle x_1 \rangle d \wedge \ldots \wedge \langle x_n \rangle d \rightarrow E$

As hereinabove:

<u>Theorem</u> The n-dimensional axiomatic system is sound and complete for n-dimensional models. #

ACKNOWLEDGEMENTS

Dimiter Vakarelov introduced me to modal logic and promoted the above investigation.

I due sincerely thanks to Solomon Passy. His ideas, advices and moral support were very helpful for me.

REFERENCES

1. Passy S., Tinchev T., PDL with Data Constants, Inf. Proc. Letters, 20(1985), 35-41.

2. Passy S., Tinchev T., Quantifiers in Combinatory PDL: completeness, definability, incompleteness, Proc. FCT'85, Springer LNCS 199, 512-519.

3. Segerberg Kr., Two-dimensional Modal Logic, Journal of Philosophical Logic, 2(1973), 76-96.

ON A NONCONSTRUCTIVE TYPE THEORY AND PROGRAM DERIVATION

Jan M. Smith

Department of Computer Science
University of Göteborg/Chalmers
S-412 96 Göteborg
Sweden

1 Introduction

Not considering philosophical arguments, the main motive for using constructive reasoning when constructing programs is that constructive proofs have computational content. For instance, formulating a specification and proving it in Martin-Löf's type theory, gives a program satisfying the specification. On the other hand, extracting programs from classical proofs is in general not possible. However, the process of deriving a program may not only involve the actual construction of the program but also the verification that an already constructed part of the program satisfies some property and it is then quite possible to use classical logic. A system for program derivation where you may use classical logic is the one developed by Manna and Waldinger [5].

I will here add the law of the excluded middle to Martin-Löf's type theory, thereby making classical logic available when reasoning about programs in type theory. In this nonconstructive type theory there are terms which cannot be computed even though they have types. However, a theorem will be proved which will give a simple syntactic requirement on a term, which will guarantee that the term is a program, satisfying the specification expressed by its type.

For me, the main drawback of introducing classical logic in type theory is that the simple semantics given by Martin-Löf [6,7] no longer works and that I am only able to have a formalistic understanding of type theory when the law of the excluded middle is added. The reason for extending type theory with the law of the excluded middle would be that it would be easier to find proofs when using classical logic. But I do not know of any example from programming where the correctness proof becomes considerably simpler when using classical logic. However, it must be admitted, the experience we have so far of formal program derivation is rather limited. In particular, the situation may change when we get more extensive experience of computer assisted programming logics, like, for instance, the system Nuprl [10] for type theory. There are, of course, also specifications which are impossible to satisfy without using classical logic, but it seems very unlikely that such a specification could be of any practical interest.

A specification in Manna and Waldinger [5] is always of the form

$$\textit{find } z \textit{ such that } R(x,z) \textit{ where } P(x)$$

where $P(x)$ is the condition on the input x and $R(x,z)$ is the condition on the output z of the desired program. The proof of finding z for a given input x must be constructive in the sense that it must be possible to extract from the proof a program $f(x)$ for computing z. However, when verifying $R(x, f(x))$ under the assumption $P(x)$, they allow classical reasoning.

A specification from Manna and Waldinger's system would in type theory be expressed by using the subset type:

$$\{z \in B \mid R(x,z)\} \ [x \in \{y \in A \mid P(y)\}]$$

where A is the type of the input, B the type of the output and the assumption is within square brackets. In type theory, because of the constructive explanation of logic, a specification can have a more general form in that an arbitrary formula of predicate logic can be interpreted as a specification [8]. However, one could think of using classical logic also when deriving programs in type theory, namely when the specification contains a part of the form $\{x \in A \mid B(x)\}$. As in Manna and Waldinger's system, one could here imagine that, by using constructive reasoning, one has obtained a program a of type A, but that classical logic may be used in the verification of $B(a)$. In practice, specifications in type theory involving subsets occur quite often [9].

One way of getting classical reasoning available in a constructive theory is to use Gödel's double-negation interpretation [1,3,4]; an approach taken, for example, by Coquand and Huet [2]. However, for Martin-Löf's type theory, this does not give the full power of the law of the excluded middle because the axiom of choice is provable in the theory. And it is well-known that the axiom of choice together with the law of the excluded middle makes it possible to prove classical set comprehension, interpreting subsets by characteristic functions. For instance, adding the law of the excluded middle to Martin-Löf's type theory without universes and well-orderings, a theory which proof theoretically is not stronger than arithmetic, we get a theory in which classical analysis can be interpreted. So, adding classical logic to Martin-Löf's type theory results in a much stronger theory which, in particular, is impredicative. For a theory with a simple type structure, these results can be found in Spector [12].

In Martin-Löf's type theory propositions are introduced as types and I assume familiarity with the idea of propositions as types as well as the rules of type theory [1,6,7,11].

2 A nonconstructive type theory

The obvious way of extending type theory with the law of the excluded middle is to introduce a new constant ? of arity $0 \to 0$ and for each type A, which may depend on assumptions, add the axiom

$$?(A) \in A \vee (\neg A)$$

where $A \vee (\neg A)$ is an abbreviation of $A + (A \to \emptyset)$, using the interpretation of propositions as types. The intuition behind this axiom is the following. By classical reasoning we have that either A is nonempty or A is empty. In the first case we choose an element a of type A and let $?(A)$ denote $\mathsf{inl}(a)$. If A is empty the identity function $\lambda((x)x)$ is an element of $A \to \emptyset$. So, in the second case we let $?(A)$ denote $\mathsf{inr}(\lambda((x)x))$.

In type theory, an object of a disjoint union $A + B$ can be computed to a value which is either of the form $\mathsf{inl}(a)$ where a is an object of type A, or of the form $\mathsf{inr}(b)$

where b is an object of type B. However, introducing the constant ? this is no longer true since, in general, we cannot know which of the disjuncts A and $\neg A$ it is that holds. So, $?(A)$ is an expression which cannot be computed and there will now be noncanonical expressions like, for example,

$$\mathsf{when}(?(A), (x)\mathsf{true}, (y)\mathsf{false})$$

of type Bool, which cannot be computed to a value. when is the noncanonical constant associated with the disjoint union, having the computation rules

$$\begin{cases} \mathsf{when}(\mathsf{inl}(a), e, d) & \rightarrow \; e(a) \\ \mathsf{when}(\mathsf{inr}(b), e, d) & \rightarrow \; d(b) \end{cases}$$

The following theorem will give a simple condition on an object of a type to have a value.

Theorem 1 *If $a \in A$ is derivable in Martin-Löf's type theory extended with the law of the excluded middle and the constant ? does not occur syntactically in a, then a can be computed to a canonical value of type A.*

Since we are now working in a theory including the law of the excluded middle, we can no longer use the semantics of Martin-Löf's type theory. For instance, the notion of canonical object has to be redefined; the new definition will be given in the proof of the theorem. Note that the theorem has as an immediate consequence the following corollary: If $a \in A$ is derivable in Martin-Löf's type theory then a can be computed to a canonical value of type A. Of course, this normalization result for closed terms, here obtained by metamathematical methods, follows immediately from the semantics of type theory. The theorem will be proved for a version of type theory which may include well-orderings but not universes. The proof is based on the method of Tait [13]; a term being computable here corresponding to a term having a canonical value.

Proof of theorem 1

The proof will be by induction on the length of the derivation of $a \in A$. We also have to consider derivations of judgements of the other forms, that is A *type*, $A = B$, and $a = b \in A$, as well as derivations of hypothetical judgements. For the first three forms I will lay down what we have to show for a derivation ending with a conclusion of that form; there will be no requirements on a conclusion of the form $a = b \in A$.

Not considering hypothetical judgements, we have to show the following:

- If we have a derivation with a conclusion of the form A *type*, we have to define what a canonical value of that type is.
- If a derivation has $A = B$ as conclusion, we must show that A and B are extensionally equal, that is, if an expression is canonical value of one of the types then it is also a canonical value of the other.
- If we have a derivation of $a \in A$ and a does not contain the constant ?, we must show that a can be computed to a canonical value of type A.

Of course, we also have to treat derivations with assumptions. The general idea in proofs of normalization is then to require that a term must have a canonical value when we substitute arbitrary terms having canonical values for the free variables; in [14] this is called computability under substitution. However, the situation here is more complicated, because terms having types may not be computable. In particular, if we have a derivation of $b \in B \; [x \in A]$ and x does not occur free in b we also have to consider substitutions for x of closed terms containing the constant ?; the crucial rule where this is needed is subset-elimination. We will prove the following three clauses by induction on the length of a derivation:

(I) If we have a derivation of

$$A(x_1, \ldots, x_n) \; type \; [x_1 \in A_1, \; \ldots, \; x_n \in A_n(x_1, \ldots, x_{n-1})]$$

and closed terms $a_1, \ldots, a_n$ such that

$$a_1 \in A_1, \; \ldots, \; a_n \in A_n(a_1, \ldots, a_{n-1})$$

then we must define what it means for an expression to be a canonical value of type $A(a_1, \ldots, a_n)$. We must also show for all closed terms $a_1, b_1, \ldots, a_n, b_n$ such that

$$a_1 = b_1 \in A_1, \; \ldots, \; a_n = b_n \in A_n(a_1, \ldots, a_{n-1})$$

that a canonical value of type $A(a_1, \ldots, a_n)$ is also a canonical value of type $A(b_1, \ldots, b_n)$ and vice versa.

(II) If we have a derivation of

$$A(x_1, \ldots, x_n) = B(x_1, \ldots, x_n) \; [x_1 \in A_1, \; \ldots, \; x_n \in A_n(x_1, \ldots, x_{n-1})]$$

and closed terms $a_1, \ldots, a_n$ such that

$$a_1 \in A_1, \; \ldots, \; a_n \in A_n(a_1, \ldots, a_{n-1})$$

then we must show that a canonical value of type $A(a_1, \ldots, a_n)$ is also a canonical value of type $B(a_1, \ldots, a_n)$ and vice versa.

(III) Suppose we have a derivation of

$$a(x_1, \ldots, x_n) \in A(x_1, \ldots, x_n) \; [x_1 \in A_1, \; \ldots, \; x_n \in A_n(x_1, \ldots, x_{n-1})]$$

where the expression $a(x_1, \ldots, x_n)$ does not syntactically contain the constant ?. Let $x_{k_1}, \ldots, x_{k_m}$ be the variables which occur free in the expression $a(x_1, \ldots, x_n)$. Let $a_1, \ldots, a_n$ be closed terms such that

$$a_1 \in A_1, \; \ldots, \; a_n \in A_n(a_1, \ldots, a_{n-1})$$

and also such that each of $a_{k_1}, \ldots, a_{k_m}$ can be computed to a canonical value of type

$$A_{k_1}(a_1, \ldots, a_{k_1 - 1}), \; \ldots, \; A_{k_m}(a_1, \ldots, a_{k_m - 1})$$

respectively. We must then show that $a(a_1, \ldots, a_n)$ can be computed to a canonical value of type $A(a_1, \ldots, a_n)$.

In order to do the induction step, we have to go through the rules of Martin-Löf's type theory extended with the law of excluded middle. Since many of the rules can be handled in the same way, I will only treat some of them.

In the proof, the assumptions on which the derivation depends will not be explicitly shown, except those which are discharged by the rule considered.

Concerning the formation rules for the various type formers, except subset-formation and Eq-formation, the wording of Martin-Löf's explanations [6,7] can be used, but here in a quite different context. Let us exemplify this by the formation rules for $+$, Π and N.

+-formation

$$\frac{A\ type \qquad B\ type}{A + B\ type}$$

A canonical value of type $A+B$ is either of the form $\mathsf{inl}(a)$ where a is a closed term which can be computed to a canonical value of type A, or of the form $\mathsf{inr}(b)$, where b is a closed term which can be computed to a canonical value of type B.

Note, that by the induction hypothesis, we know the definitions of canonical values of the types A and B.

Π-formation

$$\frac{A\ type \qquad B(x)\ type\ [x \in A]}{(\Pi\, x \in A)B(x)\ type}$$

$\lambda(b)$ is a canonical value of type $(\Pi\, x \in A)B(x)$ if for an arbitrary closed term a, which can be computed to canonical value of type A, $b(a)$ can be computed to a canonical value of type $B(a)$.

N-formation

$$\mathsf{N}\ type$$

The canonical values of type N are inductively defined by (1) 0 is a canonical value of type N and (2) If a evaluates to a canonical value of type N then $\mathsf{succ}(a)$ is a canonical value of type N.

For subset- and Eq-formation, we have to use definitions which allow classical proofs when forming canonical elements:

subset-formation

$$\frac{A\ type \qquad B(x)\ type\ [x \in A]}{\{x \in A \mid B(x)\}\ type}$$

A canonical value of type $\{x \in A \mid B(x)\}$ is a canonical value a of type A such that $B(a)$, that is, we must have a canonical value a of type A and a formal derivation of $b \in B(a)$ for some expression b.

In this definition, the law of the excluded middle may have been used in the derivation of $b \in B(a)$.

Eq-formation

$$\frac{a \in A \qquad b \in B\ type}{\mathsf{Eq}(A, a, b)\ type}$$

eq is the canonical value of type $\mathsf{Eq}(A,a,b)$ provided that $a = b \in A$ is formally derivable.

In this definition, the law of the excluded middle may have been used in the derivation of $a = b \in A$.

Note that an assumption

$$x \in A \ [x \in A]$$

trivially satisfies (III) since it is here tacitly assumed that we have a derivation of A *type*.

Let us exemplify substitution by the rule

$$\frac{a \in A \qquad c(x) \in C(x) \ [x \in A]}{c(a) \in C(a)}$$

If x occurs free in $c(x)$, then, since $c(a)$ must not contain the constant ?, neither a nor $c(x)$ can contain ?. By applying the induction hypothesis (III) on the derivation of $a \in A$, we know that a can be computed to a canonical value of type A. By applying the induction hypothesis on the derivation of $c(x) \in C(x) \ [x \in A]$ we get that $c(a)$ has a canonical value of type $C(a)$.

Also if x does not occur free in $c(x)$, the induction hypothesis (III) gives that $c(a)$ can be computed to a canonical value of type $C(a)$, because we have $a \in A$. Note that in this case a may contain the constant ?.

+-introduction

$$\frac{a \in A}{\mathsf{inl}(a) \in A + B} \qquad\qquad \frac{b \in B}{\mathsf{inr}(b) \in A + B}$$

a cannot contain the constant ? since $\mathsf{inl}(a)$ must not contain ?. By the induction hypothesis, a can be computed to a canonical value of type A. Hence, according to the definition of canonical value of type $A+B$, $\mathsf{inl}(a)$ is a canonical value of type $A+B$. The second rule is handled in the same way.

+-elimination

$$\frac{c \in A + B \qquad d(x) \in C(\mathsf{inl}(x)) \ [x \in A] \qquad e(y) \in C(\mathsf{inr}(y)) \ [y \in A]}{\mathsf{when}(c, d, e) \in C(c)}$$

Since ? must not occur in $\mathsf{when}(c, d, e)$ it cannot occur in any of the expressions c, $d(x)$ and $e(y)$. So, by the induction hypothesis, c can be computed to a canonical value of type $A+B$. There are two cases; either c has a value of the form $\mathsf{inl}(a)$ where a has a canonical value of type A, or c has a value of the form $\mathsf{inr}(b)$ where b has a canonical value of type B. In the first case, $\mathsf{when}(c, d, e)$ is computed by computing $d(a)$. By the induction hypothesis $d(a)$ has a canonical value of type $C(\mathsf{inl}(a))$, that is, a canonical value of type $C(c)$. The second case is treated in the same way.

Π-introduction

$$\frac{b(x) \in B(x) \ [x \in A]}{\lambda(b) \in (\Pi\, x \in A) B(x)}$$

Since ? must not occur in $\lambda(b)$, it cannot occur in $b(x)$. So, by the induction hypothesis, $b(x)$ can be computed to a canonical value of type $B(a)$ for every a which can be computed to a canonical value of type A. Hence, by the definition of canonical value of type $(\Pi\, x \in A) B(x)$, $\lambda(b)$ is a canonical value of that type.

Π-elimination

$$\frac{a \in A \qquad c \in (\Pi x \in A)B(x)}{\mathsf{apply}(c, a) \in B(a)}$$

Since $\mathsf{apply}(c, a)$ does not contain ? neither c nor a can contain ?. By the induction hypothesis, we then know that c can be computed to a canonical value of type $(\Pi x \in A)B(x)$. Hence, the value of c is of the form $\lambda(b)$ where we know that for each expression e having a canonical value of type A, $b(e)$ has a canonical value of type $B(e)$. According to the induction hypothesis, we know that a can be computed to a canonical value of type A. By the computation rule for apply, the value of $b(a)$ is also the value of $\mathsf{apply}(c, a)$.

subset-introduction

$$\frac{a \in A \qquad b \in B(a)}{a \in \{x \in A \mid B(x)\}}$$

Because a must not contain ?, we can use the induction hypothesis on the derivation of $a \in A$ and conclude that a evaluates to a canonical value of type A. Since we have a derivation of $b \in B(a)$, a is also a canonical value of type $\{x \in A \mid B(x)\}$.

subset-elimination

$$\frac{a \in \{x \in A \mid B(x)\} \qquad c(x) \in C(x) \ [x \in A, y \in B(x)]}{c(a) \in C(a)}$$

where y must not occur free in $c(x)$ nor in $C(x)$.

Suppose first that x occurs free in $c(x)$. Since the constant ? must not occur in $c(a)$, it can occur neither in a nor in $c(x)$. So, by the induction hypothesis, a can be computed to a canonical value of type $\{x \in A \mid B(x)\}$. By the definition of a canonical value of type $\{x \in A \mid B(x)\}$, we then know that a can be computed to a canonical value of type A and that we for some expression b have a derivation of $b \in B(a)$. Since $c(x)$ does not contain ? we can apply the induction hypothesis on the derivation of $c(x) \in C(x)$ $[x \in A, y \in B(x)]$. Hence, since y does not occur free in $c(x)$, $c(a)$ can be computed to a canonical value of type $C(a)$.

If x does not occur free in $c(x)$ we can also apply the induction hypothesis on the derivation of $c(x) \in C(x)$ $[x \in A, y \in B(x)]$ since, by subset-elimination, $a \in A$ follows from $a \in \{x \in A \mid B(x)\}$.

Eq-introduction

$$\frac{a = b \in A}{\mathsf{eq} \in \mathsf{Eq}(A, a, b)}$$

Since we have a formal derivation of $a = b \in A$ it follows directly that eq is a canonical value of type $\mathsf{Eq}(A, a, b)$.

N-elimination

$$\frac{n \in \mathsf{N} \qquad d \in C(0) \qquad e(x, y) \in C(\mathsf{succ}(x)) \ [x \in \mathsf{N},\ y \in C(x)]}{\mathsf{rec}(n, d, e) \in C(n)}$$

Since the constant ? must not occur in $\mathsf{rec}(n, d, e)$, it cannot occur in n, d and $e(x, y)$. According to the induction hypothesis, n has a canonical value of type N. We will use induction on n in order to show that $\mathsf{rec}(n, d, e)$ has a canonical value of type $C(n)$.

If n has value 0 we can use the induction hypothesis on the derivation of $d \in C$ and conclude that d has a canonical value of type $C(0)$. By the computation ru for rec this value is also a canonical value of $\text{rec}(0, d, e)$. If n has a value of the fo $\text{succ}(a)$, the computation rule for rec gives that $\text{rec}(n, d, e)$ is computed by computi $e(a, \text{rec}(a, d, e))$. By the subordinate induction hypothesis, $\text{rec}(a, d, e)$ has a canonic value of type $C(a)$. Hence, by applying the induction hypothesis on the derivation $e(x, y) \in C(\text{succ}(x))$ $[x \in \mathsf{N},\ y \in C(x)]$ we get that $e(a, \text{rec}(a, d, e))$ has a canonic value of type $C(\text{succ}(a))$.

3 Collapsing the types

Although adding the law of the excluded middle to type theory results in a muc stronger theory, no new programs not containing ? are introduced. More precisely:

Theorem 2 *If $a \in A$ is derivable in Martin-Löf's type theory extended with the law the excluded middle and the constant ? does not occur syntactically in a, then we ca find a type A' such that $a \in A'$ is derivable without the law of the excluded middle.*

Proof. In general, A and A' must be different. For instance, using subset-introductio and the law of the excluded middle, we can, for an arbitrary type B, prove

$$\text{true} \in \{x \in \text{Bool} \mid B \vee (\neg B)\}$$

Clearly, there is no hope of proving this constructively.

I will prove the theorem for type theory without universes. The proof will be b induction on the length of the derivation of $a \in A$. A' will be defined by induction o the length of the derivation of A *type*, using the definitions:

$$\begin{aligned}
\text{Bool}' &\equiv \text{Bool} \\
\emptyset' &\equiv \emptyset \\
\mathsf{N}' &\equiv \mathsf{N} \\
(A + B)' &\equiv (A' + B') \\
(\Pi\, x \in A)B(x)' &\equiv (\Pi\, x \in A')B'(x) \\
(\Sigma\, x \in A)B(x)' &\equiv (\Sigma\, x \in A')B'(x) \\
(\mathsf{W}\, x \in A)B(x)' &\equiv (\mathsf{W}\, x \in A')B'(x) \\
\{x \in A \mid B(x)\}' &\equiv A' \\
\text{Eq}(A, a, b)' &\equiv \text{Eq}(\mathsf{N}, 0, 0)
\end{aligned}$$

It is now easy to see that if

$$A(x_1, \ldots, x_n)\ type\ \ [x_1 \in A_1,\ \ldots,\ x_n \in A_n(x_1, \ldots, x_{n-1})]$$

is derivable, then

$$A(x_1, \ldots, x_n)'\ type$$

is derivable without any assumptions; in particular, $A(x_1, \ldots, x_n)'$ does not contain any free variables. It is also easy to see that if $A = B$ is derivable then $A' \equiv B'$.

In the proof we have to consider derivations depending on assumptions. So, suppose we have a derivation of

$$a(x_1, \ldots, x_n) \in A(x_1, \ldots, x_n)\ \ [x_1 \in A_1,\ \ldots,\ x_n \in A_n(x_1, \ldots, x_{n-1})]$$

where the expression $a(x_1,\ldots,x_n)$ does not syntactically contain the constant ?. Let $x_{k_1}, \ldots, x_{k_m}$ be the variables which occur free in the expression $a(x_1,\ldots,x_n)$. We then have to show that we can derive, without the law of the excluded middle,

$$a(x_1,\ldots,x_n) \in A(x_1,\ldots,x_n)' \ [x_{k_1} \in A_{k_1}(x_1,\ldots,x_{k_1-1})', \ \ldots, \ x_{k_m} \in A_{k_m}(x_1,\ldots,x_{k_m-1})']$$

The reason for deleting some of the assumptions in the induction hypothesis is that it is needed for the subset-elimination rule. The only rules where free variables may disappear without the corresponding assumptions being discharged are Eq-introduction and subset-introduction.

Eq-introduction

$$\frac{a = b \in A}{\mathsf{eq} \in \mathsf{Eq}(A, a, b)}$$

is trivial because $\mathsf{Eq}(A,a,b)' \equiv \mathsf{Eq}(\mathsf{N},0,0)$ and we can derive $\mathsf{eq} \in \mathsf{Eq}(\mathsf{N},0,0)$ without any assumptions and without the law of the excluded middle.

subset-introduction

$$\frac{a \in A \qquad b \in B(a)}{a \in \{x \in A \mid B(x)\}}$$

The derivation $b \in B(a)$ may depend on assumptions which the derivation of $a \in A$ does not, so there may be variables introduced by assumptions in the derivation of $b \in B(a)$ which do not occur free in a. However, the induction step for this rule is trivial since $\{x \in A \mid B(x)\}' \equiv A'$ and, according to the induction hypothesis, we have a derivation of $a \in A'$.

None of the remaining rules involves any problems.

Remark. In Martin-Löf [7] the fourth Peano axiom

$$(\forall x \in \mathsf{N}) \neg \mathsf{Eq}(\mathsf{N}, 0, \mathsf{succ}(x))$$

is proved by using universes. In [7] it is also conjectured that this axiom cannot be proved without universes. Using the proof of theorem 2 we can now prove this conjecture. So, assume that we can prove Peano's fourth axiom in Martin-Löf's type theory without universes, that is, that we have a closed term t and a derivation of

$$t \in (\forall x \in \mathsf{N}) \neg \mathsf{Eq}(\mathsf{N}, 0, \mathsf{succ}(x))$$

By Π-elimination we get $\mathsf{apply}(t,0) \in \neg \mathsf{Eq}(\mathsf{N},0,\mathsf{succ}(x))$, which, by theorem 2, gives $\mathsf{apply}(t,0) \in \neg \mathsf{Eq}(\mathsf{N},0,\mathsf{succ}(x))'$. Since $(\neg \mathsf{Eq}(\mathsf{N},0,\mathsf{succ}(x)))' \equiv \mathsf{Eq}(\mathsf{N},0,0) \to \emptyset$ we then get, by Eq-introduction and Π-elimination,

$$\mathsf{apply}(\mathsf{apply}(t,0), \mathsf{eq}) \in \emptyset$$

which is impossible, granted the consistency of Martin-Löf's type theory without universes.

One way of making the fourth Peano axiom derivable without universes is to define the empty type by $\emptyset \equiv \mathsf{Eq}(\mathsf{N},0,\mathsf{succ}(0))$. The definition of A' in the proof of theorem 2 then has to be changed by leaving out the clause $\emptyset' \equiv \emptyset$.

Acknowledgement
The idea of adding the law of the excluded middle to type theory emerged from a discussion with Peter Aczel and Richard Waldinger and I would like to thank them both.

References

[1] M. Beeson. *Foundations of Constructive Mathematics.* Springer, 1985.

[2] T. Coquand and G. Huet. *Constructions: A Higher Order Proof System for Mechanizing Mathematics.* Presented at Eurocal 85, April 1985, Linz, Austria.

[3] K. Gödel. *Zur intuitionistischen Arithmetik und Zahlentheorie.* Ergebnisse eines mathematischen Kolloquiums, Heft 4 (for 1931-32, pub. 1933), pp.34-38.

[4] S.C. Kleene. *Introduction to Metamathematics.* North-Holland 1952.

[5] Z. Manna and R. Waldinger. *A deductive approach to program synthesis.* ACM Transactions on Programming Languages and Systems Vol. 2, No. 1, January 1980, pp. 90-121.

[6] P. Martin-Löf. *Constructive Mathematics and Computer Programming.* In Sixth International Congress for Logic, Methodology, and Philosophy of Science, pp. 153-175. North-Holland, 1982.

[7] P. Martin-Löf. *Intuitionistic Type Theory.* Studies in Proof Theory, Lecture Notes, Bibliopolis, Napoli, 1984.

[8] B. Nordström and J.M. Smith. *Propositions, Types, and Specifications in Martin-Löf's Type Theory.* BIT Vol. 24, No. 3, 1984, pp. 288-301.

[9] K. Petersson and J.M Smith. *Program Derivation in Type Theory: A Partitioning Problem.* To appear in Journal of Computer Languages.

[10] The Prl Staff (R. Constable et al.). *Implementing Mathematics with The Nuprl Proof Development System.* Computer Science Department, Cornell University, 1985.

[11] J.M. Smith. *An Interpretation of Martin-Löf's Type theory in a Type Free Theory of Propositions.* Journal of Symbolic Logic Vol. 49, No. 3, September 1984, pp. 730-753.

[12] C. Spector. *Provable Recursive Functionals of Analysis: A Consistency Proof of Analysis by an Extension of Principles Formulated in Current Intuitionistic Mathematics.* Proceedings of Symposia in Pure Mathematics, Volume V, pp.1-27, American Mathematical Society 1962.

[13] W.W. Tait. *Intensional interpretation of functionals of finite type I.* Journal of Symbolic Logic Vol. 32, No. 2, June 1967, pp. 198-212.

[14] A.S. Troelstra. *Metamathematical Investigations of Intuitionistic Arithmetic and Analysis.* Lecture Notes in Mathematics 344, Springer-Verlag 1973.

PRIME COMPUTABILITY ON PARTIAL STRUCTURES

Ivan N. Soskov

Laboratory of Computer Science
Sofia University
1126 Sofia, Bulgaria

The notion of prime computability on abstract (unordered) domains is introduced by Moschovakis [1]. The prime computable functions are exactly those which are computable by means of deterministic (serial) procedures. In partial structures not every computable by means of nondeterministic (parallel) procedures function is prime computable. The aim of this paper is to give a generalization of the notion of prime computability in order to obtain the functions which are computable by means of parallel procedures.

This work is a part of the author's Ph. D. thesis [2]. Some of the results are announced in [3].

1. PRELIMINARIES

Let $\mathcal{A} = (B, \theta_1, \ldots, \theta_m, T_1, \ldots, T_k)$ be a partial structure. Where B is an arbitrary set, $\theta_1, \ldots, \theta_m$ are partial functions of many arguments on B, every T_i is a partial predicate of many arguments on B. There $m, k \geq 0$. The relational type of $\mathcal{A}$ is the ordered pair $\langle\langle a_1, \ldots, a_m\rangle, \langle b_1, \ldots, b_k\rangle\rangle$ where each θ_i is a_i-ary and each T_j is b_j-ary. We shall assume that every $a_i > 0$ and every $b_j > 0$. The structure $\mathcal{A}$ will be called total when all θ_i, T_j are total.

Let O be some object not in B, let $B^o = B \cup \{O\}$. The set B^* is defined by the inductive clauses:

If $s \in B^o$, then $s \in B^*$;

If $s \in B^*$, $t \in B^*$, then $\langle s,t\rangle \in B^*$.

Here $\langle s,t\rangle$ is the ordered pair of s and t and is assumed that an operation is chosen to represent the ordered pair so that no object in B^o is an ordered pair.

The natural numbers $0, 1, 2, \ldots$ are identified with the objects O, $\langle O,O\rangle$, $\langle O,\langle O,O\rangle\rangle, \ldots$ of B^*. Let N denote the set of all natural numbers.

We shall use the notation $\langle s_1, s_2, \ldots, s_n\rangle$ for $\langle s_1, \langle s_2, \ldots, \langle s_{n-1}, s_n\rangle \ldots\rangle\rangle$; x, y, z will denote natural numbers and s, t -- elements of B^*.

Let $B_{r,j}$ be the subset of B^* which consists of all elements

$\langle x_1, \ldots, x_r, s_1, \ldots, s_j \rangle$, where $x_1, \ldots, x_r \in N$ and $s_1, \ldots, s_j \in B^*$. We shall denote with $\langle x^\sim, s^\sim \rangle$ arbitrary elements of $B_{r,j}$, writing $x^\sim$ for $x_1, \ldots, x_r$ and $s^\sim$ for $s_1, \ldots, s_j$.

Let $\mathcal{F}_{r,j} = \{\varphi / \varphi : B_{r,j} -\cdot\to B\}$ and $\Sigma_{r,j} = \{\sigma / \sigma : B_{r,j} -\cdot\to N\}$.

Throughout this paper we shall identify the partial predicates in $B_{r,j}$ with the elements of $\Sigma_{r,j}$ which obtain values in $\{0,1\}$.

An element $\varphi \in \mathcal{F}_{r,j}$ ($\varphi \in \Sigma_{r,j}$) we shall call prime computable in the structure $\mathcal{A}$ iff φ is absolutely prime computable in $\{\theta_1, \ldots, \theta_m, T_1, \ldots, T_k\}$ (in the sense of Moschovakis [1]).

Many properties of the prime computable functions are proved in [1], which we shall use in this paper. Moreover we shall use and the following three propositions, which are proved in detail in [2].

<u>Theorem</u> 1 Let $r \geq 1$. The prime computable in $\mathcal{A}$ elements of $\Sigma_{r,0}$ are exactly the r-ary partial recursive functions. The unique prime computable in $\mathcal{A}$ element of $\mathcal{F}_{r,0}$ is the totally undefined function $\neg!$.

Let $j \geq 1$. The j-ary termal functions are defined by the inductive clauses:

$\lambda s_1, \ldots, s_j . s_i$, $j \geq i \geq 1$, is a termal function;

If $\alpha_1, \ldots, \alpha_a$ are termal functions, $\theta \in \{\theta_1, \ldots, \theta_m\}$, θ is a-ary, then $\lambda s^\sim . \theta(\alpha_1(s^\sim), \ldots, \alpha_a(s^\sim))$ is a termal function.

Let T_0 denote the unary total predicate $\lambda s.0$. Atomic predicates are called elements of $\Sigma_{0,j}$ of the kind $\lambda s^\sim . T(\alpha_1(s^\sim), \ldots, \alpha_b(s^\sim))$ or $\lambda s^\sim . \neg T(\alpha_1(s^\sim), \ldots, \alpha_b(s^\sim))$ where $T \in \{T_0, \ldots, T_k\}$, T is b-ary and $\alpha_1, \ldots, \alpha_b$ are j-ary termal functions.

Termal predicates are defined by the inductive clauses:

Each atomic predicate is termal;

If π_1 and π_2 are termal predicates then so is the predicate $\lambda s^\sim . (\pi_1(s^\sim) = 0 \to \pi_2(s^\sim), 1)$.

Let F_j and P_j denote the prime computable in $\mathcal{A}$ elements of $\mathcal{F}_{1,j}$ and $\Sigma_{1,j}$, respectively, which are universal for the j-ary termal functions and for the j-ary termal predicates.

Let $p_0 = 2$, $p_1 = 3$, $p_2 = 5$, ... and $(u)_i \doteq$ largest a such that p_i^a divides u.

<u>Theorem</u> 2 An element σ of $\Sigma_{r,j}$, $j \geq 1$, is prime computable in $\mathcal{A}$ iff there exists a total recursive r+1-ary function β such that for every $\langle x^\sim, s^\sim \rangle$ in $B_{r,j}$ the following conditions hold:

(i) if $\sigma(x^\sim, s^\sim)$ is defined then for every z, $P_j(\beta(z, x^\sim), s^\sim)$ is defined;

(ii) $\sigma(x^\sim, s^\sim)$ is defined iff $\exists z (P_j(\beta(z, x^\sim), s^\sim) = 0)$;

(iii) if for some z, $P_j(\beta(z, x^\sim), s^\sim) = 0$ then $\sigma(x^\sim, s^\sim) = (z)_0$.

<u>Theorem</u> 3 An element φ of $\mathcal{F}_{r,j}$, $j \geq 1$, is prime computable in $\mathcal{A}$ iff there exists a prime computable in $\mathcal{A}$ element $\sigma \in \Sigma_{r,j}$ such that for every $\langle x^\sim, s^\sim \rangle$ in $B_{r,j}$ the following conditions hold:

(i) $\varphi(x^\sim, s^\sim)$ is defined iff $\sigma(x^\sim, s^\sim)$ is defined;

(ii) $\varphi(x^\sim, s^\sim) \simeq F_j(\sigma(x^\sim, s^\sim), s^\sim)$.

Note. A result is formulated in Gordon [4] which is similar to theorem 3. It is obvious that in case of total structures if $r = 0$ both propositions are equivalent.

In his paper [5] Shepherdson introduced the so called treelike EDS which are a kind of Friedman's effectively definitional schemes (EDS) [6]. Shepherdson proved that treelike EDS, effective schemes of Kfoury and generalized Turing algorithms [6] are equivalent over all interpretations.

Using theorem 3 and theorem 2 one can prove the following:

Theorem 4 If $\varphi \in \mathfrak{F}_{0,j}$, $j \geq 1$, then φ is prime computable in $\mathfrak{A}$ iff φ is computable in $\mathfrak{A}$ by means of treelike EDS.

Proof. The proof is technical and is omitted.

2. COMPUTABLE FUNCTIONS

Theorem 4 together with the results of Shepherdson [5] show that in every partial structure the prime computable functions are exactly those which are computable by means of serial procedures. It is well known that in partial structures not any "effective" function is computable by means of serial procedures. In order to obtain the functions which are computable through parallel procedures we introduce the notion of computable function in a partial structure.

Def. An element φ of $\mathfrak{F}_{r,j}$, $r+j \geq 1$ is said to be computable in $\mathfrak{A}$ iff there exists a prime computable function $\varphi^* \in \mathfrak{F}_{r+1,j}$ such that for every $\langle x^\sim, s^\sim \rangle \in B_{r,j}$ and $t \in B$ the following condition is true:

$$\varphi(x^\sim, s^\sim) = t \longleftrightarrow \exists z(\varphi^*(z, x^\sim, s^\sim) = t).$$

Similarly, an element σ of $\Sigma_{r,j}$, $r+j \geq 1$ is said to be computable in $\mathfrak{A}$ iff there exists a function $\sigma^* \in \Sigma_{r+1,j}$, which is prime computable in $\mathfrak{A}$ and for every $\langle x^\sim, s^\sim \rangle$ and $y \in N$,

$$\sigma(x^\sim, s^\sim) = y \longleftrightarrow \exists z(\sigma^*(z, x^\sim, s^\sim) = y).$$

Let N^2 denote the p.m.v. function $\{\langle x,y \rangle / x \in N \,\&\, y \in N\}$.The following proposition, which describes the computable functions in terms of prime computability, is the main result of this section.

Theorem 5 An element $\varphi \in \mathfrak{F}_{r,j}$ $(\varphi \in \Sigma_{r,j})$ is computable in $\mathfrak{A}$ if and only if φ is absolutely prime computable in $\{\theta_1, \ldots, \theta_m, T_1, \ldots, T_k, N^2\}$.

Corollary. The computable functions are closed under prime computability.

In order to prove theorem 5 we shall use the following definition of prime computability given by Skordev in [7]. Let $\mathfrak{F}^*$ be the set of all unary p.m.v. functions in B^*. We shall consider the elements of $\mathfrak{F}^*$ as binary relations in B^* writing $\langle s,t \rangle \in \varphi$ for $t \in \varphi(s)$. Let L, R be elements of $\mathfrak{F}^*$ (in [1] L and R are denoted by π and δ) such that the following equalities hold:

$$L(0) = R(0) = 0; \quad L(s) = R(s) = 1 \quad \text{for} \quad s \in B;$$
$$L(\langle s,t \rangle) = s, \; R(\langle s,t \rangle) = t \text{ for } s,t \in B^*.$$

Let composition $\cdot$ and Π be binary operations in $\mathfrak{F}^*$, defined as follows:

$$\langle s,t \rangle \in \varphi\cdot\sigma \longleftrightarrow \exists u(\langle s,u \rangle \in \sigma \,\&\, \langle u,t \rangle \in \varphi),$$
$$\langle s,t \rangle \in \Pi(\varphi,\sigma) \longleftrightarrow \exists u_1 \exists u_2(\langle s,u_1 \rangle \in \varphi \,\&\, \langle s,u_2 \rangle \in \sigma \,\&\, t = \langle u_1,u_2 \rangle).$$

Let D be the element $(B^{o}\times\{0\}) \cup ((B^{*}\setminus B^{o})\times\{1\})$ and iteration [ ] be binary operation in $\mathcal{F}^{*}$ defined by $\langle s,t\rangle \in [\varphi,\sigma]$ iff there exists a finite sequence $u_0, u_1, \ldots, u_n$ of elements of B^{*} such that $u_0 = s$, $u_n = t$ and for $i < n$, $\langle u_i,u_{i+1}\rangle \in \varphi$ and $\langle u_i,1\rangle \in D\cdot\sigma$; $\langle u_n,0\rangle \in D\cdot\sigma$.

In [7] it is proved that if $\mathcal{B}$ is a subset of $\mathcal{F}^{*}$ then φ is absolutely prime computable in $\mathcal{B}$ if and only if φ can be obtained from elements of $\mathcal{B} \cup \{L,R\}$ by means of composition, $\sqcap$ and iteration.

Let $\mathcal{B}$ denote the set $\{\theta_1, \ldots, \theta_m, T_1, \ldots, T_k\}$.

We need the following lemma.

Lemma 1. Let $\varphi \in \mathcal{F}^{*}$ and φ be absolutely prime computable in $\mathcal{B} \cup \{N^2\}$. There exists an absolutely prime computable in $\mathcal{B}$ function φ^{*} such that for every $s,t \in B^{*}$, $\langle s,t\rangle \in \varphi \longleftrightarrow \exists x(\varphi^{*}(\langle x,s\rangle) = t)$.

Proof. Suppose φ is absolutely prime computable in $\mathcal{B} \cup \{N^2\}$. In order to prove the lemma by induction on the construction of φ, the following fou cases are considered.

1. φ is an element of $\mathcal{B} \cup \{L,R,N^2\}$. If $\varphi = N^2$ let us define φ^{*} using the following definition by cases:

$$\varphi^{*}(s) \simeq \begin{cases} L(s), & \text{if } R(s) \in N, \\ \text{undefined, otherwise.} & \end{cases}$$

If $\varphi \neq N^2$ then put $\varphi^{*} = \varphi\cdot R$.

Suppose φ_1 and φ_2 are elements of $\mathcal{F}^{*}$ and there exist elements φ_1^{*} and φ_2^{*} of $\mathcal{F}^{*}$ which are absolutely prime computable in $\mathcal{B}$ and

$$\langle s,t\rangle \in \varphi_1 \longleftrightarrow \exists x(\varphi_1^{*}(\langle x,s\rangle) = t),$$
$$\langle s,t\rangle \in \varphi_2 \longleftrightarrow \exists x(\varphi_2^{*}(\langle x,s\rangle) = t).$$

Let $\alpha = \lambda x.(x)_0$, $\beta = \lambda x.(x)_1$ and $\Gamma = \lambda x,y.(x)_y$. By theorem 1 the function α, β, Γ are prime computable in $\mathcal{B}$.

2. $\varphi = \varphi_2\cdot\varphi_1$. Put $\varphi^{*} = \varphi_2^{*}\cdot\sqcap(\beta\cdot L,\varphi_1^{*}\cdot\sqcap(\alpha\cdot L,R))$.

Then, $\langle s,t\rangle \in \varphi \longleftrightarrow \exists u(\langle s,u\rangle \in \varphi_1 \ \& \ \langle u,t\rangle \in \varphi_2) \longleftrightarrow$
$\exists u(\exists x_0(\varphi_1^{*}(\langle x_0,s\rangle)=u) \ \& \ \exists x_1(\varphi_2^{*}(\langle x_1,u\rangle) = t)) \longleftrightarrow$
$\exists x\exists u(\varphi_1^{*}(\langle (x)_0,s\rangle)=u \ \& \ \varphi_2^{*}(\langle (x)_1,u\rangle)=t) \longleftrightarrow \exists x(\varphi^{*}(\langle x,s\rangle)=t)$.

3. $\varphi = \sqcap(\varphi_1,\varphi_2)$. Put $\varphi^{*} = \sqcap(\varphi_1^{*}\cdot\sqcap(\alpha\cdot L,R),\varphi_2^{*}\cdot\sqcap(\beta\cdot L,R))$.

Then, $\langle s,t\rangle \in \varphi \longleftrightarrow \exists u_0\exists u_1(\langle s,u_0\rangle \in \varphi_1 \ \& \ \langle s,u_2\rangle \in \varphi_2 \ \& \ t = \langle u_1,u_2\rangle) \longleftrightarrow$
$\exists x\exists u_0\exists u_1(\varphi_1^{*}(\langle (x)_0,s\rangle) = u_0 \ \& \ \varphi_2^{*}(\langle (x)_1,s\rangle) = u_1 \ \& \ t = \langle u_0,u_1\rangle) \longleftrightarrow$
$\exists x(\varphi^{*}(\langle x,s\rangle) = t)$.

4. $\varphi = [\varphi_1,\varphi_2]$. Let S denote the recursive function $\lambda x.x+1$ and σ -- the element $D\cdot\varphi_2$ of $\mathcal{F}^{*}$. Put $D^{*} = D\cdot R$. Apparently D^{*} is absolutely prime computable in $\mathcal{B}$ and $\langle s,t\rangle \in D \longleftrightarrow \exists x(D^{*}(\langle x,s\rangle) = t)$. Due to the induction hypothesis and case 1 there exists a prime computable in $\mathcal{B}$ function σ^{*} such that $\langle s,t\rangle \in \sigma \longleftrightarrow \exists x(\sigma^{*}(\langle x,s\rangle) = t)$. Define the absolutely prime computable in $\mathcal{B}$ elements μ_1 and μ_2 of $\mathcal{F}^{*}$ by the equalities:

$$\mu_1 = \sqcap(S\cdot L,\sqcap(L\cdot R,\varphi_1^{*}\cdot\sqcap(\alpha\cdot\Gamma\cdot\sqcap(L\cdot R,L),R\cdot R)))$$
$$\mu_2 = \sigma^{*}\cdot\sqcap(\beta\cdot\Gamma\cdot\sqcap(L\cdot R,L),R\cdot R).$$

Let φ^* be $R\cdot R\cdot[\mu_1,\mu_2]\cdot\Pi(0^\wedge,I)$ where $0^\wedge$ and I denote the absolutely prime computable functions $\lambda s.0$ and $\lambda s.s$. We have to prove that $\langle s,t\rangle \in \varphi \longleftrightarrow \exists x(\varphi^*(\langle x,s\rangle) = t)$.

Suppose that $\langle s,t\rangle \in \varphi$. Then, there exists a finite sequence $u_0, u_1, \ldots, u_n$ of elements of B^*, such that $u_0 = s$, $u_n = t$; for $i < n$, $\langle u_i,u_{i+1}\rangle \in \varphi_1$ and $\langle u_i,1\rangle \in \sigma$; $\langle u_n,0\rangle \in \sigma$. According to the induction hypothesis there exist natural numbers $y_0, y_1, \ldots, y_{n-1}, z_0, z_1, \ldots, z_n$ for which $\varphi_1^*(\langle y_i,u_i\rangle) = u_{i+1}$ and $\sigma^*(\langle z_i,u_i\rangle) = 1$, $i < n$, and $\sigma^*(\langle z_n,u_n\rangle) = 0$.

Put $x_i = p_0^{y_i}\cdot p_1^{z_i}$, $i < n$; $x_n = p_1^{z_n}$ and $x = p_0^{x_0}\cdot p_1^{x_1}\cdots p_n^{x_n}$. It is easy to verify that $\mu_1(\langle i,x,u_i\rangle) = \langle i+1,x,u_{i+1}\rangle$ and $\mu_2(\langle i,x,u_i\rangle) = 1$, $i < n$; $\mu_2(\langle n,x,u_n\rangle) = 0$. Using these equalities one can prove by induction on i upside down $[\mu_1,\mu_2](\langle i,x,u_i\rangle) = \langle n,x,t\rangle$, $i < n$. Particularly for $i = 0$ this means that $[\mu_1,\mu_2](\langle 0,x,s\rangle) = \langle n,x,t\rangle$.

Thus it follows that $\varphi^*(\langle x,s\rangle) = t$.

Conversely, suppose that for some natural x, $\varphi^*(\langle x,s\rangle) = t$. Then $R\cdot R\cdot[\mu_1,\mu_2](\langle 0,x,s\rangle) = t$ and, therefore, there exist elements $u_0, \ldots, u_n$ of B^* such that the following equalities hold: $\langle 0,x,s\rangle = u_0$, $t = R\cdot R(u_n)$; $u_{i+1} = \mu_1(u_i)$ and $D\cdot\mu_2(u_i) = 1$, for $i < n$, and $D\cdot\mu_2(u_n) = 0$. Hence, there exist elements $t_0, \ldots, t_n$ of B^* such that for $i < n+1$, $u_i = \langle i,x,t_i\rangle$ and $t_{i+1} = \varphi_1^*(\langle((x)_i)_0,t_i\rangle)$ and $\sigma^*(\langle((x)_i)_1,t_i\rangle) = 1$, $i < n$; $\sigma^*(\langle((x)_n)_1,t_n\rangle) = 0$.

We have $t_0 = s$, $t_n = t$; for $i < n$, $\langle t_i,t_{i+1}\rangle \in \varphi$ and $\langle t_i,1\rangle \in \sigma$; $\langle t_n,0\rangle \in \sigma$.

Thus, it follows that $\langle s,t\rangle \in [\varphi_1,\varphi_2]$.

Proof to theorem 5. Let φ be a computable in $\mathfrak{A}$ element of $\mathcal{F}_{r,j}$. Let φ^* be a prime computable in $\mathfrak{A}$ element of $\mathcal{F}_{r+1,j}$ and for every $\langle x^\sim,s^\sim\rangle \in B_{r,j}$ and $t \in B$, $\varphi(\langle x^\sim,s^\sim\rangle) = t \longleftrightarrow \exists z(\varphi^*(\langle z,x^\sim,s^\sim\rangle) = t)$. Then, $\varphi = \varphi^*\cdot\Pi(N^2\cdot 0^\wedge,I)$, writing $0^\wedge$ for $\lambda s.0$ and I for $\lambda s.s$.

The vice versa follows from Lemma 1.

The case $\varphi \in \Sigma_{r,j}$ is treated in the same way.

3. SEMICOMPUTABLE SETS

Def. A subset A of $B_{r,j}$, $r+j \geq 1$ is said to be semicomputable in the partial structure $\mathfrak{A}$ iff for some prime computable in $\mathfrak{A}$ function $\sigma^* \in \Sigma_{r+1,j}$, for every $\langle x^\sim,s^\sim\rangle \in B_{r,j}$,

$$\langle x^\sim,s^\sim\rangle \in A \longleftrightarrow \exists z(\sigma^*(\langle z,x^\sim,s^\sim\rangle) = 0).$$

Using the definition above and the properties of the prime computable functions it is easy to obtain the following propositions.

<u>Theorem</u> 6 The semicomputable sets are closed under pairwise intersection and union.

<u>Theorem</u> 7 Let A be a semicomputable subset of $B_{r+1,j}$. Then either of the following sets is semicomputable in $\mathfrak{A}$:
$A_1 = \{\langle x^\sim,s^\sim\rangle \ / \ \exists y(\langle y,x^\sim,s^\sim\rangle \in A)\}$,
$A_2 = \{\langle z,x^\sim,s^\sim\rangle \ / \ \forall y(y<z \longrightarrow \langle y,x^\sim,s^\sim\rangle \in A)\}$.

Theorem 8 Let A_1 be a semicomputable subset of $B_{k,1}$. Let $\sigma_1, \ldots, \sigma_k$ be computable in $\mathcal{A}$ elements of $\Sigma_{r,j}$ and $\varphi_1, \ldots, \varphi_1$ be computable in $\mathcal{A}$ elements of $\mathcal{F}_{r,j}$. Then the following set is semicomputable in $\mathcal{A}$: $\{\langle x^\sim, s^\sim\rangle / \langle \sigma_1(x^\sim,s^\sim), \ldots, \sigma_k(x^\sim,s^\sim), \varphi_1(x^\sim,s^\sim), \ldots, \varphi_1(x^\sim,s^\sim)\rangle \in A_1\}$.

The following two propositions describe the computable functions in $\mathcal{A}$ by means of semicomputable sets.

Theorem 9 An element σ of $\Sigma_{r,j}$ is computable in $\mathcal{A}$ iff there exists a semicomputable in $\mathcal{A}$ subset A of $B_{r+1,j}$ such that

$$\sigma(x^\sim,s^\sim) = y \longleftrightarrow \langle y,x^\sim,s^\sim\rangle \in A.$$

Proof Immediate from the corresponding definitions.

Theorem 10 Let $j \geq 1$ and $\varphi \in \mathcal{F}_{r,j}$. Then φ is computable in $\mathcal{A}$ iff there exists a semicomputable subset A of $B_{r+1,j}$ for which

$$\varphi(x^\sim,s^\sim) = t \longleftrightarrow \exists y(\langle y,x^\sim,s^\sim\rangle \in A \ \& \ F_j(y,s^\sim) = t).$$

Proof Suppose φ is computable in $\mathcal{A}$. There exists a prime computable in $\mathcal{A}$ element φ^* of $\mathcal{F}_{r+1,j}$ for which

$$\varphi(x^\sim,s^\sim) = t \longleftrightarrow \exists z(\varphi^*(z,x^\sim,s^\sim) = t).$$

According to theorem 3, there exists a prime computable in $\mathcal{A}$ function $\sigma \in \Sigma_{r+1,j}$, for which $\varphi^*(z,x^\sim,s^\sim) \simeq F_j(\sigma(z,x^\sim,s^\sim),s^\sim)$.

Thus, $\varphi(x^\sim,s^\sim) = t \longleftrightarrow \exists z \exists y(\sigma(z,x^\sim,s^\sim) = y \ \& \ F_j(y,s^\sim) = t)$.

Define A by $\langle y,x^\sim,s^\sim\rangle \in A \longleftrightarrow \exists z(\sigma(z,x^\sim,s^\sim) = y)$.

In order to prove the vice versa let us suppose that A is a semicomputable subset of $B_{r+1,j}$ and

$$\varphi(x^\sim,s^\sim) = t \longleftrightarrow \exists y(\langle y,x^\sim,s^\sim\rangle \in A \ \& \ F_j(y,s^\sim) = t).$$

Let σ be a prime computable in $\mathcal{A}$ function, $\sigma \in \Sigma_{r+2,j}$ and $\langle y,x^\sim,s^\sim\rangle \in A \longleftrightarrow \exists z(\sigma(z,y,x^\sim,s^\sim) = 0)$. Then,

$$\varphi(x^\sim,s^\sim) = t \longleftrightarrow \exists z(\sigma((z)_0,(z)_1,x^\sim,s^\sim) = 0 \ \& \ F_j((z)_1,s^\sim) = t).$$

Define the function σ^* using the following definition:

$$\sigma^*(z,x^\sim,s^\sim) \simeq \begin{cases} (z)_1, & \text{if } \sigma((z)_0,(z)_1,x^\sim,s^\sim) = 0, \\ \text{undefined, otherwise.} & \end{cases}$$

Put $\varphi^*(z,x^\sim,s^\sim) \simeq F_j(\sigma^*(z,x^\sim,s^\sim),s^\sim)$. It is clear that φ is prime computable in $\mathcal{A}$ and $\varphi(x^\sim,s^\sim) = t \longleftrightarrow \exists z(\varphi^*(z,x^\sim,s^\sim) = t)$.

Using theorem 9 one can prove the following

Theorem 11 The domains and graphs of the computable in $\mathcal{A}$ elements of $\Sigma_{r,j}$ are semicomputable.

Theorem 12 The domains of the computable in $\mathcal{A}$ elements of $\mathcal{F}_{r,j}$ are semicomputable.

Proof Let φ be a computable in $\mathcal{A}$ element of $\mathcal{F}_{r,j}$. If $j = 0$ then $\mathrm{Dom}(\varphi)$ is equal to the empty set -- $\emptyset$. It is obvious that $\emptyset$ is semicomputable in $\mathcal{A}$.

Assume $j \geq 1$. If φ is prime computable in $\mathfrak{A}$ then, by theorem 3, there exists a prime computable in $\mathfrak{A}$ function $\sigma \in \Sigma_{r,j}$ such that $Dom(\varphi) = Dom(\sigma)$. Therefore $Dom(\varphi)$ is semicomputable in $\mathfrak{A}$.

Suppose that φ is computable in $\mathfrak{A}$. According to theorem 10 there exists a semicomputable subset A of $B_{r+1,j}$ for which

$$\langle x^\sim, s^\sim\rangle \in Dom(\varphi) \longleftrightarrow \exists y(\langle y, x^\sim, s^\sim\rangle \in A \ \&\ \langle y, s^\sim\rangle \in Dom(F_j)).$$

Because F_j is prime computable in $\mathfrak{A}$, $Dom(F_j)$ is semicomputable and hence $Dom(\varphi)$ is semicomputable, too.

<u>Theorem</u> 13 Let $A_1, \ldots, A_n$ be pairwise disjoint semicomputable subsets of $B_{r,j}$, $j \geq 1$, and $\varphi_1, \ldots, \varphi_n$ be computable in $\mathfrak{A}$ elements of $\mathfrak{F}_{r,j}$ ($\Sigma_{r,j}$). Let us define φ as follows

$$\varphi(x^\sim, s^\sim) \simeq \begin{cases} \varphi_1(x^\sim, s^\sim), & \text{if } \langle x^\sim, s^\sim\rangle \in A_1, \\ \ldots & \ldots \\ \varphi_n(x^\sim, s^\sim), & \text{if } \langle x^\sim, s^\sim\rangle \in A_n, \\ \text{undefined}, & \text{otherwise.} \end{cases}$$

Then φ is computable in $\mathfrak{A}$.

<u>Proof</u> The proof is rather straightforward.

Corollary. For every semicomputable subset A of $B_{r,j}$, $j \geq 1$,there exist a computable in $\mathfrak{A}$ function $\varphi \in \mathfrak{F}_{r,j}$ and a computable in $\mathfrak{A}$ function $\sigma \in \Sigma_{r,j}$ for which $Dom(\varphi) = Dom(\sigma) = A$.

The following theorem we shall call normal form theorem for semicomputable sets.

<u>Theorem</u> 14 A subset A of $B_{r,j}$, $j \geq 1$, is semicomputable in $\mathfrak{A}$ iff there exists a total recursive function δ such that

$$\langle x^\sim, s^\sim\rangle \in A \longleftrightarrow \exists z(P_j(\delta(z, x^\sim), s^\sim) = 0).$$

<u>Proof</u> The theorem follows from theorem 2 and theorem 13.

4. PRIME COMPUTABILITY, SEARCH COMPUTABILITY AND FRIEDMAN'S SCHEMES

In this section some connections between prime computability, search computability and Friedman's effectively definitional schemes are established.

It follows by theorem 5:

<u>Theorem</u> 15 Every prime computable in $\mathfrak{A}$ function is computable in $\mathfrak{A}$.

In the case of total structures vice versa is true.

<u>Theorem</u> 16 If the structure $\mathfrak{A}$ is total then every computable in $\mathfrak{A}$ function is prime computable in $\mathfrak{A}$.

<u>Proof</u> Suppose that φ is a computable in $\mathfrak{A}$ element of $\mathfrak{F}_{r,j}$. If $j = 0$ then φ is the totally undefined function $\neg!$ and hence φ is prime computable.

Suppose $j > 0$. There exists a semicomputable in $\mathfrak{A}$ subset A of $B_{r+1,j}$ for which $\varphi(x^\sim, s^\sim) = t \longleftrightarrow \exists y(\langle y, x^\sim s^\sim\rangle \in A \ \&\ F_j(y, s^\sim) = t)$.

According to theorem 14 there exists a total recursive function δ such that $\langle y, x^{\sim}, s^{\sim}\rangle \in A \longleftrightarrow \exists z (P_j(\delta(z,y,x^{\sim}), s^{\sim}) = 0)$.

Because $\mathcal{A}$ is total the functions F_j and P_j are totally defined in $B_{1,j}$.

Put $\sigma(x^{\sim}, s^{\sim}) \simeq (\mu z[P_j(\delta((z)_0, (z)_1, x^{\sim}), s^{\sim}) = 0])_1$. It is clear that the function σ is prime computable in $\mathcal{A}$ and $\varphi(x^{\sim}, s^{\sim}) \simeq F_j(\sigma(x^{\sim}, s^{\sim}), s^{\sim})$.

Thus, φ is prime computable in $\mathcal{A}$.

The case $\varphi \in \Sigma_{r,j}$ is treated in the same way.

Let $j \geq 1$. The natural number x is said to be a Goedel number of the termal function α iff $\alpha = \lambda s^{\sim}.F_j(x, s^{\sim})$. Similarly, y is said to be a Goedel number of the termal predicate π iff $\pi = \lambda s^{\sim}.P_j(y, s^{\sim})$.

Let D be a set consisting of ordered pairs $\langle \alpha, \pi \rangle$, where α is a termal function and π is a termal predicate. Using the terminology of Ershov [8], we shall call D determinant.

A determinant D is said to be recursively enumerable (r.e.) iff there exists a r.e. subset D' of N^2 such that $\langle \alpha, \pi \rangle \in D$ iff for some Goedel number x of α and some Goedel number y of π, $\langle x, y \rangle \in D'$.

Let $\varphi \in \mathcal{F}_{0,j}$, $j \geq 1$. A determinant D is said to determinate φ iff for every $s^{\sim} \in B_{0,j}$ and $t \in B$, $\varphi(s^{\sim}) = t$ iff there exists an ordered pair $\langle \alpha, \pi \rangle$ from D such that $\alpha(s^{\sim}) = t$ and $\pi(s^{\sim}) = 0$.

Using the normal form theorem for semicomputable sets and theorem 10 one can prove the following:

Theorem 17 A function $\varphi \in F_{0,j}$, $j \geq 1$, is computable in $\mathcal{A}$ iff there exists a r.e. determinant which determinates φ.

In Shepherdson [5] the notion of recursively enumerable definitional schemes (REDS) is introduced .This schemes are a modification of Friedman's EDS and pseudo EDS appropriate to their use on partial structures. In the wider notion of REDS all compatibility conditions are omitted. Our definition for computability by means of determinants is very similar to the corresponded definition of Shepherdson. Namely the following theorem is true:

Theorem 18 The computable in $\mathcal{A}$ elements of $\mathcal{F}_{0,j}$, $j \geq 1$ are exactly those which are computable in $\mathcal{A}$ by means of REDS.

We turn now to search computability.In [7] it is proved that a p.m.v. function φ is absolutely search computable in a subset $\mathcal{B}$ of $\mathcal{F}^*$ (in the sense of Moschovakis [1]) if φ is absolutely prime computable in $\mathcal{B} \cup \{(B^*)^2\}$, where $(B^*)^2 = \{\langle s,t \rangle / \ s,t \in B^*\}$.

Because N^2 is absolutely prime computable in $(B^*)^2$ the following theorem is true.

Theorem 19 Every computable in $\mathcal{A}$ function is absolutely search computable in $\{\theta_1, \ldots, \theta_m, T_1, \ldots, T_k\}$.

We conclude our paper with an example of a simple structure on the natural numbers in which not every computable function is prime computable and not every absolutely search computable function is computable.

Let us consider the structure $\mathcal{A} = (B, \theta, T_1, T_2)$, where B is the set of all natural numbers, $\theta = \lambda x.x\dot{-}1$, $T_1(0)$ = true and $T_1(s)$ is undefined for $s > 0$; $T_2(1)$ = true and $T_2(s)$ is undefined for $s \neq 1$.

We first show that every r.e. subset of B is semicomputable in $\mathcal{A}$.

Let $\pi_0, \pi_1, \ldots$ be a sequence of unary termal predicates, which are defined as follows:

$$\pi_0 = \lambda s.T_1(s); \; \pi_{n+1} = \lambda s.T_2(\theta^n(s)).$$

Let μ be a total recursive function which for every n gives a Goedel number of π_n. It is easy to prove that for every s and n,

$$P_1(\mu(n),s) = 0 \longleftrightarrow s = n.$$

Suppose that A is a r.e. set and λ is a total recursive function, such that $s \in A \longleftrightarrow \exists z(\lambda(z) = s)$. Therefore, $s \in A \longleftrightarrow \exists z(P_1(\mu(\lambda(z)),s) = 0)$. Thus A is semicomputable in $\mathcal{A}$.

Let E and $E^\sim$ be subsets of $B_{0,2}$ and

$$E = \{\langle s,t\rangle \;/\; s \in B \;\&\; t \in B \;\&\; s = t\}, \quad E^\sim = B_{0,2} \setminus E.$$

We have $\langle s,t\rangle \in E \longleftrightarrow \exists z(P_1(\mu(z),s) = 0 \;\&\; P_1(\mu(z),t) = 0)$ and $\langle s,t\rangle \in E^\sim \longleftrightarrow \exists x\exists y(x \neq y \;\&\; P_1(\mu(x),s) = 0 \;\&\; P_1(\mu(y),t) = 0)$. Hence E and $E^\sim$ are semicomputable in $\mathcal{A}$.

Using induction on the construction of the unary termal predicates one can prove that for every termal predicate π exactly one of the following is true:

(i) There exists only finite elements s of B for which $\pi(s)$ is defined.

(ii) $\pi = \lambda s.0$.

(iii) $\pi = \lambda s.1$.

Let $A \subseteq B_{0,1}$, A is infinite and $A \neq B_{0,1}$. Suppose that there exists a prime computable in $\mathcal{A}$ function φ such that $Dom(\varphi) = A$. According to theorem 2 and theorem 3 there exists a total recursive function β ,such that

(1) if $s \in A$ then for every z, $P_1(\beta(z),s)$ is defined.

(2) $s \in A \longleftrightarrow \exists z(P_1(\beta(z),s) = 0)$.

By (1), because A is infinite , for every $z \in N$,

$\forall s(P_1(\beta(z),s) = 0)$ or $\forall s(P_1(\beta(z),s) = 1)$.

Hence, by (2), $A = B_{0,1}$. The last contradicts with the choice of A.

Let $\varphi \in \mathcal{F}_{0,1}$ and $\varphi(s) = s$ if s is even and $\varphi(s)$ is undefined otherwise. It is clear that φ is not prime computable in $\mathcal{A}$ but, by theorem 13, φ is computable in $\mathcal{A}$.

Using induction on the construction of the unary termal functions one can prove that for every termal function α if $\alpha(s) = t$ then $s \geq t$. Therefore, by theorem 10, the function $S = \lambda s.s+1$ is not computable in $\mathcal{A}$. We shall prove that S is absolutely search computable in $\{\theta, T_1, T_2\}$.

Let $G = \{\langle s,t\rangle \;/\; t = s+1\}$. We have $\langle s,t\rangle \in G \longleftrightarrow \langle s,\theta(t)\rangle \in E \;\&\; \langle t,\theta(t)\rangle \in E^\sim$. Hence, G is semicomputable in $\mathcal{A}$. Let σ be a computable in $\mathcal{A}$ element of $\Sigma_{0,2}$, which obtains 0 for elements of G and is undefined otherwise. Because

$$S(s) \simeq \begin{cases} (B^*)^2(s), \text{ if } \sigma \cdot \Pi(I, (B^*)^2)(s) = 0 \\ \text{undefined, otherwise,} \end{cases}$$

S is absolutely search computable in $\{\Theta, T_1, T_2\}$.

REFERENCES

[1] Y. N. Moschovakis, Abstract first order computability I, Trans. Amer. Math. Soc. 138 ,(1969), p. 427-464

[2] I. N Soskov, Computability over partial algebraic systems. Ph. D. Thesis, Sofia, (1983)

[3] I. N. Soskov, Computability over algebraic systems. Compt. Rend. Acad. Bulg. Sci., 36:3, (1983), p. 301-303

[4] C. E. Gordon, Finistically computable functions and relations on an abstract structure (Abstract), Journal of Symbolic Logic 36 (1971), p. 704

[5] J. C. Shepherdson, Computation over abstract structures. in: "Logic Colloquium '73", H. E. Rose, J. C. Shepherdson ed., North-Holland, Amsterdam, (1975), p. 445-513

[6] H. Friedman, Algorithmic procedures, generalized Turing algorithms and elementary recursion theory, in "Logic Colloquium '69", R. O. Gandy, C. E. M. Yates ed., North-Holland, Amsterdam, (1971), p. 361-389

[7] D. G. Skordev, "Combinatory spaces and recursiviness in them" Publishing House Bulg. Acad. Sci., Sofia, (1980)

[8] A. P. Ershov, Abstract computability on algebraic structures. Lecture notes in computer science, 122, 1981, p. 397-420

COMPLEXITY BOUNDED MARTIN-LÖF TESTS

Marius Zimand

Dept. of Mathematics
Univ. of Bucharest str. Akademiei 14
70109 Bucharest, Romania

1. INTRODUCTION

One of the main ways of attacking the famous P = ?NP problem and its associates consists in the consideration of some classical tools from the recursive function theory (different kind of reducibilities, relativization, immunity, a.s.o.) in complexity bounded forms.

On this line of research, Hartmanis [5] has recently found interesting results involving complexity classes such as P, NP, co-NP and others, using time and space bounded versions of Kolmogorov complexity. It is known that the classical (i.e., unbounded) theory of Kolmogorov complexity is in a great measure equivalent to the theory of Martin-Löf tests [1,2,9]. Moreover Martin-Löf tests have already been used successfully in computation theory [4,10]. Hence it is natural to consider and to study a bounded version of Martin-Löf tests.

In this paper we define the complexity bounded Martin-Löf tests which play the same role toward the bounded Kolmogorov complexity as their unbounded homologues do toward the classical Kolmogorov complexity. In the final part some applications to the study of complexity classes are given.

2. BASIC NOTIONS AND NOTATIONS

As usual N is the set of natural numbers and R is the set of real numbers. Unless specified otherwise, all sets are languages over the finite alphabet $X = \{\emptyset, 1\}$. The length of a string x in X^* is denoted by $|x|$. X^n is the set of strings of length equal to n. The lexicographical order on X^* (i.e., $\lambda < \emptyset < 1 < \emptyset\emptyset \ldots$, where λ is the null word) will be used and denoted by $X^* = \{\alpha_1 < \alpha_2 < \ldots\}$. Thus $\alpha_1 = \lambda$, $\alpha_2 = \emptyset$, $\alpha_3 = 1$, a.s.o. For a set S, card (S) denotes the cardinality of S. Let $\mathcal{M}$ be the set of many-tape deterministic Turing transducers having $\Sigma = X \cup \{\text{special signs}\}$ as input and working alphabet and X as output alphabet. For a machine M in $\mathcal{M}$, T_M and S_M are the time and space complexity functions defined in the usual way [7]. TIME (f(n)) is the class of sets that can be accepted by a deterministic Turing machine in f(n) time. SPACE (f(n)) is the class of sets that can be accepted by a deterministic Turing machine in f(n) space.

The time and space bounded Kolmogorov complexity of a string x is defined as follows.

Definition 1

Let M be a machine in $\mathcal{M}$ and f be a recursive function from N into N.

(a) The f-time bounded Kolmogorov complexity of a string x in X^* conditioned by n in N is the value:

$$K_M^f(x|n) = \min\{|p|,\ p \in X^*,\ M(p\$n) = x \text{ and } T_M(p\$n) \leqslant f(|x|)\}.$$

(b) The f-space bounded Kolmogorov complexity of a string x in X^* conditioned by n in N is the value:

$$KS_M^f(x|n) = \min\{|p|,\ p \in X^*,\ M(p\$n) = x \text{ and } S_M(p\$n) \leqslant f(|x|)\}.$$

Observations: (i) In case there is no p in X^* such that $M(p\$n) = x$ within $f(|x|)$ time (space) then we consider $K_M^f(x|n) = \infty$ ($KS_M^f(x|n) = \infty$).
(ii) The above definition is approximatively identical to the one given by Hartmanis, only that we have considered the conditioned version of Kolmogorov complexity which is more adequate for a comparison with Martin-Löf tests theory.

The next two Propositions show the existence of universal machines for the time and space bounded Kolmogorov complexity.

Proposition 1 ([5])

There is a machine M_U in $\mathcal{M}$ (M_U will be called universal) such that for any other M in $\mathcal{M}$ there is a constant c such that for all x in X^*

$$K_{M_U}^{cf(|x|)\log f(|x|)}(x||x|) \leqslant K_M^{f(|x|)}(x||x|) + c.$$

Proposition 2 ([5])

There is a machine M'_U in $\mathcal{M}$ (M'_U will also be called universal) such that for any other M in $\mathcal{M}$ there is a constant c such that for all x in X^*

$$KS_{M'_U}^{cf(|x|)}(x||x|) \leqslant KS_M^{f(|x|)}(x||x|) + c.$$

In the sequel K_{M_U} and $KS_{M'_U}$ will be denoted shortly by K and KS, M_U and M'_U being fixed universal machines.

3. COMPLEXITY BOUNDED MARTIN-LÖF TESTS

We introduce next the main object of our study, the complexity bounded Martin-Löf test.

Definition 2

Let f be a recursive function from N into N. The set $V \subseteq X^* \times (N \setminus \{\emptyset\})$ is a Martin-Löf test (M - L test in the sequel) with time (space) bounded by f if

1. $V_{m+1} \subseteq V_m$ for all m in N where $V_m = \{x \in X^* | (x,m) \in V\}$;

2. card $(V_m \cap X^n) \leqslant 2^{n-m} - 1$;

3. there is M in $\mathcal{M}$ such that M starting with input n and m (m < n) prints on the output tape all the elements of $V_m \cap X^n$ in time (space) bounded by f(n).

If instead of (2) we have (2') card $(V_m \cap X^n) \leqslant 2^{g(n)-m} - 1$, where g is a recursive function from N into N, $g(n) \leqslant n$ for all n, then we say that the M - L test is g-density bounded.

For the motivation of this notion and as well for the statistical terminology the interested reader should read [9]. We have made one single modification: the bounding condition (3) in the above definition replaces the condition that V is recursively enumerable from Martin-Löf's original definition.

By a simple counting argument we obtain the following lemma:

Lemma 3

For every machine M in $\mathcal{M}$, for all m in N and f, g, two recursive functions from N in N we have

(a) card $\{x | x \in X^*, |x| = n, K_M^{f(n)}(x|n) < g(n) - m\} < 2^{g(n)-m}$;

(b) card $\{x | x \in X^*, x = n, KS_M^{f(n)}(x|n) < g(n) - m\} < 2^{g(n)-m}$.

The next two propositions present two examples of M - L tests which will be used throughout this paper.

Proposition 4

For every machine M in $\mathcal{M}$ and f, g two recursive functions from N into N, $g(n) \leqslant n$ for all n, there is a constant c such that the set V(M) = $= \{(x,m) | KS_M^f(x||x|) < g(|x|) - m\}$ is a M - L test with space bounded by cf and g - density bounded.

Proof (a) $V(M)_{m+1} \subseteq V(M)_m$ can be shown immediately; (b) card $(X^n \cap V(M)_m) < 2^{g(n)-m}$ follows by Lemma 3(b); (c) the enumeration of the set $(X^n \cap V(M)_m)$ can be made as follows: take in lexicographical order all the strings z with $l(z) \leqslant g(n) - m - 1$. Simulate M(z) in space f(n) and if it yields a string x of length $|x| = n$, then write x on the output tape. Then consider the next z, a.s.o. Clearly the necessary space for the above procedure is bounded by cf(n), for some constant c.

Proposition 5

For every machine M in $\mathcal{M}$ and any two recursive functions f and g from N into N, $g(n) \leqslant n$ for all n in N, there is some constant c such that the set $V(M) = \{(x,m) | K_M^f(x||x|) < g(|x|) - m\}$ is a M - L test with time bounded by $c.f(n).2^{g(n)}$ and g - density bounded.

Proof Similar to the proof of Proposition 4.

Martin-Löf tests such as those defined in Proposition 4 and Proposition 5 are called representable tests. Calude and Chitescu [2] have shown that all recursive M - L tests can be embedded into a representable one. In the next two Propositions we find the same result for the complexity bounded M - L tests.

Proposition 6

Let V be a M - L test with space bounded by f and g - density bounded, where $g(n) \leq n$ for all n in N. Then there is a machine M in $\mathcal{M}$ and some constant c such that:

$$V \subseteq \{(x,m) \mid KS_M^{c(f(n)+n)}(x \mid |x|) < g(|x|) - m\}.$$

Proof (a) We construct a Turing transducer M with the property that for each x in V we have

$$KS_M^{cf}(x \mid |x|) \leq g(|x|) - m_V(x) - 1 \qquad (1)$$

where $m_V(x) = \max\{m \mid (x,m) \in V\}$. From this relation the conclusion results immediately. We fix a machine E which for all m and n enumerates $X^n \cap V_m$ in space bounded by f(n). The machine M works as follows:

1. It starts with input $\alpha\$\bar{n}$, where $\alpha \in X^*$ and n is the binary code for n. (On inputs of any other form, M will stop without writing anything on the output tape.)

2. M computes the rank r of α in the lexicographical order (i.e., $\alpha = \alpha_r$).

3. Consider the following ordering of strings x in $V \cap X^n$: $x < y$ iff $m_V(x) > m_V(y)$ or $m_V(x) = m_V(y)$ and in the enumeration E of $X^n \cap V_{m_V(x)}$ x precedes y.

Let $V \cap X^n = \{x_1 < x_2 < \ldots < x_k\}$. Then M writes on the output tape x_r, if $r \leq k$, otherwise it stops with the specification that no output has been produced.

We show that relation (1) holds true. Indeed the rank of x in $V \cap X^n$ (corresponding to the ordering introduced above) is less or equal than card $(V_{m_V(x)} \cap X^n) < 2^{g(n)-m_V(x)} - 1 = 1 + 2 + 2^2 + \ldots + 2^{g(n)-m_V(x)-1}$ and it follows that there is some α in X^*, $|\alpha| \leq g(n) - m_V(x) - 1$ such that the rank of α in the lexicographical order equals the rank of x in $V \cap X^n$ and in consequence $M(\alpha\$n) = x$.

(b) We show that M can be constructed in such a manner so that there is some constant c such that for every x in $V \cap X^n$ for which $M(\alpha\$\bar{n}) = x$, we have $S_M(x\$\bar{n}) \leq cf(n)$. Such a machine M acts as follows:

1. It computes the rank r of α (remark that $|\alpha| \leq n - m_V(x) - 1 \leq n - 2$) and writes r on some work tape.

2. M produces one by one the elements of $V_{n-2} \cap X^n$. Keeping in mind that V is a M - L test with space bounded by f, this operation can be done in less than f(n) tape cells. When some x of $V_{n-2} \cap X^n$ is generated, M decrements r by 1. If $r = \emptyset$ the last x produced is also written on the output tape.

3. If r has not become $\emptyset$, M generates one by one the elements of $V_{n-3} \cap X^n$. Each generated element is written on a special work tape. Then the elements of $V_{n-2} \cap X^n$ are generated again and M verifies if the element written on the special work tape has not been produced already in the

precedent step (i.e., the step of generating the elements of $V_{n-2} \cap X^n$). If the element on the special work tape is new then r is decremented by 1. Then the next element of $V_{n-3} \cap X^n$ is considered. The procedure continues in an obvious way (always verifying if some element in $V_p \cap X^n$ has not been produced already during the enumeration of $V_{p+1} \cap X^n$), until $r = \emptyset$. The last x is written on the output tape. By an easy examination of the space required by the above procedure, we find a constant c so that $K_M^{c(f(n)+n}(x||x|) \leqslant g(|x|) - m_V(x) - 1$.

Proposition 7

Let V be a M - L test with time bounded by f and g - density bounded, where $g(n) \leqslant n$ for all n in N. Then there is a machine M in $\mathcal{M}$ such that

$$V \subseteq \{(x,m) \mid K_M^{f'}(x||x|) < g(|x|) - m\},$$

where $f' \in O(n(f(n) + 2^{2g(n)}))$.

Proof (a) The same as part (a) of the proof of Proposition 6; and (b) the machine M acts as follows:

1. It computes the rank r of α in $O(n)$ time (remember that $|\alpha| < n$).

2. It produces one by one the elements of $V_{n-2} \cap X^n$, $V_{n-3} \cap X^n$, ..., $V_1 \cap X^n$ (in this order) and each element is compared with the elements previously written on the output tape. If it is a new element, then it is written on the output tape (after those generated earlier) and r is then decremented by 1.

Producing the elements of $V_{n-2} \cap X^n$, ..., $V_1 \cap X^n$ takes no more than $(n-2)f(n)$ steps of computation and comparing the jth generated element with the previous $j - 1$ elements takes $(j-1)kn$ steps, where k is some constant, so that all the comparisons take less than

$$kn \sum_{j=1}^{\text{card } V_1 \cap X^n} (j-1) \leqslant kn \sum_{i=0}^{2^{g(n)-1}} i = O(n \cdot 2^{2g(n)}) \text{ steps.}$$

Hence the time needed by M on the input $\alpha\$\bar{n}$ for the computation of x is bounded by $(n-2)f(n) + O(n\cdot 2^{2g(n)}) = O(n(f(n) + 2^{2g(n)}))$.

Combining the last propositions we derive simply the existence of universal complexity bounded M - L tests, in the sense that if some word x is rejected by such an universal M - L, then x is rejected by any other M - L test eventually at a higher level.

Proposition 8

Let V be a M - L test with space bounded by f and g - density bounded, where $g(n) \leqslant n$ for all n in N. There is a constant c such that

$$V_{m+c} \subseteq V(M_U)_m \text{ for all m in N,}$$

where $V(M_U)$ is the M - L test generated by the universal machine M_U with space bounded by cf and g - density bounded.

Proof From Proposition 6 we get a machine M in $\mathcal{M}$ and a constant c_1 in N such that

$$V \subseteq \{(x,m) \mid KS_M^{c_1 f}(x \mid |x|) < g(|x|) - m\}.$$

From Proposition 2, there is a constant c_2 in N such that $\{(x,m) \mid KS_M^{c_1 f}(x \mid |x|) < g(|x|) - m\} \subseteq \{(x,m) \mid KS_M^{c_2 f}(x \mid |x|) < g(|x|) - (m-c_2)\}$. We take $c = \max\{c_1, c_2\}$. Then $V_m \subseteq V(M_U)_{m-c}$ for all $m \geq c$. So $V_{m+c} \subseteq V(M_U)_m$ for all m in N.

Proposition 9

Let V be a M - L test with time bounded by f and g - density bounded. Then there is some constant c such that $V_{m+c} \subseteq V(M_U)_m$ for all m in N, where $V(M_U)$ is the M - L test generated by the universal machine M_U with time bounded by some function $f \in O(n \log n(f(n) + 2^{2g(n)}))$ and g - density bounded.

Proof Similar to the proof of Proposition 8.

It is known that the set of strings having "high" classical Kolmogorov complexity (the random strings in Kolmogorov's sense) is immune [3]. We shall prove similar results for the complexity bounded version of Kolmogorov complexity, which may have interesting consequences for the study of complexity classes, since immunity is a property intensively investigated in the last time in connection with the classical open problems involving complexity classes.

Lemma 10

Let f be a function from N into X^* such that $|f(n)| = n$ infinitely often (i.o.), and g, h two functions from N into N, h being an increasing and unbounded function, and suppose that

$$KS^g(f(n) \mid n) \geq h(n) \text{ for all n with } |f(n)| = n.$$

Then $f(n) \notin SPACE[p(n)g(|f(n)|)]$ for all functions p and N into R such that $\lim_{n \to \infty} p(n) = 0$.

Proof Suppose there is a function p from N into R such that $\lim_{n \to \infty} p(n) = 0$ and there is a transducer M in $\mathcal{M}$ which computes f and $S_M(n) \leq p(n)g(|f(n)|)$ for all n. We can modify M into a transducer M' such that $M'(x\$n) = M(n)$, for all x in X^* and $S_{M'}(x\$n) = S_M(n)$ (M' simply neglects x and recalls that the space on the input tape does not count for S_M). Then keeping in mind that $|f(n)| = n$, i.o., we get that $KS_{M'}^{cS_M}(f(n) \mid n) = 0$, i.o. Hence there is a constant c such that

$$KS_{M_U}^{cS_{M'}}(f(n) \mid n) \leq c, \text{ i.o.}$$

Recalling the hypothesis it follows that $c.S_{M'}(n) \geq g(|f(n)|) = g(n)$, i.o. But $S_{M'}(n) \leq p(n)g(n)$. So $p(n)g(n) > c^{-1}g(n)$ which means $p(n) > c^{-1}$, i.o. is a contradiction (remember that $\lim_{n \to \infty} p(n) = 0$).

Lemma 11

Let f be a function from N into X^* such that $|f(n)| = n$, i.o., and g, h two functions form N into N, h being an increasing and unbounded function and suppose that $K^g(f(n)|n) \geq h(n)$ for all n. Then $f(n) \notin \mathrm{TIME}[(p(n) \times g(n))^{1/2}]$ for every function p from N into R such that $\lim_{n\to\infty} p(n) = 0$.

Proof Suppose there is a function p from N into R with $\lim_{n\to\infty} p(n) = 0$ and a transducer M in $\mathcal{M}$ which computes f and $T_M^2(n) \leq p(n)g(n)$ for all n in N. We modify M into M' as in Lemma 10. It follows that

$$K_{M'}^{T_{M'}(n)}(f(n)|n) = 0 \text{ for all n such that } |f(n)| = n.$$

Hence $K_{M_U}^{cT_{M'}(n)\log T_{M'}(n)}(f(n)|n) \leq c$ for all n such that $|f(n)| = n$ and for some c. It follows that $c.T_{M'}(n)\log T_{M'}(n) > g(n)$. So $p(n) > c^{-1}$, i.o., contradicting the fact that $\lim_{n\to\infty} p(n) = 0$.

Proposition 12

Let $A \subseteq \{x | KS^g(x||x|) \geq |x|\}$ be an infinite set. The $A \in \mathrm{SPACE}[p(n) \times g(n)n^{-1}]$ for all functions p from N into R with $\lim_{n\to\infty} p(n) = 0$.

Proof Suppose that the characteristic function of A is in $\mathrm{SPACE}[p(n)g(n)n^{-1}]$ for some function p from N into R, with $\lim_{n\to\infty} p(n) = 0$. Let f be the function

$$f(n) = \begin{cases} \text{the least (lexicographically) x in A with } |x| = n, \\ \qquad\qquad \text{if there is such a x;} \\ \lambda, \text{ otherwise;} \end{cases}$$

then clearly $f \in \mathrm{SPACE}[cp(n)g(n)]$. But f satisfies the conditions of Lemma 10, a contradiction.

Proposition 13

Let $A \subseteq \{x \in X^* | K^g(x||x|) \geq |x|\}$. Then $A \notin \mathrm{TIME}[(p(n)g(n))^{1/2}n^{-1}2^{-n}]$ for all p with $\lim_{n\to\infty} p(n) = 0$.

Proof Similar to the proof of Proposition 12.

4. APPLICATIONS TO COMPLEXITY CLASSES

Complexity bounded M - L tests could be useful when studying some properties of the complexity classes. Complexity bounded M - L have natural interpretations especially when sets having certain kinds of densities (sparse or co-sparse sets) are investigated. We present in this section two results exemplifying the way complexity-bounded M - L are used when studying space or time complexity of some important sets. For the definition of P, NP, co-NP, a.s.o., [7] is a good reference. We also use the following definition.

Definition 3

A set $V \subseteq X^*$ is sparse if there exists a constant k such that for all n

$$\mathrm{card}\,\{x \in V, |x| = n\} \leq n^k + k.$$

A set $V \subseteq X^*$ is co-sparse if its complementary is sparse.

Theorem 14

Let f be a function from N into N such that $\lim_{n\to\infty} n^k/f(n) = 0$ for all k in N. Any co-sparse set V in PSPACE contains a subset A such that $A \notin SPACE[f(n)]$.

Proof. Let V be a co-sparse set in PSPACE. It follows that its complementary $\bar{V}$ is in PSPACE (PSPACE is closed under complementation) and $\bar{V}$ is sparse, i.e., there is a constant k such that card $\{x \in \bar{V} \mid |x| = n\} \leqslant n^k + k$ for all n in N. Let $W = \{(x,m) \mid x \in \bar{V}, |x| \geqslant n_o, 1 \leqslant m \leqslant |x|^{1/2}\}$ where n_o is taken sufficiently large so that for every $n \geqslant n_o$, $2^{n-n^{1/2}} > n^k + k$. As W is in PSPACE, it follows immediately that W is a M - L test with space bounded by some polynomial q(n). Then by Proposition 8, there exists a constant c such that $W_{m+c} \subseteq V(M_U)_m$ for all m, where $V(M_U)$ is an universal M - L test with space bounded by any function $f(n) \geqslant t.q(n)$ where t is some constant. We have co-$V(M_U)_m \subseteq$ co-W_{m+c} for all m (co-A is the notation for the complementary of A). In consequence co-$V(M_U)_1 \subseteq$ co-W_{1+c}. Hence if $|x| \geqslant (1 + c)^2$ and $x \in$ co-$V(M_U)_1$, it follows that $x \in$ co-$\bar{V} = V$. In conclusion the set $A = \{x \in \text{co-}V(M_U)_1, |x| \geqslant (1 + c)^2\} \subseteq V$. But the set A $\notin SPACE[p(n)f(n)n^{-1}]$ for every function p from N into R with $\lim_{n\to\infty}(n) = \emptyset$, because co-$V(M_U)_1 \notin SPACE[p(n)f(n)n^{-1}]$, by Proposition 12. But as f can be chosen arbitrarily large, the conclusion of the theorem follows immediately.

Remark that Theorem 14 can be used for every co-sparse set V from the polynomial hierarchy PH, since PH $\subseteq$ PSPACE. In particular V can be in P, NP or co-NP. Also note that the "arbitrary complex" subset A of V contains more that half the elements of V. Indeed card $(X^n \cap \text{co-}V(M_U)_1) \geqslant 2^n - 2^{n-1} = 1/2.2^n$.

The next result concerns the p-printable sets.

Definition 4 ([6])

A set $A \subseteq X^*$ is p-printable if there is a deterministic Turing machine M which starting with input n prints all the elements of A of length n in polynomial time in n.

The importance of p-printable sets follows from a result of Hartmanis and Yesha [6] which states that "EXPTIME = NEXPTIME" if "any sparse set in NP is p-printable" if "there are no sparse sets in NP-P". We show that the hypothesis that any spare set in NP is p-printable has interesting consequences.

Theorem 15

Let $V \subseteq X^*$ be a p-printable set in NP. Then V contains at most a finite set of random strings.

Proof Recall that a string x in X^* is called random if $K(x \mid |x|) \geqslant |x|$ [1,8], where K is the unbounded Kolmogorov complexity. From the condition that V is p-printable we get immediately that V is sparse. Let k be such that card $(V \cap X^n) \leqslant n^k + k$ for all n and $W = \{(x,m) \mid x \in V, |x| \geqslant n_o, 1 \leqslant m \leqslant |x|^{1/2}\}$, where n_o is sufficiently large so that $2^{n-n^{1/2}} > n^k + k$ for all $n \geqslant n_o$. We observe that W is a M - L test with time bounded by some polynomial q(n) and with density less than $n^k + k$. It follows that $W_{m+c} \subseteq V(M_U)_m$ for all m and for some constant c, where $V(M_U)$ is the universal M - L test with time bounded by a function $f \in O(n(q(n) + (n^k + k)^2)$. Hence f has the growth order of a polynomial. In particular $W_{1+c} \subseteq V(M_U)_1$, but $W_{1+c} = V$ with the exception of a finite set. In consequence for every x with length x sufficiently large, we have $x \in V(M_U)$, i.e., $K^f_{M_U}(x||x|) \leqslant |x| - 1$, which clearly implies $K(x||x|) < |x|$.

It is interesting to note that the above reasoning, made more tightly, leads us to the conclusion that for every x in V of sufficiently large length we have $K^f(x||x|) \leqslant O(\log(|x|))$ where f has the growth order of a polynomial. Hence if we find a sparse set in NP containing an infinity of strings x with $K^g(x||x|) > O(\log(|x|))$ with g larger than any polynomial, then we would have EXPTIME $\neq$ NEXPTIME and from here, by a standard downward translation, P $\neq$ NP.

REFERENCES

1. C. Calude and I. Chitescu, Random strings according to A. N. Kolmogorov and P. Martin-Löf: classical approach, Found. Control. Engrg., 7:73-85 (1982).
2. C. Calude, I. Chitescu and L. Staiger, P. Martin-Löf tests: representability and embeddability, Rev. Roumaine Math. Pures Appl., 30: 719-732 (1985).
3. C. Calude and I. Chitescu, Strong noncomputability of random strings, Internat. J. of Comput. Math., 11:43-45 (1982).
4. C. Calude and M. Zimand, A relation between correctness and randomness in the computation of probabilistic algorithms, Internat. J. of Comput. Math., 13:47-53 (1984).
5. J. Hartmanis, Generalized Kolmogorov complexity and the structure of feasible computations, Cornell University, TR 14853 (1984).
6. J. Hartmanis and Y. Yesha, "Computation times of NP sets of different densities", LNCS 154:319-330, Springer-Verlag (1983).
7. J. Hopcroft and J. Ullman, "Introduction to Automata Theory, Languages and Computation", Addison-Wesley (1979).
8. A. N. Kolmogorov, Three Approaches to the Definition of the Concept Quantity of Information, Peredachi Informatsii, 1:3-11 (1964).
9. P. Martin-Löf, The Definition of Random Sequences, Information and Control, 9:602-619 (1966).
10. M. Zimand, On the topological size of sets of random strings, Z. Math. Logik Grundl. Math., vol. 32:81-88 (1986).

PARTICIPANTS AND CONTRIBUTORS

Johan van Benthem
Dept. of Mathematics
Univ. of Amsterdam
Roetersstraat 15
1018 WB Amsterdam
HOLLAND

Douglas S. Bridges
Univ. of Buckingham
Buckingham, MK18 1EG
ENGLAND

Cristian Calude
Dept. of Mathematics
Univ. of Bucharest
str. Academiei 14,
70109 Bucharest
ROMANIA

*Curt C. Christian
Inst. für Logistik der Univ. Wien
Universitätsstr. 10 / 2 / 1
1090 Wien
AUSTRIA

Cheryl Dawson
Dept. of Mathematics
Stanford Univ.
Stanford, CA 94305
U.S.A.

John W. Dawson, Jr.
Dept. of Mathematics
Stanford Univ.
Stanford, CA 94305
U.S.A.

Osvald Demuth
Na vrcholu 17
13 000 Praha 3
CZECHOSLOVAKIA

Elka Dimitrova/ temporary at:
Sector of Mathematical Logic
Faculty of Math. & Mech.
boul. Anton Ivanov 5
Sofia 1126
BULGARIA

Angel V. Ditchev
boul. Simeon Veliki 19/II, ap.33
Shumen 9704
BULGARIA

Albert G. Dragalin
Szegfü u. 9, fszt. 2
4027 Debrecen
HUNGARY

Yurij Ershov
Novosibirsk State Univ.
630090 Novosibirsk
U.S.S.R.

*Pavel Filipec
Inst. of Economics
Czechoslovak Acad. of Sci.
Prague
CZECHOSLOVAKIA

Svetla Gandeva
Inst. of Philosophy
boul. P. Evtimi 6
Sofia 1000
BULGARIA

George Gargov
Sector of Mathematical Logic
Faculty of Math. & Mech.
boul. Anton Ivanov 5
Sofia 1126
BULGARIA

Miroslav Genov
Sector of Mathematical Logic
Faculty of Math. & Mech.
boul. Anton Ivanov 5
Sofia 1126
BULGARIA

Nadezhda Georgieva
Mladost I, 16, vh. E
Sofia 1199
BULGARIA

Sergej Goncharov
Inst. of Mathematics
Soviet Acad. of Sci./Siberian Div.
630090 Novosibirsk
U.S.S.R.

Valentin Goranko
Sector of Mathematical Logic
Faculty of Math. & Mech.
boul. Anton Ivanov 5
Sofia 1126
BULGARIA

Hans-Jürgen Höehnke
Karl-Weierstraß-Inst.
Mohrenstr. 39 (Pf.1304)
Berlin - 1086
DDR

Kornelia Ilieva
Inst. for Bulgarian Language
Chapaev 52, bl.17
Sofia 1113
BULGARIA

Ljubomir L. Ivanov
Sector of Mathematical Logic
Faculty of Math. & Mech.
boul. Anton Ivanov 5
Sofia 1126
BULGARIA

Mitko Janchev
Sector of Mathematical Logic
Faculty of Math. & Mech.
boul. Anton Ivanov 5
Sofia 1126
BULGARIA

Herman R. Jervell
Inst. of Informatics
P.O.Box 1080 Blindern
0316 Oslo 3
NORWAY

Vladimir Jotsov
boul. Kl. Gotvaldt 58
Sofia
BULGARIA

Borislav Jurukov
Pedagogical Inst.
Blagoevgrad
BULGARIA

Krasimir Kirov
Corporation SPS
compl. Bukston 207 A
Sofia
BULGARIA

*Andrej N. Kolmogoroff
Mech. Math. Faculty
Moscow State Univ.
117234 Moscow
U.S.S.R.

Violeta Koseska
Inst. for Slavonic Languages
Polish Acad. of Sciences
PKiN XVIIp.
00-901 Warsaw
POLAND

Wojciech Kowalczyk
Inst. of Mathematics
Univ. of Warsaw
PKiN, IX p.
00-901 Warsaw
POLAND

*Evangelos Kranakis
Centre for Mathematics & Comp. Sci.
P.O.Box 4079
1009 AB Amsterdam
HOLLAND

Andreana Madgerova
boul. Tolbukhin 24
Sofia 1000
BULGARIA

Antoni Mazurkiewicz
Inst. for Computer Science
Polish Acad. of Sciences
PKiN Xp.
00-901 Warsaw
POLAND

Christoph Meinel
Sektion Mathematik
Humboldt Univ.
PSF 1297
1086 Berlin
DDR

Pierangelo Miglioli
Dipart. di Sci. dell'Informazione
via Moretto da Brescia 9
20133 Milano
ITALY

Marion Mircheva
Inst. of Philosophy
boul. P. Evtimi 6
Sofia 1000
BULGARIA

Juan Jose Moreno-Navarro
Dpto. de Algoritmica
Facultad de Informatica
Univ. Politechnica de Madrid
Carretera de Valencia, Km. 7
28031 Madrid
SPAIN

Ugo Moscato
Dipart. di Sci. dell'Informazione
via Moretto da Brescia 9
20133 Milano
ITALY

Yiannis Moschovakis
Dept. of Mathematics
UCLA
405 Hilgard Ave.
Los Angeles, CA 90024
U.S.A.

Marcin Mostowski
Dept. of Logic
Warsaw Univ./ Bialistok Branch
Swierkowa 20
Bialystok
POLAND

Nikolaj Nagornyj
Comp. Center of Soviet Acad. Sci.
ul. Vavilova 40
117967 GSP Moscow
U.S.S.R.

Valerij A. Nepomniaschy
Computing Centre
Soviet Acad. of Sci./Siberian Div.
pr. Lavrentieva 6
630090 Novosibirsk
U.S.S.R.

Stela Nikolova
Sector of Mathematical Logic
Faculty of Math. & Mech.
boul. Anton Ivanov 5
Sofia 1126
BULGARIA

*Mario Ornaghi
Dipart. di Sci. dell'Informazione
via Moretto da Brescia 9
20133 Milano
ITALY

Solomon Passy
Sector of Mathematical Logic
Faculty of Math. & Mech.
boul. Anton Ivanov 5
Sofia 1126
BULGARIA

Liliana Pavlova
Sector of Mathematical Logic
Faculty of Math. & Mech.
boul. Anton Ivanov 5
Sofia 1126
BULGARIA

Assen Petkov
Inform. Comp. Centre
Research & Technologies Committee
ul. Chapaev 55A
Sofia 1547
BULGARIA

Petio P. Petkov
Sector of Mathematical Logic
Faculty of Math. & Mech.
boul. Anton Ivanov 5
Sofia 1126
BULGARIA

Vesselin Petrov
Studentski grad, 19 bl., 704
Sofia 1100
BULGARIA

*Alberto Pettorossi
IASI-CNR
Viale Manzoni 30
00185 Roma
ITALY

Jan Plaza
Madalinskiego 21, m. 29
02-513 Warszawa
POLAND

Vassil Plougchiev / photographer
boul. Georgi Traikov 99
Varna 9002
BULGARIA

Slavian Radev
Sector of Mathematical Logic
Faculty of Math. & Mech.
boul. Anton Ivanov 5
Sofia 1126
BULGARIA

Ivan Rajchev
blvd. Lenin, bl. 6, ap. 9
Smoljan 4700
BULGARIA

Helena Rasiowa
Wiejska 9, m 5
00-480 Warszawa
POLAND

Mario Rodriguez-Artalejo
Dpto. de Algoritmica
Facultad de Informatica
Univ. Politechnica de Madrid
Carretera de Valencia, Km. 7
28031 Madrid
SPAIN

Giovanni Sambin
Dept. of Mathematics
Univ. of Siena
via del Capitano 15
53100 Siena
ITALY

Krister Segerberg
Dept. of Philosophy
Univ. of Auckland
Private Bag
Auckland
NEW ZEALAND

Aleksej L. Semenov
Lab. on Theory of Algorithms
Inst. for Problems in Cybernetics
Soviet Acad. of Sciences
ul. Vavilova 37
117 319 Moscow
U.S.S.R.

Nikolaj A. Shanin
LOMI
Fontanka 27
191011 Leningrad
U.S.S.R.

*Nikolaj V. Shilov
Computing Centre
Soviet Acad. of Sci./Siberian Div.
pr. Lavrentieva 6
630090 Novosibirsk
U.S.S.R.

Dirk Siefkes
TU Berlin
Fb. 20 (Informatik)
Franklinstr. 28/29
1000 Berlin 10
WEST BERLIN

Dimiter Skordev
Sector of Mathematical Logic
Faculty of Math. & Mech.
boul. Anton Ivanov 5
Sofia 1126
BULGARIA

Andrzej Skowron
Inst. of Computer Science
Polish Acad. of Sciences
PKiN, p. IX
00-901 Warsaw
POLAND

Jan M. Smith
Dept. of Computer Science
Univ. of Göteborg/Chalmers
412 96 Göteborg
SWEDEN

Ivan N. Soskov
LICS
boul. Anton Ivanov 5
Sofia 1126
BULGARIA

Alexandra Soskova
Sector of Mathematical Logic
Faculty of Math. & Mech.
boul. Anton Ivanov 5
Sofia 1126
BULGARIA

Vladimir Sotirov
Sector of Mathematical Logic
Faculty of Math. & Mech.
boul. Anton Ivanov 5
Sofia 1126
BULGARIA

Svetlana Stoyanova
Inst. for Bulgarian Language
ul. Chapaev 52, bl. 17
Sofia 1113
BULGARIA

Martin Tabakov
Inst. of Philosophy
boul. P. Evtimii 6
Sofia 1000
BULGARIA

Helmuth Thiele
Sektion Mathematik
Humboldt Univ.
Unter den Linden 6
1086 Berlin, PSF 1297
DDR

Tinko Tinchev
Matematika magazin
boul. Vl. Zaimov 62
Sofia 1504
BULGARIA

Antoni Tonchev
GD IIZ of ATU
ul. Sveta Sofia 5
Sofia 1000
BULGARIA

Boris A. Trakhtenbrot
Dept. of Comp. Sci.
Tel Aviv Univ.
Ramat Aviv
ISRAEL

Aldo Ursini
Dept. of Mathematics
Univ. of Siena
via del Capitano 15
53100 Siena
ITALY

Vladimir A. Uspensky
Dept. of Math. Logic
Mech. Math. Faculty
Moscow State Univ.
117234 Moscow
U.S.S.R.

Dimiter Vakarelov
Sector of Mathematical Logic
Faculty of Math. & Mech.
boul. Anton Ivanov 5
Sofia 1126
BULGARIA

Krastjo Vakarelov
Inst. of Philosophy
boul. P. Evtimi 6
Sofia 1000
BULGARIA

Mars Valiev
VNIISI
prosp. 60-letija Oktjabria 9
117312 Moscow
U.S.S.R.

*Veli Valpola
Tähdenlennontie 12-14
02240 Espoo 24
FINLAND

Vasil Vasilev
Inst. for Tech. Cyber. & Robotics
ul. Acad. G. Bonchev 2
Sofia 1113
BULGARIA

Ognjan Zahariev
Sector of Mathematical Logic
Faculty of Math. & Mech.
boul. Anton Ivanov 5
Sofia 1126
BULGARIA

Ivo Zapletal
Czech. Acad. of Sciences
Jilska 1
11000 Prague 1
CZECHOSLOVAKIA

Jordan Zashev
Sector of Mathematical Logic
Faculty of Math. & Mech.
boul. Anton Ivanov 5
Sofia 1126
BULGARIA

*Marius Zimand
Dept. of Mathematics
Univ. of Bucharest
str. Akademiei 14
70109 Bucharest
ROMANIA

Neli Zlatareva
Inst. for Tech. Cyber. & Robotics
ul. Akad. G.Bonchev 2
Sofia 1113
BULGARIA

* did not attend

INDEX